future arquitecturas s.l. 编

# 未来建筑竞标 中国 第4辑

# 经济适用房样板

# *FUTURE ARQUITECTURAS COMPETITIONS CHINA IV*
# *SOCIAL HOUSING PROTOTYPES*

图书在版编目 (CIP) 数据

未来建筑竞标. 中国. 第4辑：经济适用房样板：汉英对照 / 西班牙未来建筑出版社编. 一 杭州：浙江大学出版社，2011.12
ISBN 978-7-308-09289-0

Ⅰ. ①未… Ⅱ. ①西… Ⅲ. ①建筑设计—世界—现代—图集 Ⅳ. ①TU206

中国版本图书馆CIP数据核字（2011）第229470号

浙江省版权局著作权合同登记图字：11-2011-205号



**未来建筑竞标 中国 第4辑 经济适用房样板**

*FUTURE ARQUITECTURAS COMPETITIONS CHINA IV SOCIAL HOUSING PROTOTYPES*

future arquitecturas s.l. 编

责任编辑 张凌静 李峰伟
封面设计 future arquitecturas s.l. 未来建筑
出版发行 浙江大学出版社
（杭州天目山路148号 邮政编码：310007）
（网址：http://www.zjupress.com）
印 刷 杭州多丽彩印有限公司
印 张 30
开 本 889mm×1194mm 1/8
字 数 384千
版印次 2011年12月第1版 2011年12月第1次印刷
书 号 ISBN 978-7-308-09289-0
定 价 300.00元

**主编** · DIRECTORS AND PUBLISHERS
Gerardo Mingo Pinacho · Gerardo Mingo Martínez (西)

**联合主办单位** · CO-SPONSORS
浙江大学建筑工程学院 · ZUCCEA
浙江大学建筑设计研究院 · ADRZU
西班牙未来建筑 · future arquitecturas s.l.

**联合出版** · CO-PUBLISHERS
浙江大学出版社 Zhejiang University Press

**执行编辑 · MANAGING EDITOR**
Gabriela Vélez Trueba (西) · gabriela@arqfuture.com

**图形 · LAYOUT**
Carlos de Navas Paredes (西)

**图形制作 · GRAPHIC PRODUCTION**
Tamara García Garrido (西) · tamara@arqfuture.com
左雯莎 Zuo Wensha · news@arqfuture.com

**中国地区公司合伙人 · CORPORATE PARTNER IN CHINA**
赵磊 Zhao Lei · leizhao@arqfuture.com
地址 address: 中国杭州下城区文晖路303号
浙江交通集团大厦11楼
邮编 postal code: 310014
电话 telephone: +86 571 85303277
手机 cell phone: +86 13706505166

**美洲地区公司合伙人 · CORPORATE PARTNER IN AMERICA**
Santiago Vélez (厄) · svelez@arqfuture.com
地址 address: c/ Portugal E10-276 y 6 de Diciembre, 2do piso
Quito · Ecuador
电话 telephone: +593 98036427

**行政人员 · ADMINISTRATION**
Belén Carballedo (西) · belen@arqfuture.com

**销售部 · DISTRIBUTION DEPARTMENT**
曾江福 Zeng Jiangfu
手机 cell phone: 13564489269
电话 telephone: 20 65877188

**广告 · ADVERTISING**
china@arqfuture.com

***future*** arquitecturas
Rafaela Bonilla 17, 28028
Madrid, Spain

**www.arqfuture.com**

# 序言

## MASQUE
## 如何设计未来城市？

Masque最早起源于18世纪，是音乐与文本、旅程记录的集合体，阐述了所参观地方的本质内涵及精神、与居住者的关系，以及古代与当今之间态度的差异，所有这些都给我们带来了惊喜。其中，人物、景物与意象和象征元素相融合，甚至还包括超现实主义的作品，如希罗尼穆斯·波希的三连画 *Bosch* 、《乐园》，融合了完美的绘画技术和质量、极佳的作品及建筑风格，使当代经典文化继承了15世纪时的元素。

中国大多数作品源于个人回忆、日常故事以及片段记录，将其拼接成一个完整作品，因此处处充斥着时间的痕迹。文化的多样性丰富了城市内涵，而遗产的继承又促进了社会的发展：识别度、真实性、完整性及内涵。

“20世纪建筑遗产保护办法”国际会议（CAH20thC）于2011年6月通过了《马德里文件》。该文件与之前1993年通过的《雅典：都会宣言》、1942年勒·柯布西耶出版的《巴黎2号文件》、1964年威尼斯、2000年克拉科夫等文件一起，帮助人们认识、了解建筑遗产及其文化意义，并建立有关优胜景观的真实性、完整性和内涵的研究，保护环境及其可持续发展等方面的方法和标准。

**遗产的保护标准是否与以往有所不同?**

如何设计未来城市？革新旧时的审美观，还是保护城市文化遗产？

大城市历史与未来之间的冲突在所难免，21世纪，这些矛盾将一一凸显。

# Preface

## MASQUE
## How do we imagine the cities of the future?

Masque, a word of the 18th century, represents a work with music and text which, together with a journey notebook that examines the substantial beliefs of the places which are visited, the soul of a place and its relationship with its inhabitants, in ancestral modern attitudes realities, leads us to the logic of the happy surprise, where the characters and the landscape intermingle with imaginary and symbolic elements even surrealistic, like Hieronymus Bosch's triptych Bosch, The Garden of Earthly Delights which saves the contemporary classical culture from the 15th century, with technical perfection and quality of the drawing, the excellent work, with good architecture.

China is full of monuments which come from personal memories, everyday stories and ephemeral moments, to retrieve a collective story. The diversity of cultures has enriched cities, improving the development of society with a necessary maintenance of the heritage: identification, authenticity, integrity and meaning.

The International Conference "Approaches for Intervention in the Architectural Heritage of the 20th century, CAH20thC", adopted on June 2011 "Madrid Document", joining to other previous: Athens, urban manifesto written in 1933 aboard the "Patris II", journey notebook published in 1942 by Le Corbusier; Venice 1964, Krakow 2000... to advance towards the knowledge, understanding and cultural meaning of architectural heritage and establish methodologies and criteria regarding research, conservation and environmental sustainability, respecting the authenticity, integrity and meaning of good landscape.

**Is the theory of preservation different from other periods of the heritage now?**

How do we imagine the cities of the future? Aesthetic reclamation of the past or concern for the cultural legacy of the city?

The 21st century personifies the inevitable conflict between past and future of the big cities.

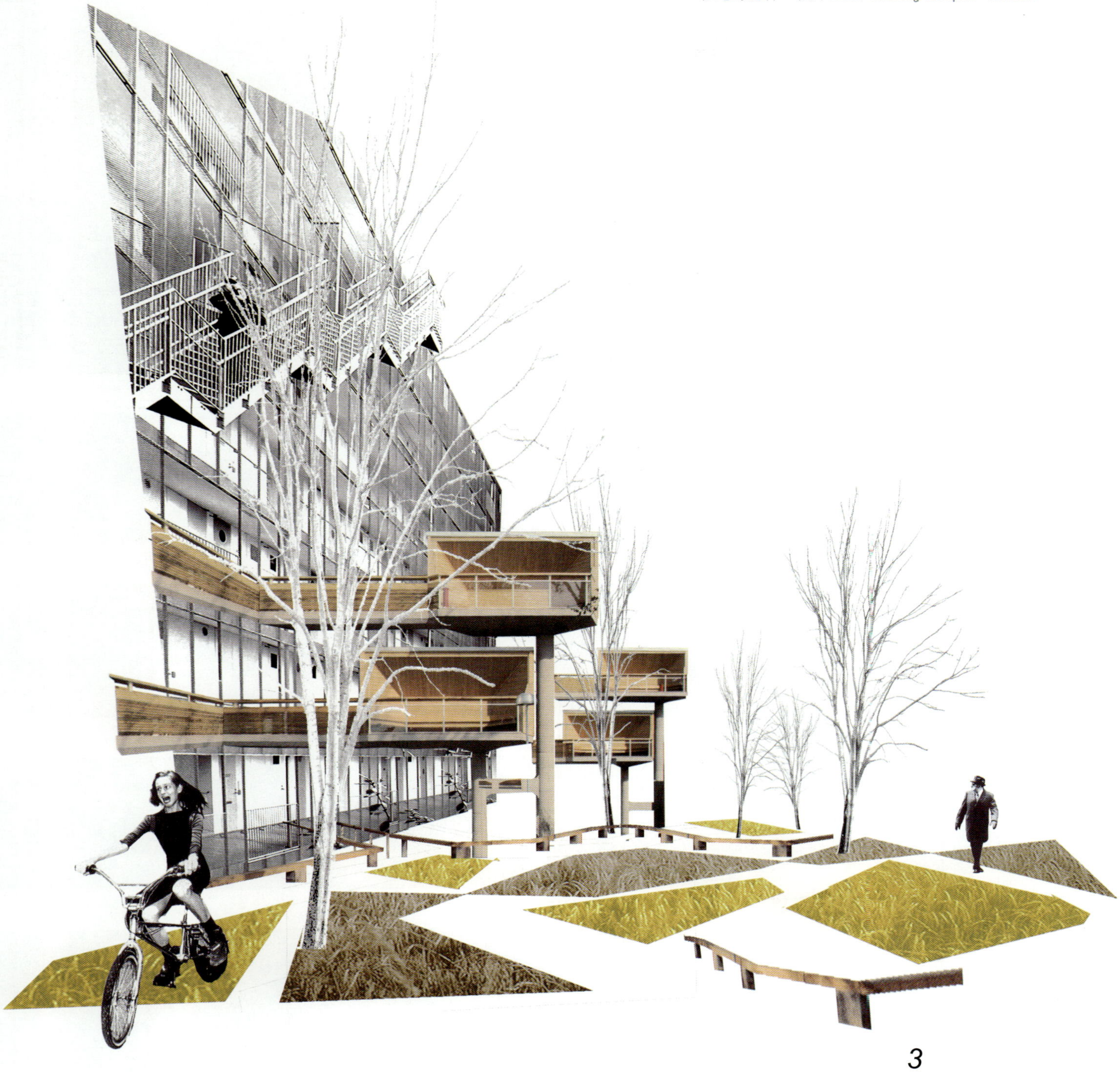

**浙江大学建筑工程学院**现有规划、建筑、土木、水利四大学科，基本涵盖了国家基本建设领域的全部学科，涉及到建筑、市政、交通、水利、铁道、港口与海洋工程等主要产业领域。在学科上具有良好的互补性和交叉性，为学科发展奠定了良好的基础。

**The College of Civil Engineering and Architecture of Zhejiang University** has a wide coverage of the fields of national capital construction including: building design and construction, municipal engineering, transportation, water conservancy, railway and harbor and offshore engineering. And it forms a construction with Regional and Urban Planning, Architecture, Civil Engineering and Hydraulic Engineering to have the integration of production, learning and research.

## 七堡小学 · 杭州

### Qibao Primary School

王晖 Wang Hui · 胡频飞 Hu Pinfei · 包望韬 Bao Wangtao · 杜丽婷 Du Liting · 姚立夫 Yao Lifu
范庄媛 Fan Zhuangyuan · 赵小青 Zhao Xiaoqing

## 天目湖娱乐中心 · 溧阳

### Tianmu Lake Leisure Centre

浦欣成 Pu Xincheng · 孙炜玮 Sun Weiwei · 吴璟 Wu Jing · 李澍田 Li Shutian · 陈吉 Chen Ji

## 富阳东大道城市设计 · 富阳

### Urban Design of Fuyang East Avenue

罗卿平 Luo Qingping · 高小妹 Gao Xiaomei · 钱逸卿 Qian Yiqing

## 江苏美术馆 · 江苏

### Jiangsu Art Gallery

高裕江 Gao Yujiang · 卢挺 Lu Ting · 章文杰 Zhang Wenjie · 高庶三 Gao Shusan · 高生皓 Gao Shenghao

浙江大学建筑设计研究院始建于1953年，是国家重点高校中最早成立的甲级设计研究院之一。业务范围有高层、超高层的大型办公、宾馆、商业综合体、行政办公楼；学校校园规划与设计；影剧院、图书馆、博物馆等文化建筑；居住区规划与设计；体育建筑；医院类建筑；城市设计；智能建筑设计、室内设计；风景园林与景观设计；市政公用工程；岩土工程；幕墙设计；古建筑和近现代建筑的维修保护、文物保护规划等。

**Architectural Design and Research Institute of Zhejiang University** was founded more than half a century ago in 1953, it has been one of the earliest Grade-A design institutes established among state key universities. The business covers high-rise or super high-rise large office, hotel, business complex and administrative office building; planning and design of campus; cinema, library, museum and other cultural buildings; residential community planning and design; construction of sports facilities; construction of hospital; urban design; intelligence architectural design, indoor design; landscape architecture and landscape design; municipal public engineering; geotechnical engineering; curtain walling design; maintenance and protection of ancient building and modern building, planning of cultural relics protection, etc.

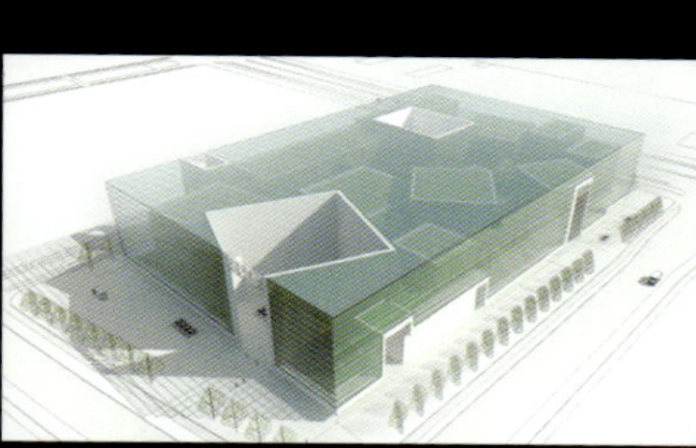

## 中国国家美术馆新馆 · 北京

### Expansion of China National Art Museum

陈瑜 Chen Yu · 邱文晓 Qiu Wenxiao · 朱睿 Zhu Rui

## 中国国家美术馆新馆 · 北京

### Expansion of China National Art Museum

方涛 Fang Tao

## 中国国家美术馆新馆 · 北京

### Expansion of China National Art Museum

陈 冰 Chen Bing

## 中国国家美术馆新馆 · 北京

### Expansion of China National Art Museum

霍飞 Huo Fei

## 中国国家美术馆新馆 · 北京

### Expansion of National Art Museum

郑茂恩 Zheng Maoen

## 中国国家美术馆新馆 · 北京

### Expansion of China National Art Museum

叶长青 Ye Changqing · 张舒文 Zhang Shuwen · 郑筇 Zheng Kun · 朱君 Zhu Jun

## 中国国家美术馆新馆 · 北京

### Expansion of China National Art Museum

施明化 Shi Minghua

七堡小学 · 杭州

## Qibao Primary School · China

王晖 Wang Hui · 胡频飞 Hu Pinfei · 包望韬 Bao Wangtao · 杜丽婷 Du Liting · 姚立夫 Yao Lifu · 范庄媛 Fan Zhuangyuan · 赵小青 Zhao Xiaoqing

设计方案 design proposal

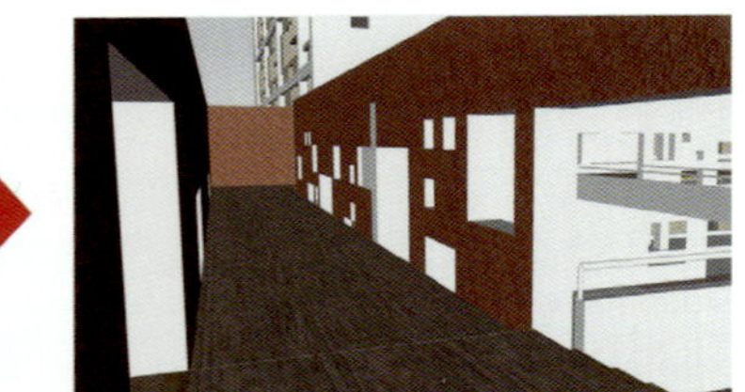

园林手法 LANDSCAPE GARDEN MODEL

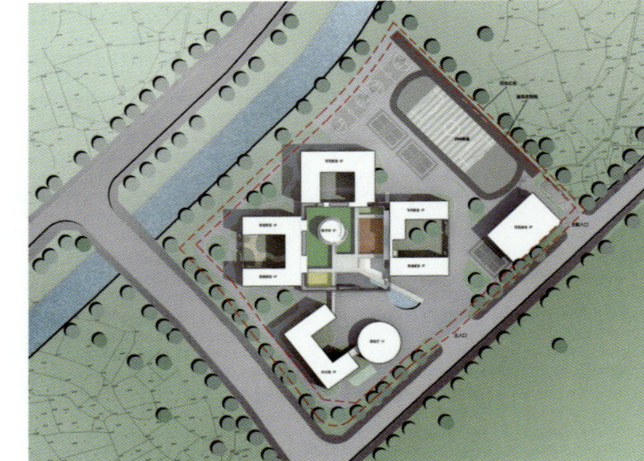

总平面图 OVERALL PLAN

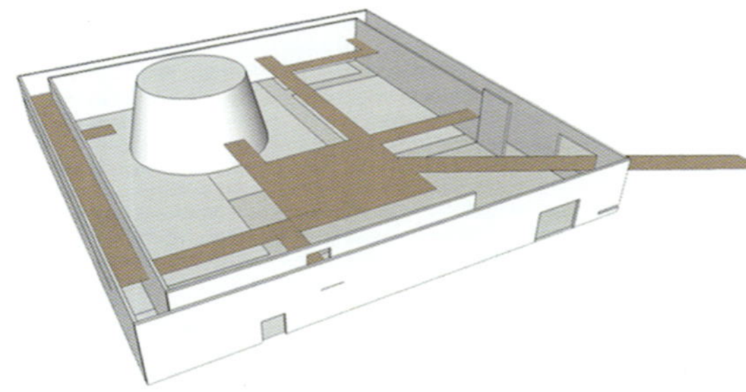

中式院落 CHINESE COURTYARD

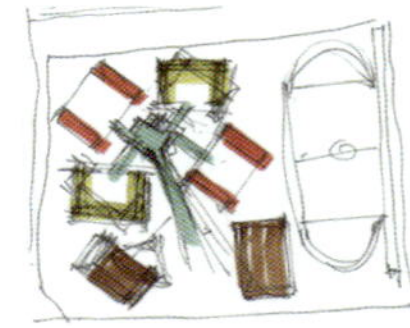
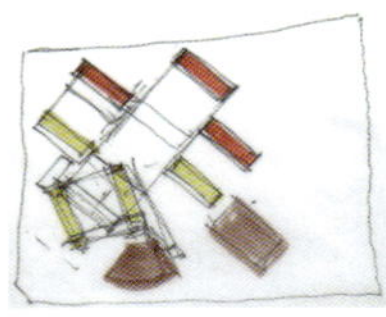
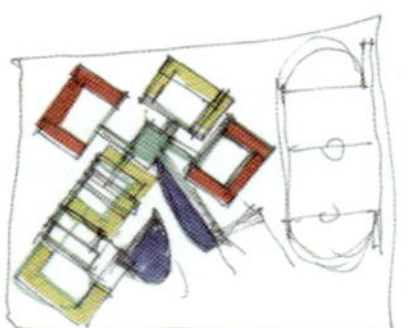
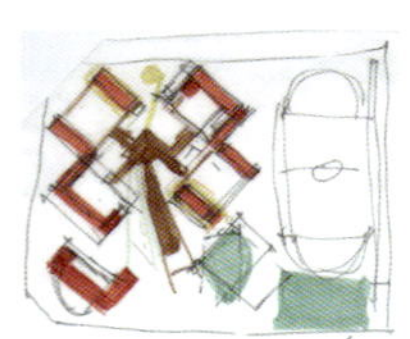
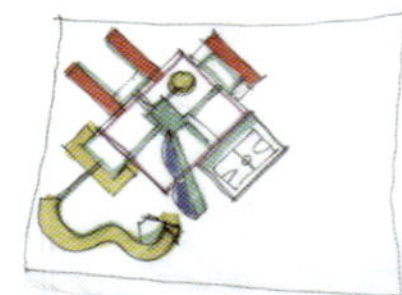
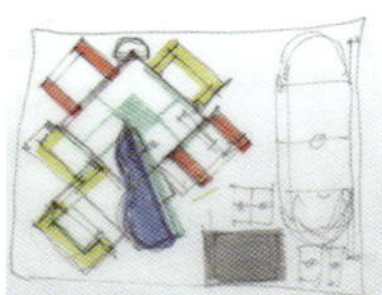
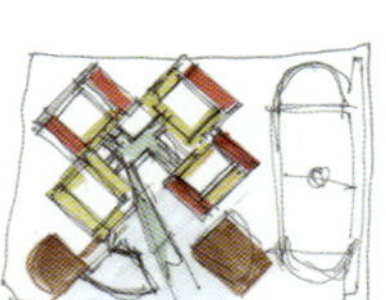
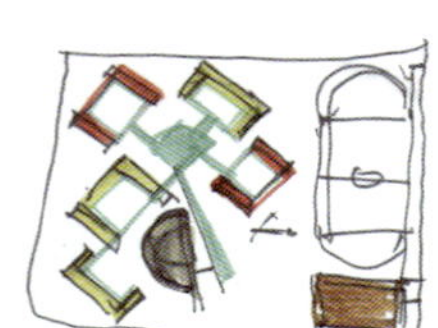
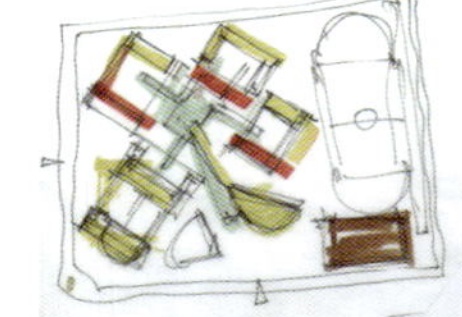
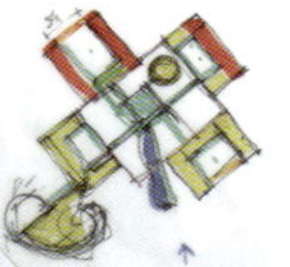

布局推敲 PROCESS LAYOUT

鲁迅先生说起私塾里"四角的天空"的时候，是羡慕在海边劳作的闰土，其实闰土也会羡慕他。如今多数小学校的平面，像串成一串的烤肠，一成不变。原因不光因为规范，更因为成见。我们担心那里的童年会像一阵风吹过空旷的校园，只剩下单调乏味。

因此我们回顾了传统的合院式教育空间，看看能否同时得到鲁迅和闰土的生活。我们认为在一个大院子里成长的小孩是幸福的。我们期望这样的校园里有安全感，有向心性，能肆意玩耍，偶尔也能有一点沉思默想。

Mr. Lu Xun envies Run Tu laboring at the seaside when talking of the four-cornered sky in the old-style private Chinese school, while Run Tu is also jealous of Lu Xun. The plane in a great deal of schools nowadays is just the cookie-cutter and prosaic type, just like a bunch of sausages linked together. The reason for this is not only for standardization, but more for prejudices. It is a matter of concern that the childhood spent there is just a gust of wind blowing through the vast campus, with only dullness left in its trail.

We look back on the traditional four-walled school, seeking for the signs of the lives of Mr. Lu Xun (writer, ideologist, revolutionist, and educationist in contemporary China) and Run Tu. We suppose the kids must be happier living in this kind of courtyard. Our will is that this kind of school would provide more security and cohesion, giving full vent to the children when they play, and sometimes also an opportunity for them to be deep in thought.

立面处理——古典元素应用 TREATMENT OF FAÇADE-APPLICATION OF CLASSICAL ELEMENTS

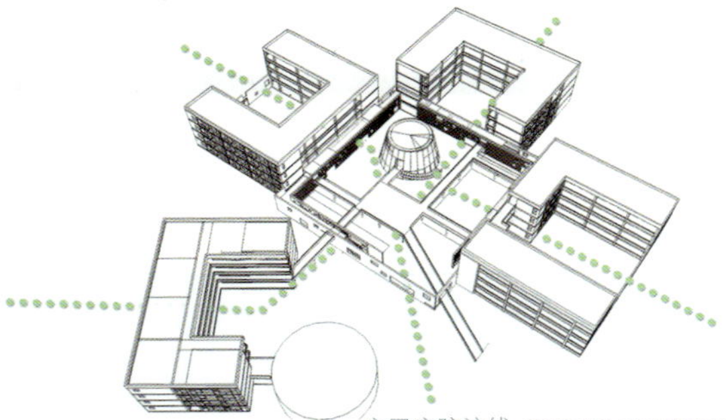

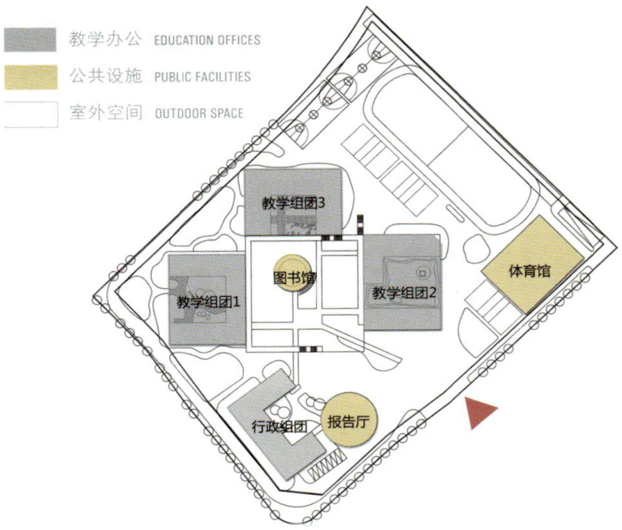

教学办公 EDUCATION OFFICES
公共设施 PUBLIC FACILITIES
室外空间 OUTDOOR SPACE

教学组团3
图书馆
教学组团1
教学组团2
体育馆
行政组团
报告厅

底层庭院流线 COURTYARD STREAMLINE FOR THE GROUND FLOOR

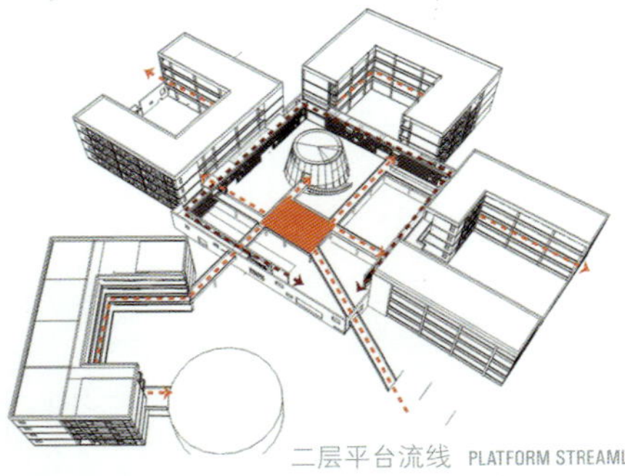

二层平台流线 PLATFORM STREAMLINE FOR THE FIRST FLOOR

流线分析 ANALYSIS OF STREAMLINE

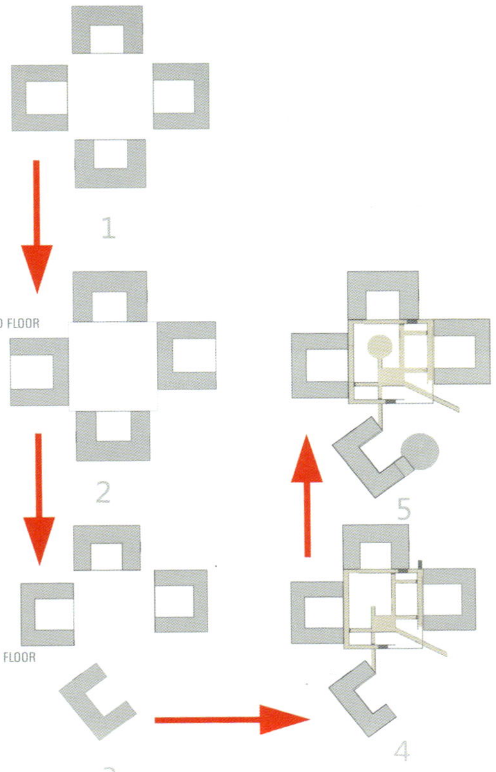

1 2 3 4 5

生成过程 GENERATION PROCESS

二层平台流线 PLATFORM STREAMLINE FOR THE FIRST FLOOR

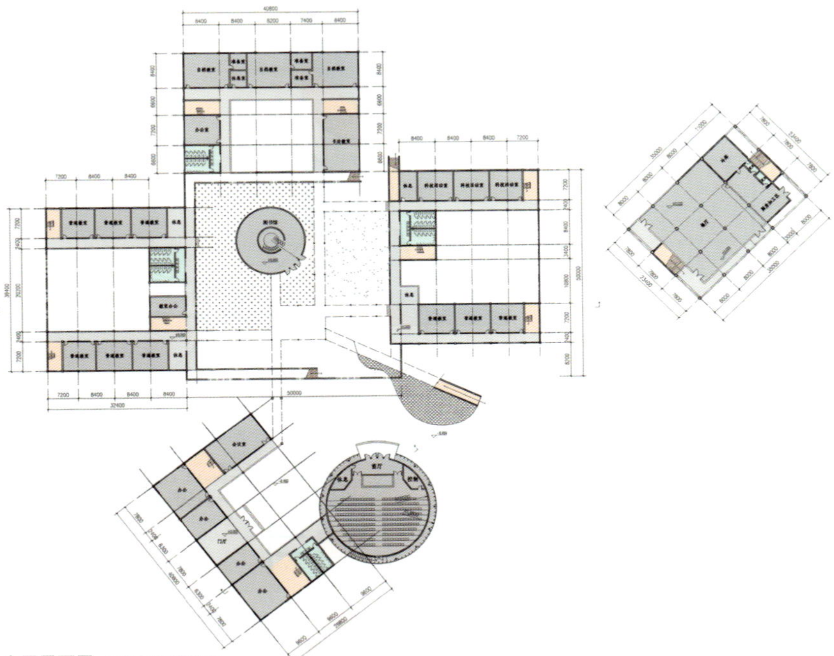

底层平面图 GROUND FLOOR PLAN

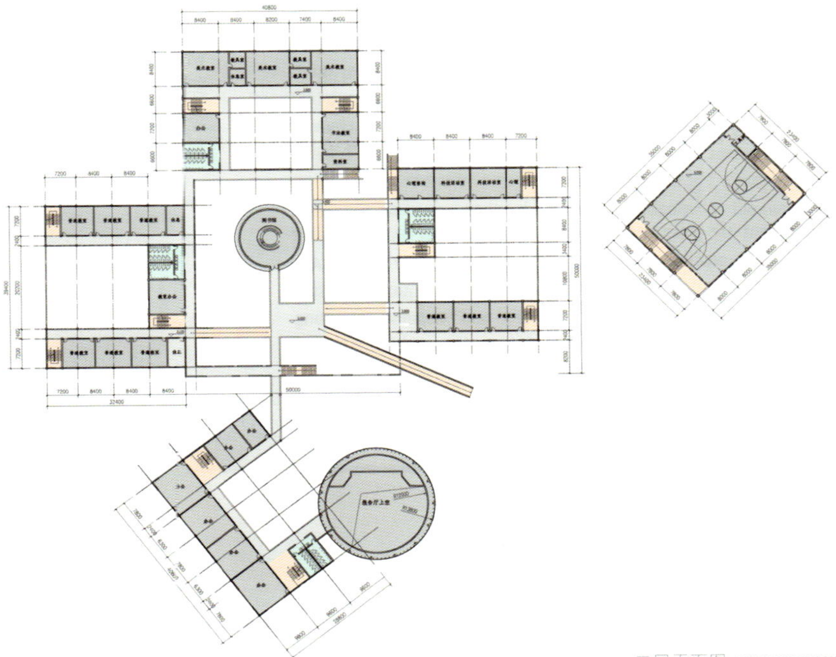

二层平面图 FIRST FLOOR PLAN

# 富阳东大道城市设计·富阳

## Urban Design of Fuyang East Avenue · China

罗卿平 Luo Qingping · 高小妹 Gao Xiaomei · 钱逸卿 Qian Yiqing

设计方案 design proposal

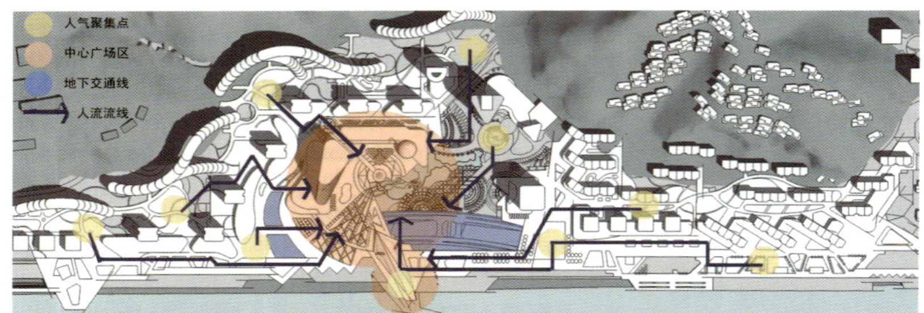


内部人流—由地面而独立的步行道路系统进入中心广场区

INTERNAL CROWD - ENTERING INTO THE CENTRAL SQUARE ALONG THE INDEPENDENT WALKWAY FROM

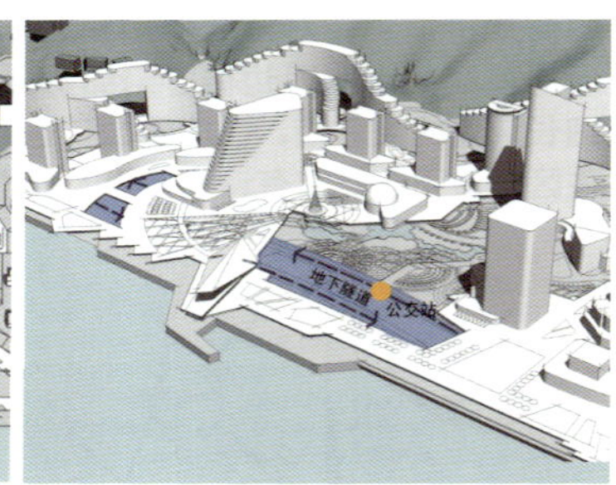


外部人流—由地下公交站进入中心广场区

EXTERNAL CROWD - ENTERING INTO THE CENTRAL SQUARE FROM THE SUBWAY STATION

该地块良好的景观价值决定了设计的出发点，拓展和重塑这个价值同等重要。因而无论是空间的布局、建筑形体、竖向设计还是色彩的确定，都是在力争景观价值最大化的同时去构筑一个能与自然相融的城市片区，并且其自身是一个充满活力、自由丰富的生活场所。设计中“大平台”的提出，综合解决了防洪及堤坝阻挡景观视线、地面停车、城市交通等问题，是设计中的一个亮点。

1. 良好的景观 2. 独立的步行系统带来稳定而安逸的场所感 3. 充分的休闲场地

1. GREAT LANDSCAPE; 2. STEADY AND PLEASANT FEELINGS WITH AN INDEPENDENT WALKWAY SYSTEM; 3. BROAD RECREATION SPACE

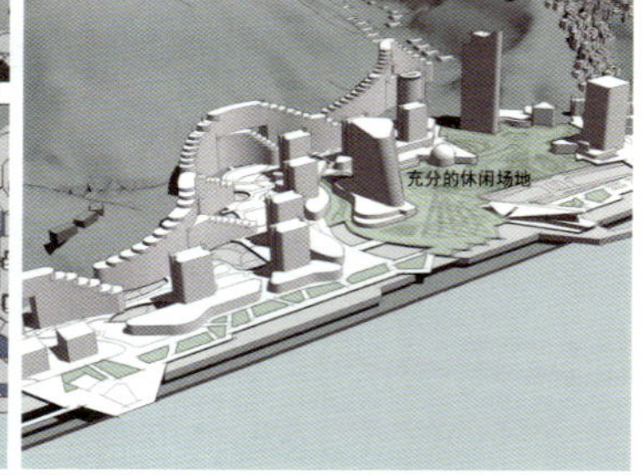


场所形式上的吸引

ATTRACTION FROM SPACE FORM

The starting point of the design is decided by the good landscape value of the plot, and it's equally important to expand and remold the value. Therefore, on the premise of maximizing the landscape value, best efforts are exerted to construct an urban block in harmonious coexistence with nature and enable the building to be a living place full of vigor, freedom and variety, no matter from the aspects of layout, architecture form, vertical design or colors. The proposal of "Big Platform" is a highlight of the design, solving such problems as flood control, dykes hindering people's vision for landscape, ground parking, and urban transportation.

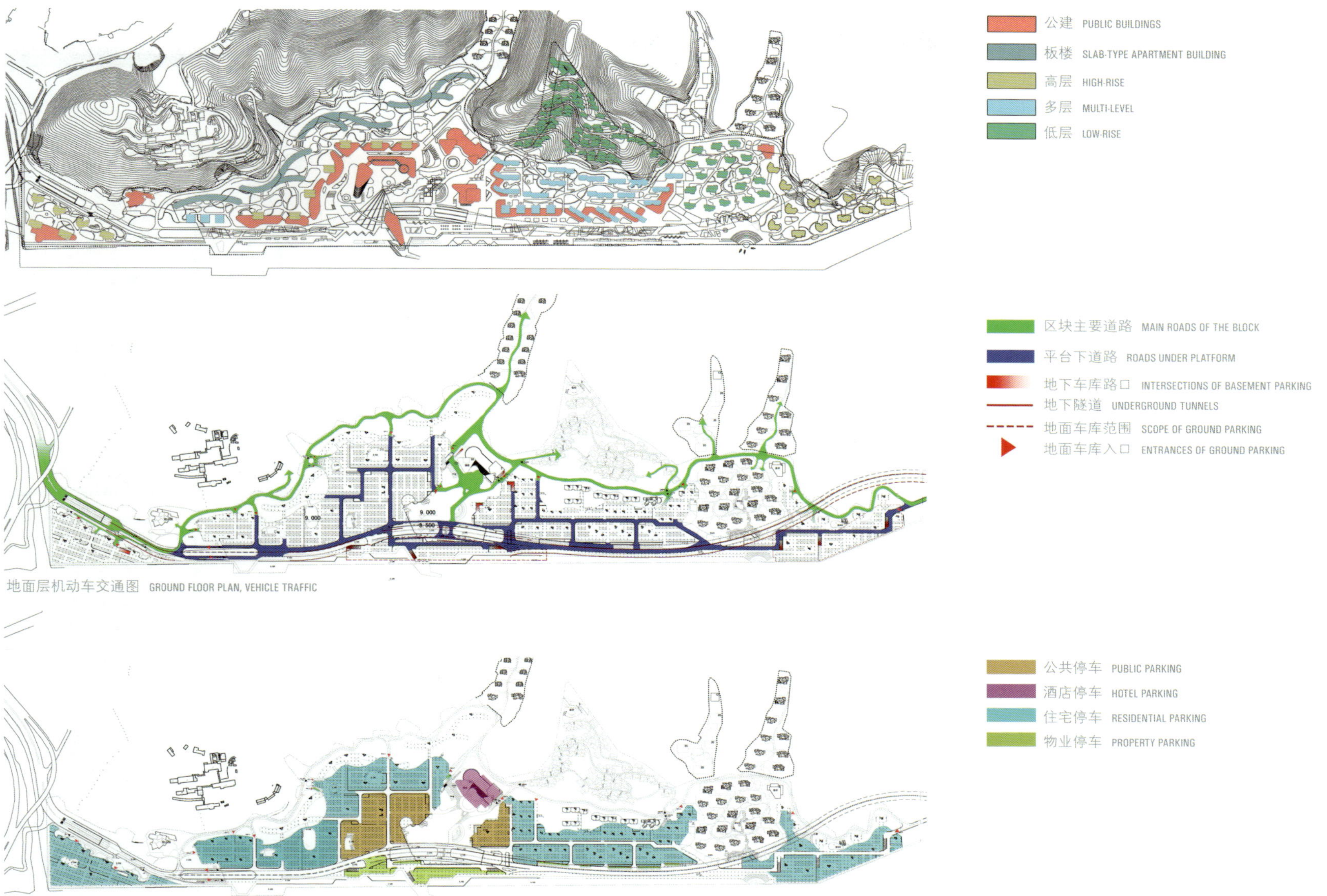


地面层机动车交通图 GROUND FLOOR PLAN, VEHICLE TRAFFIC

平台层停车图 PLATFORM FLOOR PLAN, PARKING

# 江苏美术馆·江苏

## Jiangsu Art Gallery · China

高裕江 Gao Yujiang · 卢挺 Lu Ting · 章文杰 Zhang Wenjie · 高庶三 Gao Shusan · 高生皓 Gao Shenghao

国际竞标方案 international competition proposal

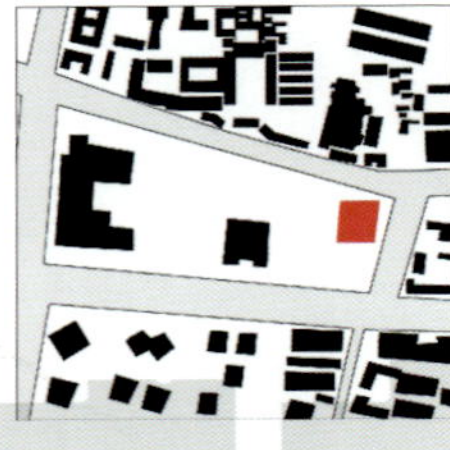

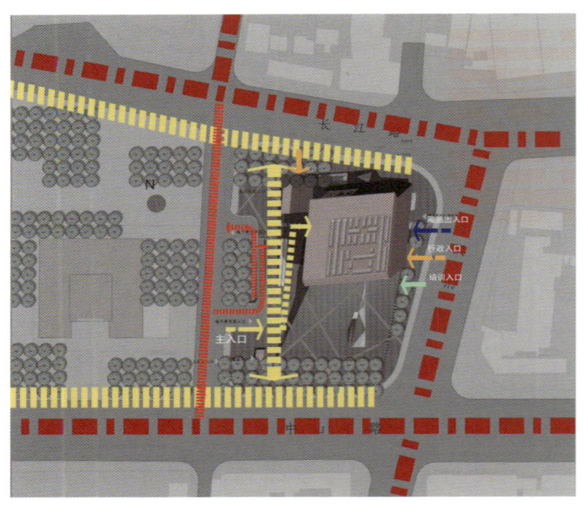

源自中国传统印鉴——方印的造型，以及出自国文综艺体“艺”的数字化处理后的窗格体面肌理的有机整构，通过倾斜、扭动的动势整合，一座现代而古悠、简约而意趣的城市标志性建筑呼之而出。它是建筑化的雕塑，也是雕塑化的建筑。俗称“南京红”的暗红色调，具有强烈的地域文化色彩及浓郁的人文精神气息。

Derived from the traditional Chinese seal - the project transforms a square seal, and organically reconstructs the volume-surface texture as a consequence of the digitalization of Chinese Mixed Art Font "Yi", with the integration of motive forces: sloping and twisting. We create a city's symbolic architecture which is modern and ancient, simple and interesting, is vividly portrayed. It is not only an architecturalized sculpture but also a sculpturalized architecture. The dark red commonly called as "Nanjing Red", features intensive regional cultural characters and a distinctive spirit of humanism.

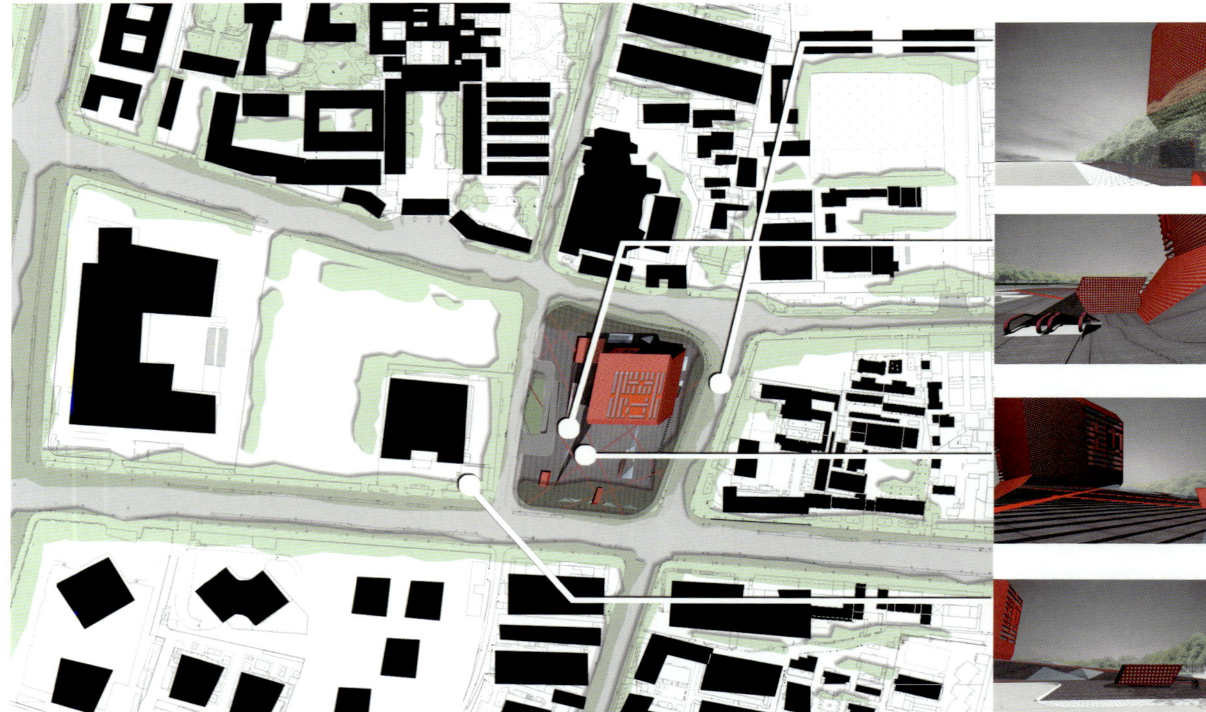

坡地与微微倾斜的建筑成了独特而宜人的城市景观
SLIGHTLY TILTED BUILDING, COMBINED WITH RAMP, BECOMES A UNIQUE AND PLEASANT URBAN LANDSCAPE

下沉的入口将参观者自然地引入建筑
LOWERED ENTRANCE GUIDES VISITORS INTO THE BUILDING NATURALLY

缓缓向上的台阶烘托出场地的宁静和建筑的雕塑感
STEPS RISE GRADUALLY, ENHANCING TRANQUILLITY OF THE PLOT AND THE SCULPTURAL APPEARANCE OF THE BUILDING

建筑的退让带来了连续的绿化和舒适的广场
THE BUILDING BACKS DOWN, SHAPING CONTINUOUS GREENING AND A PLEASANT SQUARE

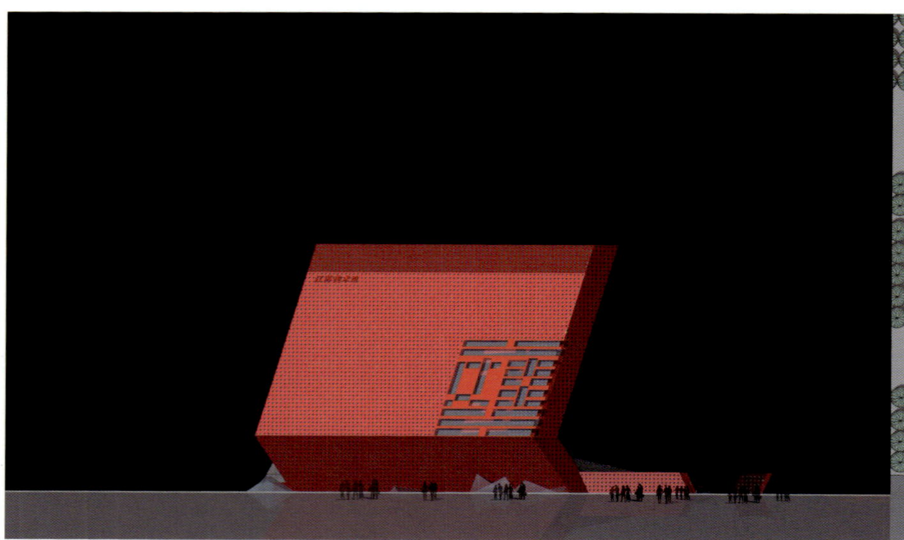

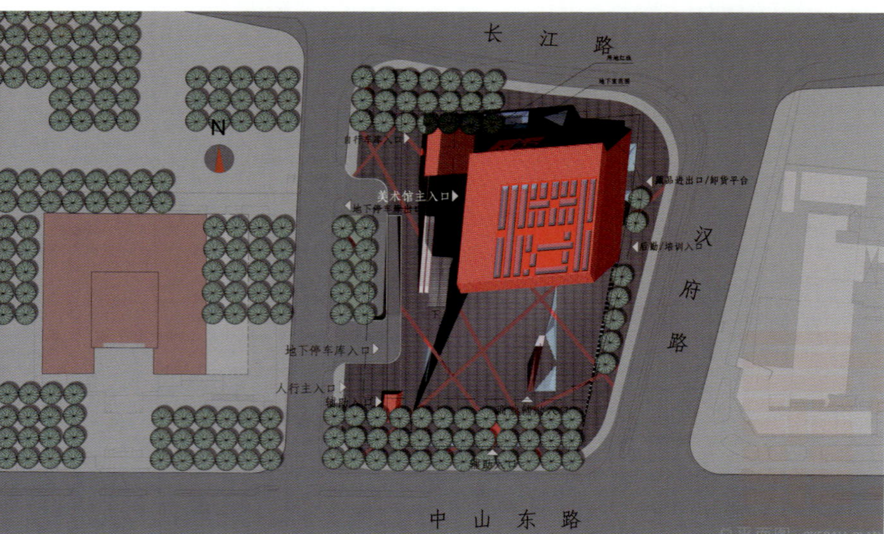


总平面图 OVERALL PLAN

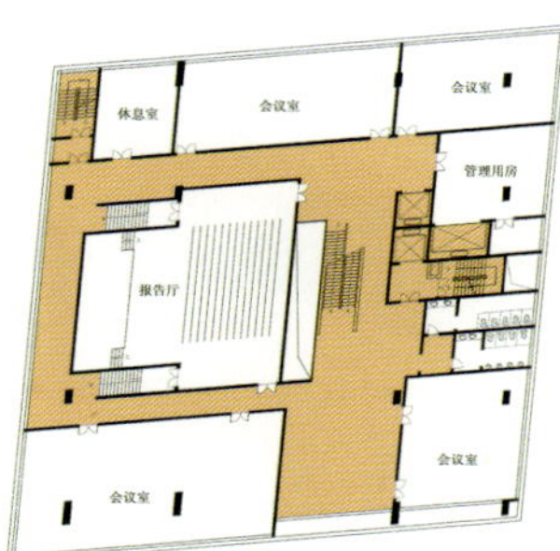

四层平面图 FLOOR PLAN 3

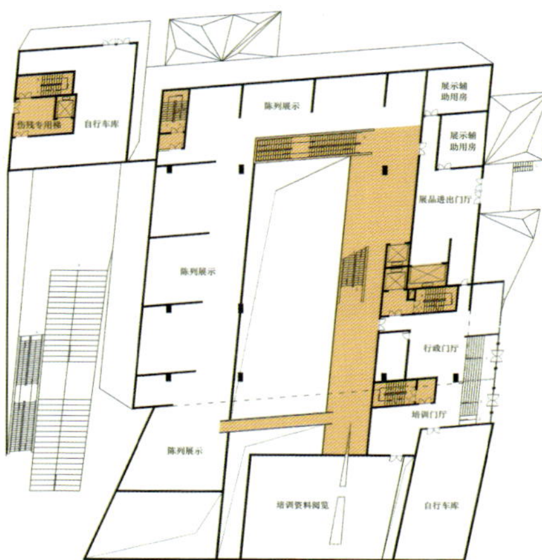

底层平面图 GROUND FLOOR PLAN

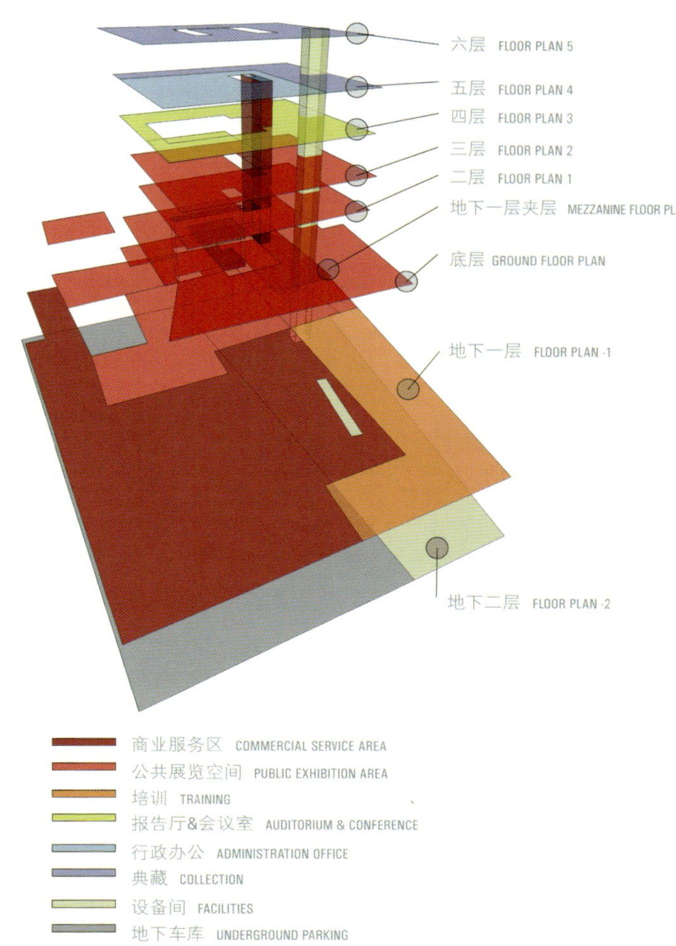


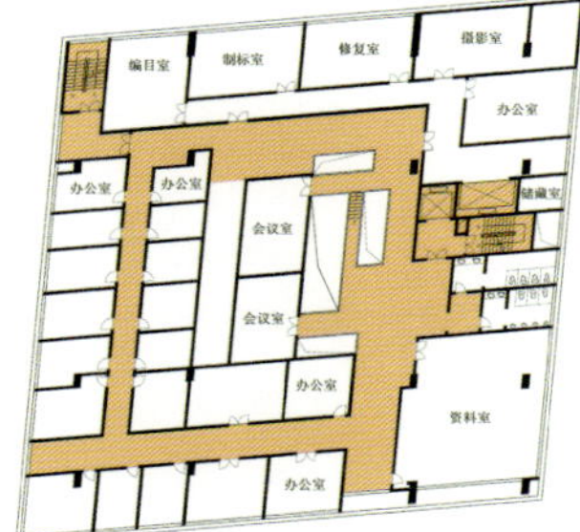

五层平面图 FLOOR PLAN 4

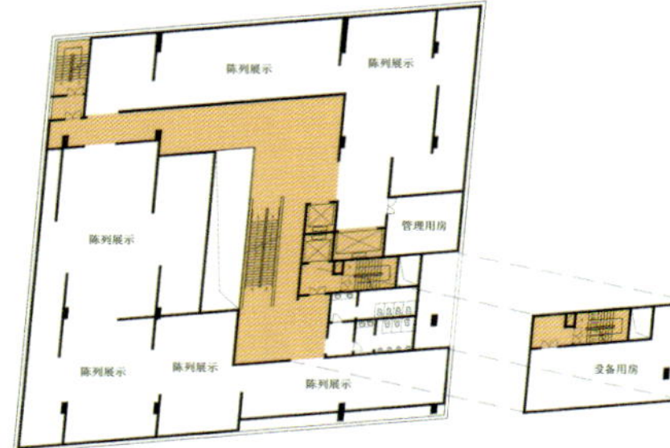

二层平面图 FLOOR PLAN 1

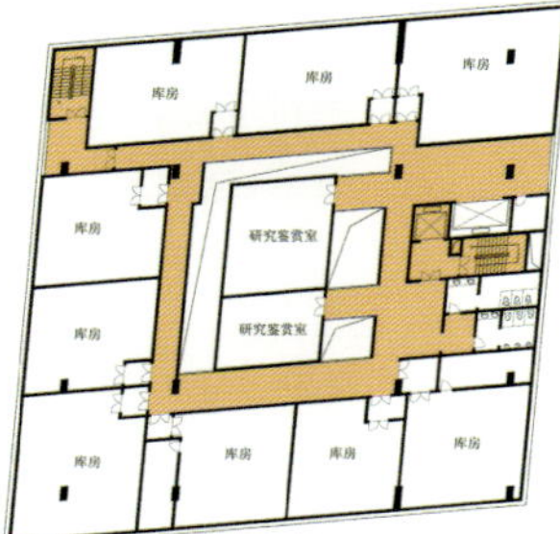

六层平面图 FLOOR PLAN 5

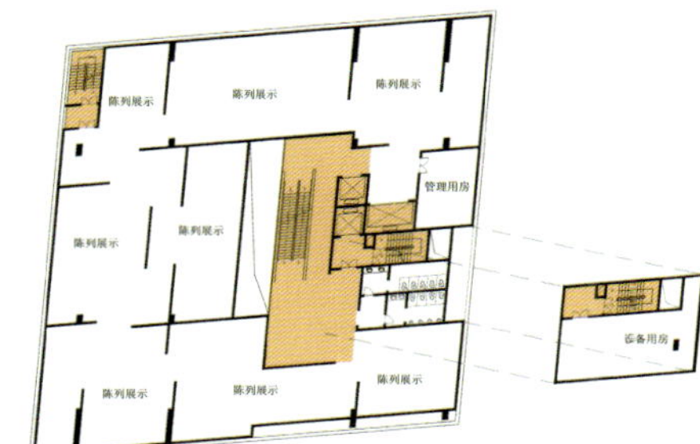

三层平面图 FLOOR PLAN 2

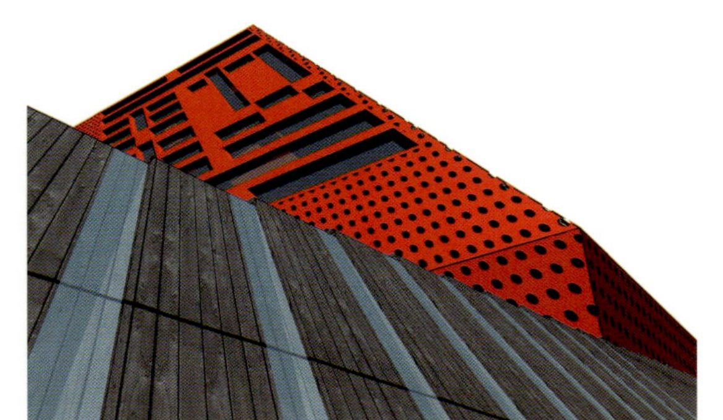

# 中国国家美术馆新馆·北京

## Expansion of China National Art Museum · China

陈瑜 Chen Yu · 邱文晓 Qiu Wenxiao · 朱睿 Zhu Rui

竞标方案 competition proposal

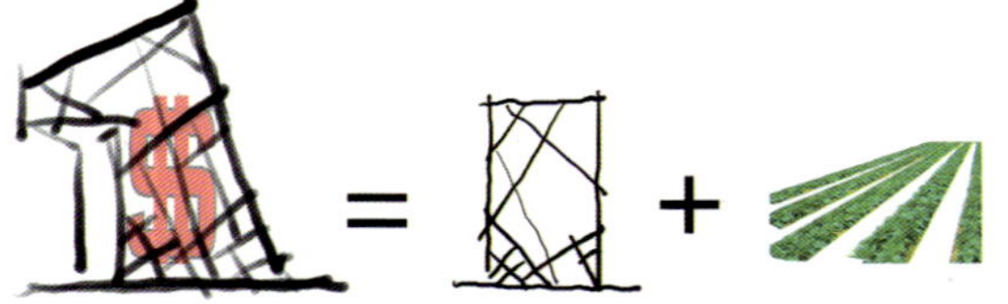

应该寻求高代价的标新立异，
One should aim to be outstanding which carries a high prize,

还是承担更多的社会责任，
or should accept more social responsibility,

亦或将二者统一，实现土地的重生。
or accept the combination for the revival of the arable land.

如果运用一些智慧，我们甚至可以使土地加倍
With some intelligence, the land could take advantage

而且不受季节的局限。
without limitation of seasons.

80亩

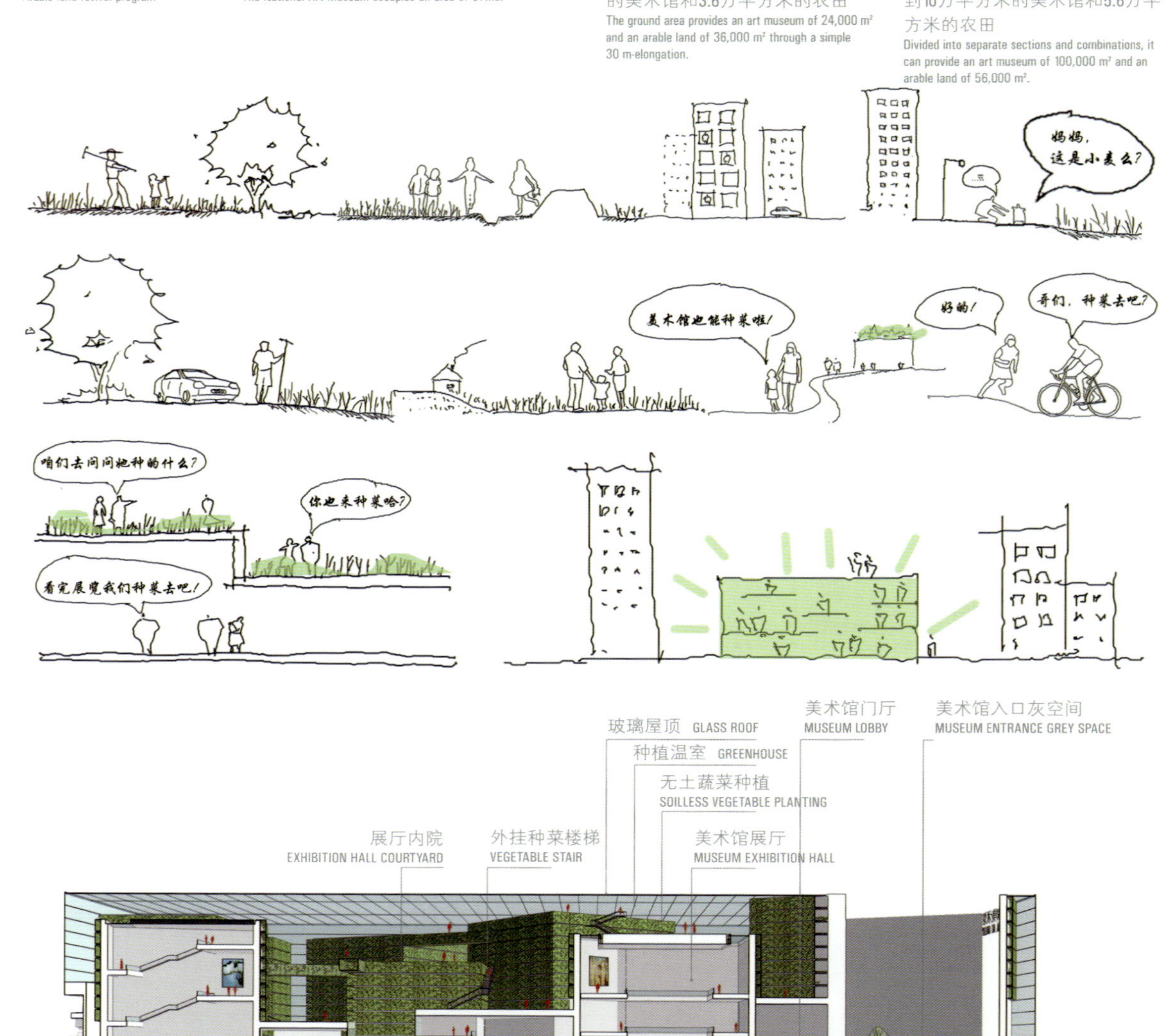


我们的立意：建设项目中的土地重生探索。

我们的策略：美术馆使用建筑容积，耕地使用建筑表皮。

种菜，被植入美术馆中，成为农耕生活的艺术。种菜行为与参观行为在这里碰撞，交流，契合了国家美术馆的公众参与职能。

我们希望通过在美术馆上种菜唤起人们心中的乡野情节，从而使国家美术馆成为北京城绿意盎然的世外桃源。

设计通过对基本立方体的扭转、拉伸、切割和咬合，实现了美术馆空间丰富性与种菜表面积最大化的完美统一。参观流线与种菜流线相互交织，却又各自独立；展览空间与种菜空间相互咬合，却又不期而遇……

Conception: revival of the arable land for the construction projects!
Strategies: the art museum uses the volume size and the arable land for the building skin.
Vegetable-growing has become an art for the agricultural tillage in the art museum. The vegetable-growing interacting with the sighting of visitors conveys with the function of audience participation in the National Art Museum of China.
It is our hope that the vegetable-growing in the art museum can arouse the appeal of countryside in the hearts of the visitors and make the art museum the "Arcadia" in Beijing.
The torsion, elongation, cutting and meshing for the architectural cubes in the design represents the unity of the versatile use of the art museum and maximization of areas for growing vegetables. The areas for visiting and vegetable-growing are crisscrossed yet independent; the exhibition space and the vegetable-growing area combine organically and naturally...

温室屋顶 GREENHOUSE ROOF
温室空间 GREENHOUSE SPACE
三维菜地 3 D VEGETABLE FIELD
结构板 STRUCTURAL SLAB
室内空间 INDOORSPACE

空间构成分析 SPACE COMPOSITION ANALYSIS

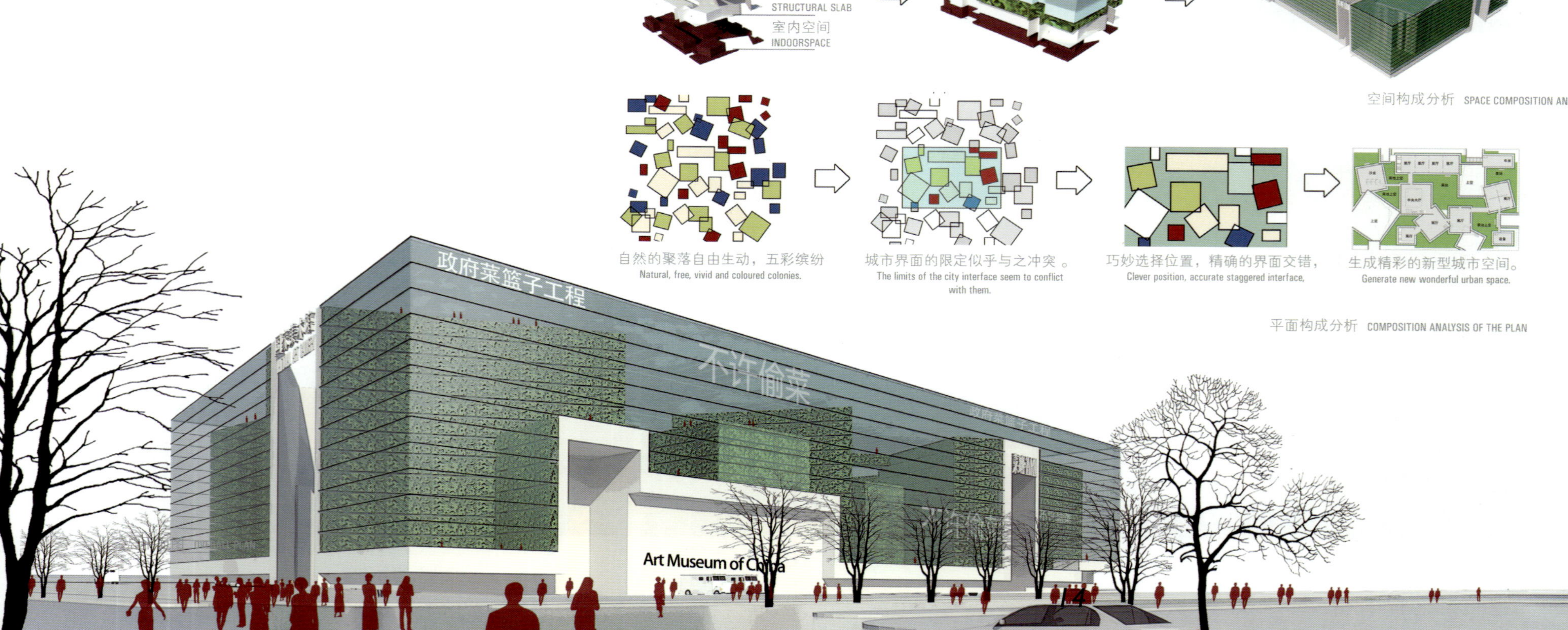


自然的聚落自由生动，五彩缤纷
Natural, free, vivid and coloured colonies.

城市界面的限定似乎与之冲突。
The limits of the city interface seem to conflict with them.

巧妙选择位置，精确的界面交错，
Clever position, accurate staggered interface,

生成精彩的新型城市空间。
Generate new wonderful urban space.

平面构成分析 COMPOSITION ANALYSIS OF THE PLAN

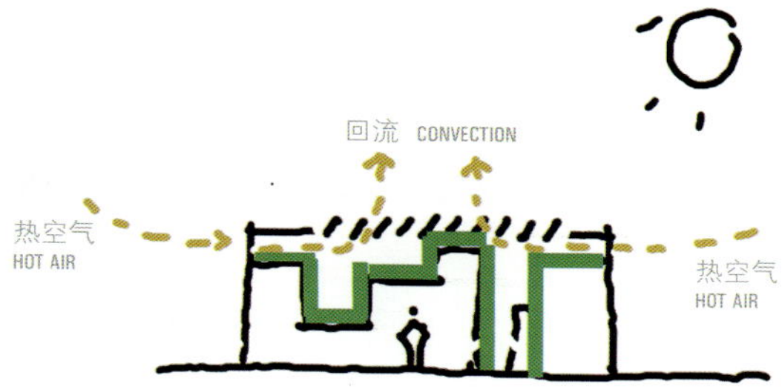


夏季，玻璃幕墙开启，外界热空气从暖房屋顶流出，形成回流风，与种植屋顶和墙面共同作用，保证美术馆外墙温度不因受到辐射而剧烈上升。

In summer, the outside hot air flows out through the roof of the greenhouse when the glass curtain wall is open, stopping, with the combination of rooftop planting and façade, the temperature of the exterior wall of the art museum from shooting up.

冬季，玻璃幕墙关闭，外界冷空气经过暖房和蔬菜加热增氧，导入美术馆室内，形成清新温暖的新风。

In winter, the glass curtain wall will be closed and the outside cold air, after being warmed up by the greenhouse and filtered by the vegetables growing, will be more fresh and warm.

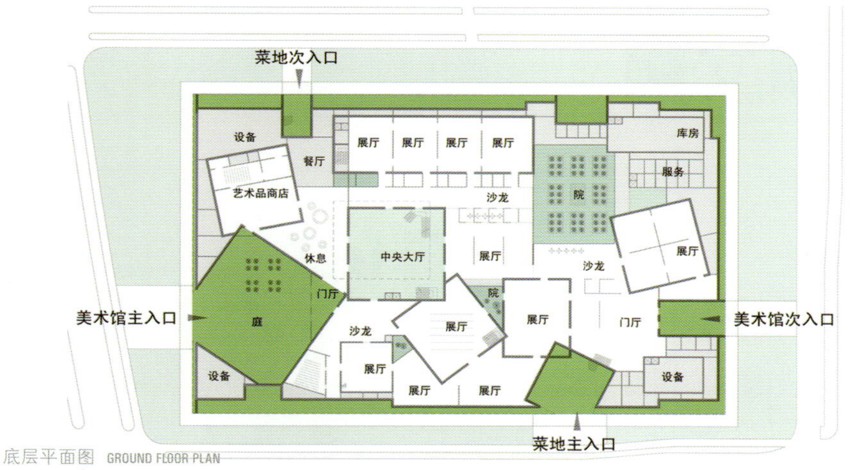

菜地次入口
美术馆主入口
美术馆次入口
菜地主入口

底层平面图 GROUND FLOOR PLAN

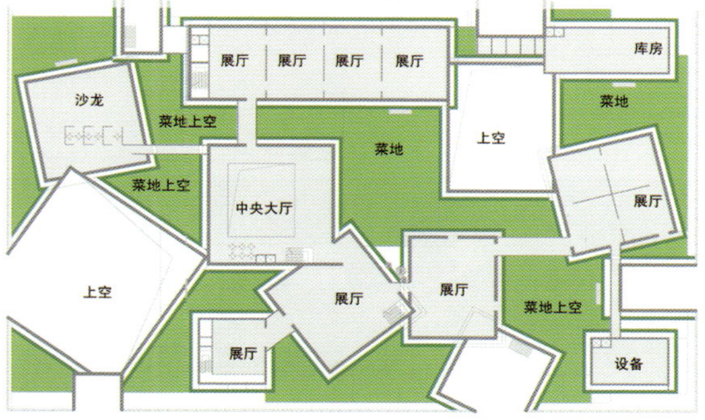

二层平面图 FIRST FLOOR PLAN

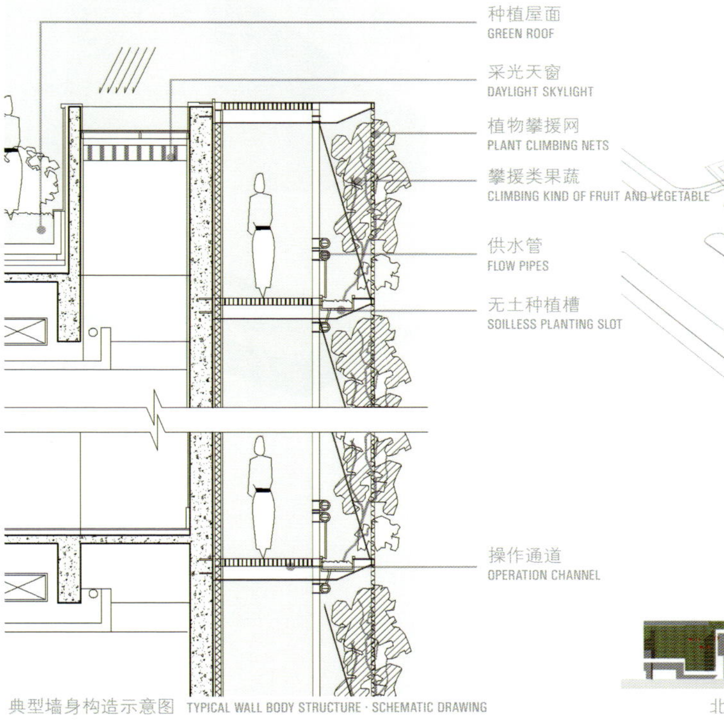

种植屋面
GREEN ROOF
采光天窗
DAYLIGHT SKYLIGHT
植物攀援网
PLANT CLIMBING NETS
攀援类果蔬
CLIMBING KIND OF FRUIT AND VEGETABLE
供水管
FLOW PIPES
无土种植槽
SOILLESS PLANTING SLOT
操作通道
OPERATION CHANNEL

典型墙身构造示意图 TYPICAL WALL BODY STRUCTURE · SCHEMATIC DRAWING

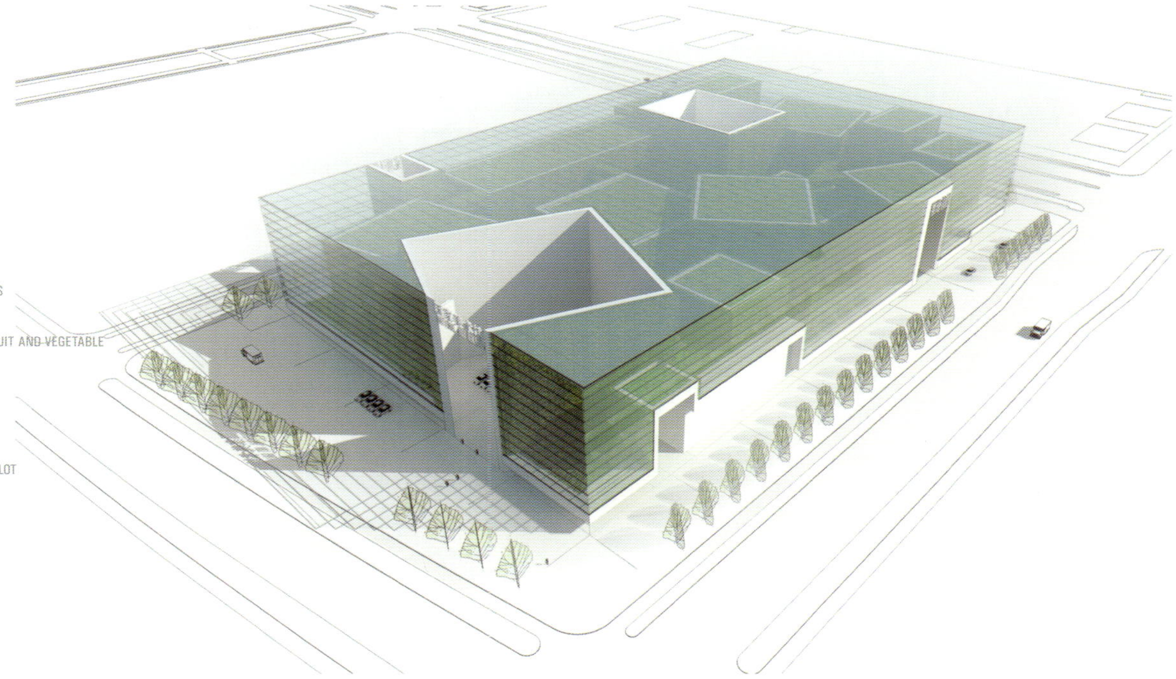

北向立面图 NORTH ELEVATION　西向立面图 WEST ELEVATION　南向立面图 SOUTH ELEVATION　东向立面图 EAST ELEVATION

# 中国国家美术馆新馆 · 北京

## Expansion of China National Art Museum · China

方涛 Fang Tao

竞标方案 competition proposal

山水
MOUNTAIN AND WATER

光线
LIGHT

树木
WOOD

林荫道
BOULEVARD

参观人流
VISITORS FLOW

穿行人流
CROSS FLOW

广场山水
MOUNTAIN AND WATER ON SQUARE

庭院的光线
LIGHT IN THE COURTYARD

立体的绿化
STEREO GREENING

农田的路径
PATH TO FARMLAND

穿行人流
CROSS FLOW

参观人流
VISITORS FLOW

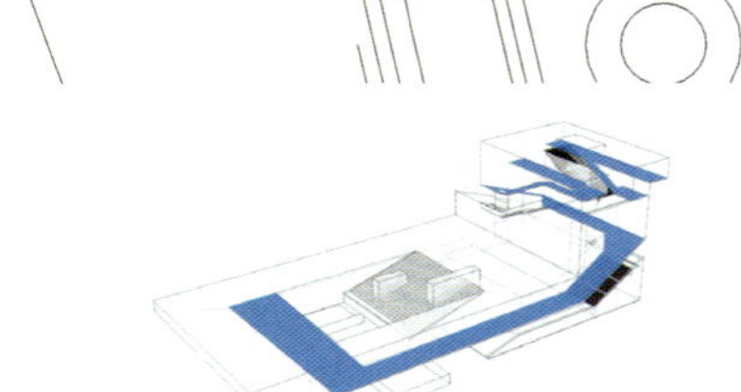
穿行流线 TRAVERSE STREAMLINE

+

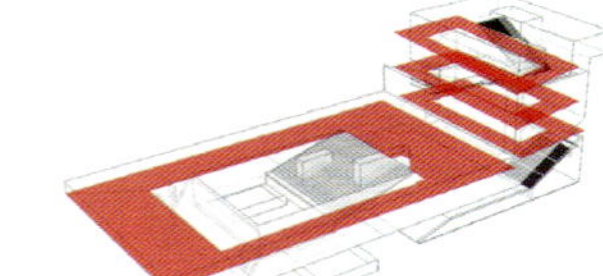
参观流线 VISITING STREAMLINE

+

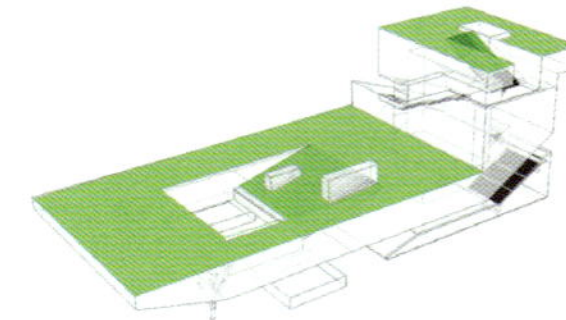
沿途景观 LANDSCAPE ALONG THE ROUTE

=

国家美术馆 CHINA NATIONAL ART MUSEUM

传统意义上的美术馆是一个艺术的圣殿。中国国家美术馆的立意之初在于展示一种开放的姿态，让人们可以不经意间接触到艺术，使美术成为人们生活文化的一部分。回归自然的生活和高于生活的艺术之间的相互融合便是此次方案的立意。

攀登，是很多人喜欢和选择的一种运动。攀登可以让你锻炼身体、磨练意志，而更多时候能让你亲近自然，饱览美景。本次方案将这一行为演化成在都市中行走，体会攀登的乐趣，将穿行路线设置为环绕美术馆逐渐向上。

In a traditional sense, an art museum is a hallowed ground for fine arts. The goal of the National Art Museum of China since its conception, is to convey its openness and make more chances for people to get inadvertently close to art, to make art a part of their cultural life at a subconscious level. The program is based on the interaction between life and nature.

As favorite sport, climbing is a sport for exercising the body and tempering the will power, while offering a chance to get closer to nature and sight-seeing. The program takes the form of a pathway which develops around the art museum and then upwards as it was climbing, as if it walked through the city enjoying the pleasure of climbing.

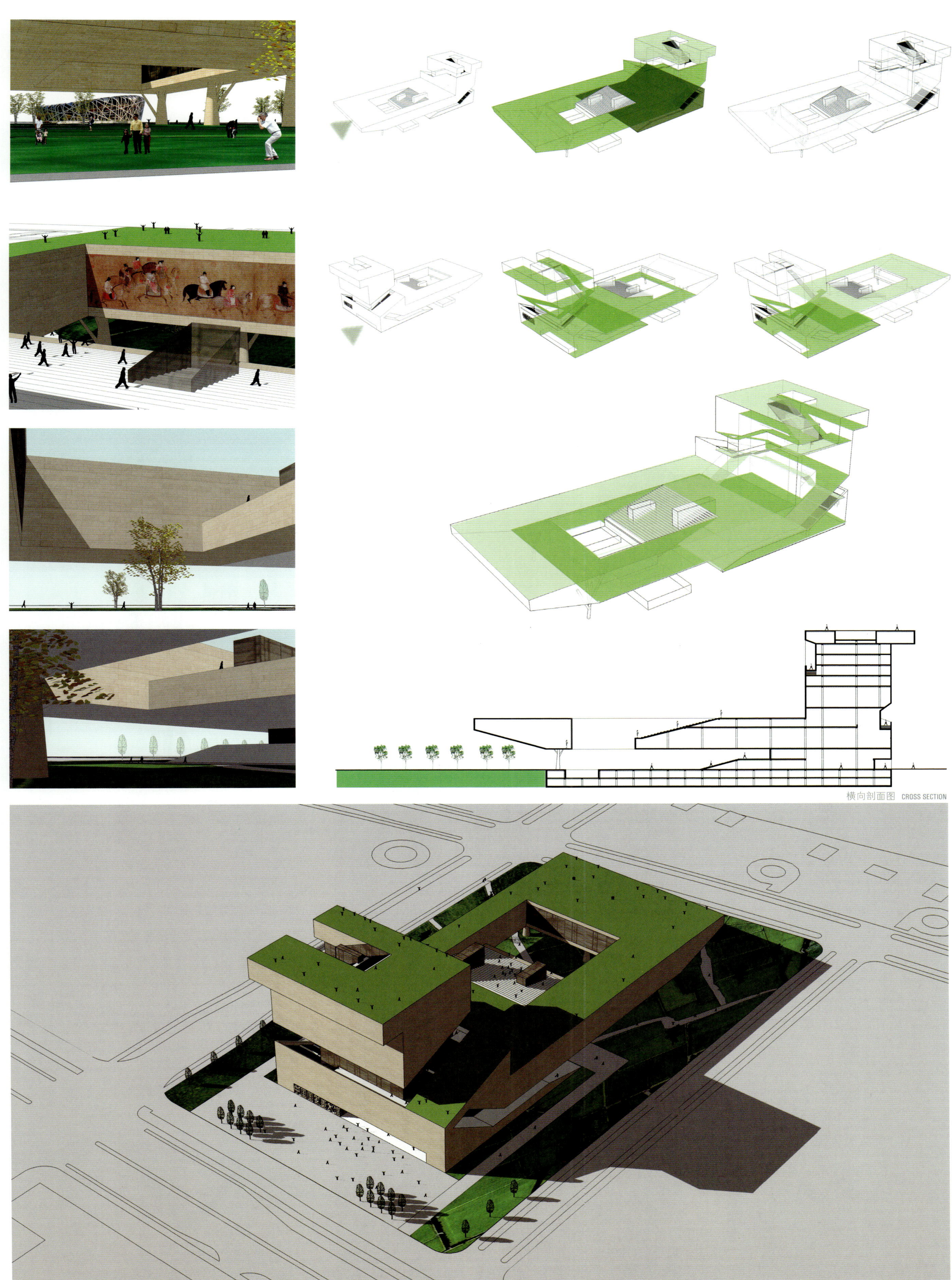
横向剖面图 CROSS SECTION

# 中国国家美术馆新馆·北京

## Expansion of China National Art Museum · China

陈 冰 Chen Bing

竞标方案 competition proposal

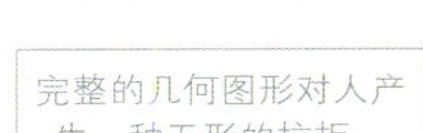
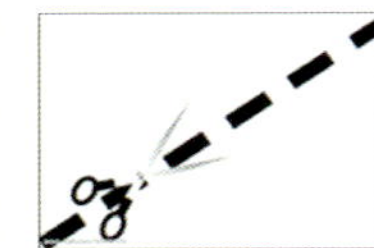

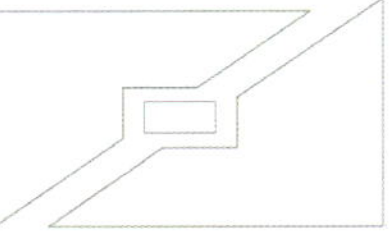
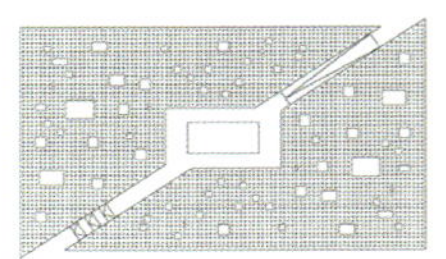

完整的几何图形对人产生一种无形的抗拒，
The complete geometric figure holds an irrespirable appeal to people.

通廊的引入使建筑呈现出一种开放的姿态。
with the vestibules signifying the open nature of the building

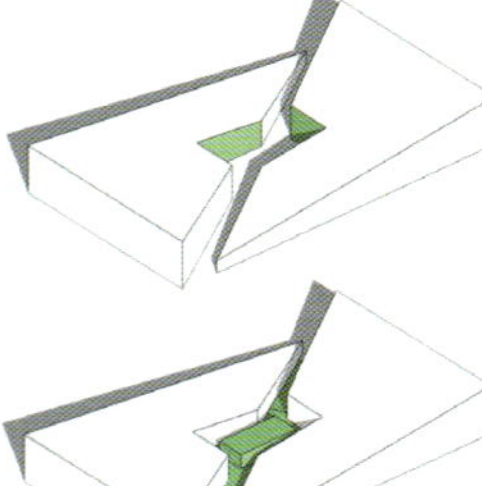
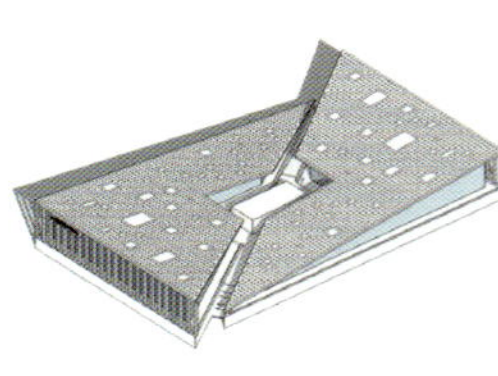
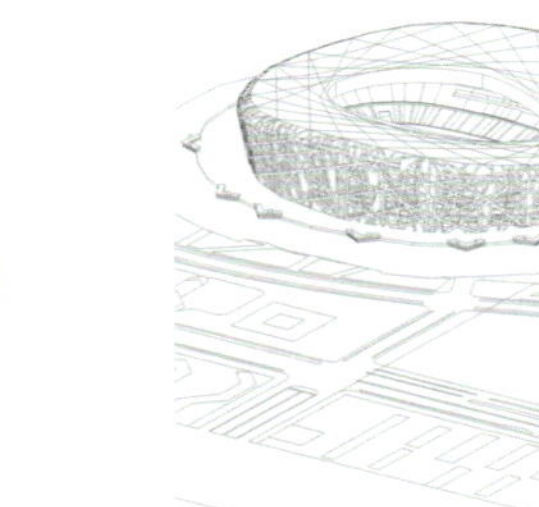
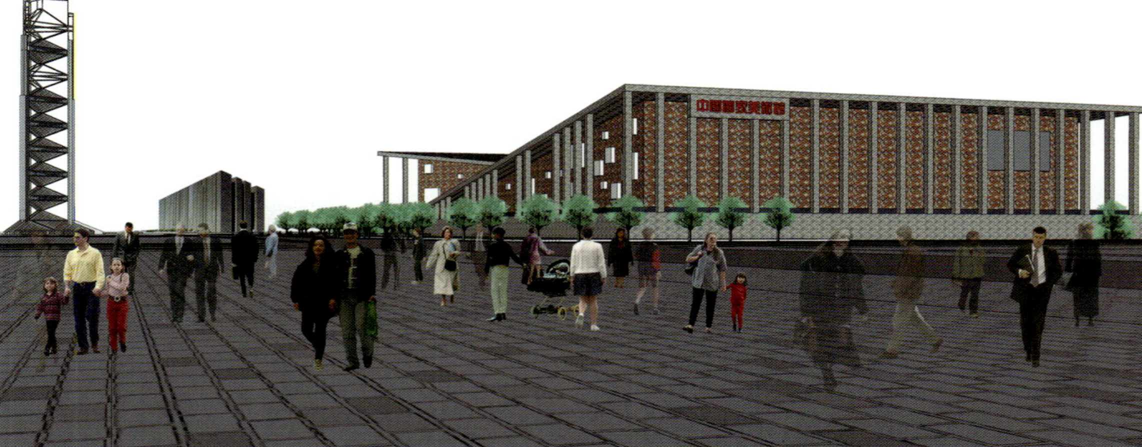

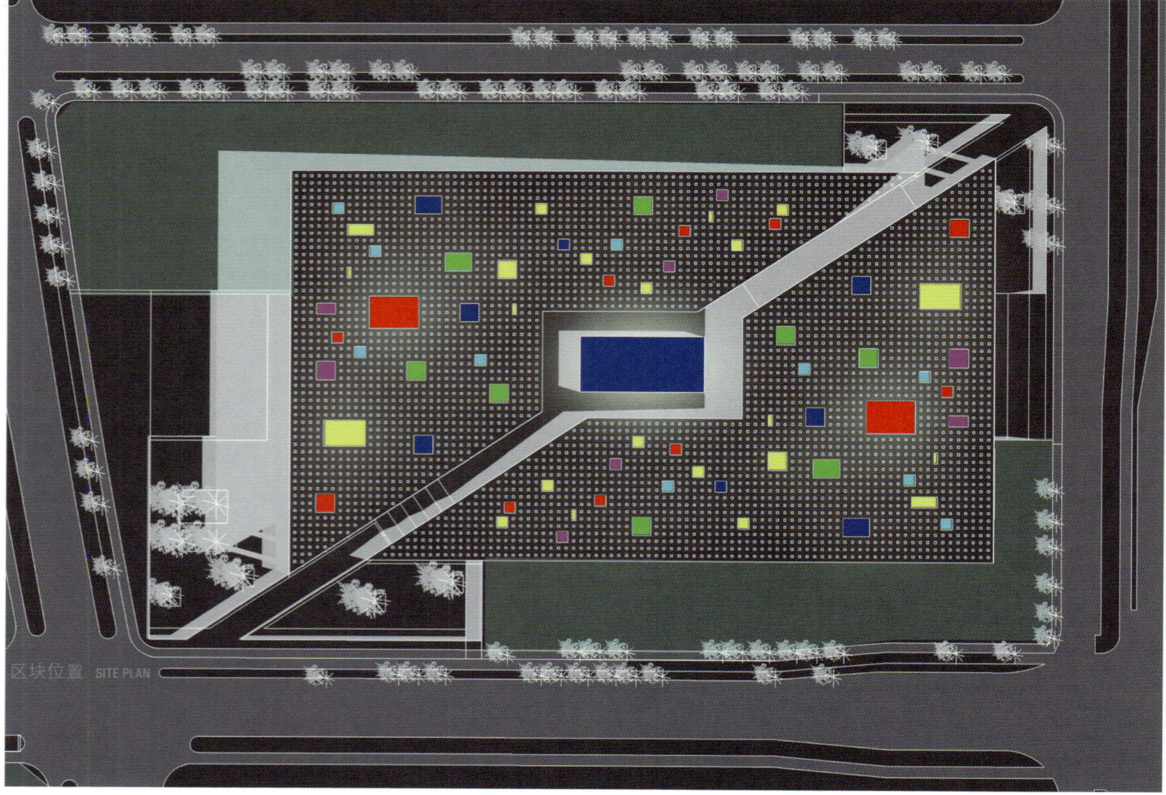

区块位置 SITE PLAN

当艺术插上翅膀之后……本案从整个奥体公园区位、国家美术馆建筑自身所需形象以及所承载的功能出发，以“翼”为设计意象，强调整个建筑稳重又不失活泼、简洁又不显得简单。

When the art is equipped with wings... Its displays its image and functions as a museum. The National Art Museum of China takes "wings" as its design image within the large picture of the Olympic Park, with an emphasis on being stately yet lively, simple yet not simplistic.

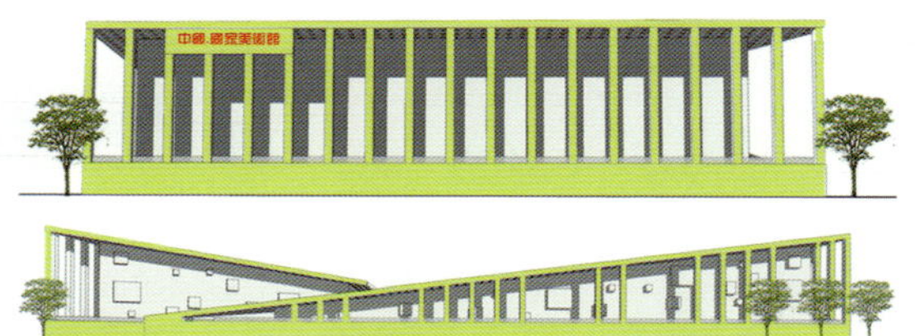

外界面：建筑沿街立面比较活跃，呈现一种非对称状态，显得较为灵动、活泼，体现亲民的理念。
Exterior: The street front side of the building is lively with its asymmetric design to convey the images of vitality, vividness and affability.

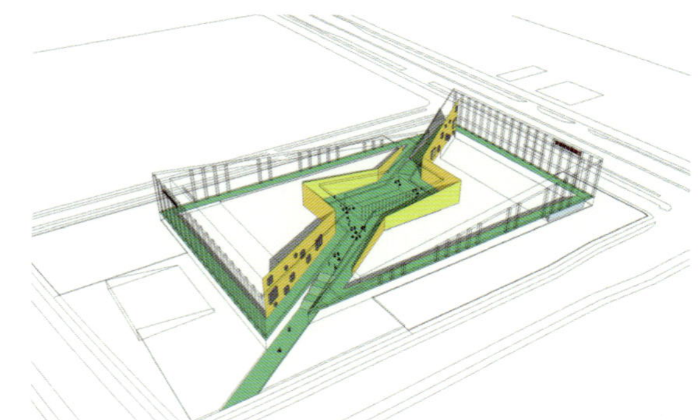

通廊的引入形成了非常富有戏剧感的内界面。
The introduction of vestibules provides a dramatic feeling to the interior.

表皮一：屋顶LED灯 FAÇADE ONE: ROOFTOP LED LIGHTS

设计尝试将现代技术结合到美术馆建筑中，将LED灯融合到美术馆斜屋面中，犹如一块“画板”，白天跟建筑整体外墙融为一体，夜晚则呈现出绚烂的灯光效果。

With the use of the modern technology, the design incorporates LED lights into the sloping roof of the art museum, just like a drawing board, which blends into the surrounding exterior during the day and shows all the colours of the rainbow at night.

表皮二：格子窗花 FAÇADE ¯WO: LATTICE WINDOW PAPER

设计从中国传统建筑汲取设计元素，以格子窗花作为美术馆表皮肌理，具有极高的可识别性，也极易被大众所认知，整个美术馆建筑以一种极其现代的姿态来诠释中国特色。

用格子窗花包裹美术馆内外两个界面之后，再结合建筑中间内廊及周边外廊设计，在格子窗表皮上嵌入一个个画框，加强内外界面的层次感，同时人民大众不需要进入美术馆，通过内外走廊就能够欣赏到珍贵的艺术品。

The lattice window paper shrouding the exterior and interior of the art museum blends with the inner lobbies and outside corridors with a frame on each lattice window considering the effects of gradation; thus the visitors can appreciate the precious works of art through the inner and outside corridors without entering the art museum.

# 中国国家美术馆新馆·北京

## Expansion of China National Art Museum · China

霍飞 Huo Fei

竞标方案 competition proposal

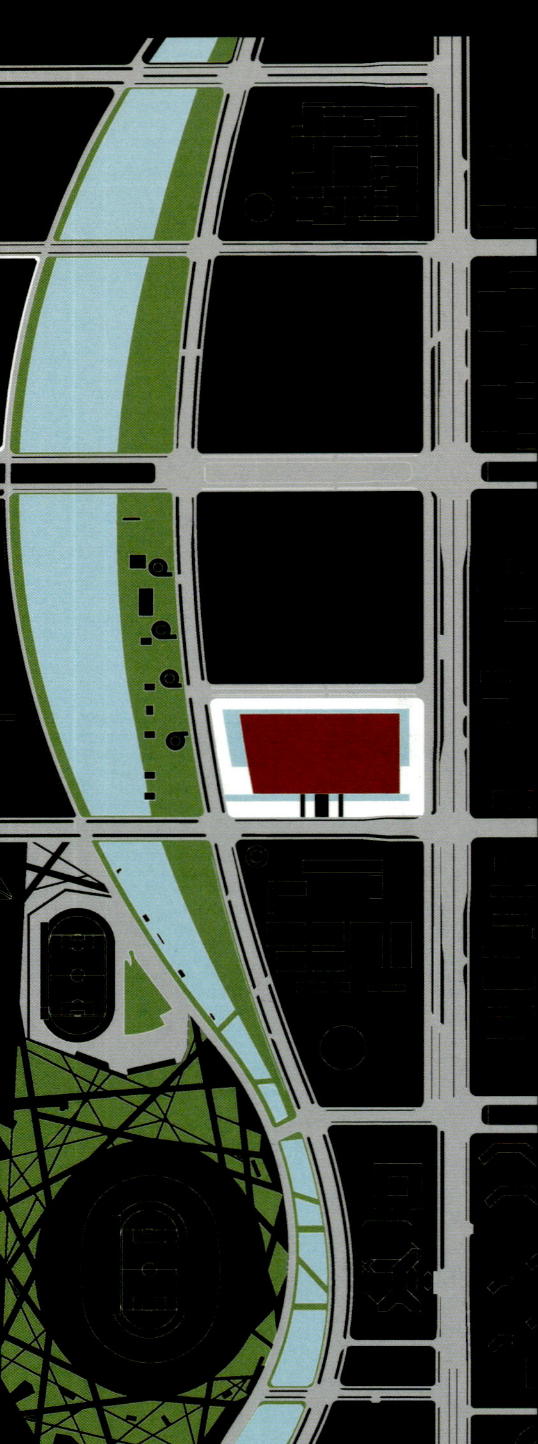

区块位置 SITE PLAN

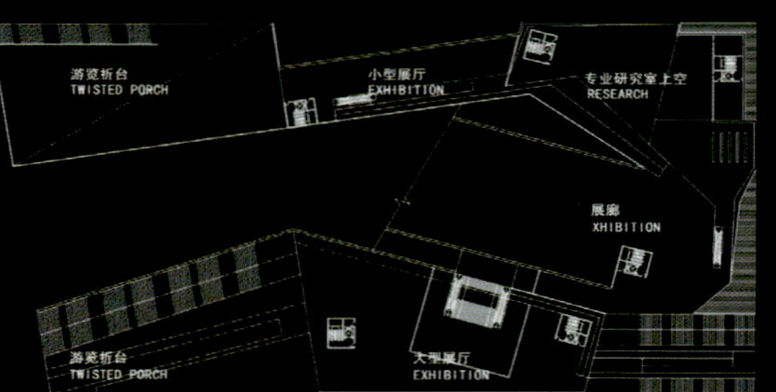


二层平面图 FIRST FLOOR PLAN

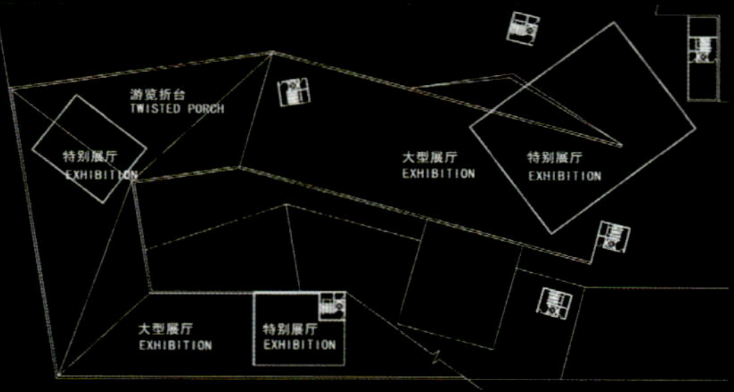


三层平面图 SECOND FLOOR PLAN

方案着眼于展示方式的探询。

在这里外形的因素被放置于一个并不显眼的位置。建筑本体对自身差异性的顽固坚持带来的过多冲突显现的实质，有时会是某种建筑“失忆”的引导。视觉疲倦在这里要尽可能避免。

展示是一个内外结合的过程。“表面的影幕”展示的是表面化趋向；“错置的盒子”展示的是集中性引导；“翻折的游廊”展示的是趣味感提升。建筑的整个表皮空间都只是展示的装置。装置艺术的灵活性涵盖本次设计的主旨，让整个建筑为展示服务，人们关注的焦点不再是室内单纯的展品。

The program focuses on exploring ways for mode of expression. The exterior elements are placed in an inconspicuous position. The marked contrast resulting from the structure's insistence on diversity act as a beacon for the amnesia of the building. Visual fatigue is beyond the realms of possibility.

Exhibition is a process for exterior and interior expression. "Surface screen" is to convey what is obvious; "Dislocated boxes" is for the central guidance role; "Folded corridor" is to add joy and pleasure. The whole surface area of the building is for expression purposes. The flexibility of the installation art embraces the goal of this design, namely the whole building as an exhibition piece, just like the items on display inside the building.

城市银幕 SURFACE BEAUTIFULLY GRACED

城市展厅 EXHIBITION CITY

旋转的盒子 ROTATED BOXES

横向剖面图 CROSS SECTION

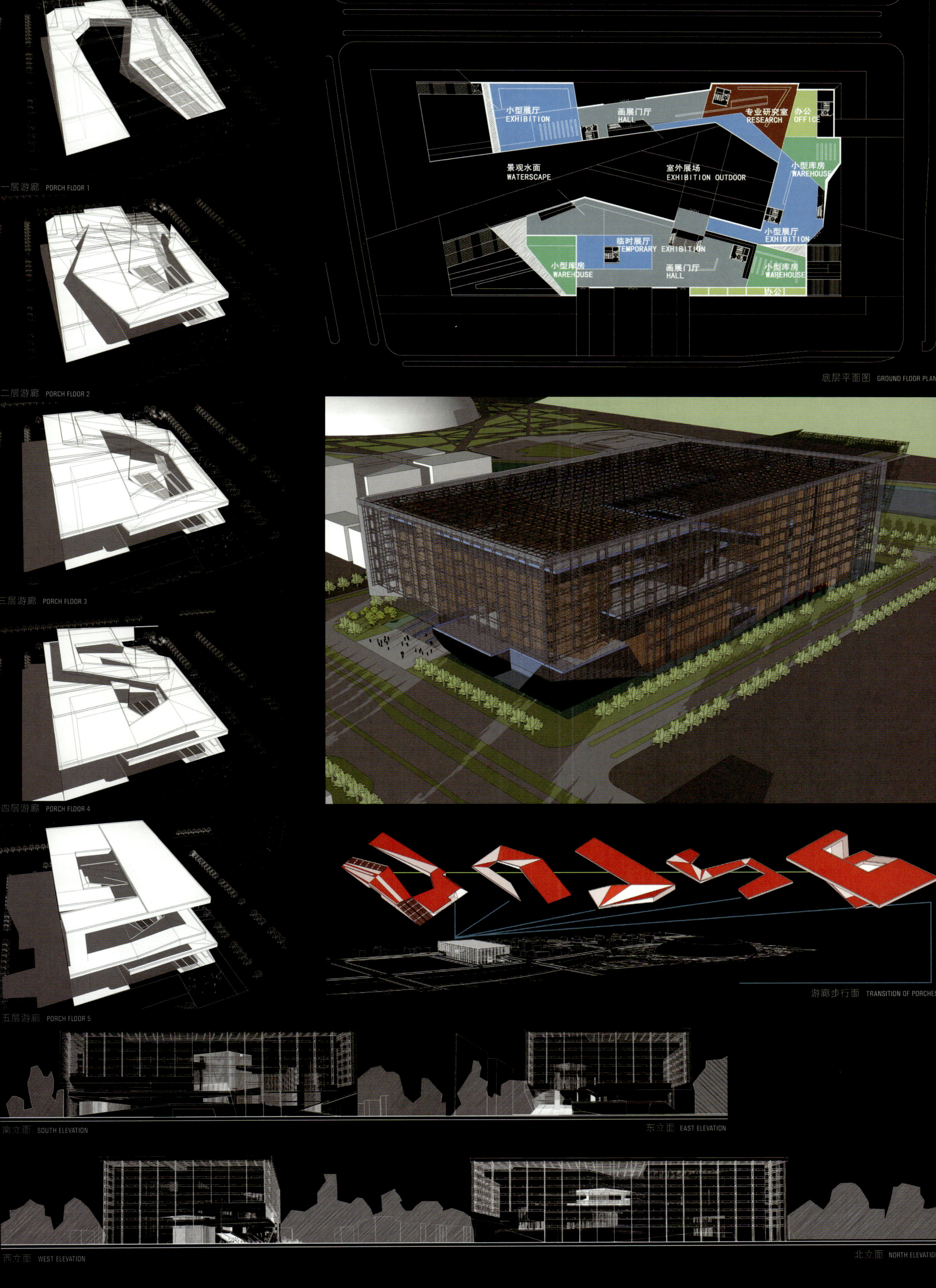
一层游廊 PORCH FLOOR 1
二层游廊 PORCH FLOOR 2
三层游廊 PORCH FLOOR 3
四层游廊 PORCH FLOOR 4
五层游廊 PORCH FLOOR 5
小型展厅 EXHIBITION
画展门厅 HALL
专业研究室 RESEARCH
办公 OFFICE
景观水面 WATERSCAPE
室外展场 EXHIBITION OUTDOOR
小型库房 WAREHOUSE
小型展厅 EXHIBITION
临时展厅 TEMPORARY EXHIBITION
小型库房 WAREHOUSE
画展门厅 HALL
小型库房 WAREHOUSE
办公
底层平面图 GROUND FLOOR PLAN
游廊步行面 TRANSITION OF PORCHES
南立面 SOUTH ELEVATION
东立面 EAST ELEVATION
西立面 WEST ELEVATION
北立面 NORTH ELEVATION

# 中国国家美术馆新馆·北京

## Expansion of China National Art Museum · China

郑茂恩 Zheng Mao'en

竞标方案 competition proposal

中庭——为防止大型美术馆“幽闭症”的产生，观看画展的人们从一个展厅出来，带着沉甸甸的思考准备到下一个展厅，其间必须穿越由绿色植物构成的中庭空间，这样的空间通风透雨，可以给人们带来片刻的轻松和愉悦，彼此交流、稍作休息。参观者在不同展厅之间来来回回的穿越行为构成了美术馆独特的体验经历。

Courtyard - a space for preventing the "claustrophobia" in the large-scale art museum. The visitors, deep in thought and fresh from one exhibition hall and to the next exhibition hall, will inevitably pass through a courtyard dotted with greenery; a grey area like this, open to the elements, will provide a space for relaxation and joy and communication for visitors. The confrontation between different exhibition halls constitutes a unique experience for the visitors to the art museum.

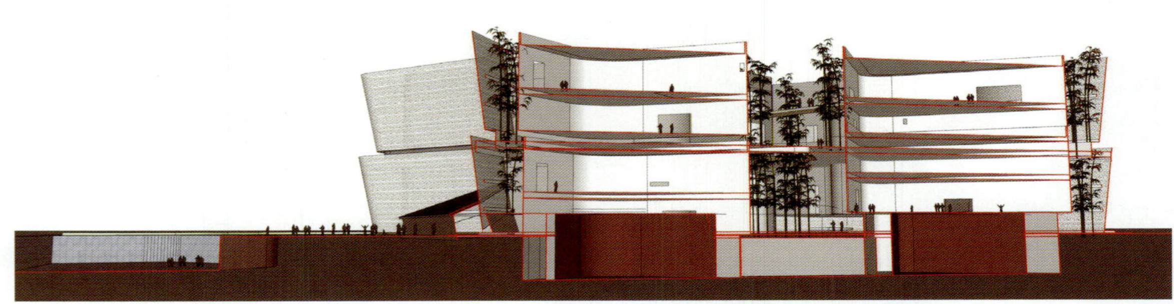

根据功能用房的封闭开放程度划分空间，简单、实用。
The museum is simple yet functional with varying degrees of openness as to different purposes.

与已有的巨型建筑“鸟巢”、“水立方”相比，美术馆外形并不显得柔弱。
The art museum compares favorably against the existing gigantic "the Bird's Nest" stadium and the Water Cube (national swimming center).

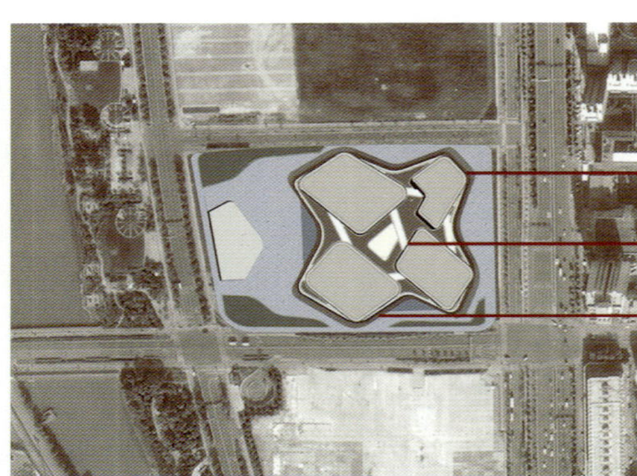


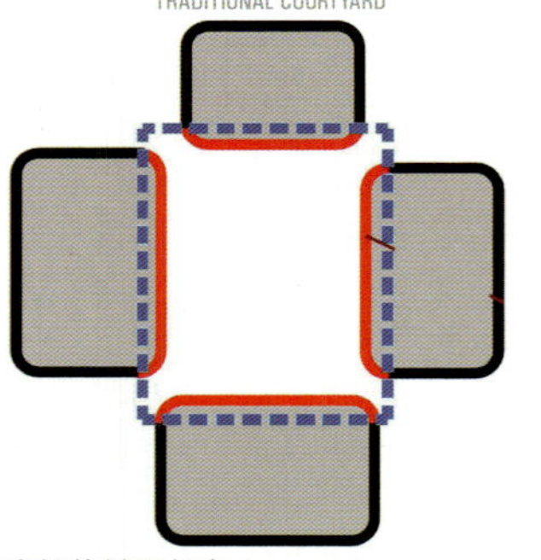


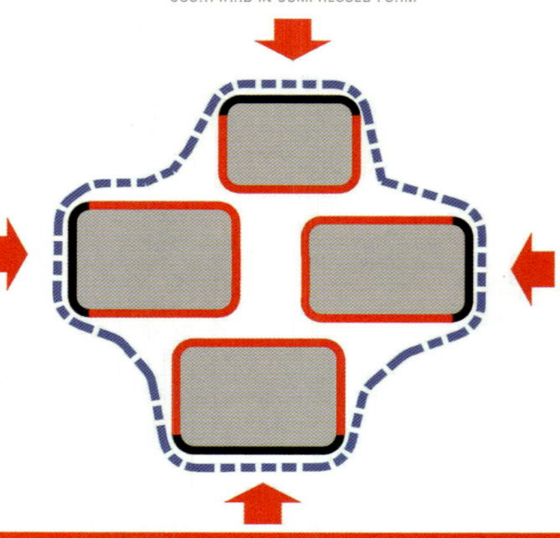

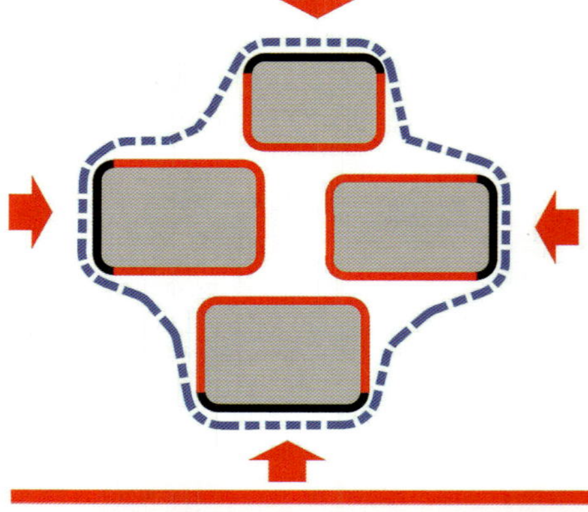


曲折的中庭，各封闭空间与共享空间的接触面长度约为传统中庭的2.5倍，效能增加。
THE CONVOLUTED COURTYARD - THE LENGTH OF THE CONTACT AREA BETWEEN THE ENCLOSED SPACE AND SHARED SPACE IS NEARLY 2.5 TIMES THAT OF THE COURTYARD WITH ENHANCED EFFICIENCY

地下一层平面图 UNDERGROUND FLOOR PLAN

# 中国国家美术馆新馆·北京

## Expansion of China National Art Museum · China

叶长青 Ye Changqing · 张舒文 Zhang Shuwen · 郑堃 Zheng Kun · 朱君 Zhu Jun

竞标方案 competition proposal

山 MOUNTAIN

砚 INKWELL

墨 INK

石 STONE

形与神 THE BODY AND THE SPIRIT

明式罗汉床 ARHAT BED

米芾 山水局部 MI DI – DETAIL OF A LANDSCAPE PAINTING

简与繁 SIMPLE AND COMPLEX

法国巴黎罗浮宫 LOUVRE MUSEUM IN PARIS

英国国家美术馆 THE NATIONAL GALLERY IN UK

含与放 EMBRACING AND EXCLUDING

天与地 HEAVEN AND EARTH

道家—庄子 TAOIST ZHUANGZI

李可染—山水局部 LI KERAN – DETAIL OF A LANDSCAPE PAINTING

水与墨 WATER AND INK

范宽—临流独坐图 FAN KUAN – SITTING ALONE BY THE STREAM

黄公望—山水立轴 HUANG GONGWANG – WALL SCROLL OF A LANDSCAPE PAINTING

黄宾虹—山水立轴 HUANG BINHONG – WALL SCROLL OF A LANDSCAPE PAINTING

陆俨少—山水立轴 LU YANSHAO – WALL SCROLL OF A LANDSCAPE PAINTING

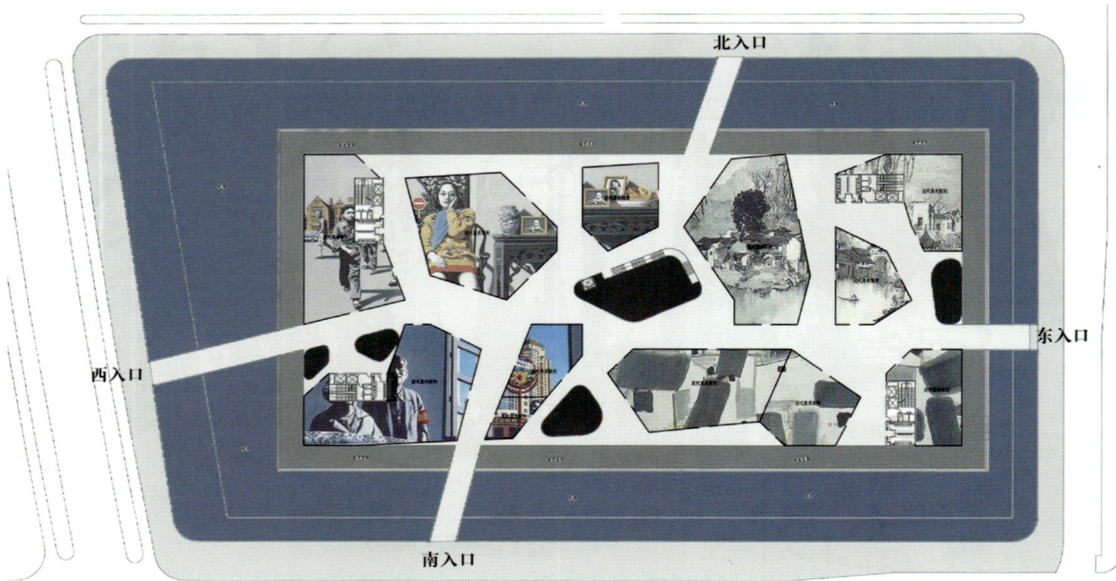

底层平面 GROUND FLOOR PLAN

中国国家美术最重要、最有特质的形式是中国书法和中国画。而溶有诗、书、画、印于一炉的中国山水画无疑是其中最有代表性者。而国画(书法亦同)无不是以咫尺空间描绘大自然和大宇宙，以艺术的形式去实践中国传统“天人合一”的根本精要。

The most crucial and unique form for Chinese national fine arts is Chinese calligraphy and Chinese paintings. The Chinese landscape painting, embracing the poetry, calligraphy, is the epitome of its kind. The Chinese painting (calligraphy alike), capturing the great nature and cosmos in its tiny dimensions, reflects the quintessence of the traditional Chinese philosophy of “human is integral to nature” in the form of art.

底层平面 GROUND FLOOR PLAN

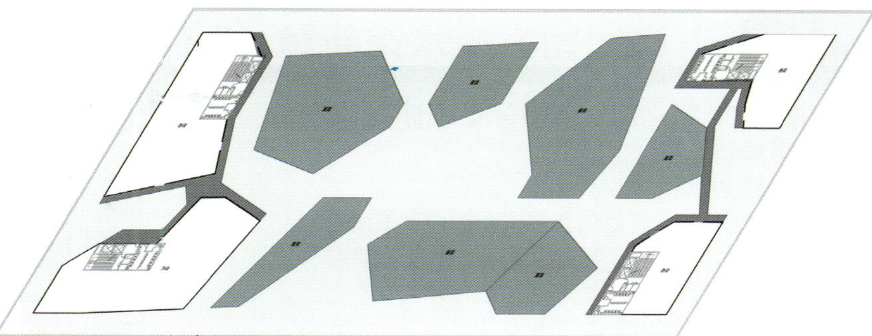

四层平面 THIRD FLOOR PLAN

地下一层平面 FLOOR PLAN -1

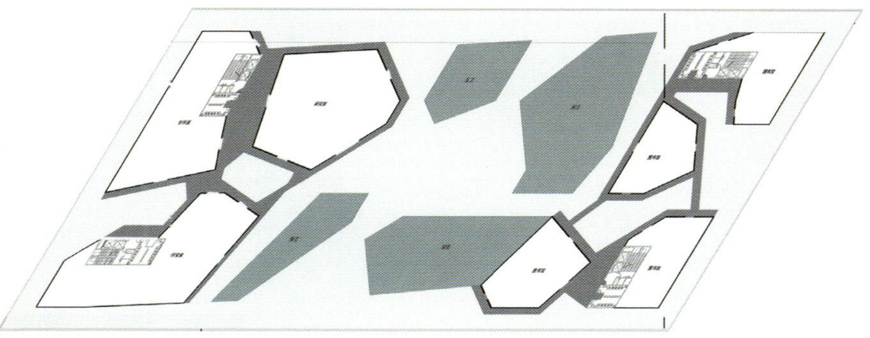

三层平面 SECOND FLOOR PLAN

地下二层平面 FLOOR PLAN -2

二层平面 FIRST FLOOR PLAN

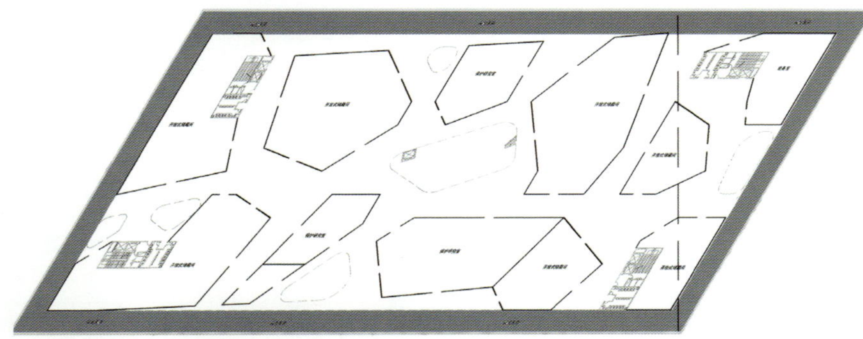

地下三层平面 FLOOR PLAN -3

# 中国国家美术馆新馆·北京

## Expansion of China National Art Museum · China

施明化 Shi Minghua

竞标方案 competition proposal

“‘国字号’建筑如何真正成为人们易于到达的场所、市民的客厅、城市空间的有机组成部分？”是对于设计起点的思考。

1.空间的尺度：空间的尺度决定了空间的活力。在大尺度建筑中引入传统建筑尺度，创造一种适宜的尺度。

2.公共性：在建筑中创造一个足够开放的公共空间，无论美术馆开放与否，人们都可以随意达到的空间，属于城市市民的空间。

3.交通核：建筑通过桥联系城市周边地块，而桥的交点就是美术馆的入口空间。这里是城市的节点，也是重要的交通核。

当年，张择端在绘制《清明上河图》时，是否也有相同的思考？我们只有在画卷中细细解读。

How can a state-sponsored building act as an accessible place and be an integral part of the city? This question sets the designers thinking.

1. Space dimensions: the vitality of the space depends on space dimensions. The inclusion of traditional buildings in the large-scale building takes the dimensions into account.

2. Public serving function: in every building, big public spaces, enough for the public, shall be provided - a place for the citizens where the public can visit at their own will, whether the art museum is open or closed.

3. Communications hub: the building is linked to the surrounding area through the bridges, which act as the entrance to the art museum. These are the highlights of the city, and are also the core of the communications hub.

Did these thoughts arise to Zhang Zeduan while drawing the Along the River during the Qingming Festival? The answer only lies in our deep reading of this paintings.

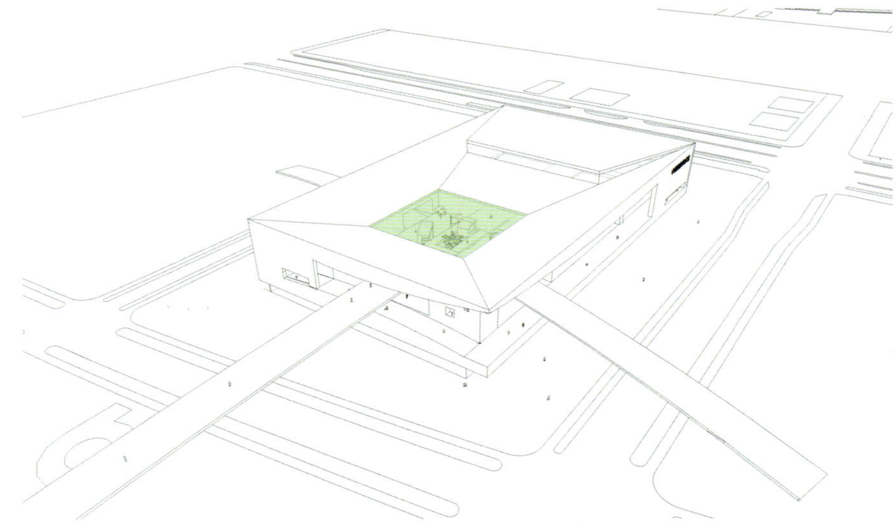

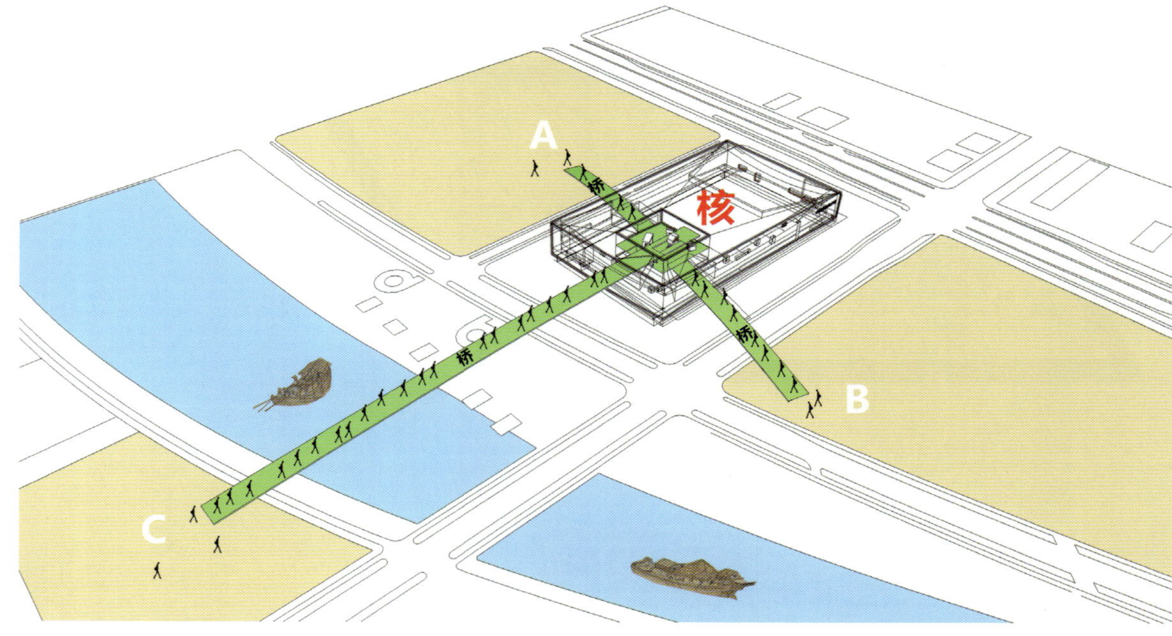

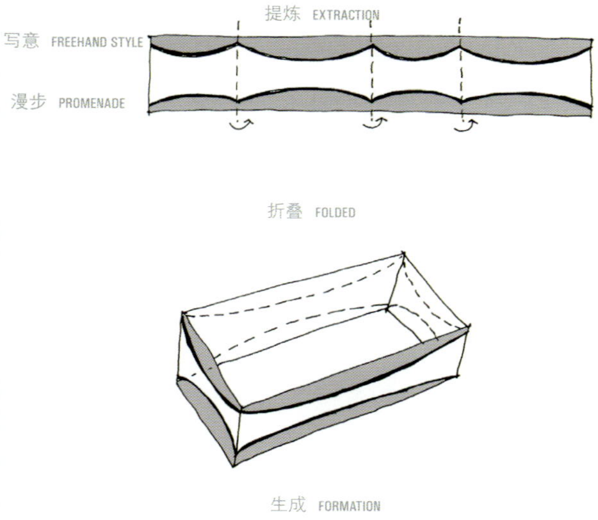
提炼 EXTRACTION
写意 FREEHAND STYLE
漫步 PROMENADE
折叠 FOLDED
生成 FORMATION

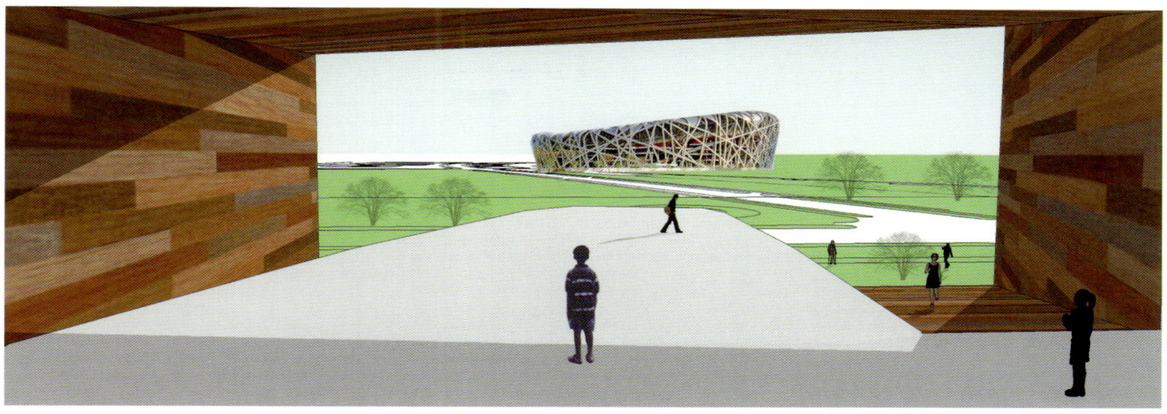

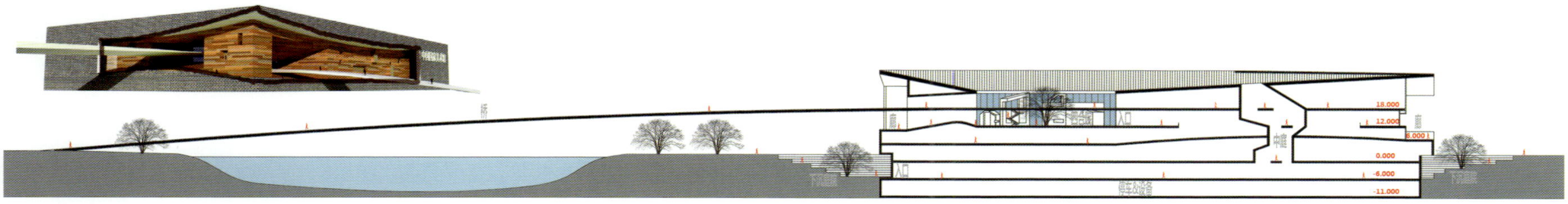

剖面图 SECTION

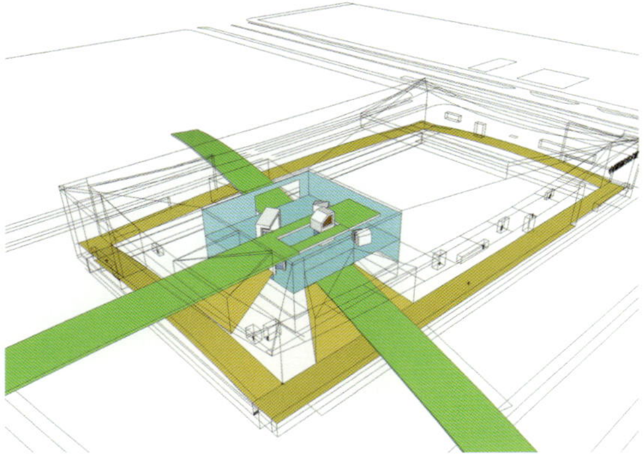

四合院 CHINESE QUADRANGLES

桥 BRIDGES

副阶周匝 SURROUNDED BY AUXILIARY STEPS

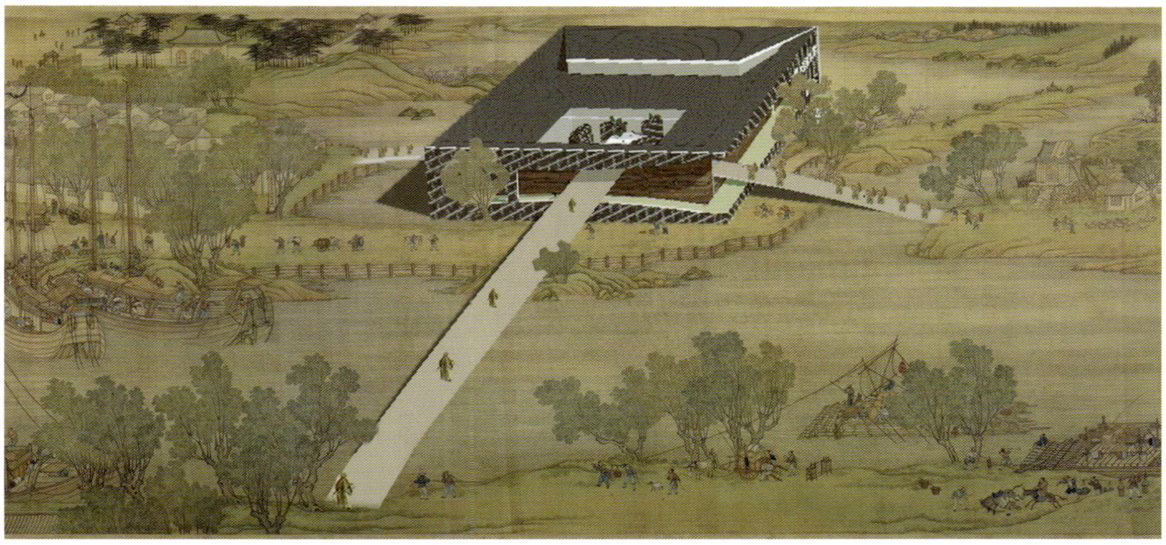

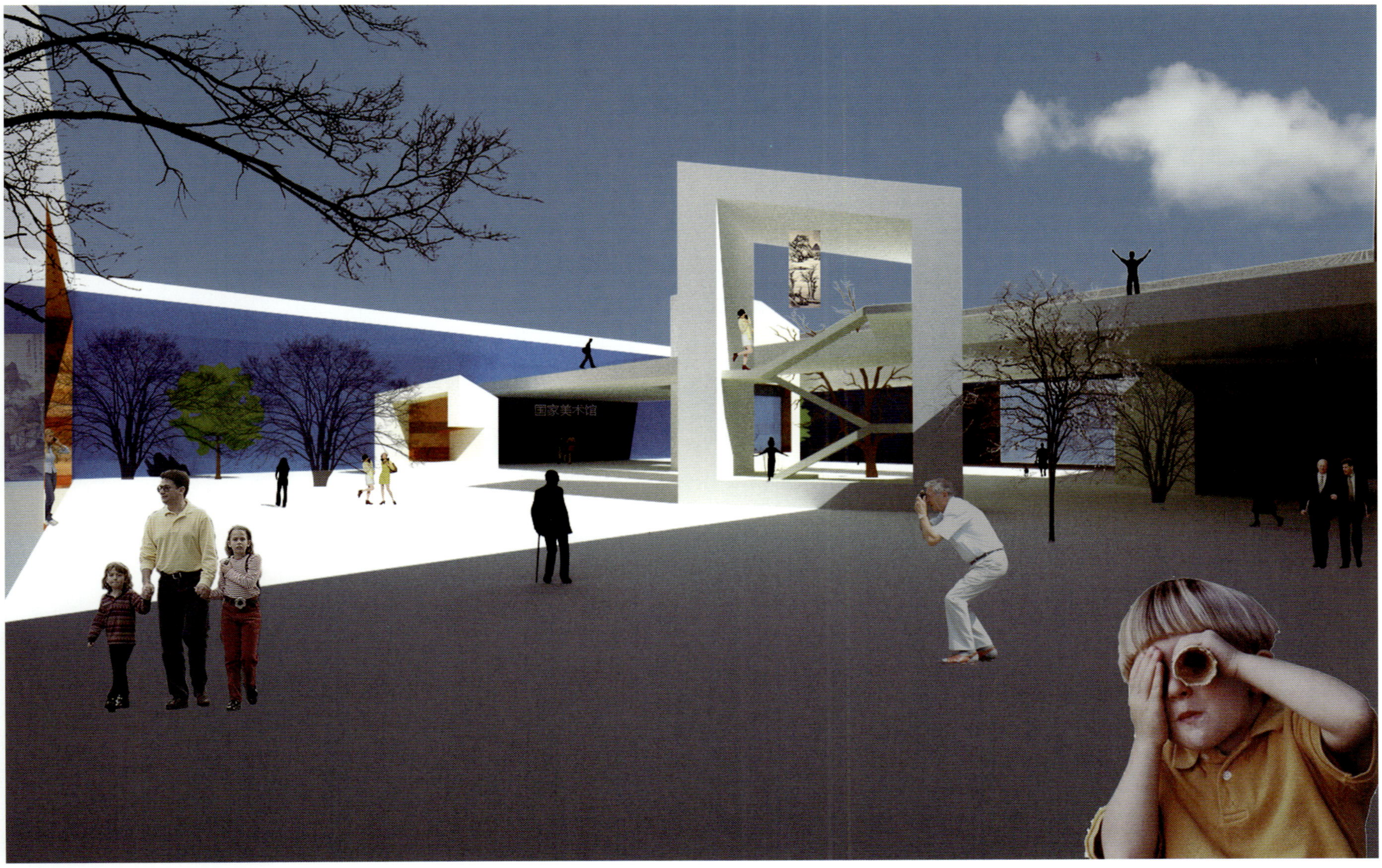
国家美术馆

# 竞标项目 competitions

# 新高速铁路列车站（AVE）· 乌艾尔瓦
## New high-speed rail station · Spain

**竞标 · competition**
新高速铁路列车站 (AVE)
New high-speed rail station

**竞标类型 · competition type**
两阶段公开竞标
two-stage open competition

**项目地点 · site area**
乌艾尔瓦 · 西班牙 Huelva · Spain

**主办方 · promoter**
铁路基建管理局 Administrador de Infraestructuras Ferroviarias (ADIF)

**日程安排 · schedule**
招标 · Announcement 01.2011
评审结果 · Jury´s results 04.2011

**评审团 · jury**

Antonio González Marín, 铁路基建管理局局长 President of Adif
José Juan Díaz Trillo, 安达卢西亚自治区政府环保局顾问 Minister for the Environment of the council of Andalucía
Petronila Guerrero Rosado, 乌艾尔瓦省议会主席 President of the Provincial Deputation of Huelva
Pedro Rodríguez González, 乌艾尔瓦省长 Mayor of Huelva
José Abraham Carrascosa Martínez, 道路桥梁隧道港口学院院长 (安达卢西亚地区) Dean of the College of Civil Engineers, Channels and Ports (Demarcation of Andalucía)
Gonzalo Prieto Rodríguez, 安达卢西亚官方建筑学院院长 Dean of the official College of Architecture of Andalucía

**获奖者 · awards**

一等奖 · first prize
**Rafael de La-Hoz Arquitectos (建筑师事务所)**
Rafael de La-Hoz Castanys (建筑师)
**Acciona Ingeniería (结构师事务所)**

入围 · finalist
**Cruz y Ortiz arquitectos (建筑师事务所)**
Antonio Cruz Villalón · Antonio Ortiz García
Blanca Sánchez Lara (建筑师)

合作 (c) Teresa Cruz · Héctor Salcedo García · Alejandro Álvarez · Rocío Peinado Mercedes Pérez
渲染 renderings · Alejandro Álvarez
结构 engineering · EUROESTUDIOS
模型 model · Jorge Queipo

入围 · finalista
**Rogers Stirk Harbour & Partners + Vidal y asociados arquitectos (建筑师事务所)**
团队 team · Rogers Stirk Harbour & Partners: Simon Smithson · Juan Laguna · Jason García Pablo Codesido · Mariola Merino · Carmen Marquez
合作团队 team vidal y asociados arquitectos · Luis Vidal · Francisco Sanjuán · David Avila · David Fernandez · Julio Isidro Lozano · Eva Couto · Naira Pérez

入围 · finalista
**AZPA/FOA (建筑师事务所)**
**Alejandro Zaera-Polo (partner in charge) (合伙负责人)**

设计团队 design team · Ravi Lopes · Guillermo Fernandez-Abascal · Cecilia de Marinis · Giuseppe Giordano · Pere Raventós · Anelie Seaman · Elisabetta Cabras · Pep Wennberg
结构 engineers · GPO Engineers

入围 · finalista
**AJN · Ateliers Jean Nouvel · COOT Arquitectos · INES Ingenieros (建筑师事务所)**
建筑师 architect · Jean Nouvel
项目主管 project directors · Federico Sotomayor + Alberto Medem
团队 team · Eloisa Siles · Juan Galbis · Athina Faraut · Mathieu Puyaubreau · Ikbal Bouaita · Rafael Renard · Rafaëlle Ishkinazi · Jugulta Le Clerre
模型 model · Jorge Queipo · Sergio Haring

乌艾尔瓦高速铁路火车站需要修建**两座相得益彰的建筑体**：一边是面积3,000平方米的旅客候车楼，另一边是火车调控指挥的配楼。

The construction of the high-speed rail station of Huelva requires **two complementary projects**: on one side, the construction of the travelers' building with 3,000 m² and the execution of all train annexes.

新高速铁路列车站 · 乌艾尔瓦

New high-speed rail station · Spain

一等奖 · First Prize

# puerta umbría (竞标代码)

Rafael de La-Hoz Arquitectos (建筑师事务所)

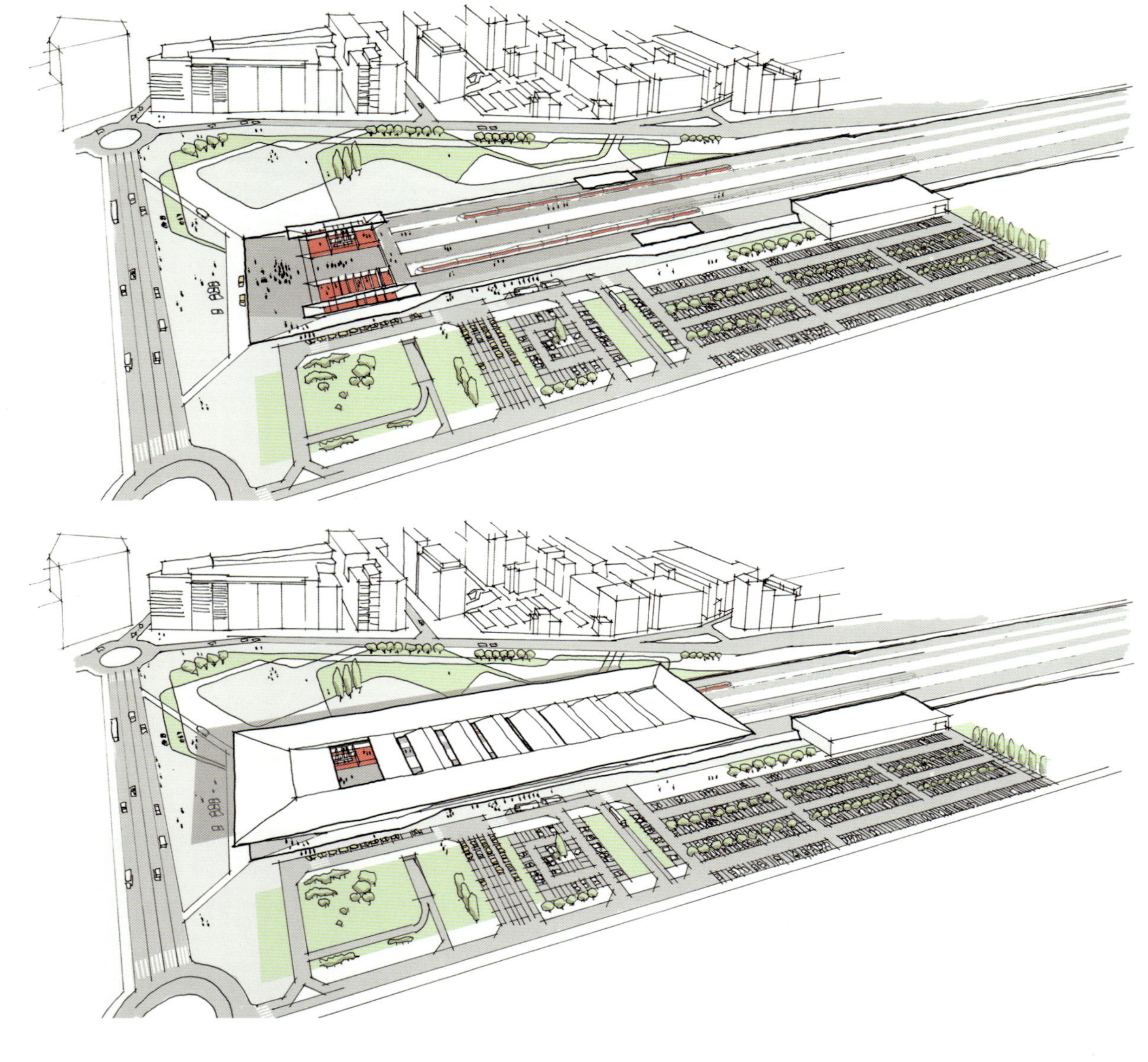

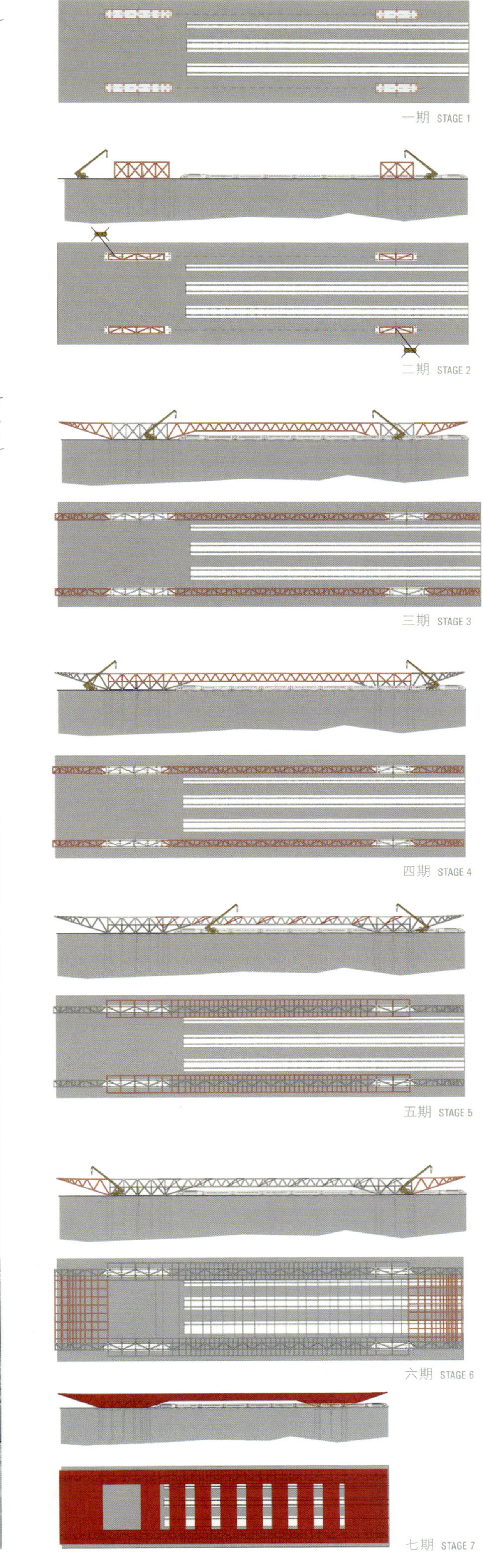

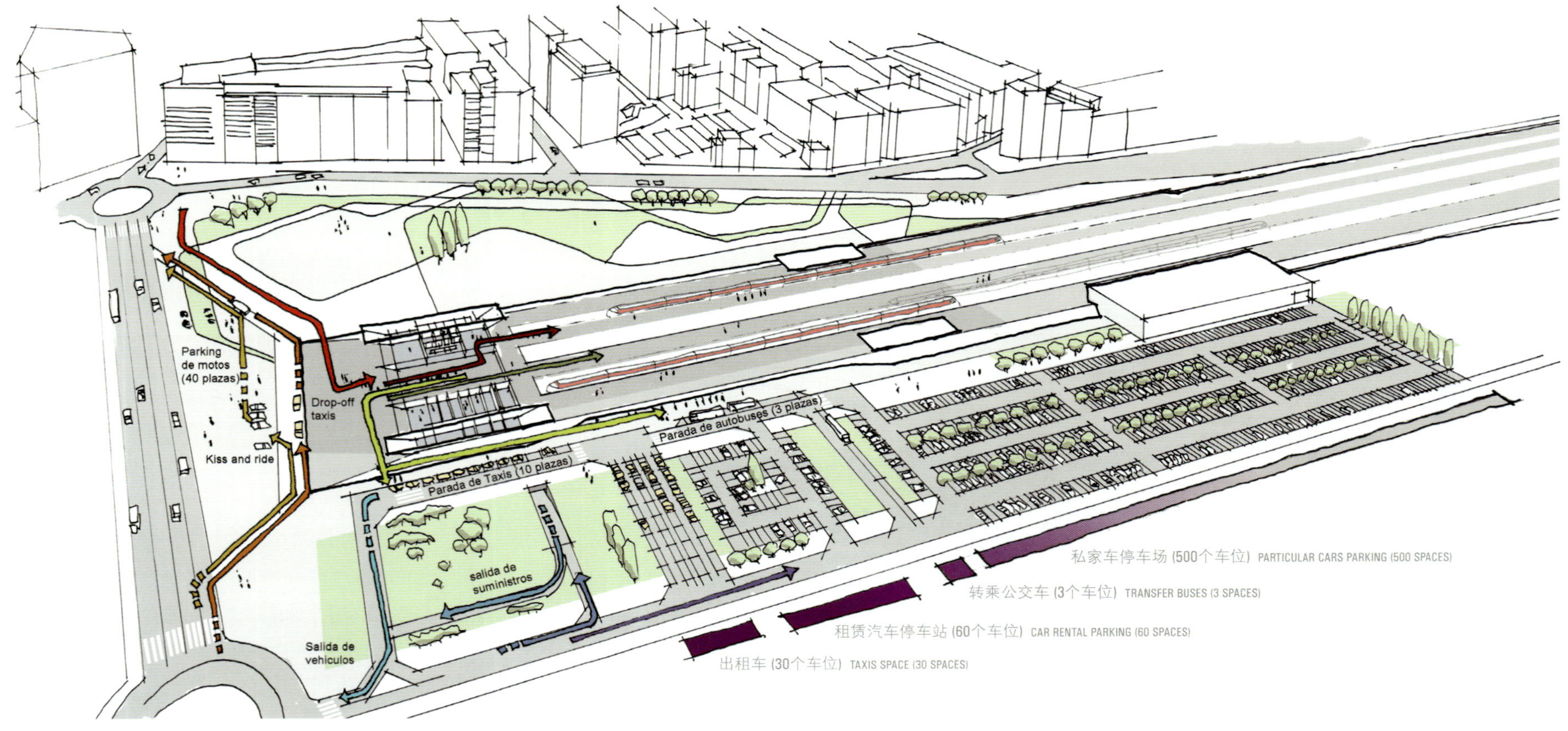


功能 PROGRAMS

行人入口 PEDESTRIAN ACCESS

个人入口 PARTICULAR ACCESSES

出租车入口 TAXI ACCESS

高铁入口 ENTRANCE TO HIGH-SPEED RAIL

城际列车区入口 REGIONAL ENTRANCE

后勤供应入口 SUPPLIES ENTRANCE

停车场入口 ENTRANCE TO THE PARKING

通向出租车、公交车、汽车租赁或停车场的出口 EXIT TO TAXIS, BUS, RENTAL OR PARKING

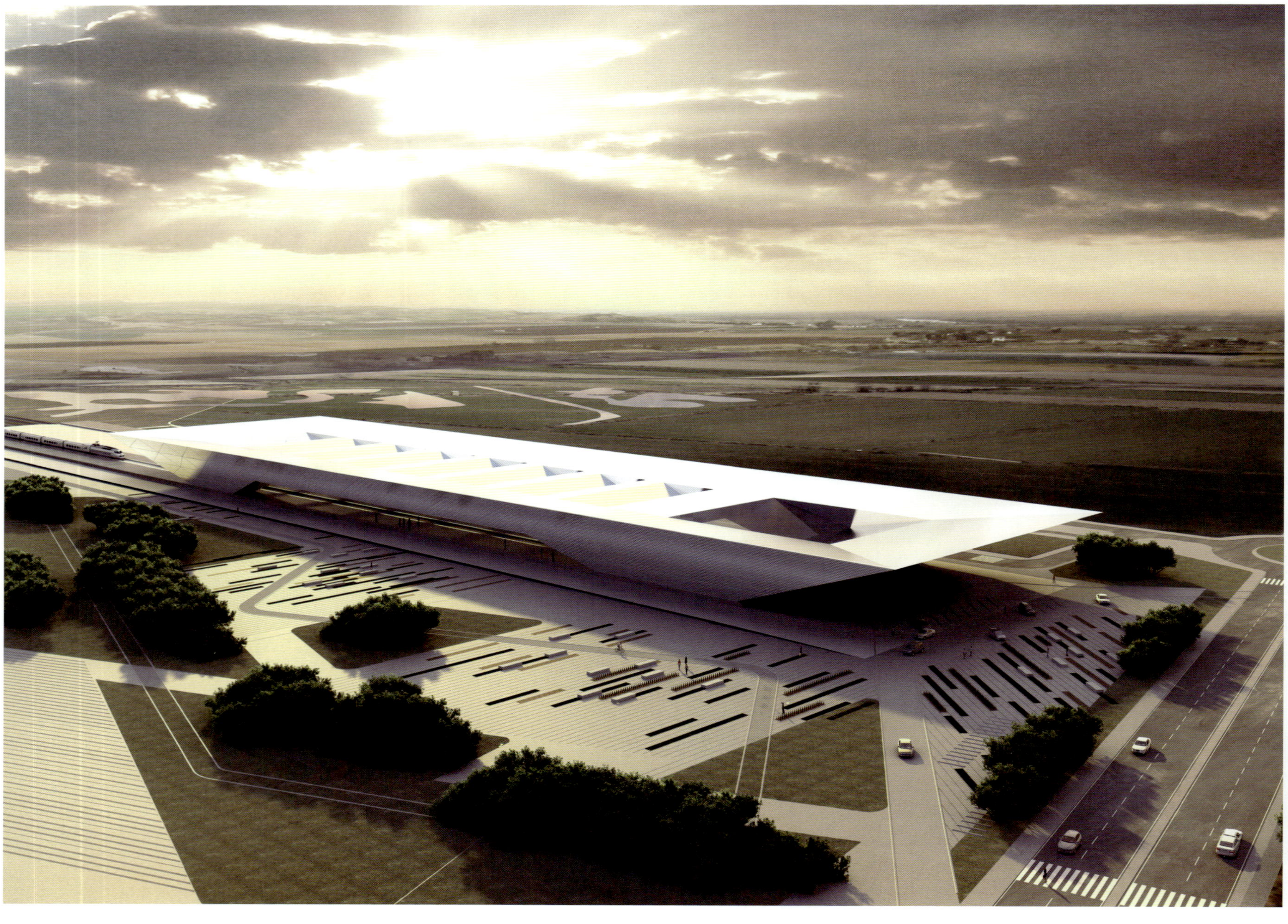

平面图 FLOOR PLAN

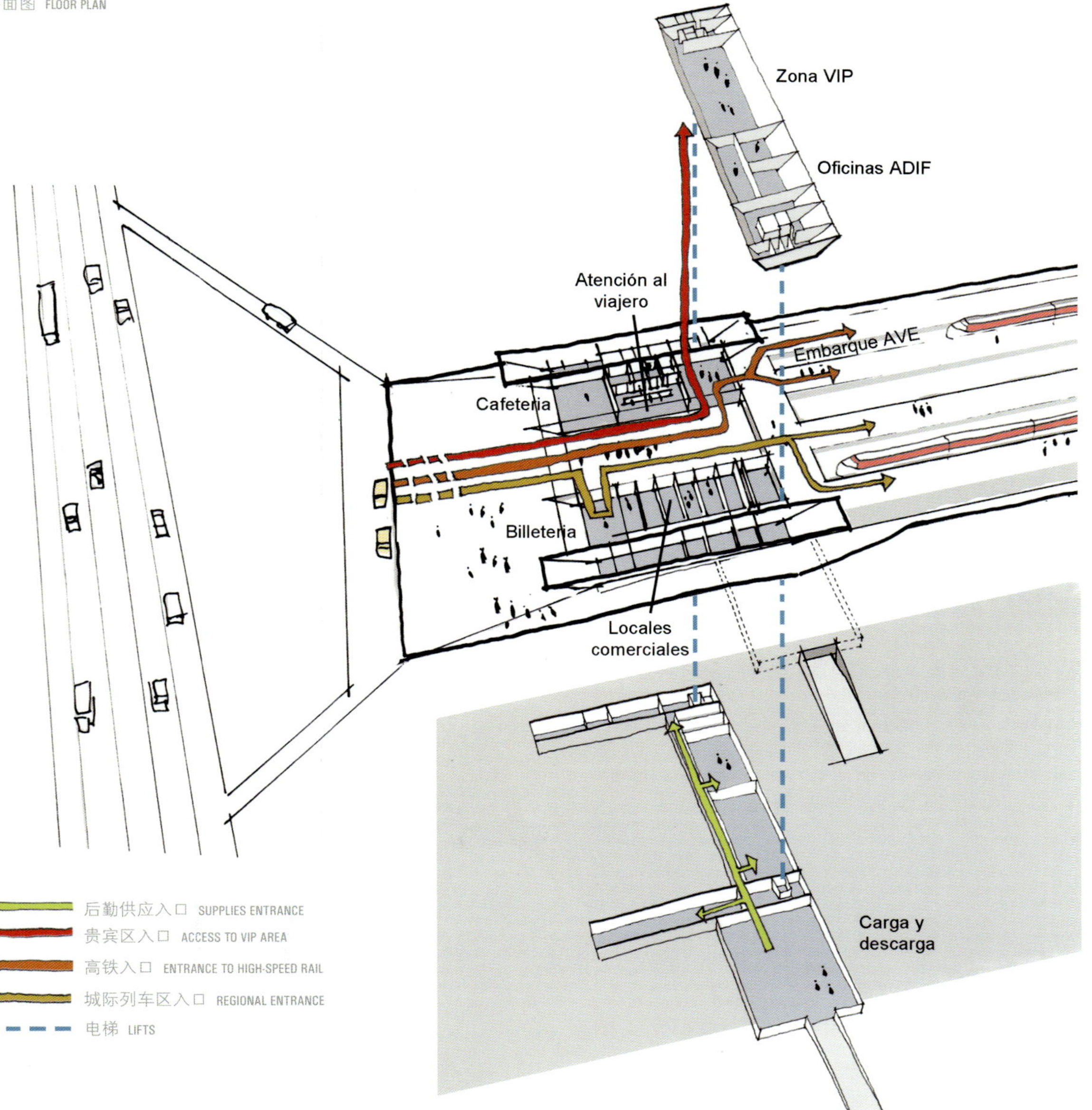


# 白板

拉丁语中的“白板”一词可能最能概括该项目的双重概念。看多了强调繁缛结构和装饰的大量建筑，厌烦了为表现宏大气势而过度运用的曲线以及夸张的弧形和符号，我们提出了修建一幢回归本源的建筑结构体的设想，一张设计平面图纸、一个穹顶、一个平板：一个白板。四大结构支撑起了一个宽敞、水平的穿孔穹顶。

# TABULA RASA

Perhaps the Latin term Tabula Rasa is the best word to synthesize the dual concept of our project. Overwhelmed by so many architectures based on superfluity and excess, with extravagant curves and arrogant arches and semiotics in the attributes of power, we propose a basic architectural gesture, a plan, a roof, a tabula: A Tabula Rasa. Four elements support a board, a plane, a perforated roof.

SECCIÓN SECTION

新高速铁路列车站 · 乌艾尔瓦

New high-speed rail station · Spain

入围 · Finalist

# dunas (竞标代码)

Cruz y Ortiz arquitectos (建筑师事务所)

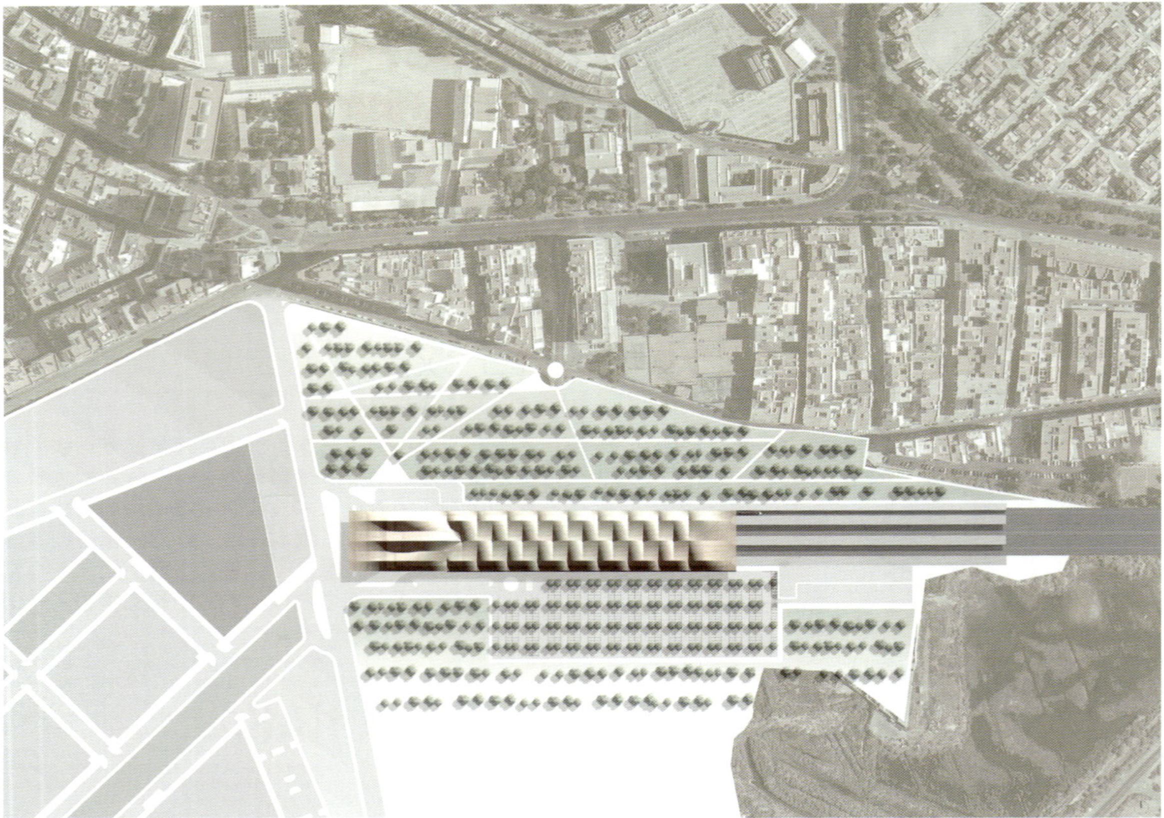

地块位置 SITE PLAN

顶部构造 ROOF STRUCTURE

1 高铁站 HIGH-SPEED RAIL STATION

2 金属构造（拱和桁架） METALLIC STRUCTURE (ARCHES AND TRUSS)

3 织物网 TEXTILE GRID

4 太阳能板 SOLAR PANELS

## 曲线

总体而言，车站可能就是一个穹顶。我们粗略设计的穹顶由聚四氟乙烯（四氟乙烯的含氟聚合物）材料制成。该材料非常轻便，使得支撑物之间可留出较宽的间距。此外，作为车站内部结构的一部分，它还可以节省许多次要构件。聚四氟乙烯可取代玻璃，并通过其不透明度自如地控制光线。

## CURVY LINES

Maybe, a station is, above all, a roof. We propose a stretched roof made up of Teflon (synthetic fluoropolymer of tetrafluoroethylene), very light, capable of providing considerable distances among supports, inherent in the stations and make unnecessary many secondary structural elements. Teflon can replace the glass and allow a good control of light through its opacity.

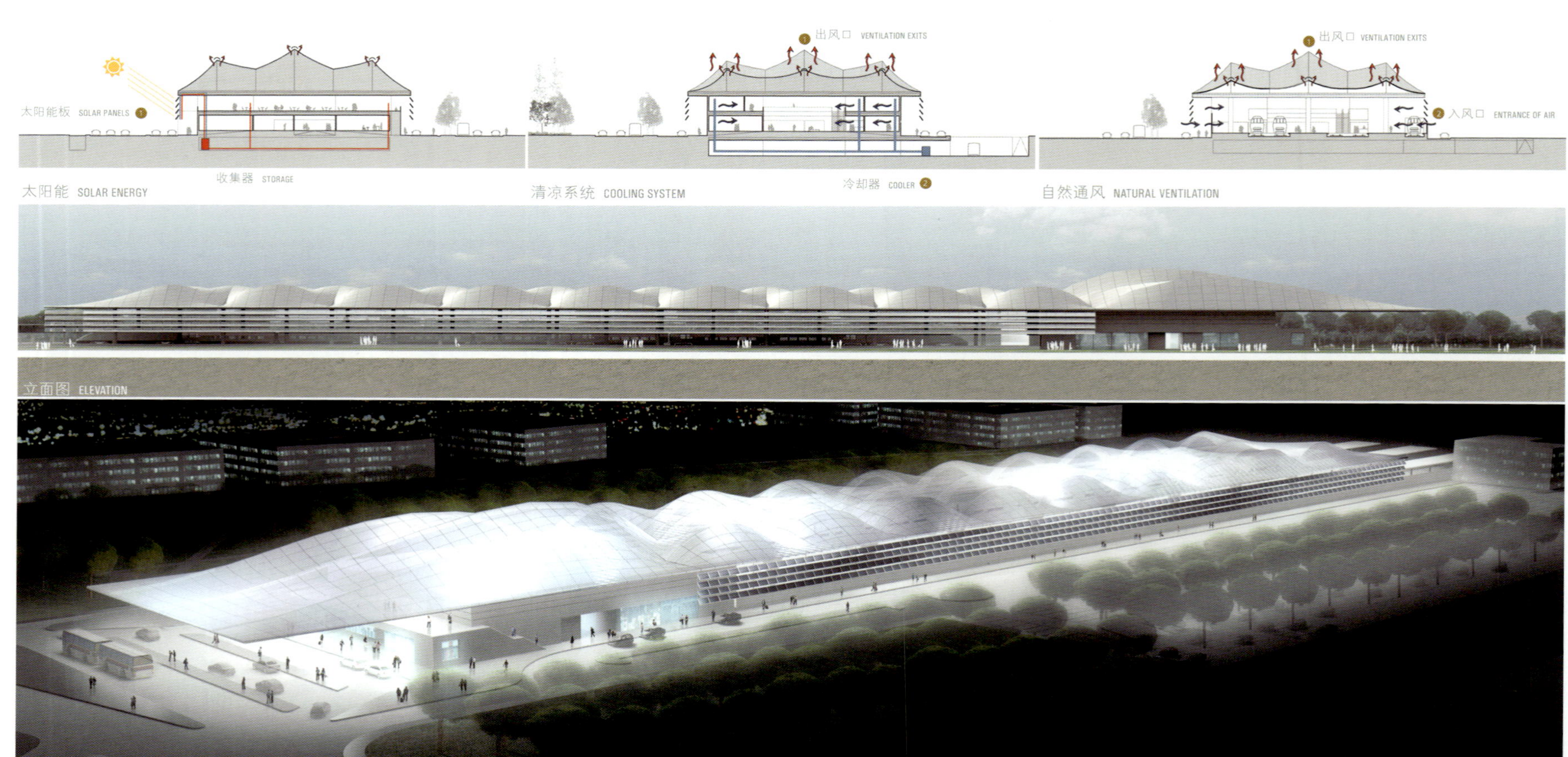

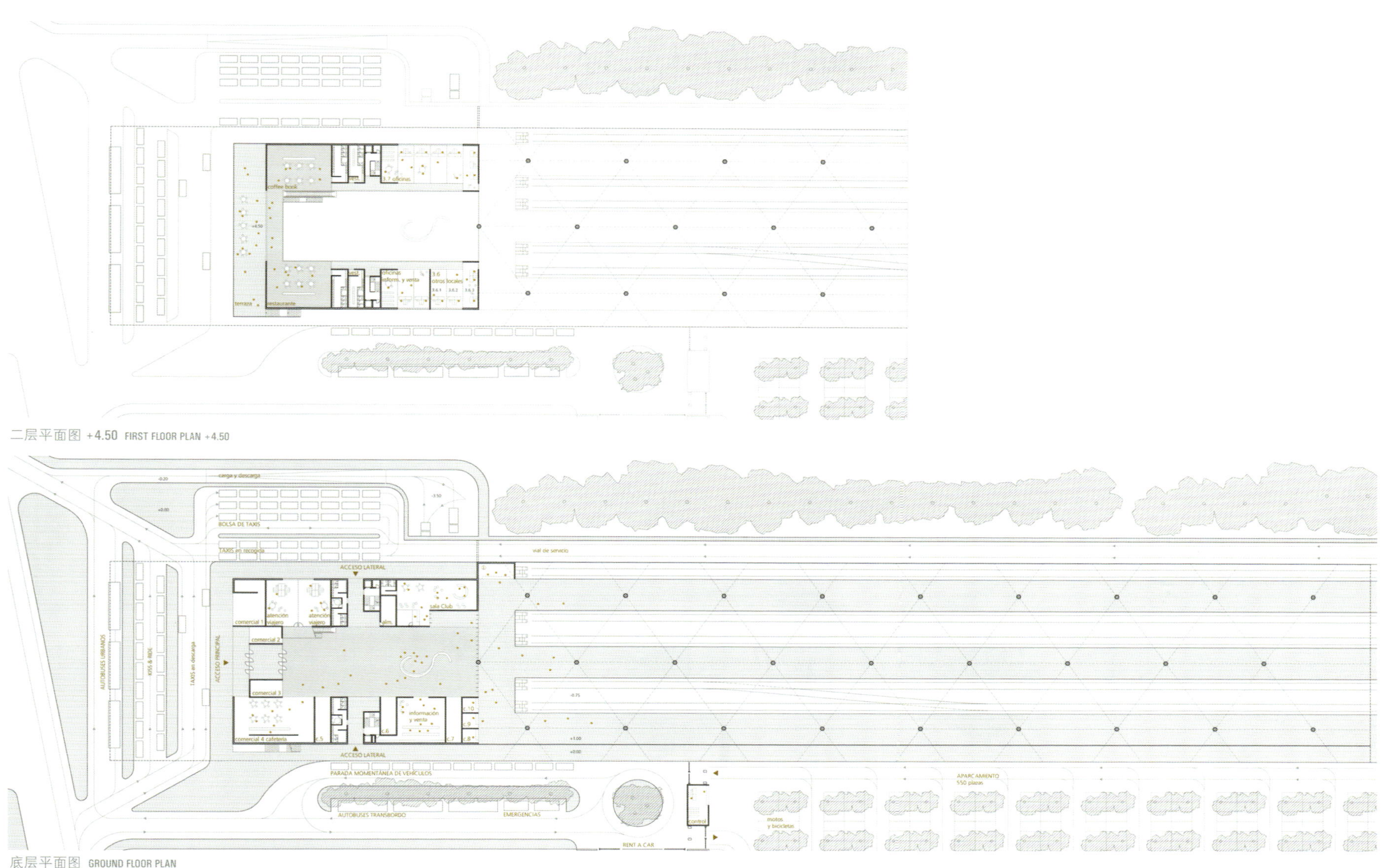
二层平面图 +4.50 FIRST FLOOR PLAN +4.50

底层平面图 GROUND FLOOR PLAN

剖面图 SECTION

新高速铁路列车站 · 乌艾尔瓦

New high-speed rail Station · Spain

入围 · Finalist

# huelv/a/ve (竞标代码)

Rogers Stirk Harbour & Partners · Vidal y asociados arquitectos (建筑师事务所)

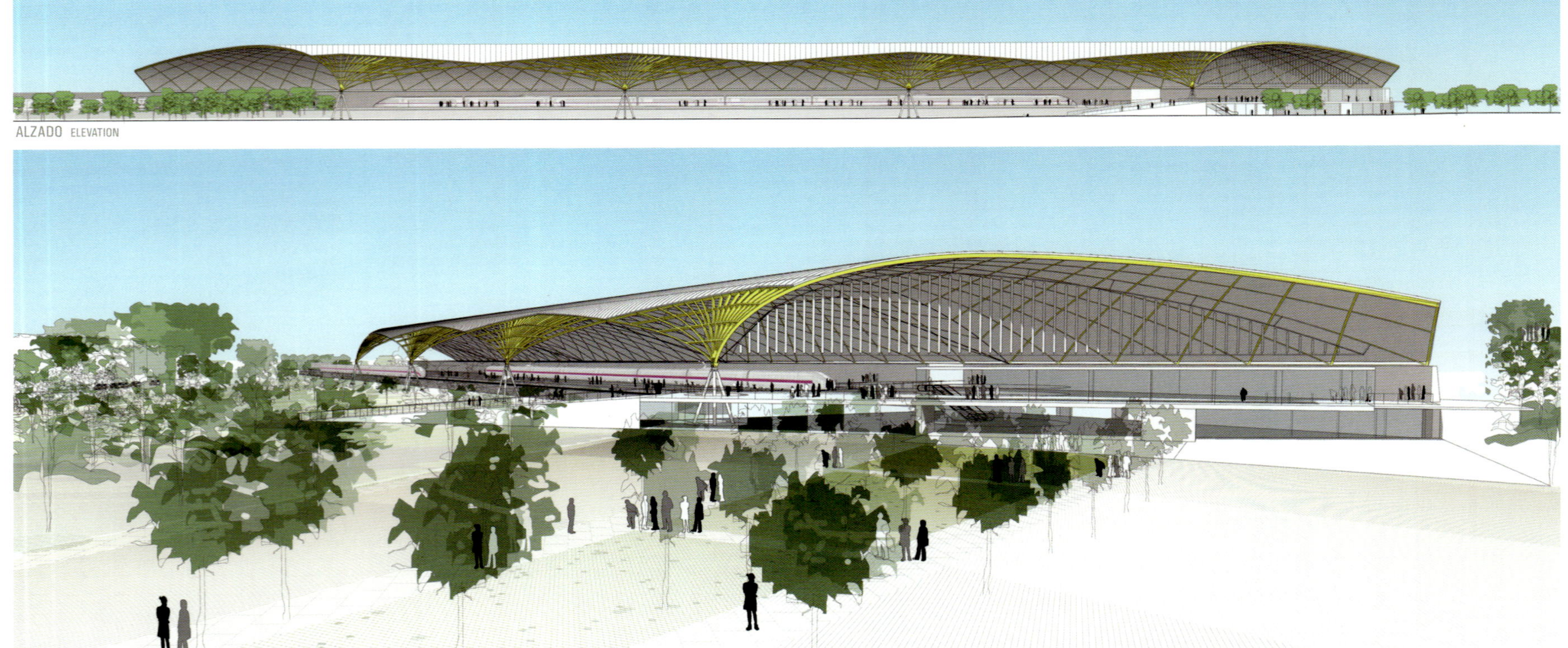


## 穹顶之下的微气象

我们为新建AVE火车站提出的设计蓝图旨在每日迎送往来的火车，并充满自豪地展示各火车的风采。因此，我们选择将站台和火车通过透明化的设计展现在公众眼前，这样，抵达的乘客便可饱览城市的风景。与此同时，城市中人们也可看到往来不息的火车。站台和其遮护结构——穹顶，是建筑的基本构造要素。我们设计的穹顶外形活力十足、极富表现力，这个轮廓清晰的单一结构将车站包裹起来，是庇护车站的一扇穹顶。

## A MICROCLIMATE UNDER THE ROOF

Our proposal for the new AVE train station aims to celebrate daily arrivals and departures of trains as well as to show, with pride, the trains. Therefore, we chose to make the platforms and trains visible, so that passengers arriving can see the city and, in turn, trains can be seen from the city. The platforms and their protection element, the roof, are the basic elements of composition. Our roof has a dynamic and expressive shape, which with a single bold gesture, protects and wraps everything.

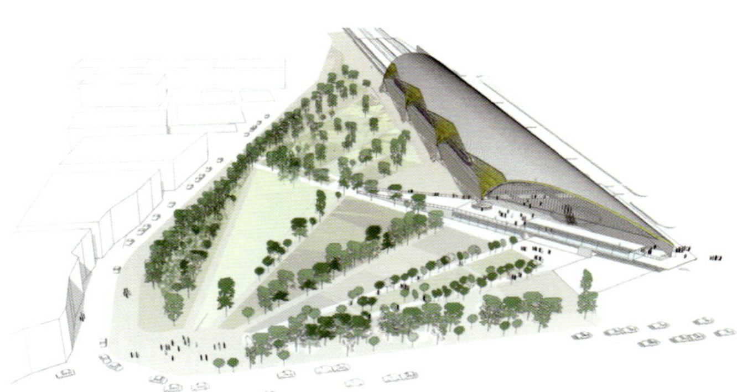

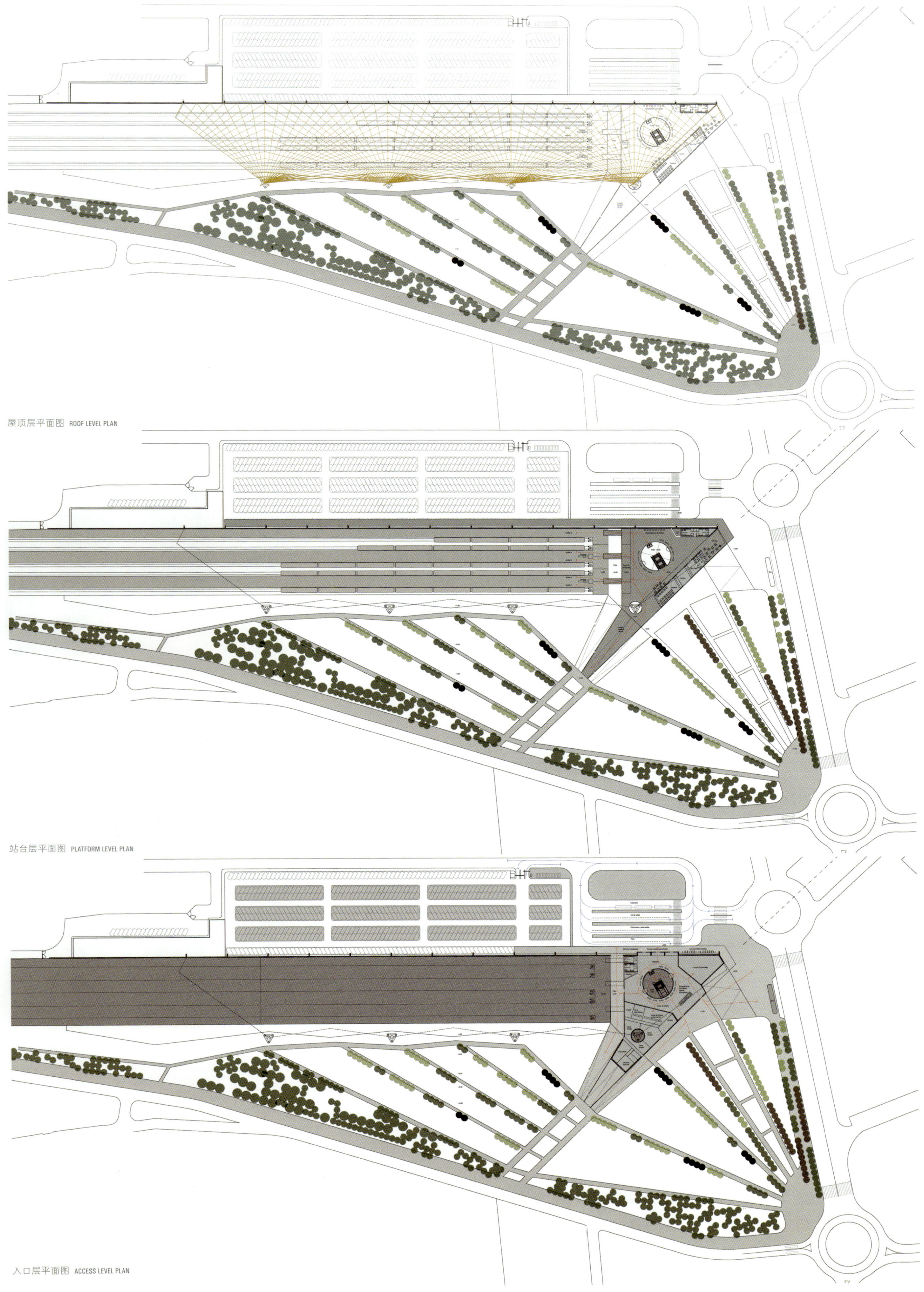

屋顶层平面图 ROOF LEVEL PLAN

站台层平面图 PLATFORM LEVEL PLAN

入口层平面图 ACCESS LEVEL PLAN

新高速铁路列车站 · 乌艾尔瓦

New high-speed rail station · Spain

入围 · Finalist

# flujos de cobre (竞标代码)

AZPA/FOA (建筑师事务所)

Alejandro Zaera-Polo (建筑师)

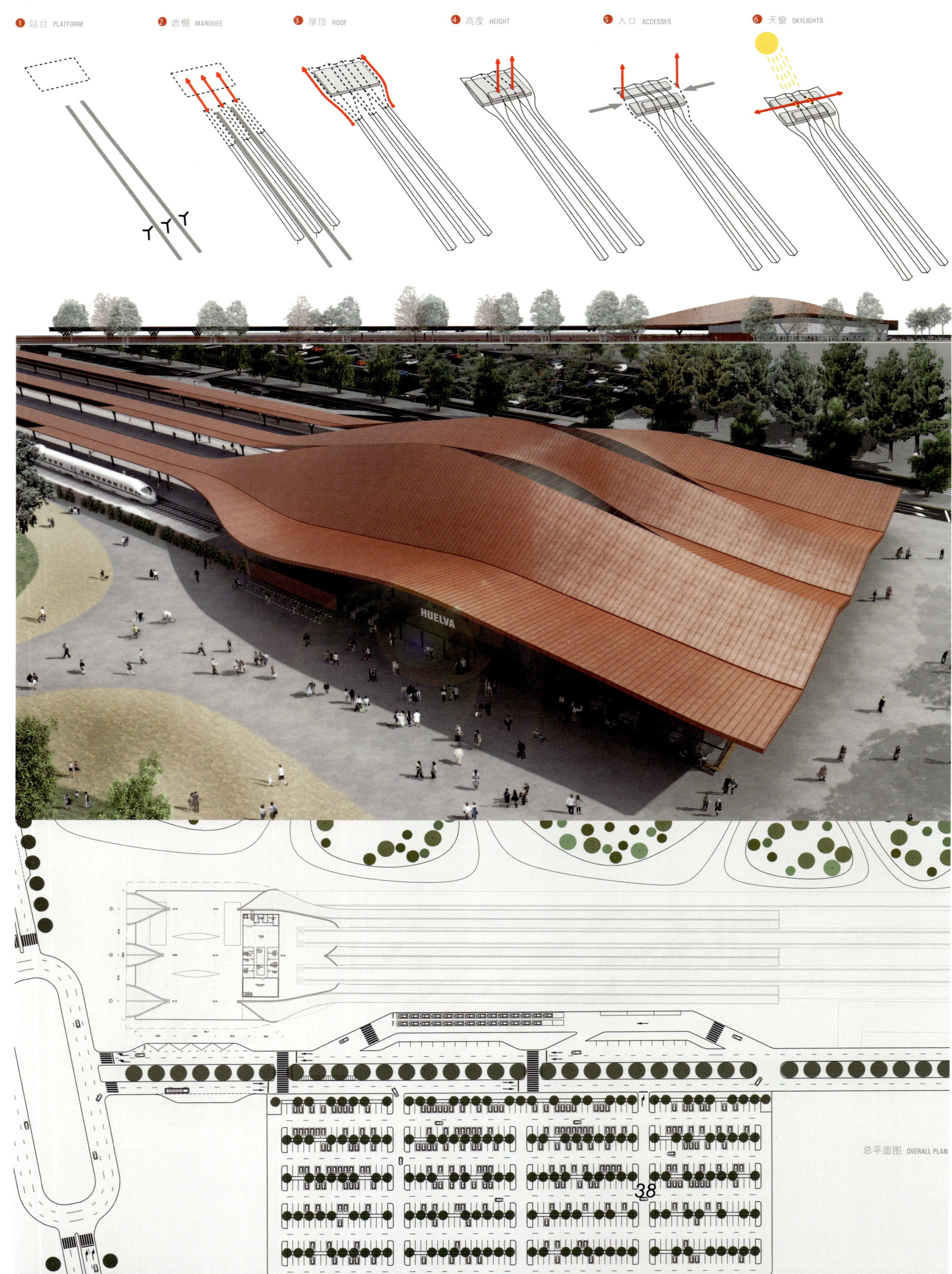


总平面图 OVERALL PLAN

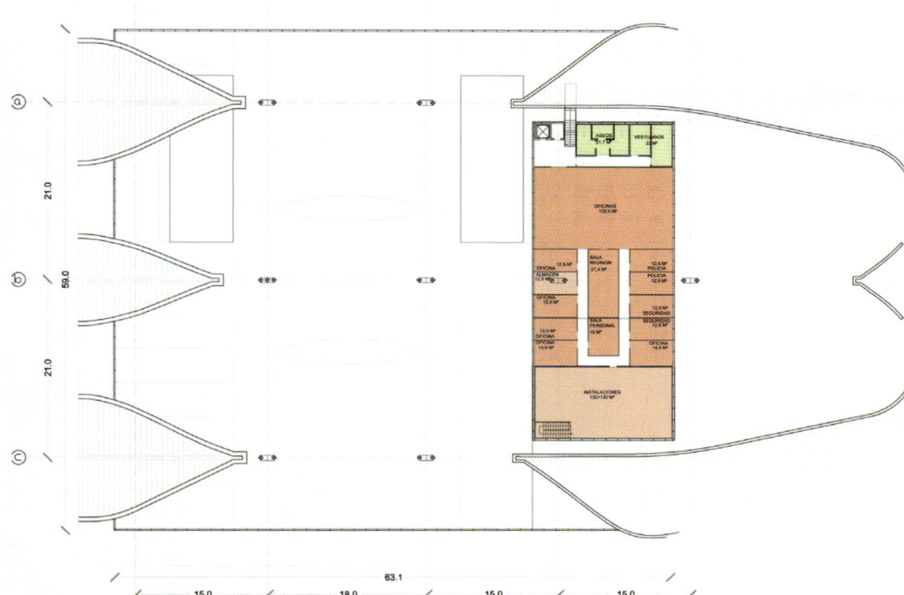

办公层平面图 +4.00 OFFICES FLOOR PLAN +4.00

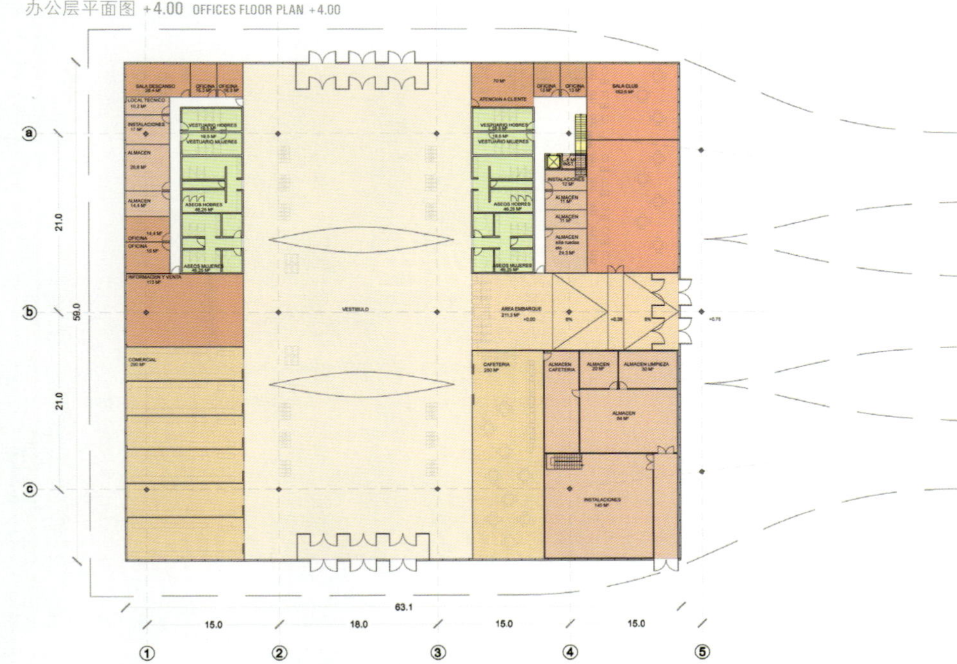

入口层平面图 ENTRANCE FLOOR PLAN

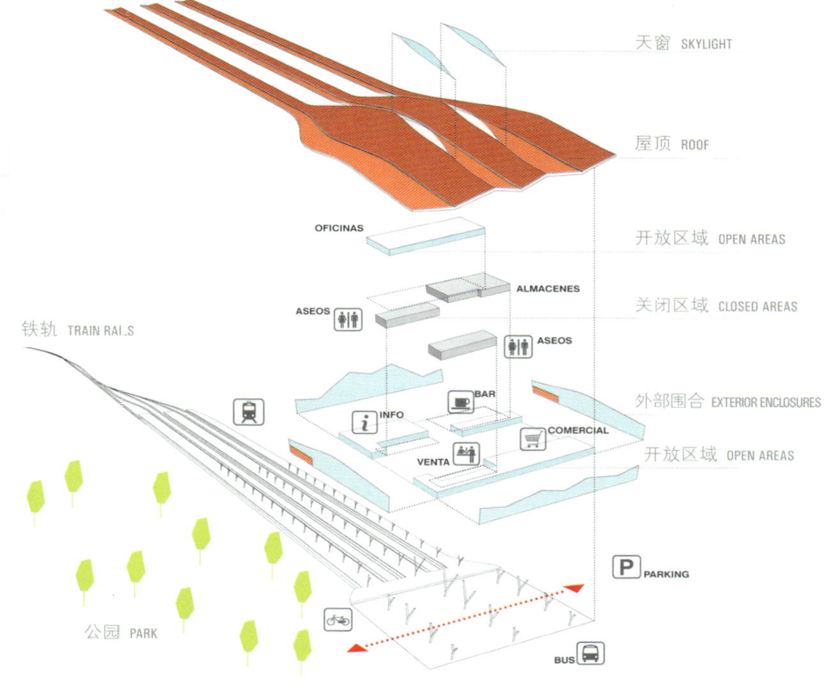


# 动态和流畅

该建筑是一个大型的封闭结构，形似一块天幕，在垂直方向上流畅地由东站台一直展开到西面的拟建地块边缘。该建筑的设计灵感是为车站打造一块地标式的巨大穹顶，因此将移除墙体，使建筑的所有周边视野都能够一览无余。

# DYNAMIC AND FLUID

The project generates a large envelope shaping a marquee in section which fluidly and longitudinally unfolds from the eastern platforms to the edge of the buildable plot to the west. The idea to produce a large roof as model for a station is precisely the attempt to remove the walls, as if the entire perimeter of the building was potentially open to passenger traffic.

新高速铁路列车站 · 乌艾尔瓦

New high-speed rail station · Spain

入围 · Finalist

# la llegada a la estacion de huelva (竞标代码)

AJN · Ateliers Jean Nouvel · COOT Arquitectos · INES Ingenieros (建筑师事务所)

Jean Nouvel (建筑师)

地块位置 SITE PLAN

## 矿物色

我们想设计一个有钢制穹顶的火车站,一个风格低调且面积适中的火车站。站台穹顶可映照出一列列火车,而大厅和通道均铺设彩色的氧化铁物质，打造出乌艾尔瓦火车站标志性的地形和景观。

## MINERAL COLOURS

We want a train station with steel canopies that reveal the trains. A train station which is not pompous not oversized. The platform canopies are mirrors from the trains; we pave the hall and the access with colorful iron oxide elements which create landforms and landscaped materials that address the identity of the station of Huelva.

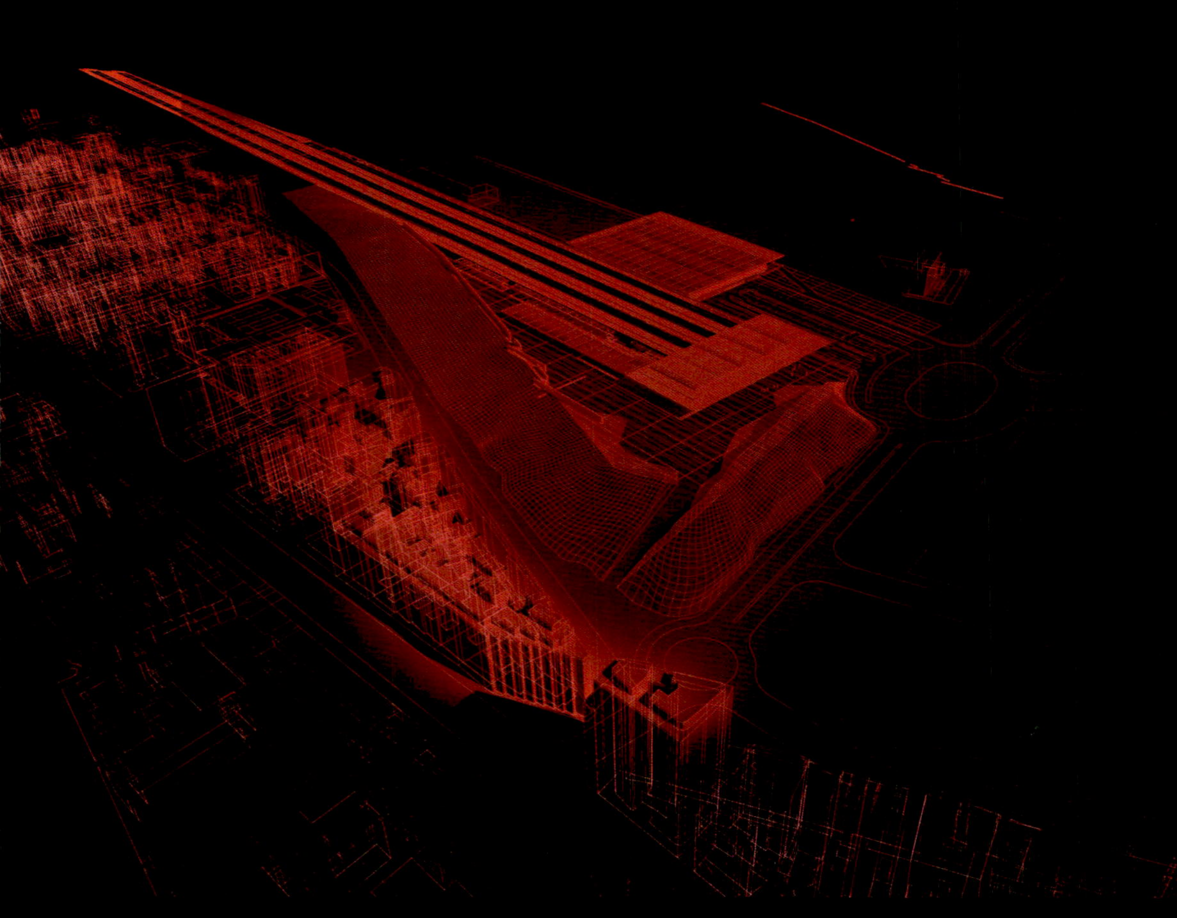

剖面图 AA SECTION AA

剖面图 BB SECTION BB

入口层平面图 FLOOR PLAN ACCESS LEVEL

剖面图 CC SECTION CC

# 华沙交响乐音乐厅 · 华沙
## Sinfonia Varsovia Concert Hall · Poland

**竞标 · competition**
华沙交响乐音乐厅
Sinfonia Varsovia Concert Hall

**竞标类型 · competition type**
两阶段国际竞标
two-stage architectural competition

**项目地点 · site area**
华沙 · 波兰 Warsaw · Poland

**主办方 · promoter**
华沙当地政府文化局 Warsaw Local Government Institution of Culture

**日程安排 · schedule**
招标 · Announcement 06.2010
评审结果 Jury's results 11.2010

**评审团 · jury**

Eckhard Kahle · 物理学家－声学家，音乐家（布鲁塞尔）Physicist – Acoustician, Musician (Brussels)
Tomasz Konior · 建筑师（卡托维兹）Architect (Katowice)
Marek Kraszewski · 华沙首都文化局局长 Director of Culture Bureau of the Capital City of Warsaw
Andrzej Krzyżanowski · 音乐家（华沙）Musician (Warszawa)
Janusz Marynowski · 音乐家 Musician, 华沙交响乐管弦乐主管（华沙）Director of Orkiestra Sinfonia Varsovia (Warsaw)
Bohdan Paczowski · 建筑师（卢森堡） Architect (Luxemburg) · 竞标评审主席 Chairman of Competition Jury
Rudy Ricciotti · 建筑师（班多，卢森堡）Architect (Bandol, France)
Małgorzata Rozbicka · 建筑师 Architect, 建筑史学家（华沙）Historian of architecture (Warsaw)
Jerzy Szczepanik-Dzikowski · 建筑师（华沙）Architect (Warsaw)
Piotr Śmierzewski · 建筑师（科斯林）Architect (Koszalin)
Hubert Trammer · 建筑师（卢柏林）Architect (Lublin)

**获奖者 · awards**

一等奖 · **first prize**
**Atelier Thomas Pucher（建筑师事务所）**

设计团队 design team · Thomas Pucher · Stephan Brugger · Klaus Hohsner · Manuel Konrad · Robert Lamprecht · Erich Ranegger · Jan Schrader · Dominik Troppan · Elisabeth Maria Weber
渲染 renderings · Jan Schrader · Bernhard Luthringshausen

并列二等奖 · **second prize ex-aequo**
**Zaha Hadid Architects（建筑师事务所）**

并列二等奖 · **second prize ex-aequo**
**Hermanowicz Rewski Architekci（建筑师事务所）**

Baltazar Brukalski · Radosław Tabor · Wojciech Hermanowicz · Dariusz Brzeziński · Luiza Anyszka · Małgorzata Grzegorzewska · Magdalena Palmowska · Joanna Orkisz · Błażej Hermanowicz · Stanisław Rewski（建筑师）

并列二等奖 · **second prize ex-aequo**
**Maka Sojka Architekci（建筑师事务所）**

荣誉提名奖 · **honorable mention**
**RE（建筑师事务所）**
Piotr Michalewicz · Mateusz Tański（建筑师）

结构工程 structural engineer · ARUP Poland
声学 acoustic · 奥雅纳 伦敦 ARUP London

荣誉提名奖 · **honorable mention**
**Nieto Sobejano Arquitectos（建筑师事务所）**
Fuensanta Nieto · Enrique Sobejano（建筑师）

合作 (c) Alfredo Baladrón · Patricia Grande · Lourenço van Innis · Sebastian Sasse
模型 models · Juan de Dios Hernández-Jesús Rey
摄像 photographs · Diego Hernandez

参标 · **proposal**
**SO – IL (Solid Objectives – Idenburg Liu)（建筑师事务所）**
Florian Idenburg · Jing Liu（建筑师）

竞标任务包括为**原兽医学院建筑群**设定一个建筑和城市设计概念，作为华沙交响乐管弦乐团的选址。由于该城市被选定为“2016年欧洲文化之都”的候选城市，因此该建筑的建造至关重要。

The Competition task consists in developing an architectural and urban concept in the **former Veterinary Institute complex** for the purpose of the seat of Orkiestra Sinfonia Varsovia. The Center is an important project to this city - as it was candidate for the European Capital of Culture 2016.

华沙交响乐音乐厅 · 华沙

Sinfonia Varsovia Concert Hall · Poland

一等奖 · First Prize

# 浮墙

该浮墙围绕整个建筑物，将公园置为新的中心，并打造了一个幽静地带——这是管弦乐队进行演奏的基本要求——周遭富有一种戏剧性的氛围。该公园成为一个对外开放的公共场所。音乐厅融合了“鞋盒厅”和舞台的特点。传统“鞋盒厅”具有卓越的音响效果，但视觉效果却不尽如人意；而舞台在视觉方面具有优势，但在音响效果方面却达不到效果。

# FLOATING WALL

By enveloping the entire site, the floating wall serves as an indication to the park as its new centre and creates a distinctive place of silence - the basic principle for an orchestra to perform - full of ambience and drama. The park becomes an open public place. The Concert Hall is a fusion of a Shoebox Hall and an Arena. While the traditional shoe box hall is known for its excellent acoustics but often offers poor visual conditions, the Arena has its advantages with regards to visibility and is hardly applicable when it comes to its acoustic conditions.

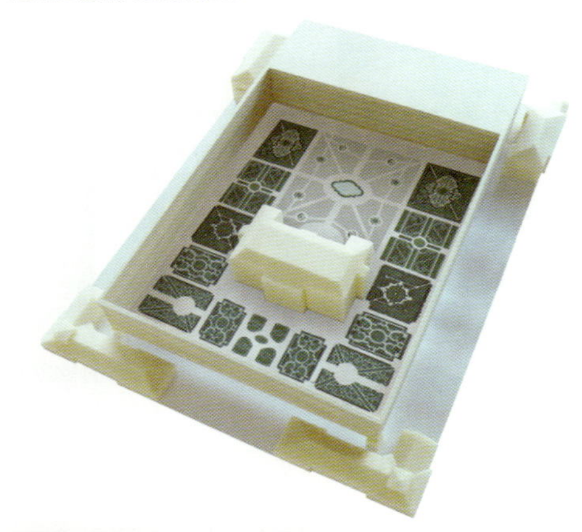

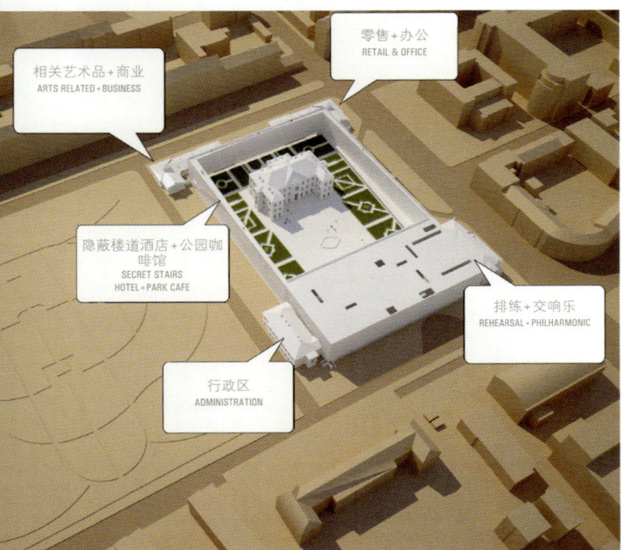


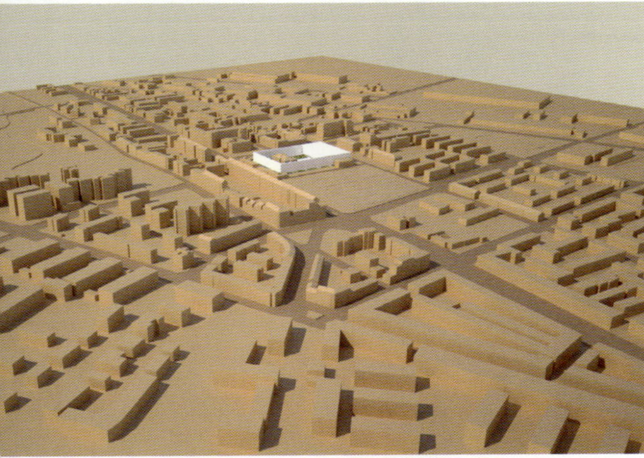

# fch013 (竞标代码)

Atelier Thomas Pucher (建筑师事务所)

Thomas Pucher (建筑师)

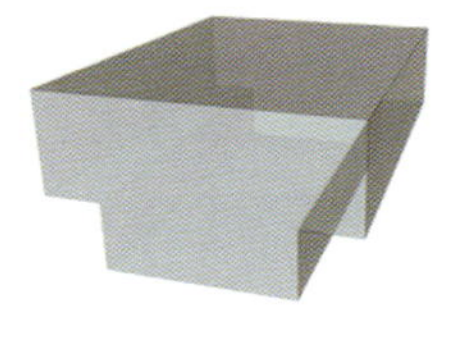

理想声学效果 IDEAL ACOUSTICS
鞋盒 SHOEBOX

理想视觉效果 IDEAL VIEW
环绕大厅 SURROUNDING HALL

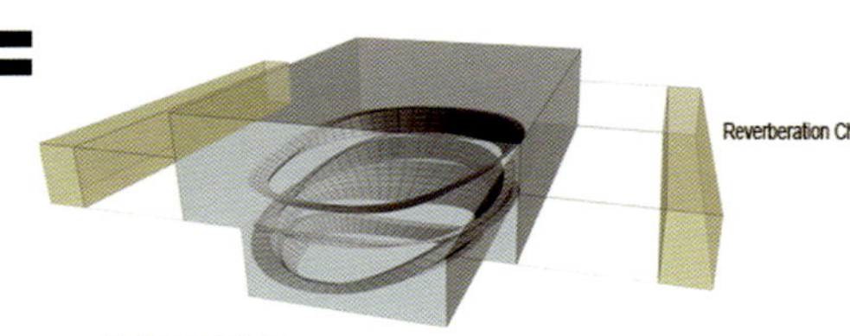


整体环境效果 FULL DRAMA AND ATMOSPHERE
华沙交响乐音乐厅 SINFONIA VARSOVIA

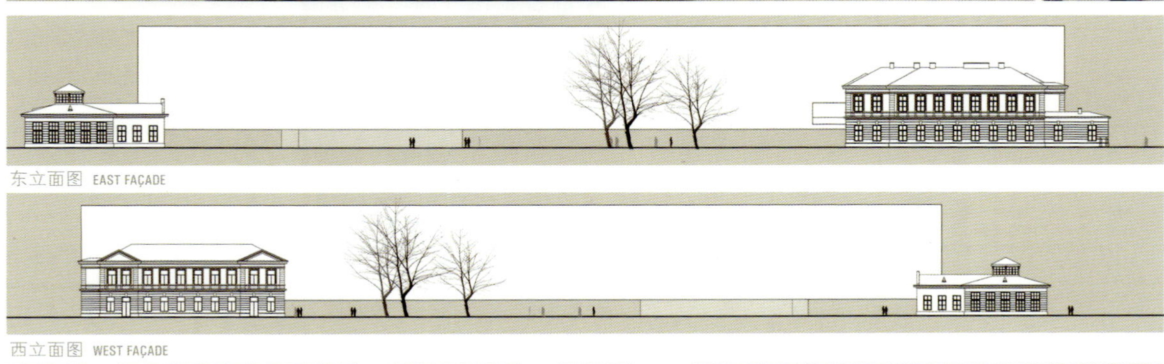

东立面图 EAST FAÇADE

西立面图 WEST FAÇADE

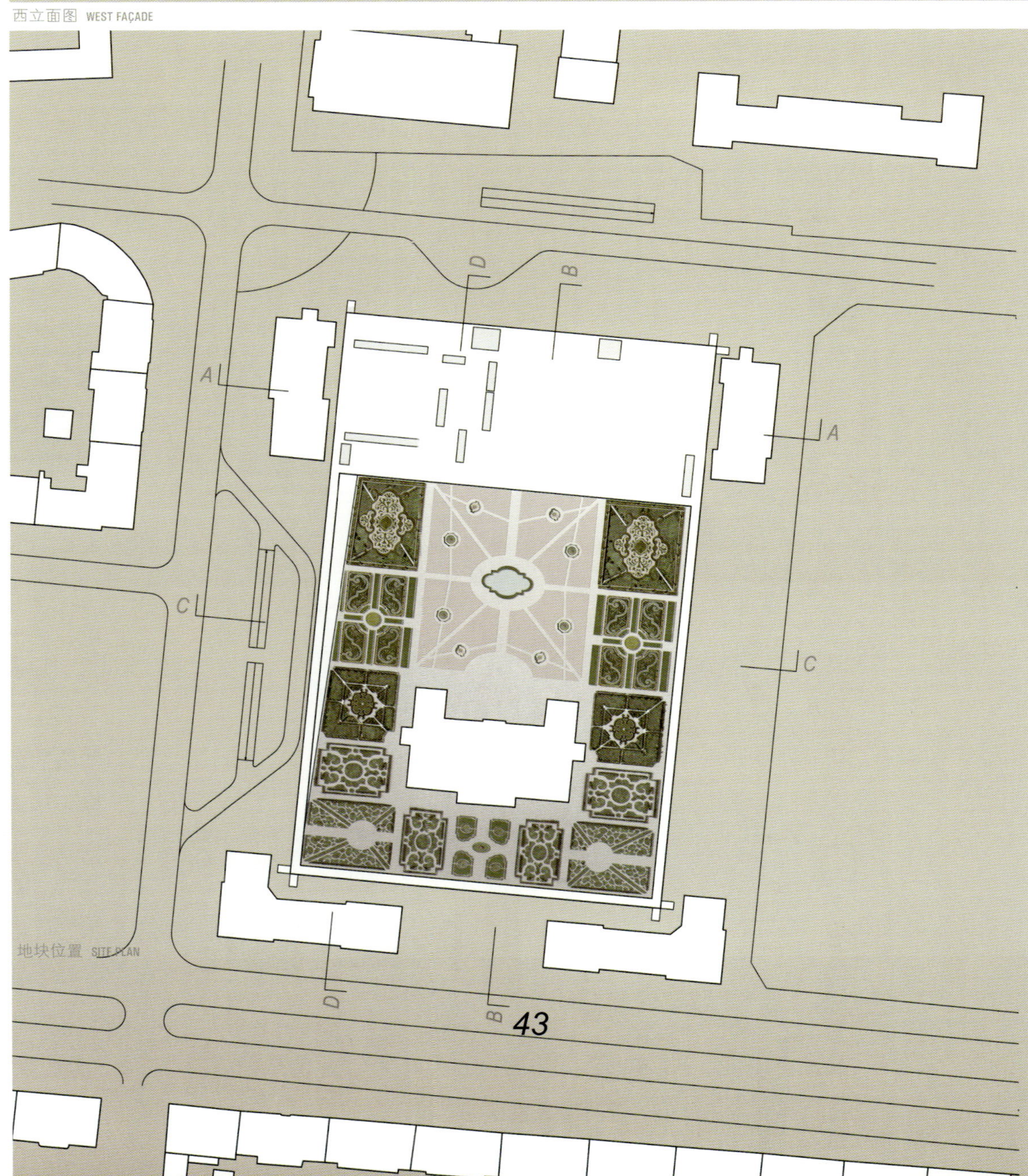


地块位置 SITE PLAN

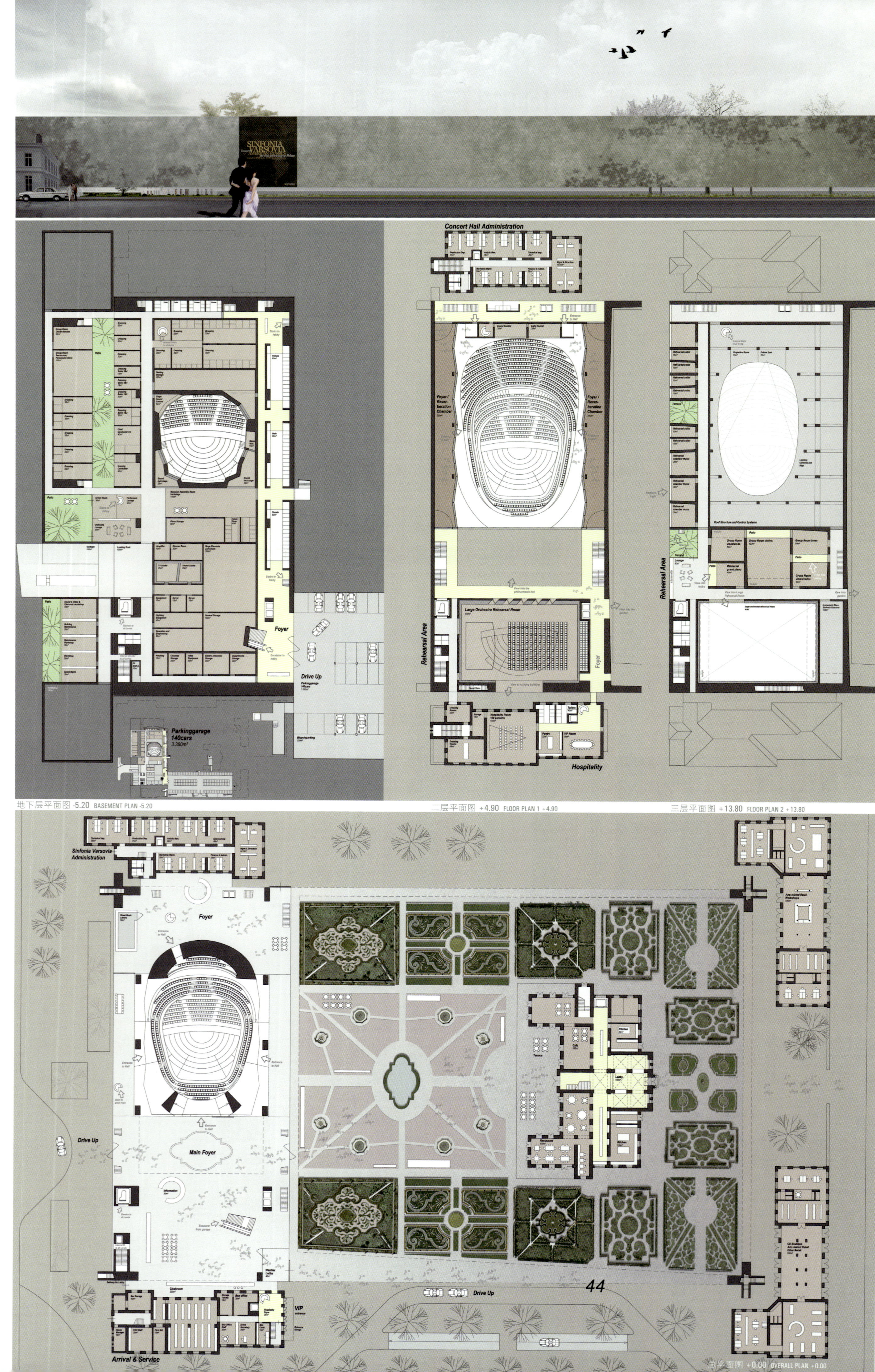


地下层平面图 -5.20 BASEMENT PLAN -5.20

二层平面图 +4.90 FLOOR PLAN 1 +4.90

三层平面图 +13.80 FLOOR PLAN 2 +13.80

总平面图 +0.00 OVERALL PLAN +0.00

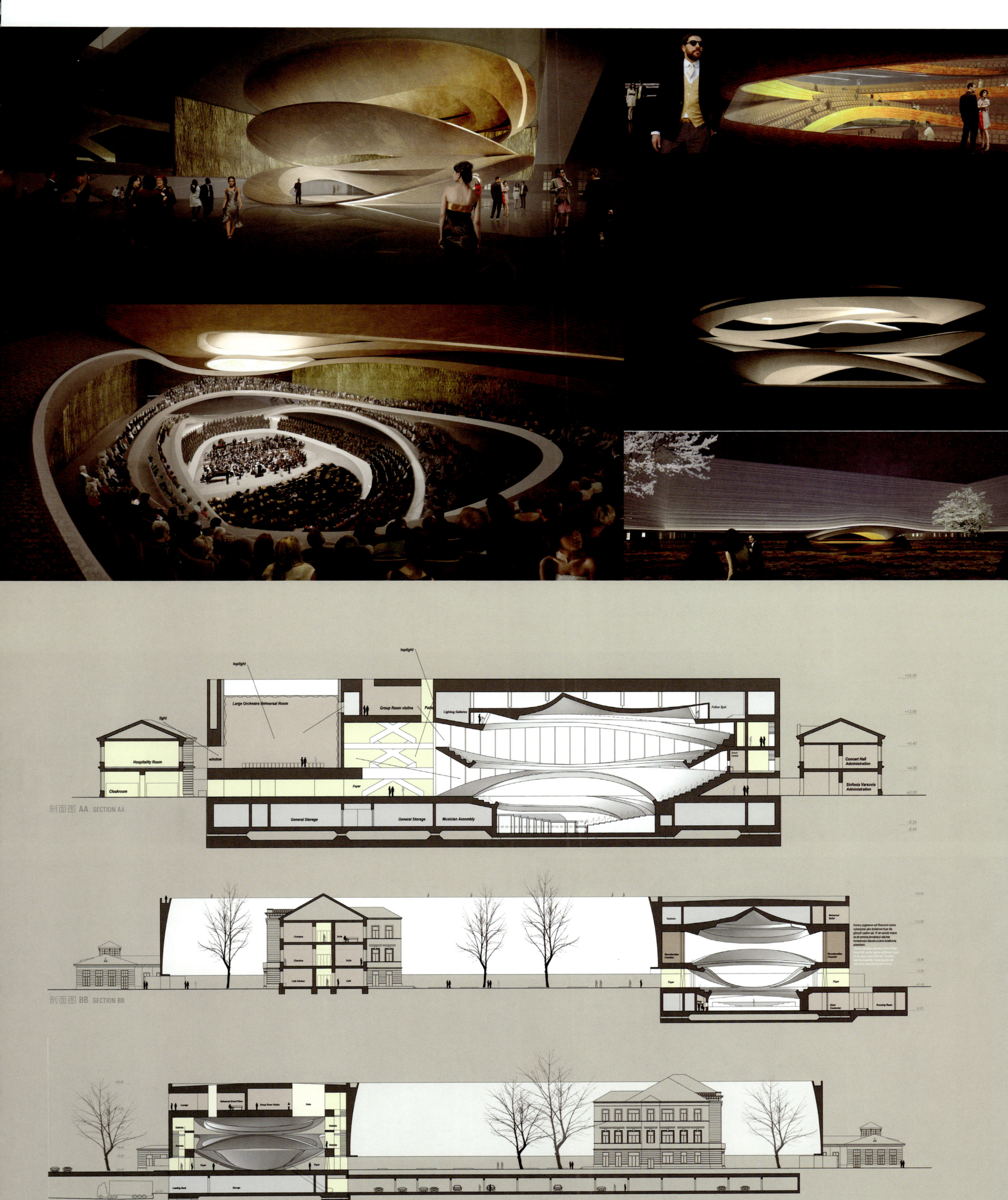

剖面图 AA SECTION AA
剖面图 BB SECTION BB
剖面图 CC SECTION CC

华沙交响乐音乐厅 · 华沙

Sinfonia Varsovia Concert Hall · Poland

并列二等奖 · Second Prize Ex-Aequo

# svu937 (竞标代码)

Zaha Hadid Architects (建筑师事务所)

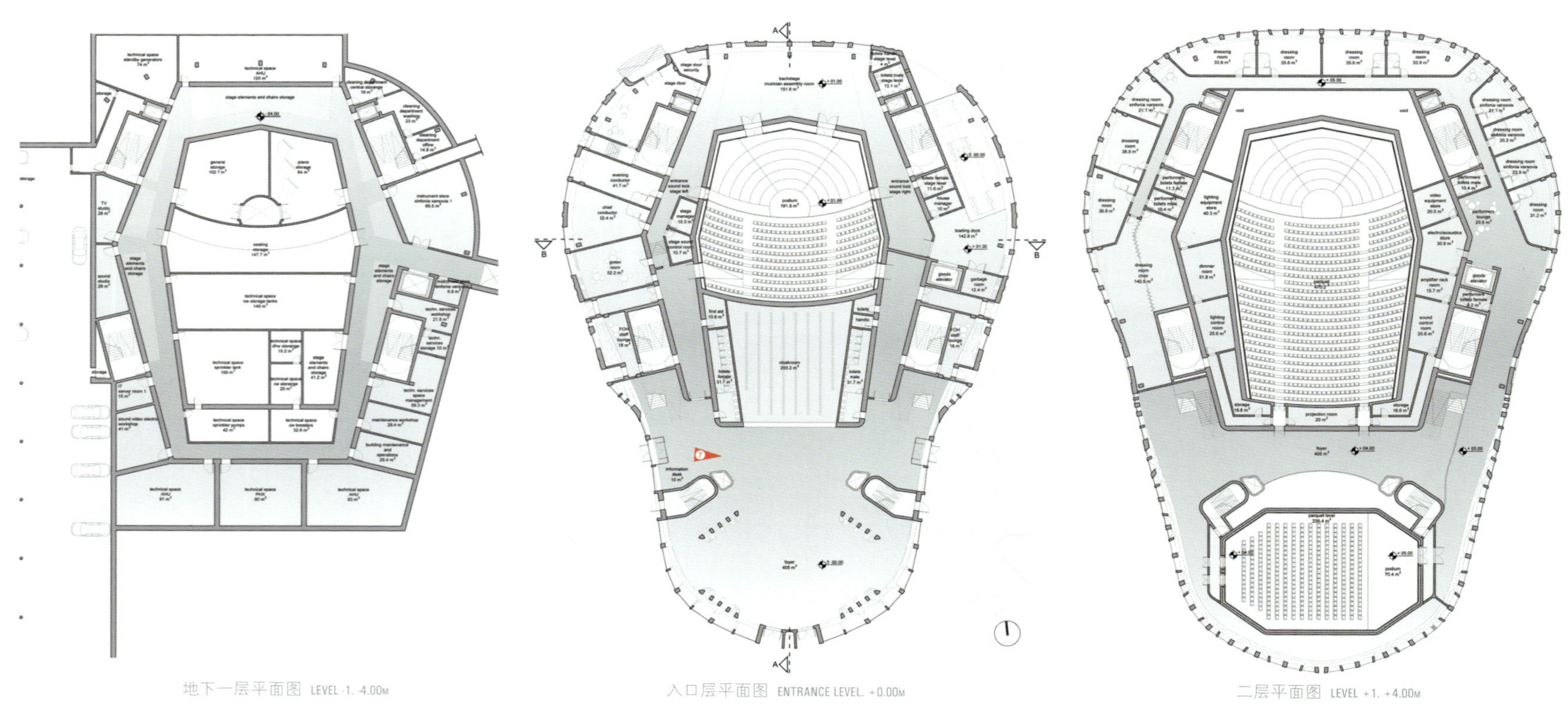

地下一层平面图 LEVEL -1. -4.00M　　入口层平面图 ENTRANCE LEVEL. ±0.00M　　二层平面图 LEVEL +1. +4.00M

地块位置 SITE PLAN

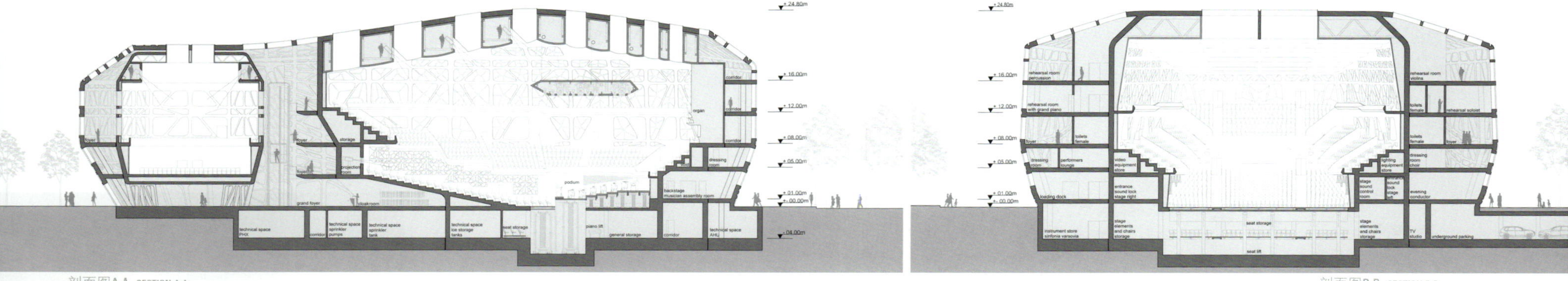

剖面图A-A SECTION A-A

剖面图B-B SECTION B-B

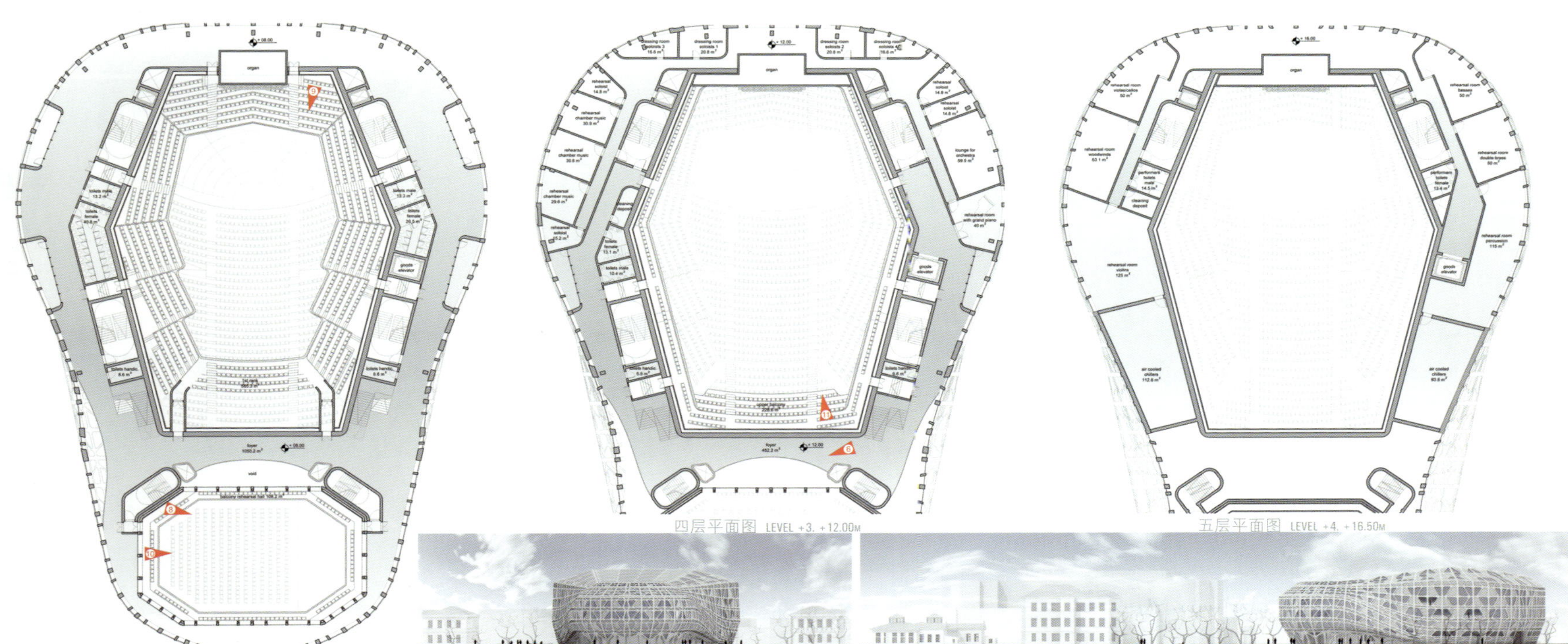

四层平面图 LEVEL +3, +12.00M

五层平面图 LEVEL +4, +16.50M

三层平面图 LEVEL +2, +8.00M

地块剖面图C-C SITE SECTION C-C

地块剖面图D-D SITE SECTION D-D

# 凹凸有致

建筑物间的走道设计通过整合既有建筑的对称性将其融合到周围的城市布局中。在凸显场地中心轴的同时，还将原先长方形的格局变得更加富有流动性。对五座建筑物进行修缮，并通过可持续的方式使得新旧建筑相融合，为总体规划区域注入活力。整个场所在空间上采用格网形状式布局，以营造新文化中心的独特氛围。该建筑建于中轴线上，以凸显其独立的地标性，同时，还通过一个通透大厅与周围相连接，这样公园和大厅即可有机地连接起来，它们的界线也因此而弱化。

# CONCAVE AND CONVEX

The passages through the site incorporate the existing symmetrical layout, extending it to strategic points to the surrounding city fabric. The sites central axis is registered and reinforced while transforming its original rectangular grid into a more fluid and continuous circulation system. Five buildings are maintained and refurbished connecting the old with the new in a sustainable manner and activating the overall planning area. A lattice pattern as a new layer of spatial organization is stretched over the whole site in order to enhance the unique atmosphere of the new cultural centre. While the positioning of the building on the central axis pronounces its solitary landmark appearance, the building is linked to the surroundings via a large transparent foyer which blurs the inside outside boundary between park and foyer space.

华沙交响乐音乐厅 · 华沙

Sinfonia Varsovia Concert Hall · Poland

并列二等奖 · Second Prize Ex-Aequo

# bml468 (竞标代码)

Hermanowicz Rewski Architekci (建筑师事务所)

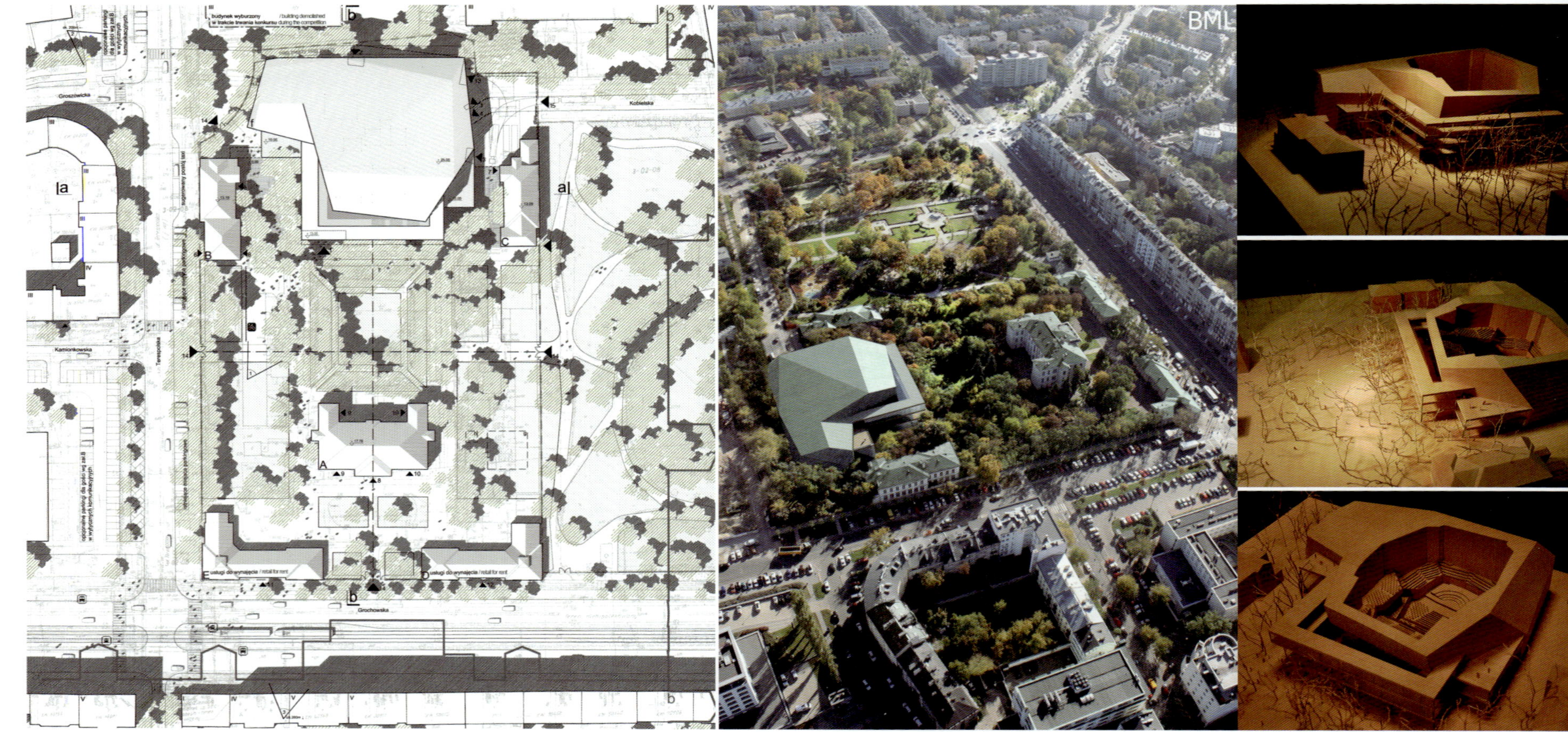

地块位置 SITE PLAN

东立面图 EAST ELEVATION

西立面图 WEST ELEVATION

四层平面图 FLOOR PLAN +3

三层平面图 FLOOR PLAN +2

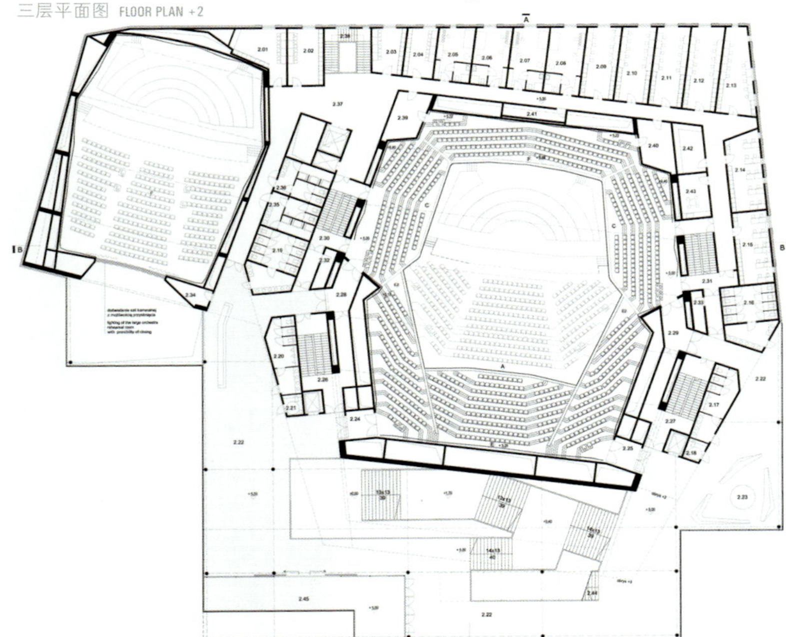

二层平面图 FLOOR PLAN +1

底层平面图 FLOOR PLAN 0

# 不规则外形

该建筑的主建筑体为椭圆形，与乐器相似，具有良好的触觉效果。它既不与遮天蔽日的树冠相抗衡，也不与历史悠久的城市环境相抵牾。它可调整其规模和位置以与树木和既有的建筑相融合。该音乐厅巧妙地打破了历史氛围的对称性，以便与绿地面积有机融合于一体。

# IRREGULAR BODY

Similarly to a musical instrument, the body of the main building is oval and pleasant to the touch. It competes neither with the wonderful tree stand nor with the historical urban surroundings. It flows between the trees and the existing buildings adjusting to both scale and location. The concert hall building subtly breaks the symmetry of historical atmosphere, organically blending with the green areas.

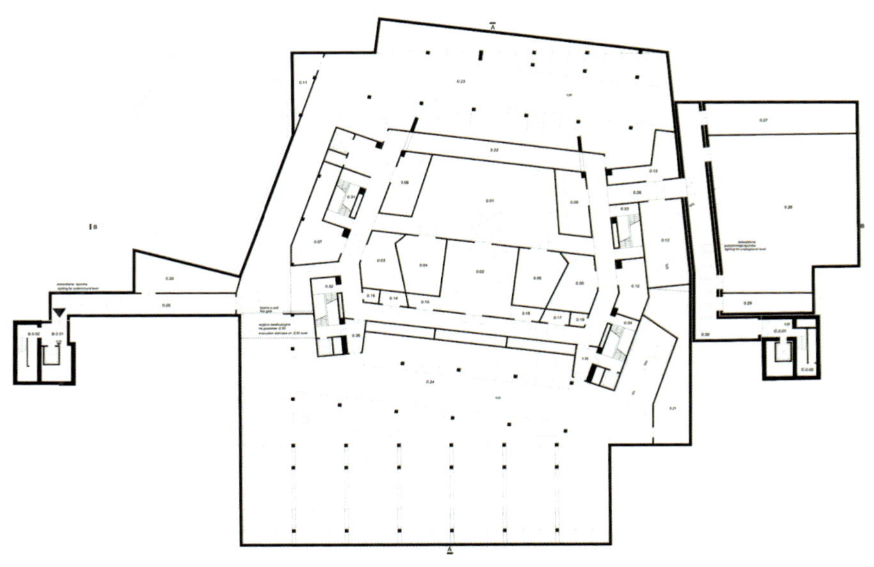

地下一层平面图 FLOOR PLAN -1

南立面图 SOUTH FAÇADE

北立面图 NORTH FAÇADE

华沙交响乐音乐厅 · 华沙

Sinfonia Varsovia Concert Hall · Poland

荣誉提名奖 · Honorable Mention

# svr900 (竞标代码)


RE (建筑师事务所)

Piotr Michalewicz · Matusz Tański (建筑师)


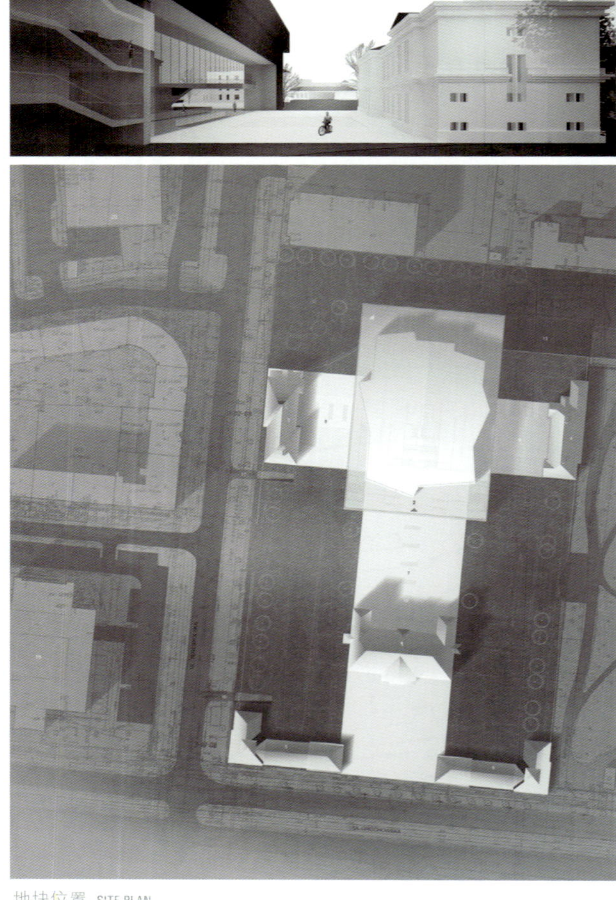

地块位置 SITE PLAN

## 简洁

该项目旨在与既有的五座建筑物经典、对称的布局相辅相成。在建筑群体内的各个建筑物相互衔接，将提供一个舒适、方便的通行环境。

## SIMPLICITY

The project is meant to be a complimentary element to the classical, symmetrical layout of five historical buildings. All the buildings comprising the complex will be connected to provide a comfortable and obstacle-free communication.

东立面图 EAST ELEVATION

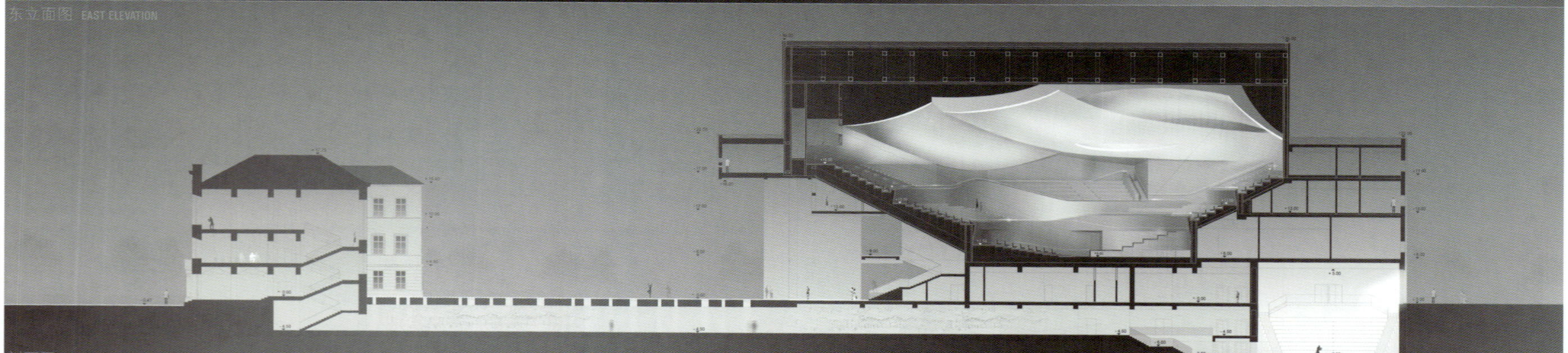

剖面图 AA SECTION AA

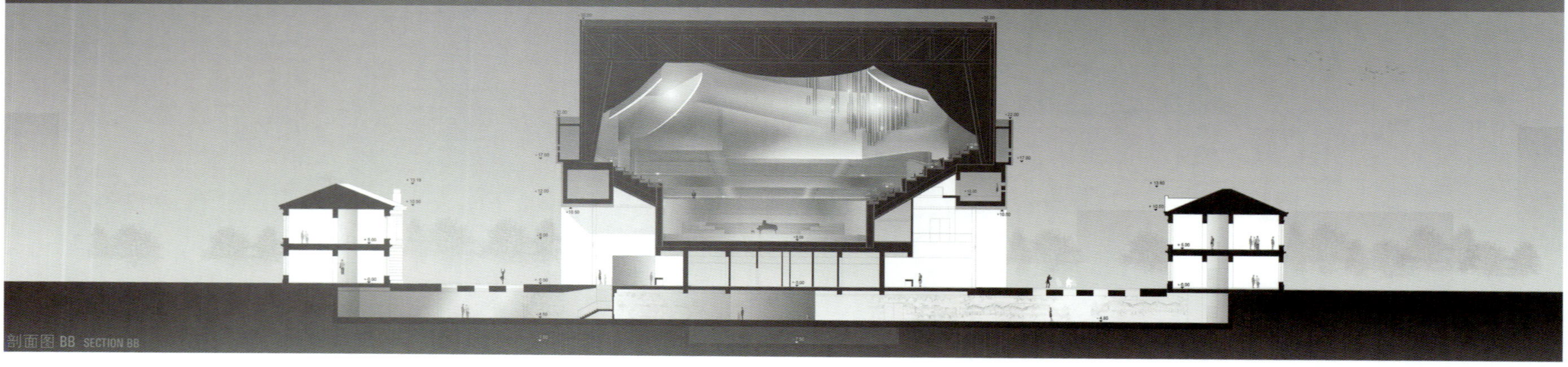

剖面图 BB SECTION BB

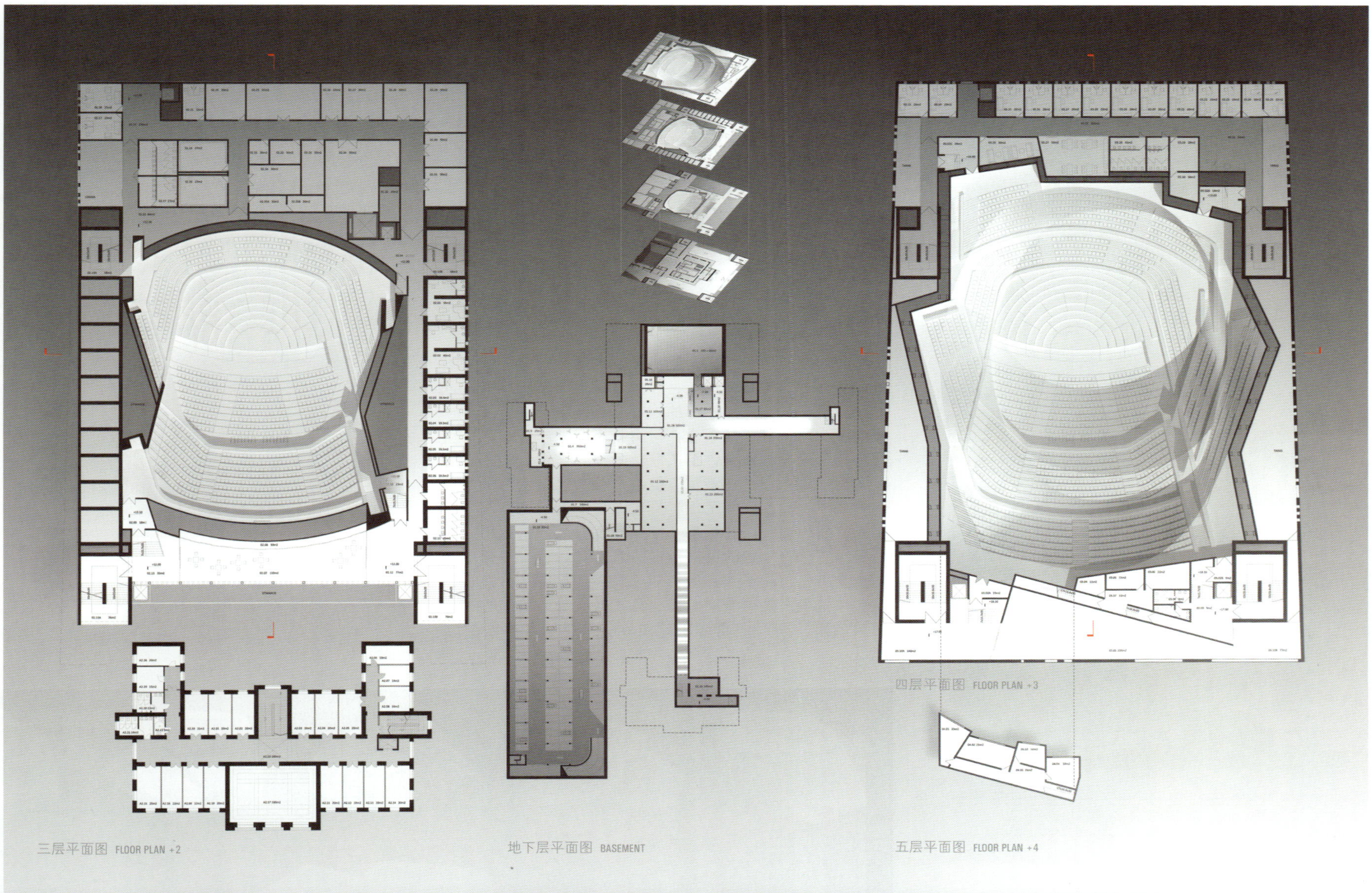

三层平面图 FLOOR PLAN +2

地下层平面图 BASEMENT

四层平面图 FLOOR PLAN +3

五层平面图 FLOOR PLAN +4

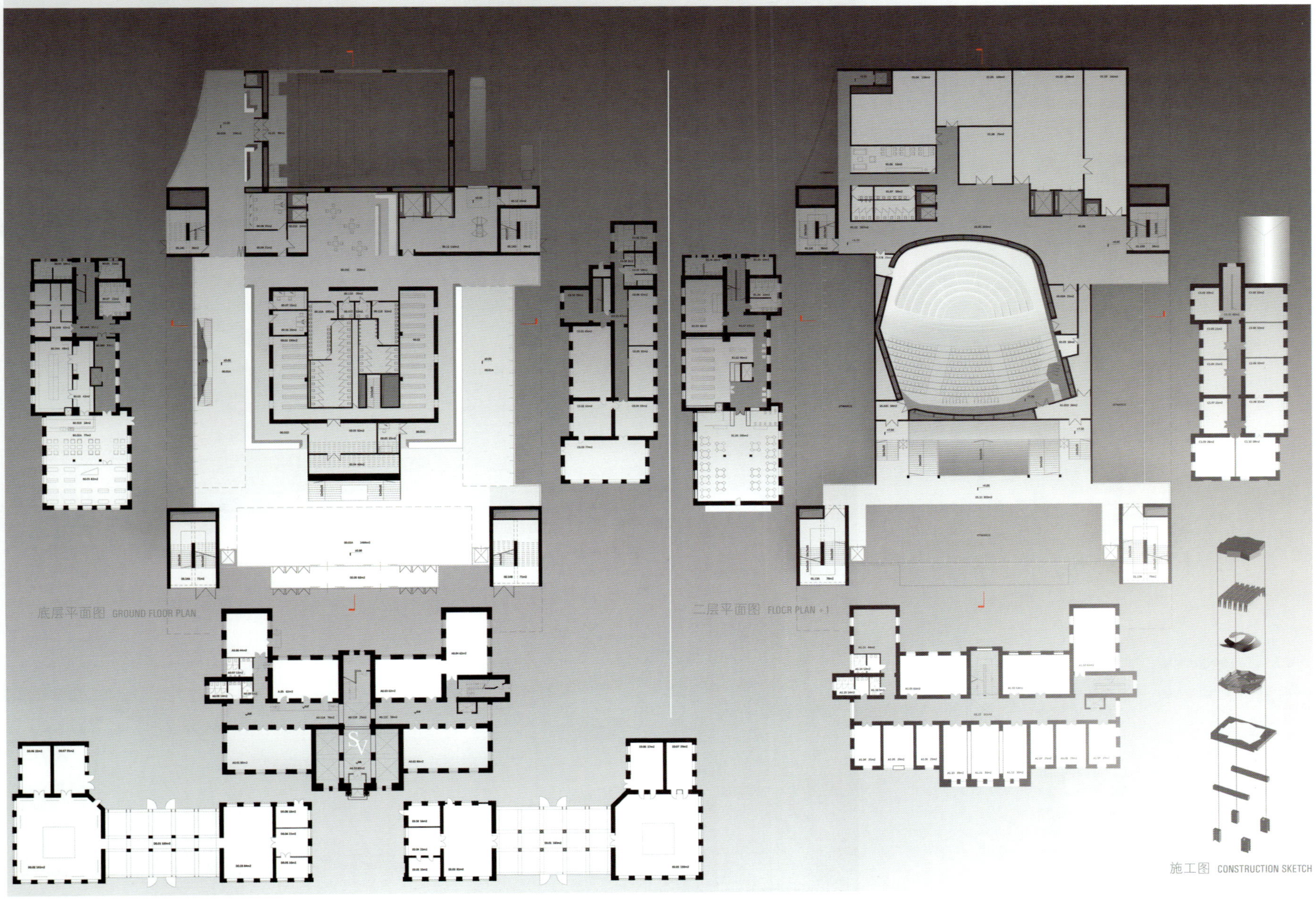

底层平面图 GROUND FLOOR PLAN

二层平面图 FLOOR PLAN +1

施工图 CONSTRUCTION SKETCH

华沙交响乐音乐厅 · 华沙

Sinfonia Varsovia Concert Hall · Poland

荣誉提名奖 · Honorable Mention

# cdg690 (竞标代码)

Nieto Sobejano Arquitectos (建筑师事务所)

Fuensanta Nieto · Enrique Sobejano (建筑师)

## 螺旋形状、音符和树木

全新的音乐厅由完全不相关联的部分组成：螺旋形状——一种几何形状，通过其打造出音乐厅的内部空间，并让大大小小的螺旋空间连为一体；音符——不同音符排列组合，可得到不同的乐音，我们将其与建筑的重复性相关联，并在建筑设计中予以体现；树木——在该建筑的各个公共区域均可看到园林中的树木。为了保存具有重要价值的五座既有建筑物，该音乐厅基于一个螺旋上升的多边形植物而修建。该项目的关键点在于其实体设计：其建筑表皮由一系列不间断的透明塑膜玻璃构成，并在组合中采用三角形布局，以达到光线波动和映射的独特效果，同时还可缩小建筑物在周围环境中的体积。

## THE SPIRAL, THE MUSICAL NOTES AND THE TREES

The new auditorium came about from apparently unconnected starting points: -the spiral- a geometrical shape that forms the space of the hall and that continuously reappears on different scales; -the musical notes-, whose combinatory variations we associate with the repetition are implied in all architectural works; -the trees-, whose presence in the gardens is made visible in all of the public parts of the building. Respecting and preserving the five existing buildings of greatest value, the new auditorium arises on the basis of a polygonal plant which is derived from a spiral. A key argument of the project is its materiality, defined by a continuous skin made up of pieces of extra clear molded glass whose triangular layout which is repeated in combinatorial series and provokes unexpected light vibrations and reflections, at the same time as reducing the scale of the building in relation to the context.

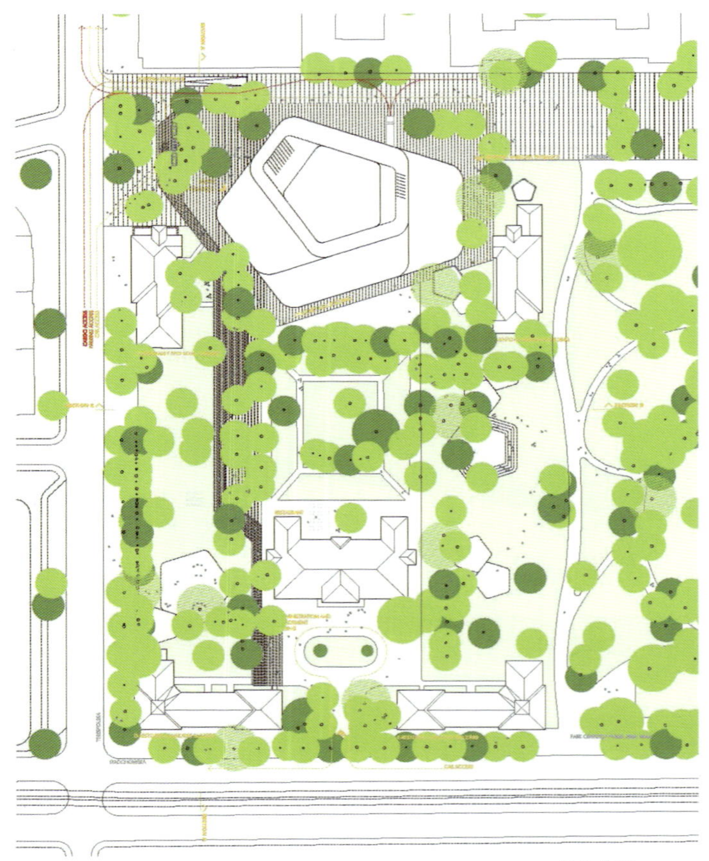

地块位置 SITE PLAN

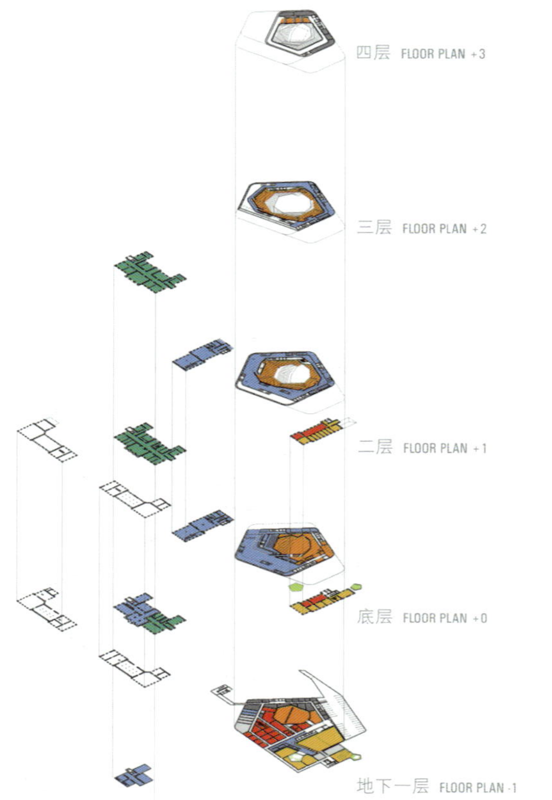


功能分布图 FUNCTIONAL SCHEME

- 公共区域 PUBLIC AREAS
- 音乐厅 CONCERT HALL
- 仓储+服务·维护 STORAGE+SERVICES·MAINTENANCE
- 演艺厅 PERFORMER'S AREA
- 排练厅 REHEARSAL ROOM
- 行政区+后勤操作 MANAGEMENT+ADMINISTRATION AREAS
- 维护+中央仓储+设备区 MAINTENANCE+CENTRAL STORAGE+FACILITY MANAGEMENT

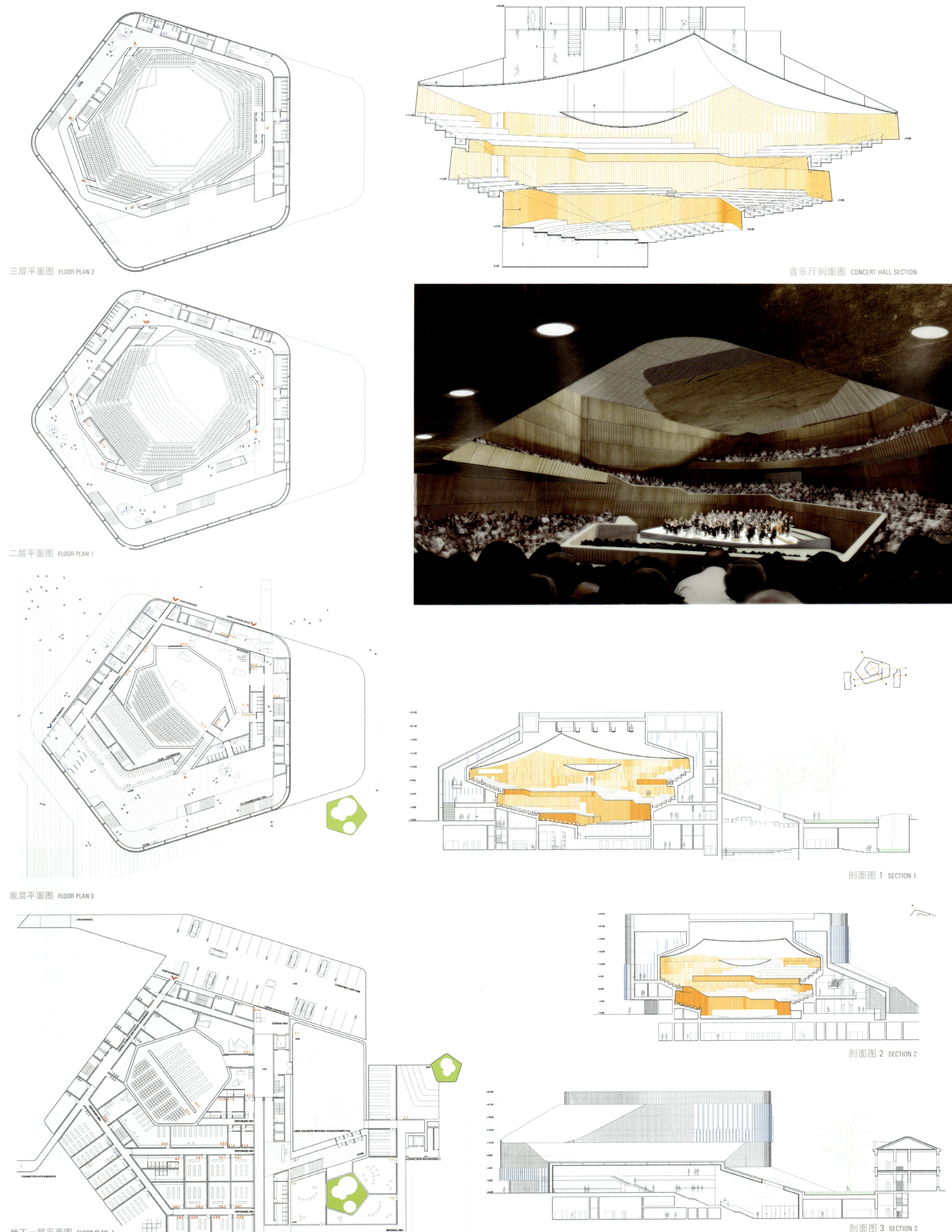
三层平面图 FLOOR PLAN 2
二层平面图 FLOOR PLAN 1
底层平面图 FLOOR PLAN 0
地下一层平面图 FLOOR PLAN -1
音乐厅剖面图 CONCERT HALL SECTION
剖面图 1 SECTION 1
剖面图 2 SECTION 2
剖面图 3 SECTION 3

华沙交响乐音乐厅 · 华沙

Sinfonia Varsovia Concert Hall · Poland

参标 · Proposal

# xyz000 (竞标代码)

so-il (建筑师事务所)

Florian Idenburg · Jing Liu (建筑师)

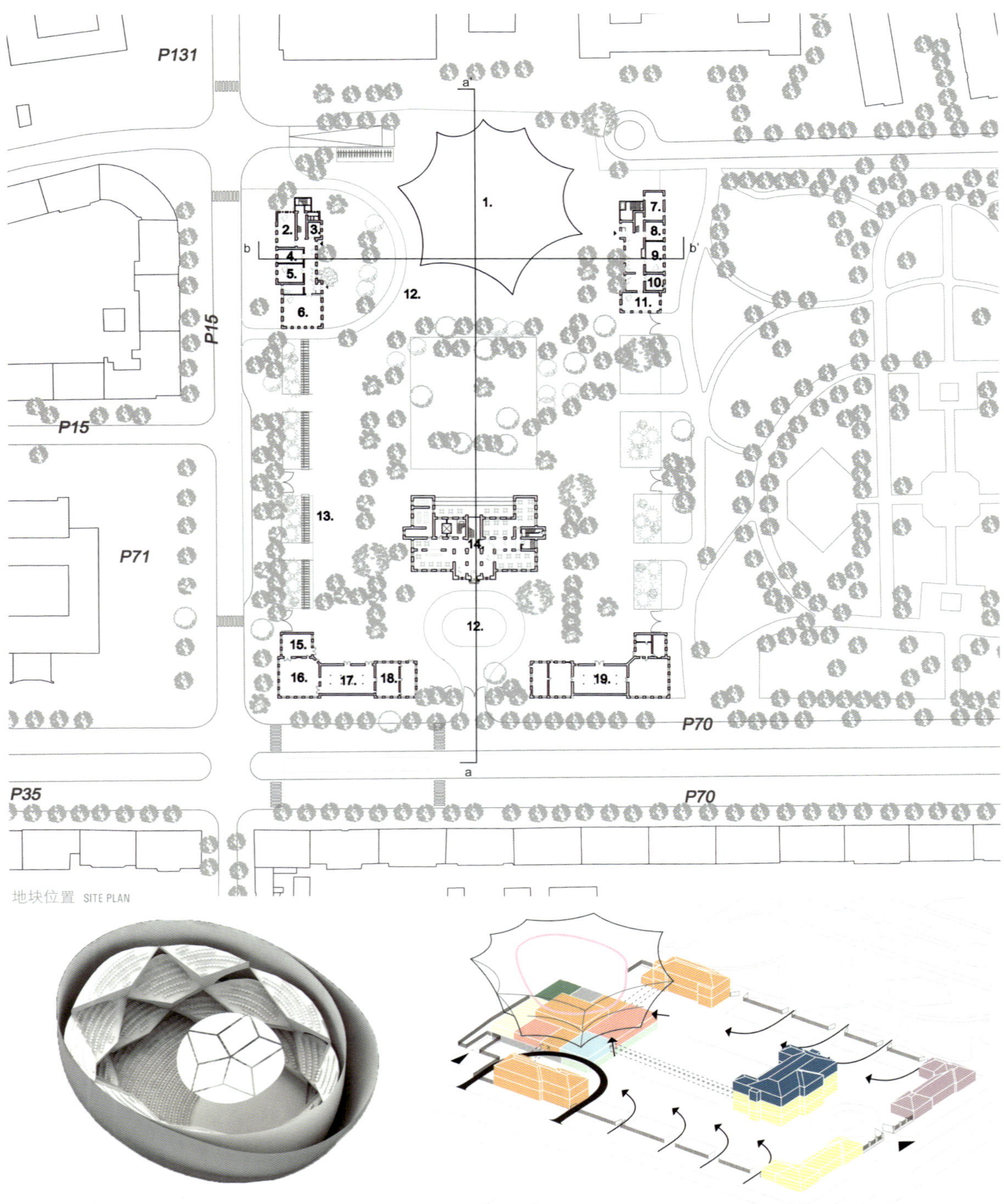


地块位置 SITE PLAN

## 中心轴位置

我们的设计目标是改造该场地，以便于大众进出。首先，我们在四周均设置入口，以示对大众的光临表示欢迎，让音乐声传到街邻四舍，甚至到更远处的市中心。为了与该建筑的整体设计概念相吻合，将演艺厅设置在中心轴上。通过其设计方案以及外观造型，该建筑物不仅与当地环境相融合，而且将其功能延伸至该建筑群体外的活动及大众。

## ON A CENTRAL AXIS

With our proposal, this site is made accessible and permeable. First, we would introduce entrances on all sides, thus creating a welcoming air, nestling the institute and the music performed therein within the neighborhood and the city beyond. In line with the overall concept for the site, the performance hall is positioned on a central axis. Through its plan and form, the building not only responds to the monuments on-site but also transcends these and opens up the site to activities and audiences beyond these buildings and borders.

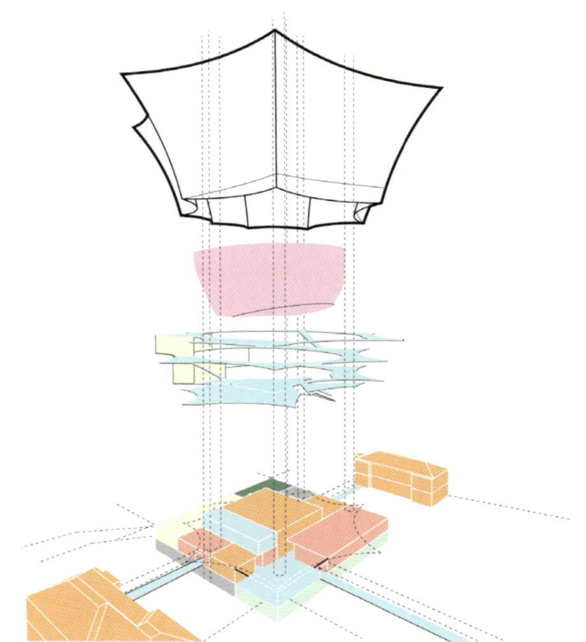

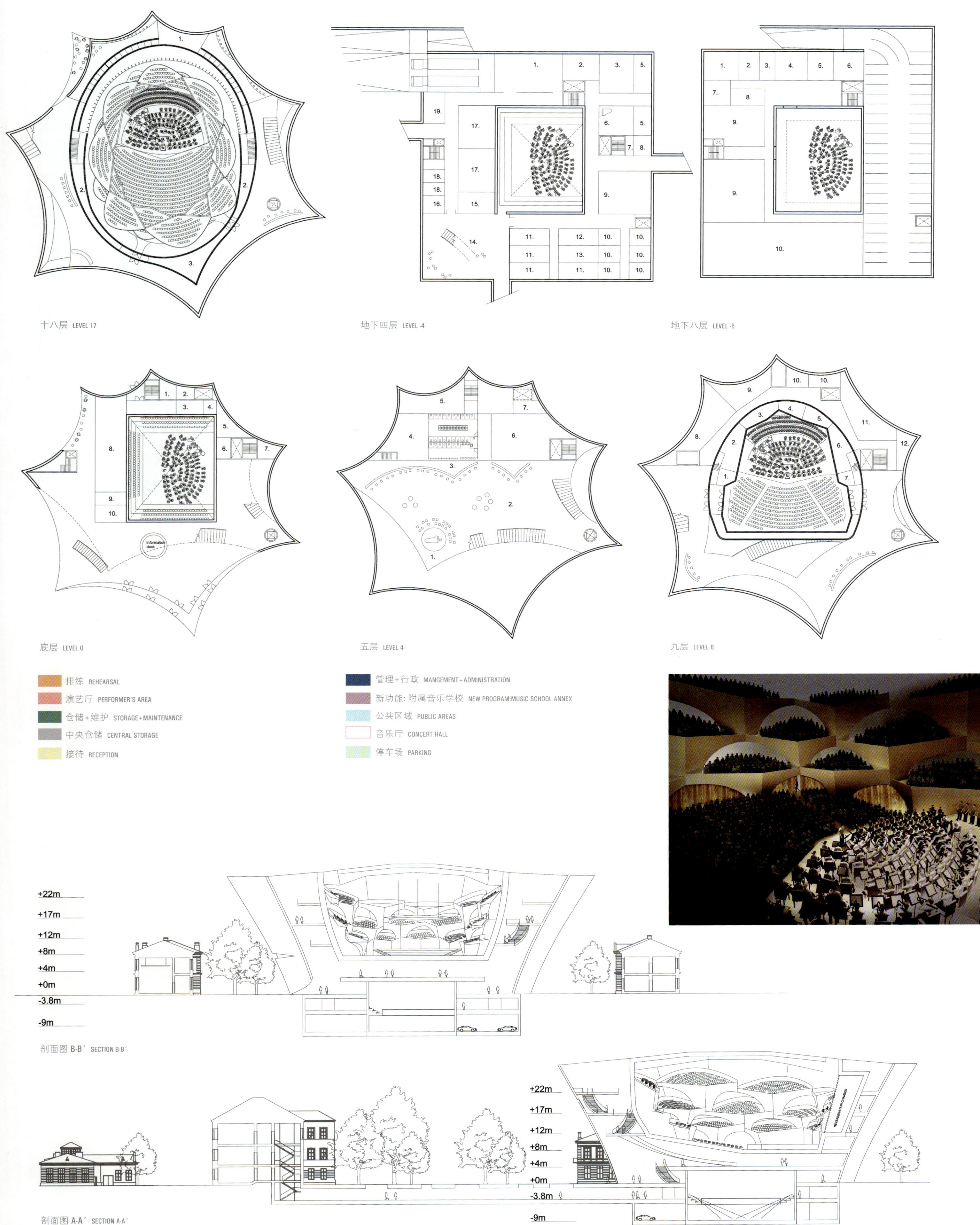


十八层 LEVEL 17

地下四层 LEVEL -4

地下八层 LEVEL -8

底层 LEVEL 0

五层 LEVEL 4

九层 LEVEL 8

剖面图 B-B´ SECTION B-B´

剖面图 A-A´ SECTION A-A´

# 动漫博物馆 · 杭州
# Comic and Animation Museum · China

**竞标 · competition**
动漫博物馆
Comic and Animation Museum

**竞赛类型 · competition type**
邀标
invited competition

**项目地点 · site area**
杭州 · 中国 Hangzhou · China

**主办方 · promoter**
杭州城市规划局 Hangzhou Urban Planning Bureau

**日程安排 · schedule**
招标 · Announcement 01.2011
评审结果 · Jury's results 04.2011

**评审团 · jury**

崔愷 Cui Kai
程泰宁 Cheng Taining
刘克成 Liu Kecheng
王澍 Wang Shu
刘辉石 Liu Huishi
Felix Zwoch
刘晓东 Liu Xiaodong
阮文静 Ruan Wenjing
汪小玫 Wang Xiaomei

**premios · awards**

一等奖 · **first prize**
**MVRDV**
Winy Maas · Jacob van Rijs · Nathalie de Vries (建筑师)

Renske van der Stoep · Stefan de Koning · Hui Hsin Liao · Rune Veile · Doris Strauch · Pepijn Bakker · Maria Lopez · Arjen Ketting · John Tsang · Aser Gimenez · Juras Lasovsky · Suchi Vora
合作建筑师 co-architect · Zhubo Architectural & Engineering Design
结构顾问 advisors-consultants · structure-services · 奥雅纳 Arup (Mitsuhiro Kanada · Andrew Luong · Peter Mensinga)
展览设计 exhibition design · Kossmann.deJong (Herman Kossmann Michel de Vaan)
图形设计 graphic design · Jonge Meesters (Jeroen Bruijn · Serge Scheepers)
艺术设计/三维建模 artist Impressions/ 3D modeling · MVRDV · Kossmann.deJong · Unlimited CG

二等奖 · **second prize**
**Miralles Tagliabue EMBT (建筑师事务所)**

建筑师 architect · Benedetta Tagliabue
方案主管 project director · Daniel Rosselló
合作 (c) Francesca Origa · Gabriele Rotelli · Felipe Pecegueiro · Francesca Neus Frontera · Tomas Montis · William Mathers · Silvia Rio · Ana Catalina Villareal Yarto · Roberto Palau Tamez Vaiva Simliunaite · Iker Alzola · Alice Puleo · Ana Redondo · Mar Flores · Diego Javier Parra Beatriz Cabanillas Macias

入围 · **finalist**
**Atelier Bow-Wow (塚本由晴事务所)**
Yoshiharu Tsukamoto · Momoyo Kaijima (建筑师)

合作 (c) Takahiko Kurabayashi · Reika Tatekawa · Mio Sekimoto · Yuki Chida · Hiroaki Goto Motoo Chiba · Shirley Woo
结构 structure · 奥雅纳日本 Arup Japan · Mitsuhiro Kanada · Kengo Takamatsu
三维多媒体及渲染 3D multi media & renderings · Atelier Bow-Wow + crystal CG Shanghai

入围 · **finalist**
**同济大学建筑设计研究院 Architectural Design & Research Institute of Tongji University**
主案建筑师 project architect · 曾群 Zeng Qun

入围 · **finalist**
**王路工作室 · 清华大学建筑设计研究院 Studio Wang Lu · Architectural Design & Research Institute of Qinghua University**

主案建筑师 project architect · 王路 Wang Lu
团队 team · 李坚 Li Jian · 徐杰 Xu Jie · 郑小东 Zheng Xiaodong · 孙德龙 Sun Delong · 钟铮 Zhong Zheng · 孟宁 Meng Ning

杭州是一座大都市，西北毗邻上海，相距约180千米，人口达640万。钱塘江南面人口较为稀少，该博物馆可成为该地区的新焦点。CCAM 可巩固杭州作为中国**漫画产业之都**的地位。这座新博物馆由一系列“山”型建筑组成，内含办公室、酒店和会议中心，其第一期即将完工，建成后将成为杭州城快速发展的新地标。

Hangzhou is a metropolis with 6.4 million inhabitants 180 km southwest of Shanghai. The Museum will become a new focal point on the less populated southern side of Qiantang river. The CCAM will consolidate the city's leading position as China's **capital of the animation industry**. The new Museum will be the icon of a larger development. It comprises a series of hill-shaped buildings containing offices, a hotel and a conference centre of which the first phase is close to completion.

动漫博物馆·杭州

Comic and Animation Museum · China

一等奖 · First Prize

MVRDV (建筑师事务所)

Winy Maas · Jacob van Rijs · Nathalie de Vries (建筑师)

地块位置 SITE PLAN

## 气球状对白圈

如何设计，使得该建筑能告诉人们这里是一个漫画的世界、动画的世界？在设计中运用漫画中的主要标志之一——气球状对白圈，使得读者看到这个建筑造型，即可知道它一定与卡通、漫画和动画相关。

该项目共分六个区域：入口、典藏区、互动区、图书馆、剧院和教育馆。各个区域呈现独立的三维对白圈状，六个区域排列组合形成一组连续的逻辑空间序列。

## SPEECH BALLOON

How could a new building specifically designed to house the worlds of comics and animations express this universal language? By using one of its prime characteristics - the speech balloon - the building will instantly be recognized as the place for cartoons, comics and animations.

The requested program has been organized into six zones: Entrance, Collection, Interaction, Library, Cinema and Education. Each zone is organized in its own 3-dimensional speech balloon shaped volume and the six of them together are arranged in such a way that they form a continuous logical spatial sequence.

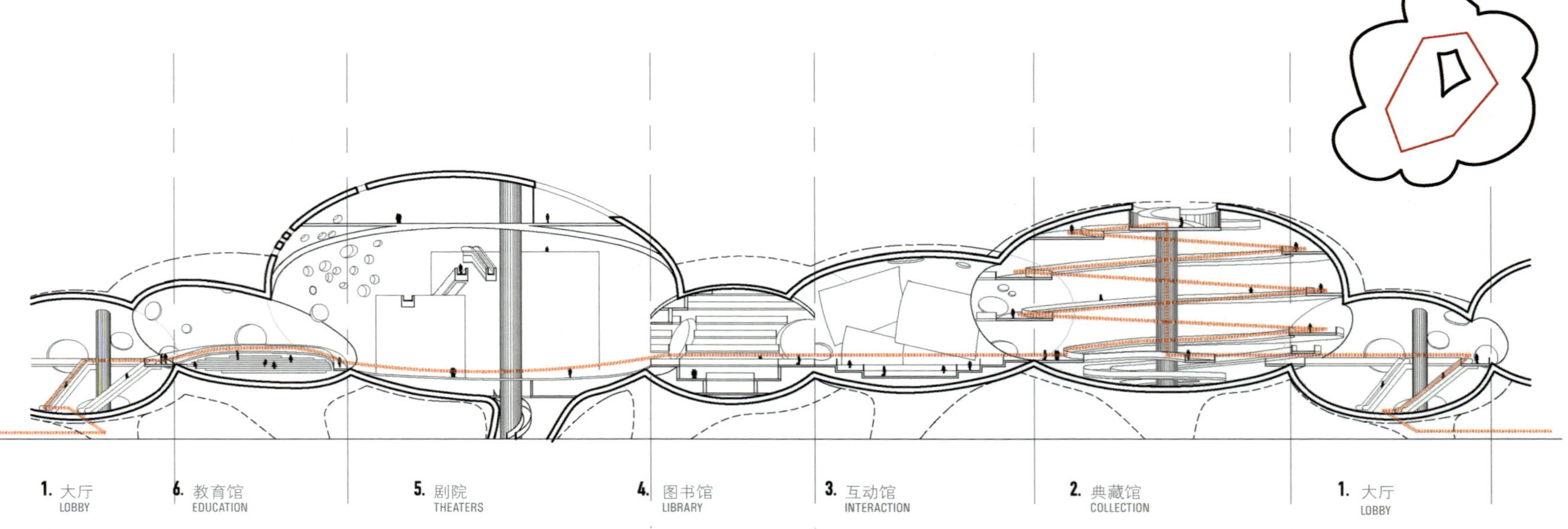


主剖面 MAIN SECTION

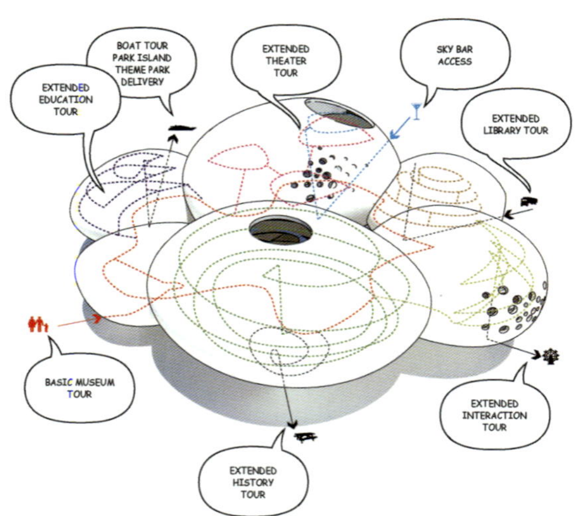


流线 CIRCULATION

- 博物馆基本流线 BASIC MUSEUM TOUR
- 博物馆历史馆扩展流线 EXTENDED HISTORY TOUR
- 互动馆扩展流线 EXTENDED INTERACTION TOUR
- 图书馆扩展流线 EXTENDED LIBRARY TOUR
- 剧院扩展流线 EXTENDED THEATER TOUR
- 教育馆扩展流线 EXTENDED EDUCATION TOUR
- 屋顶吧入口 SKY BAR ACCESS
- 游船流线·岛体公园·主体公园·货物装卸 BOAT TOUR · PARK ISLAND · THEME PARK · DELIVERY

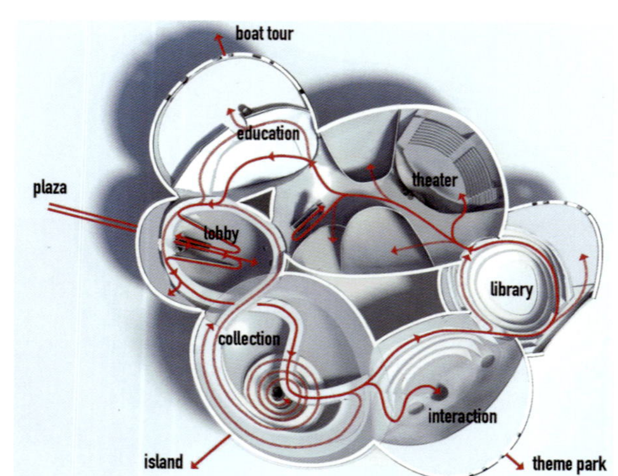


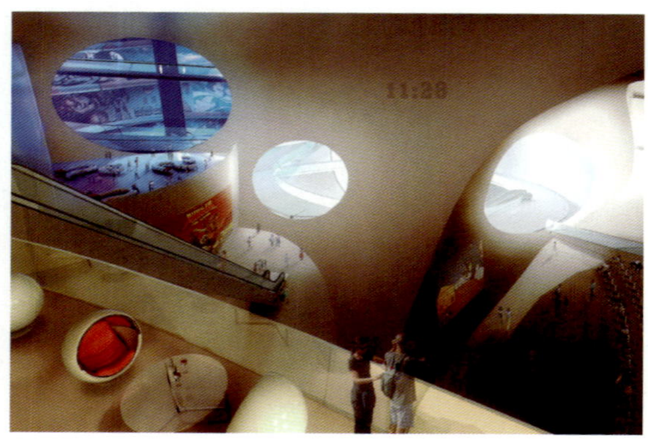

平面图 +18M FLOOR PLAN +18M

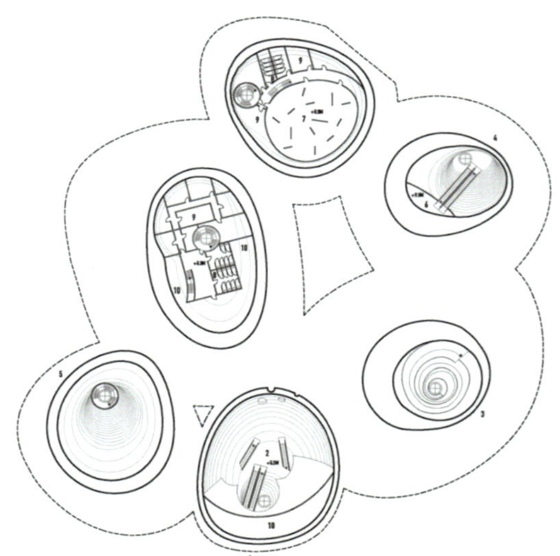

平面图 +9M FLOOR PLAN +9M

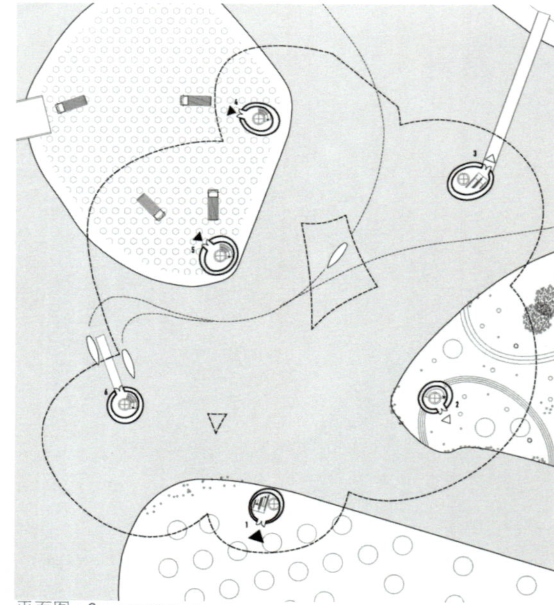

平面图 +0M FLOOR PLAN +0M

屋顶层平面图 ROOF PLAN

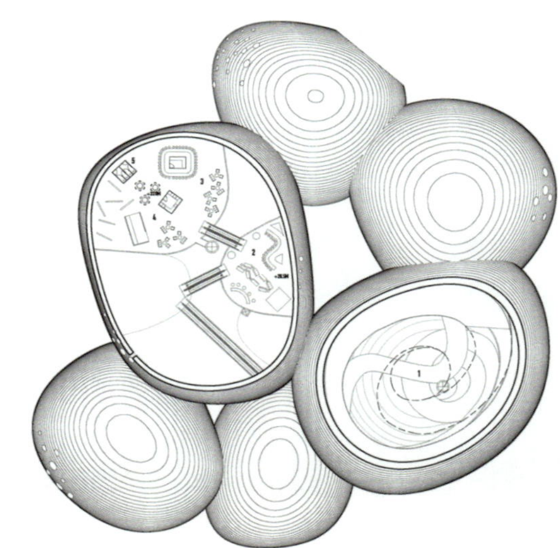

平面图 +36M FLOOR PLAN +36M

平面图 +27M FLOOR PLAN +27M

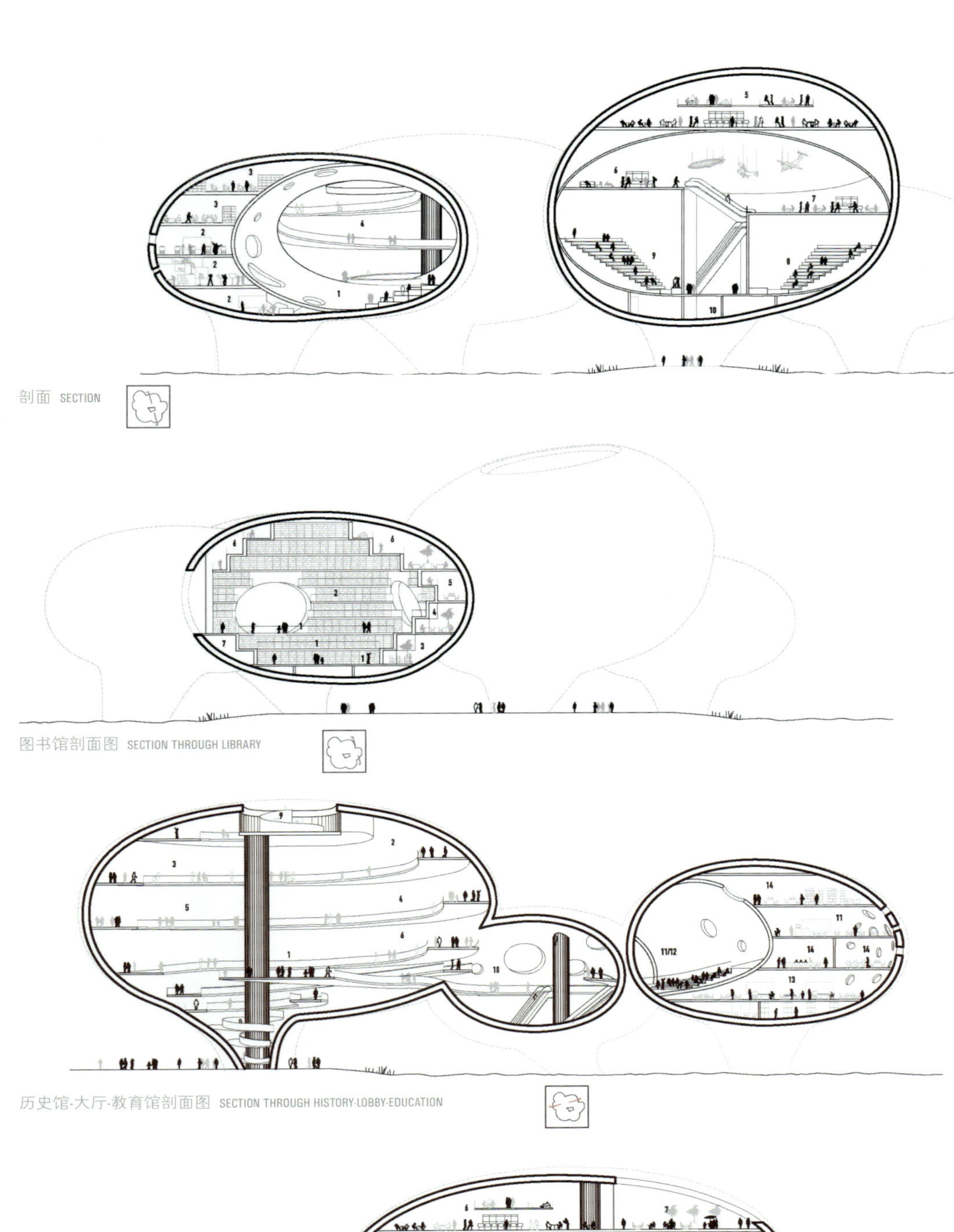

剖面 SECTION

图书馆剖面图 SECTION THROUGH LIBRARY

历史馆-大厅-教育馆剖面图 SECTION THROUGH HISTORY-LOBBY-EDUCATION

大厅-剧院剖面图 SECTION THROUGH LOBBY-THEATER

动漫博物馆 · 杭州

Comic and Animation Museum · China

二等奖 · Second Prize

Miralles Tagliabue EMBT (建筑师事务所)

Benedetta Tagliabue (建筑师)

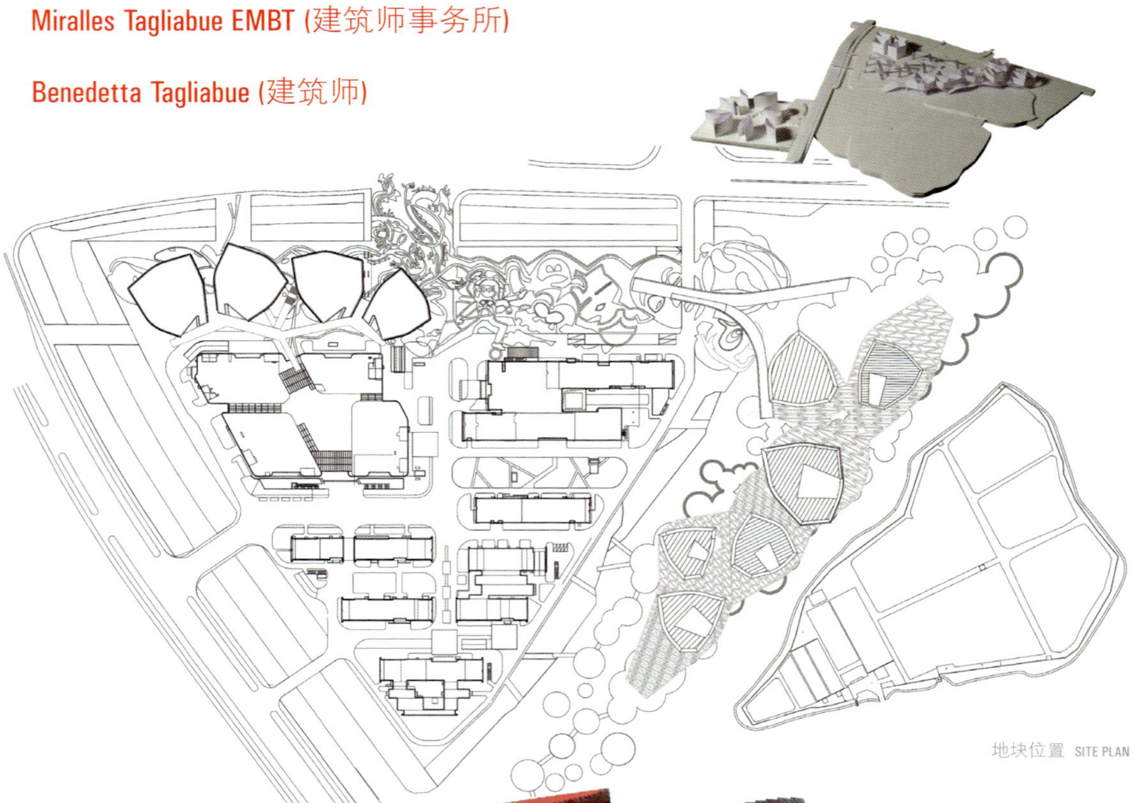

地块位置 SITE PLAN

# 惊喜连连

与环境相融合的最佳方式是，将漫画博物馆视为一种景观策略，而不仅仅是一座建筑物，使其最大化地融合于周边环境中。如果一开始即将其建成一座可漂移的小岛，效果会怎样呢？我们可利用一块地来实践我们的想法。

# BASKETS FULL OF SURPRISES

The best way to relate to these amazing surroundings is to think the comic museum not as a building but as a landscape strategy. It will dialogue with its surroundings at a larger scale. What if the first action was to bend a piece of island and make it float? We would automatically have a piece of land to shelter our activities.

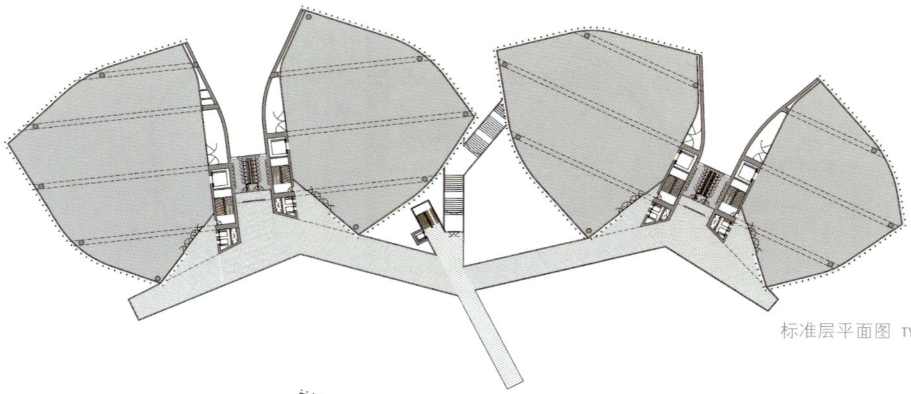

标准层平面图 TYPICAL FLOOR PLAN

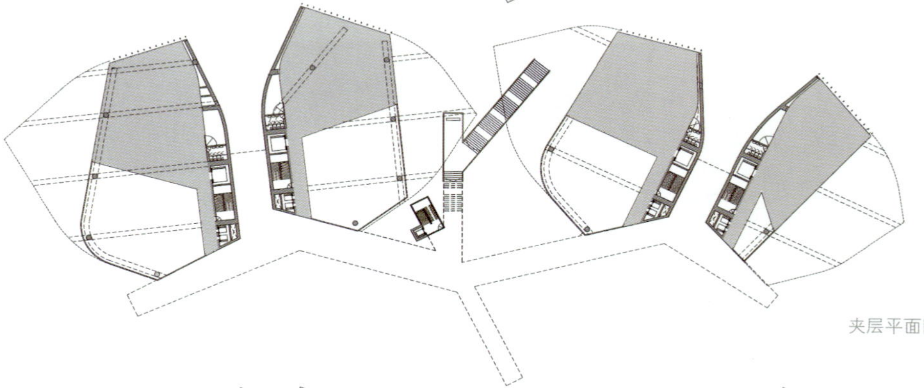

夹层平面图 MEZZANINE

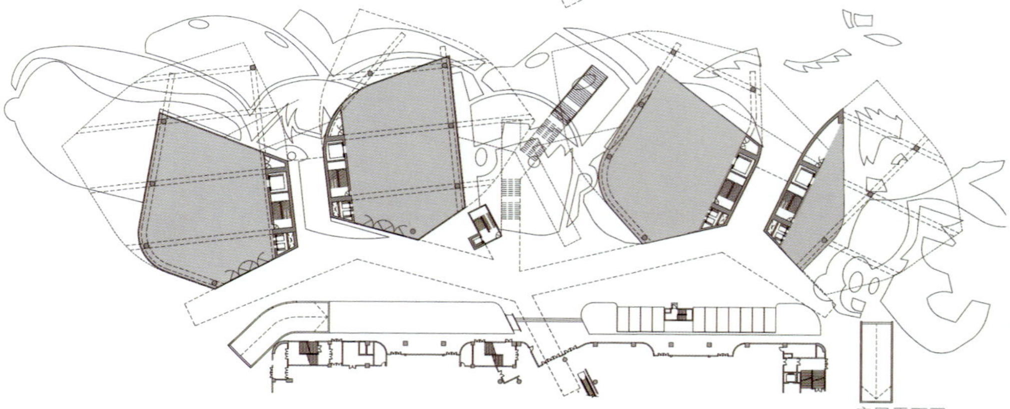

底层平面图 GROUND FLOOR PLAN

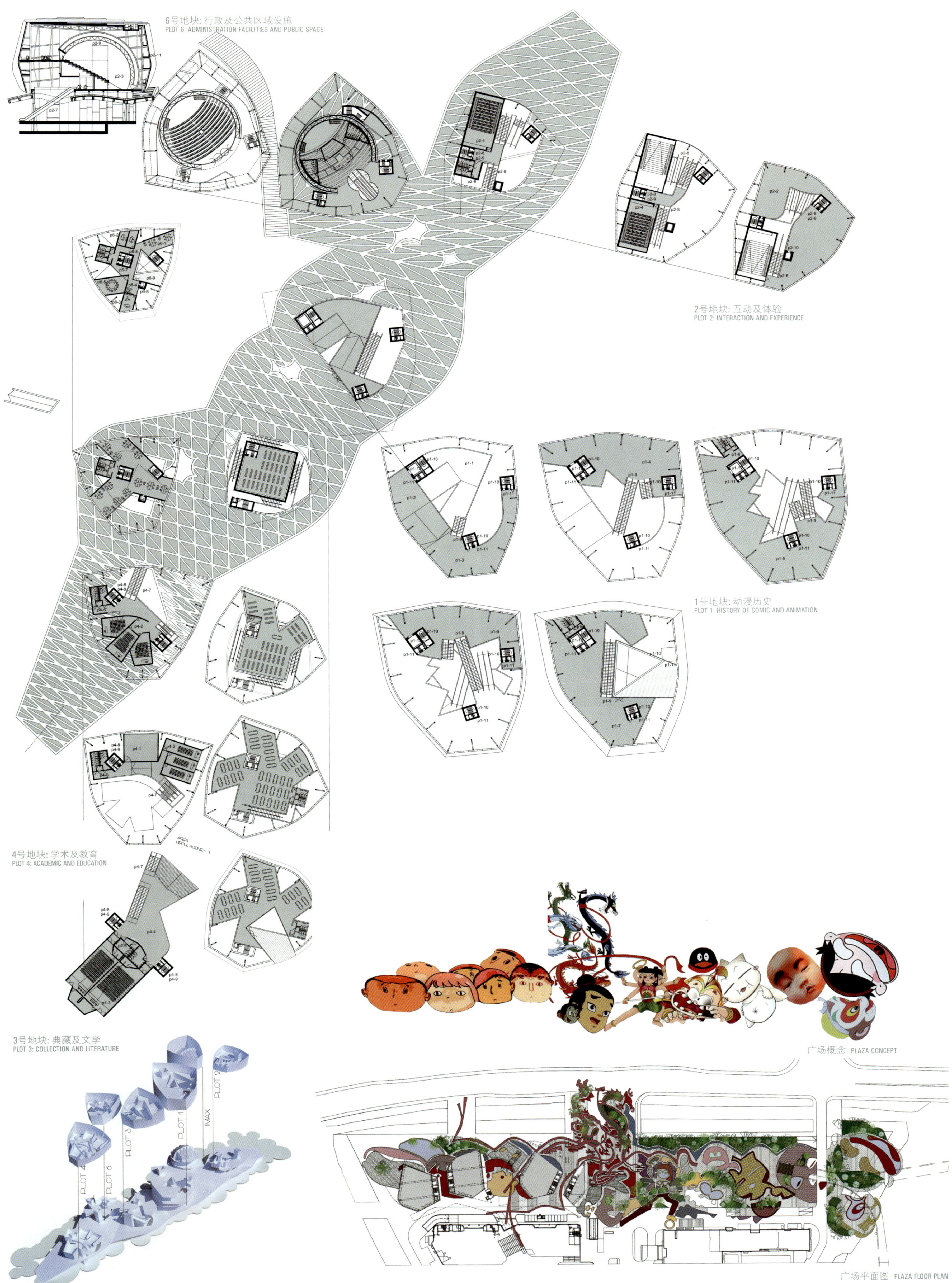
6号地块: 行政及公共区域设施
PLOT 6: ADMINISTRATION FACILITIES AND PUBLIC SPACE
2号地块: 互动及体验
PLOT 2: INTERACTION AND EXPERIENCE
1号地块: 动漫历史
PLOT 1: HISTORY OF COMIC AND ANIMATION
4号地块: 学术及教育
PLOT 4: ACADEMIC AND EDUCATION
3号地块: 典藏及文学
PLOT 3: COLLECTION AND LITERATURE
PLOT 4
PLOT 6
PLOT 3
PLOT 1
IMAX
PLOT 2
广场概念 PLAZA CONCEPT
广场平面图 PLAZA FLOOR PLAN

## 动漫博物馆 · 杭州

## Comic and Animation Museum · China

入围 · Finalist

Architectural Design & Research Institute of Tongji University (同济大学建筑设计研究院)

曾群 Zeng Qun (建筑师)

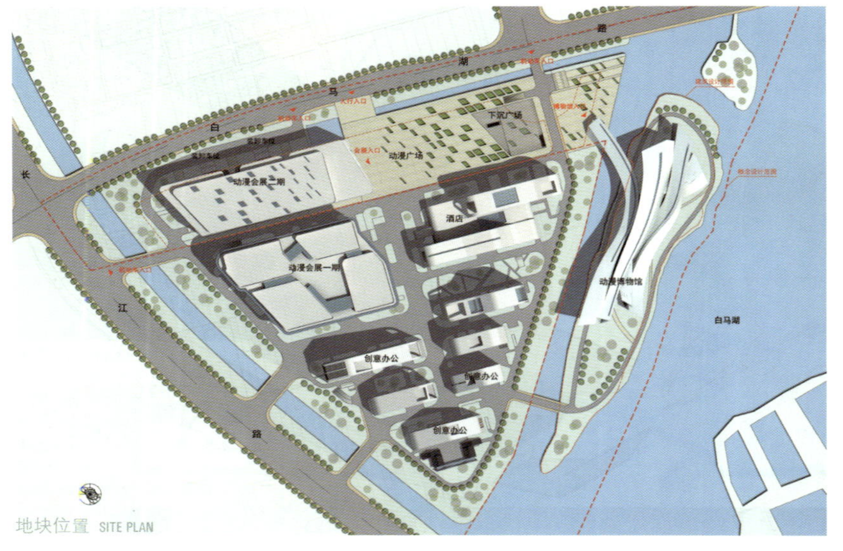

地块位置 SITE PLAN

地下层平面图 UNDERGROUND FLOOR PLAN

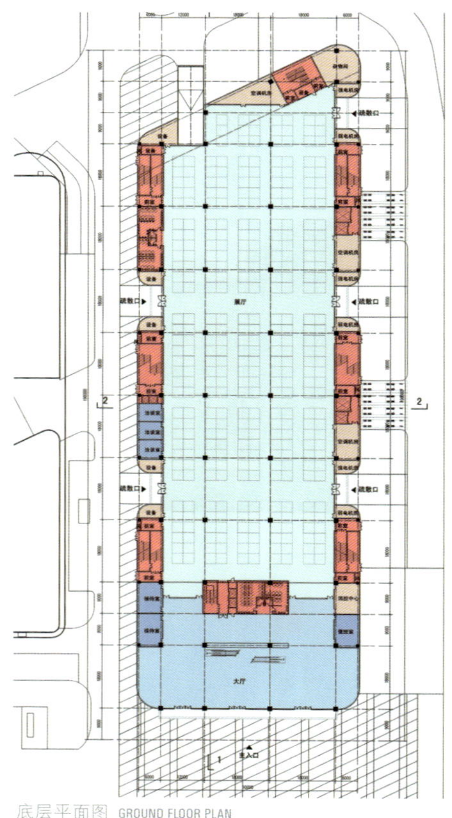

底层平面图 GROUND FLOOR PLAN

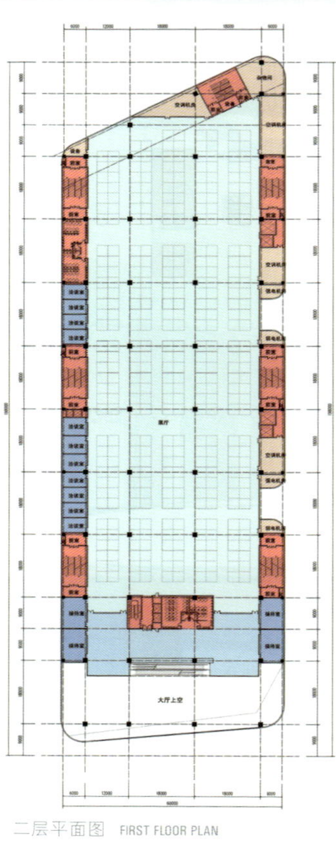

二层平面图 FIRST FLOOR PLAN

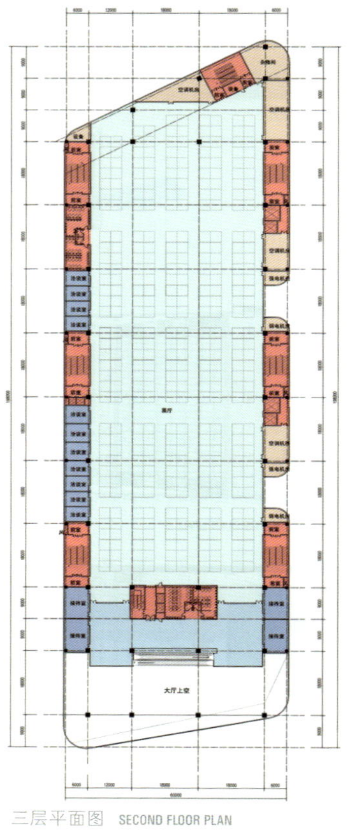

三层平面图 SECOND FLOOR PLAN

# 幻想

我们根据天成的地理环境，提出“蛟龙出水”的设计概念。因为无论从哪个角度看，该建筑的整体外形都犹如蛟龙出水，呈现出一种震撼的视觉效果。

# FANTASY

Relying on an ideal geographical environment, we propose the design concept of "a dragon that jumps out of the water". From any angle of the site, the integral form of the building is just like a dragon jumping out of the water with an impressive effect.

南立面图 SOUTH ELEVATION

北立面图 NORTH ELEVATION

西立面图 WEST ELEVATION 东立面图 EAST ELEVATION

展览中心二期 PHASE II OF EXHIBITION CENTER

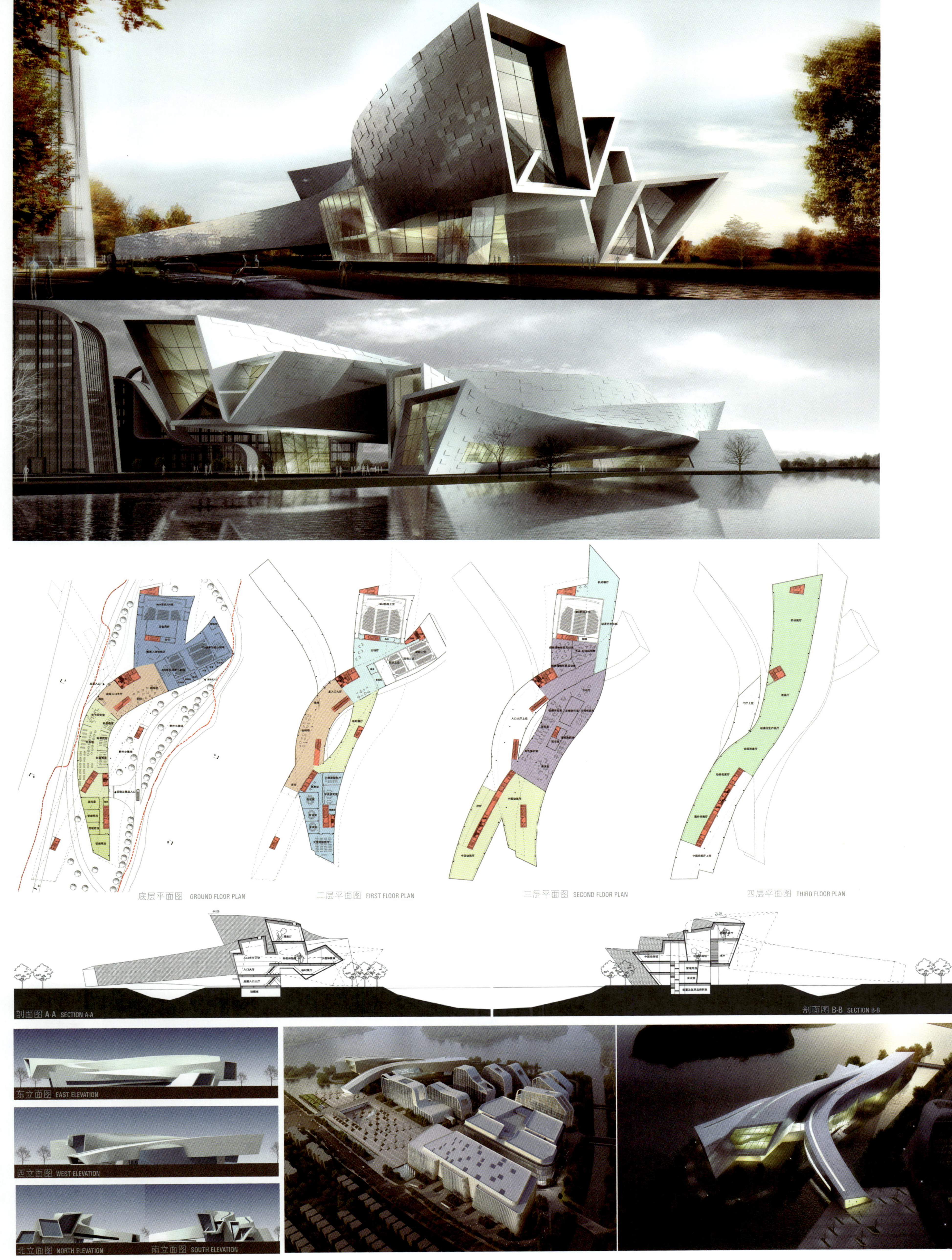
底层平面图 GROUND FLOOR PLAN
二层平面图 FIRST FLOOR PLAN
三层平面图 SECOND FLOOR PLAN
四层平面图 THIRD FLOOR PLAN
剖面图 A-A SECTION A-A
剖面图 B-B SECTION B-B
东立面图 EAST ELEVATION
西立面图 WEST ELEVATION
北立面图 NORTH ELEVATION
南立面图 SOUTH ELEVATION

动漫博物馆·杭州

Comic and Animation Museum · China

入围 · Finalist

王路工作室 Studio Wang Lu

清华大学建筑设计研究院 Architectural Design & Research Institute of Qinghua University

王路 Wang Lu (建筑师)

概念 CONCEPT

## 一体化

我们设计的关键点之一就是如何让该建筑与周边环境相融合。它可有机地将城市空间与自然环境相连接，诠释建筑的含义，并传送其空间意义。该建筑外形旨在反映动漫产业的精神。

## INTEGRATION

One of the key points of our design is how to integrate architecture into the environment. It organically links the city space with the natural environment, interprets the site and completes its spatial meaning. The form tries to reflect the animation and comic spirit.

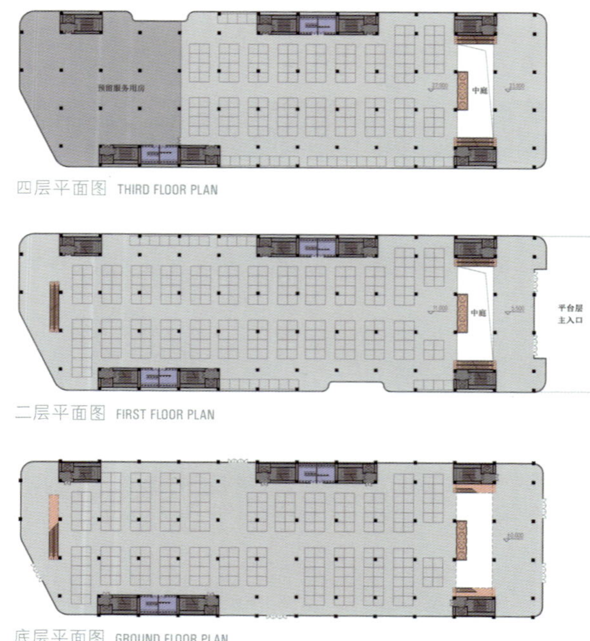

四层平面图 THIRD FLOOR PLAN

二层平面图 FIRST FLOOR PLAN

底层平面图 GROUND FLOOR PLAN

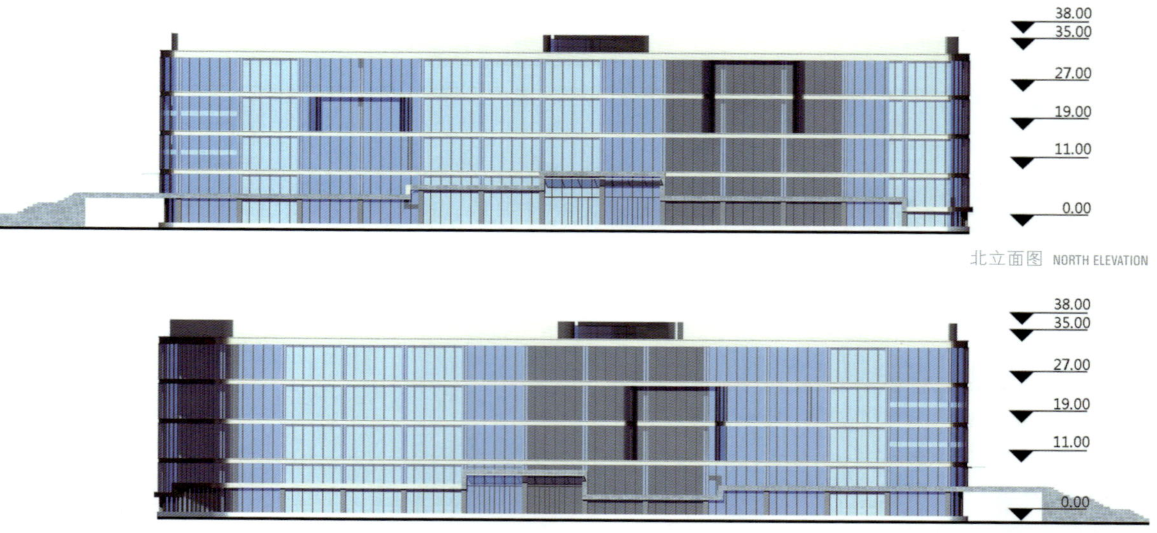



北立面图 NORTH ELEVATION

南立面图 SOUTH ELEVATION

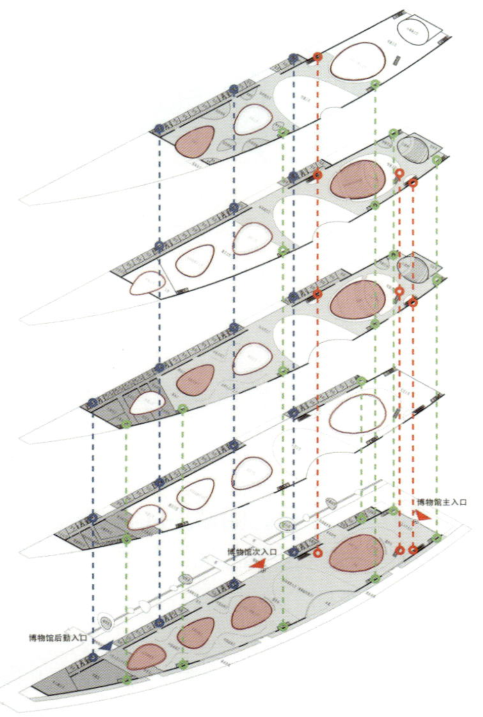

主要游客垂直流线
MAIN VERTICAL CIRCULATIONS FROM VISITORS
次要游客垂直流线
SECONDARY VERTICAL CIRCULATIONS FROM VISITORS
内部垂直流线
INTERNAL VERTICAL CIRCULATIONS

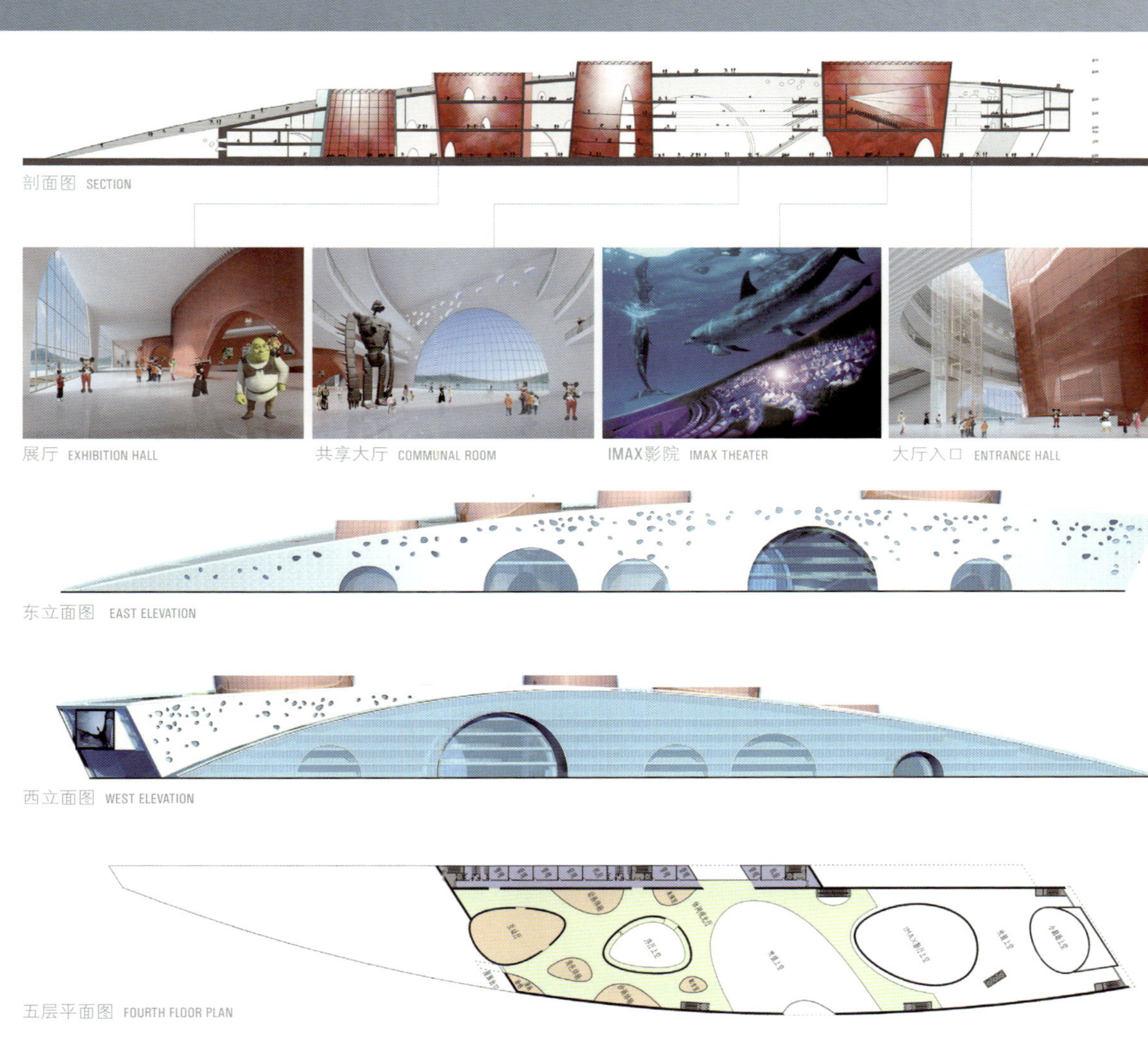

剖面图 SECTION

展厅 EXHIBITION HALL

共享大厅 COMMUNAL ROOM

IMAX影院 IMAX THEATER

大厅入口 ENTRANCE HALL

东立面图 EAST ELEVATION

西立面图 WEST ELEVATION

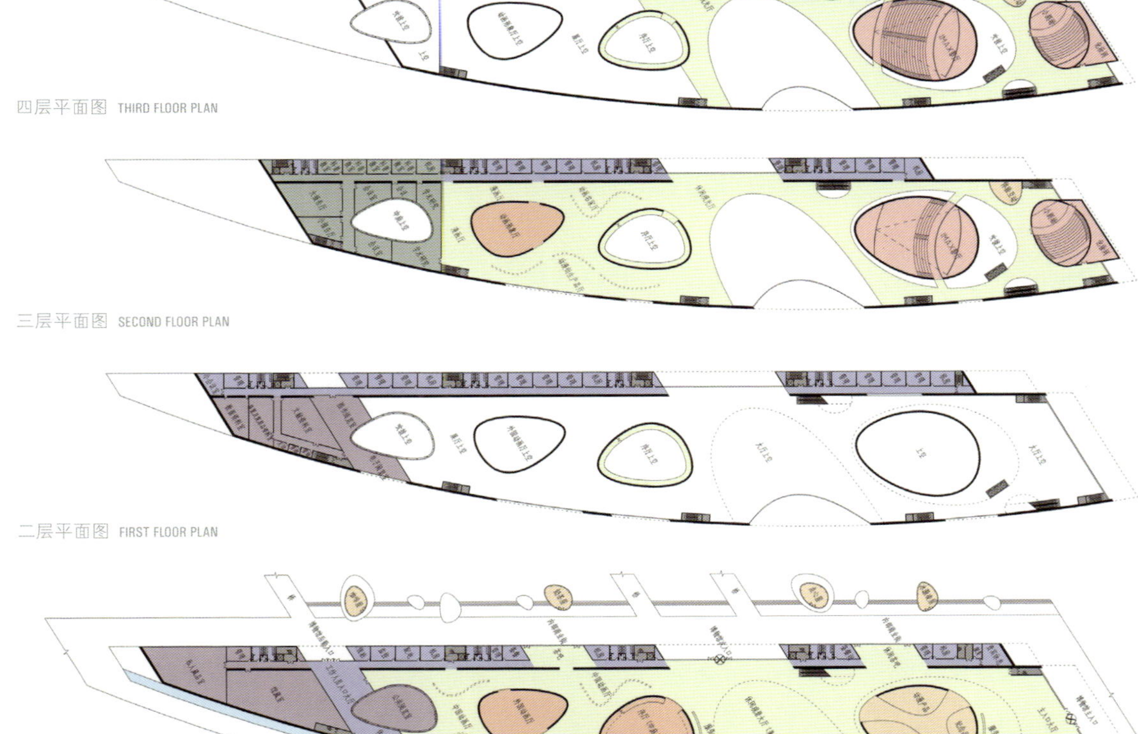

五层平面图 FOURTH FLOOR PLAN

四层平面图 THIRD FLOOR PLAN

三层平面图 SECOND FLOOR PLAN

二层平面图 FIRST FLOOR PLAN

底层平面图 GROUND FLOOR PLAN

# 西蒙·波利瓦国际音乐交流综合体·加拉加斯

## Simon Bolivar International Complex for Social Action through Music · Venezuela

**竞标 · competition**
西蒙 · 波利瓦国际音乐交流综合体 (CIASMSB)
Simon Bolivar International Complex for Social Action through Music

**竞标类型 · competition type**
公开竞标
open competition

**项目地点 · site area**
加拉加斯 · 委内瑞拉 Caracas · Venezuela

**主办方 · promoter**
CAF + Fesnojiv

**日程安排 · schedule**
招标 · Announcement 04.2010
评审结果 · Jury´s results 07.2010

**评审团 · jury**

José Antonio Abreu, 委内瑞拉国立青少年管弦乐团系统主管 Director of Fesnojiv
Yasuhisa Toyota, 日本声学工程师 Acoustic Engineer
Iñaqui Abalos, 西班牙建筑师 Spanish architect
Anita de la Rosa, 景观设计师 Landscape architect
Lorenzo González Casas, 规划师 Planning architect
Eduardo Guzmán, 波利瓦市政厅代表 Representative of the Bolivar City Hall
Omar Seijas, 委内瑞拉建筑学院副院长 Vice President of the Architects Association of Venezuela

**获奖者 · awards**

一等奖 · **first prize**
**ADJKM (建筑师事务所)**

Khristian Ceballos · Alejandro Méndez · Mawari Núñez · Daniel Otero · Jean-Marc Río (建筑师)

二等奖 · **second prize**
**SLIK STEINEMANN LEMMERZAHL KUENG ARCHITEKTEN (建筑师事务所)**
**URBAN-THINK TANK / ARQUITECTOS, URBANISTAS, C.A. (规划师事务所)**
Slik团队 team Slik · Lukas Kueng · Steffen Lemmerzahl ·Ramias Steinemann · Ivo Piazza · Thomas Vermeulen
Urban Think-Tank团队 team Urban Think-Tank · Alfredo Brillembourg · Hubert Klumpner Michael Contento · Lindsey Sherman · Rafael Machado · Melissa Ramos · Willem Boning

三等奖 · **third prize**
**SULLKA LIMA (建筑师事务所)**

合作 (c) Nicolas Labropoulos

该大赛旨在为未来的音乐建筑征集最佳的建筑方案，建造一座集音乐和社交活动于一体的综合建筑体。**目前这里是青少年乐团的所在场所**，位于加拉加斯圣罗莎的Boulevard Amador Bendayan。

The competition looked for the best architectural option for the future musical axis that will complement the Complex for Social Action through Music, **current venue of juvenile orchestras**, located in the Boulevard Amador Bendayan in the sector of Santa Rosa in Caracas.

## 国际音乐交流综合体 · 加拉加斯

## International Complex for Social Action through Music · Venezuela

一等奖 · First Prize　　adjkm (建筑师事务所)

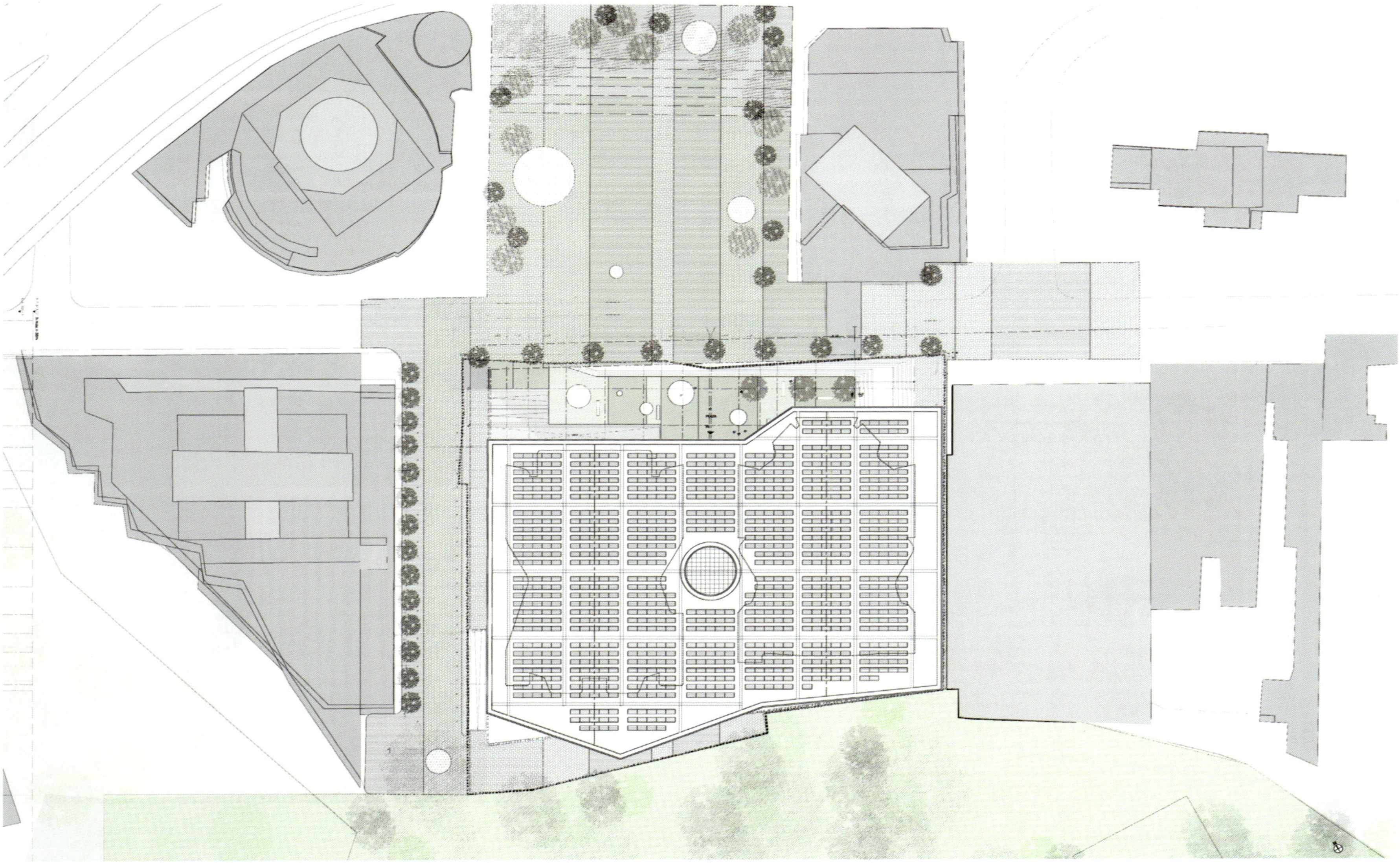

底层平面图 GROUND FLOOR PLAN

# 悬浮建筑

该建筑呈一个架空空间，通过一条水平缝隙将建筑一分为二，而该水平缝隙面向Caobos公园，并将景观融入建筑设计中。该空间内含音乐学院（一个拔地而起的密实结构体）以及音乐厅（一个跃于公园上方的轻巧悬浮结构体）。建筑外形仿大树而建，象征着树一般的诗情画意：该建筑的底部透出一股音乐气质，而其顶部彰显了天才音乐家们的奇思妙想。该建筑的总体造型呈水平层状，其水平缝隙可用于建筑体的采光和通风。

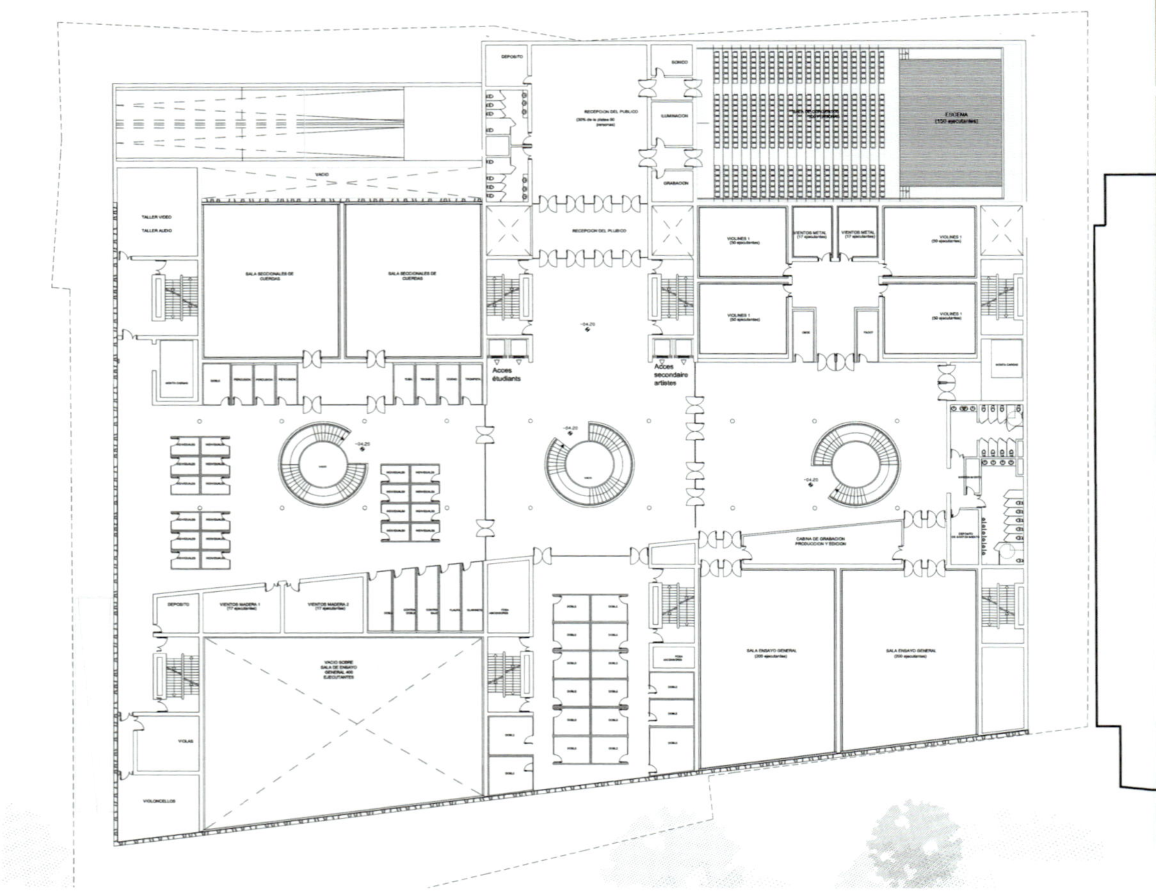

音乐学院层平面图 FLOOR PLAN CONSERVATORY LEVEL

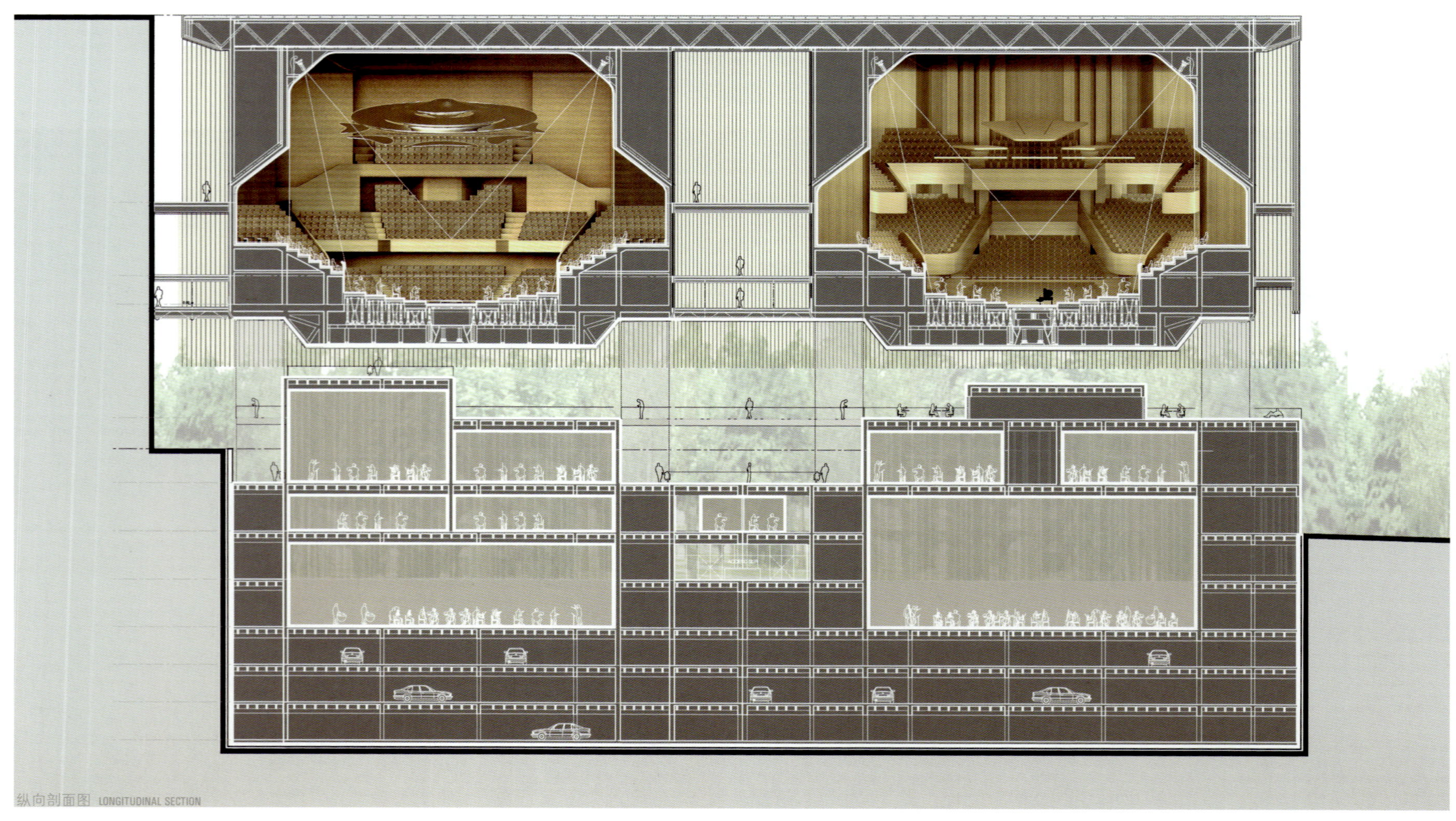

纵向剖面图 LONGITUDINAL SECTION

# SUSPENDED BLOCK

This empty fragment divides the building into two units by means of a horizontal crack that opens the view towards Caobos park and integrates the notion of landscape into the project. This space connects the music conservatory, a compact block anchored to the ground, and the concert halls, a light and suspended block, as a precious levitation of object over the park. The shape mimics the poetic symbolism of the tree: the musical knowledge is born on the root of the project and the creative genesis arises at the top from the talent of musicians. The general operation of the project is organized by horizontal layers. The horizontal crack allows for lighting and ventilation of the complex.

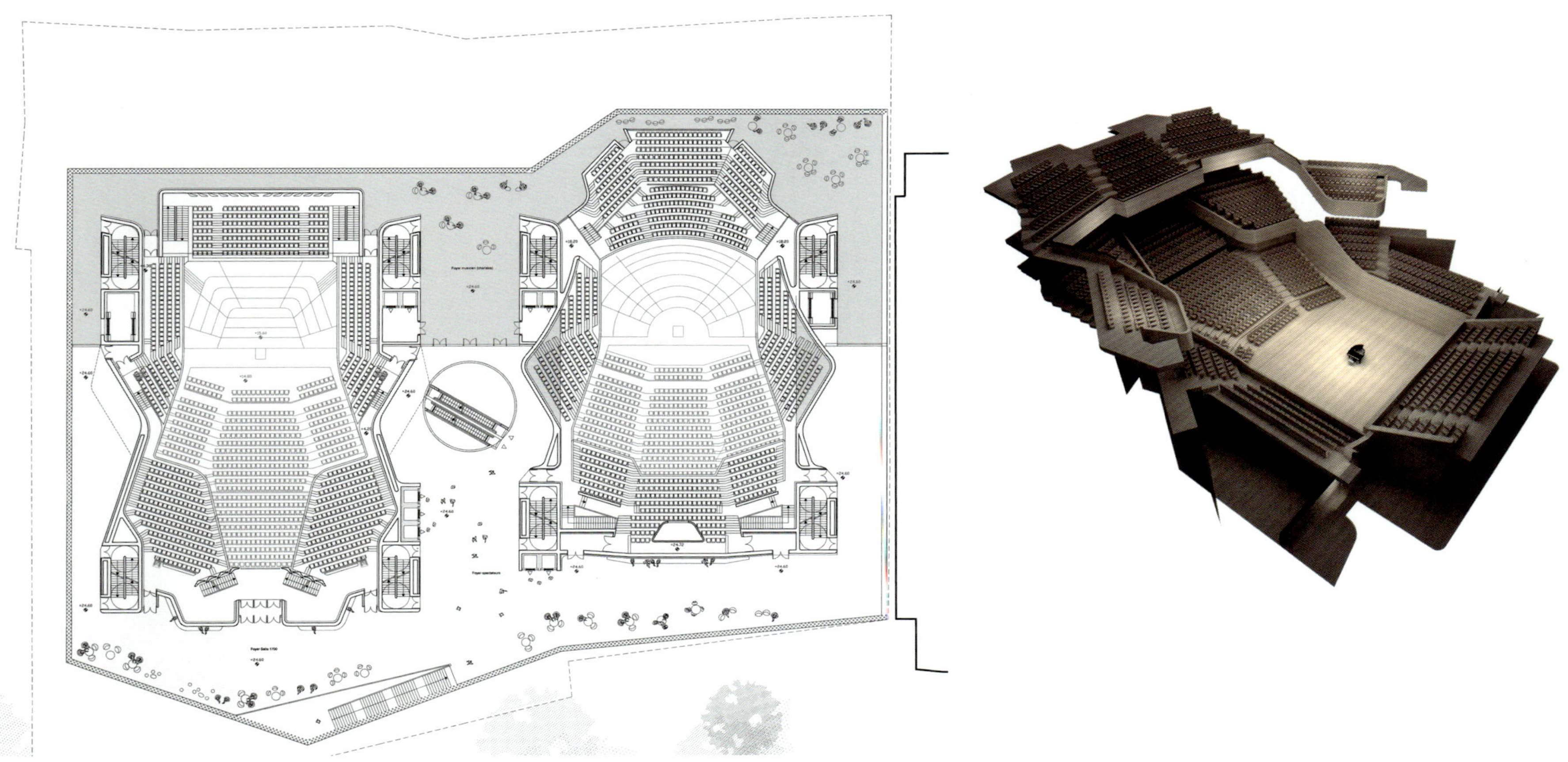

交响乐厅层平面图 FLOOR PLAN SYMPHONY HALLS LEVEL

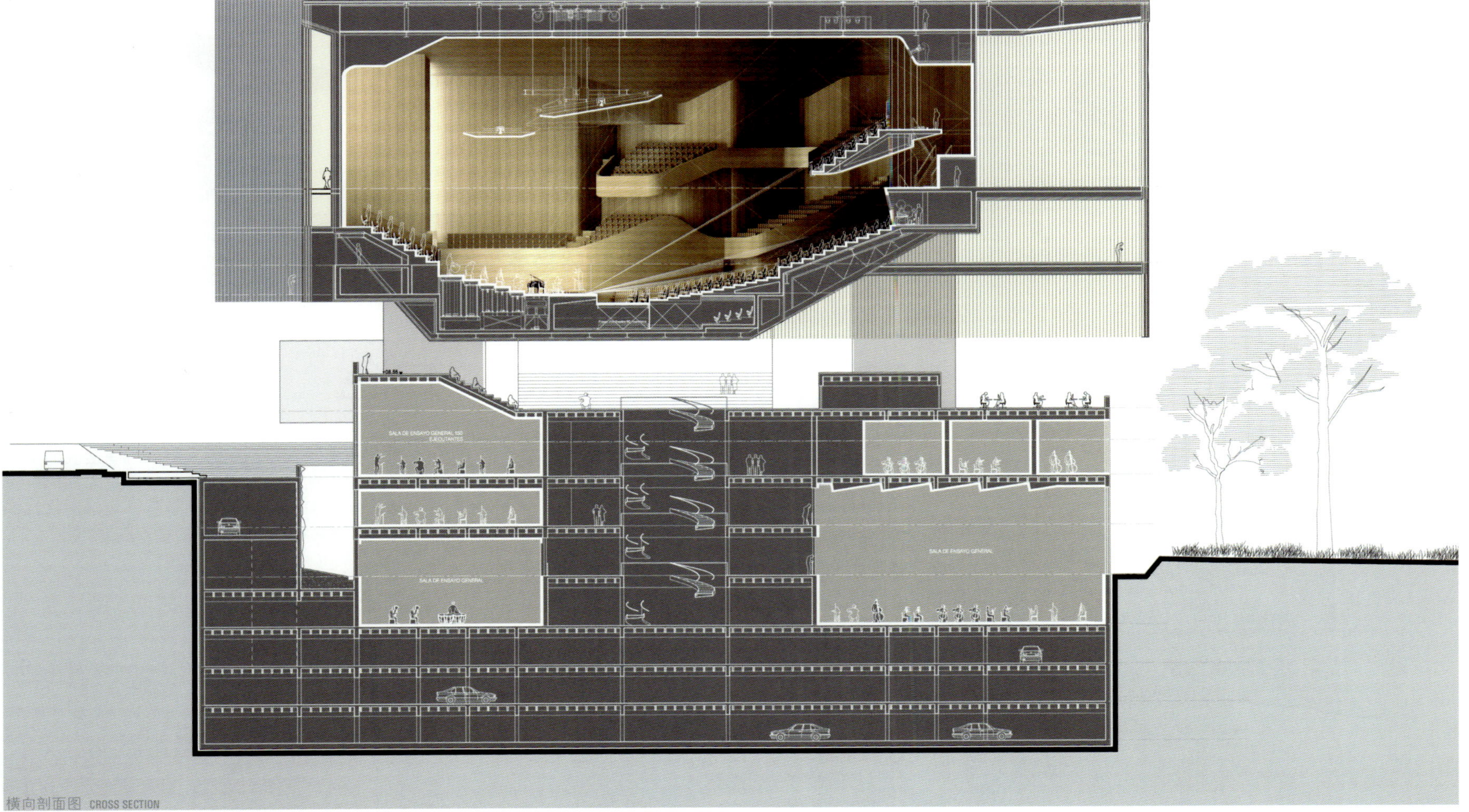

横向剖面图 CROSS SECTION

国际音乐交流综合体 · 加拉加斯

International Complex for Social Action through Music · Venezuela

二等奖 · Second Prize

# on753 (竞标代码)


SLiK Steinemann Lemmerzahl Kueng Architekten (建筑师事务所)

Urban-Think Tank / Arquitectos, Urbanistas, C.A. (规划师事务所)


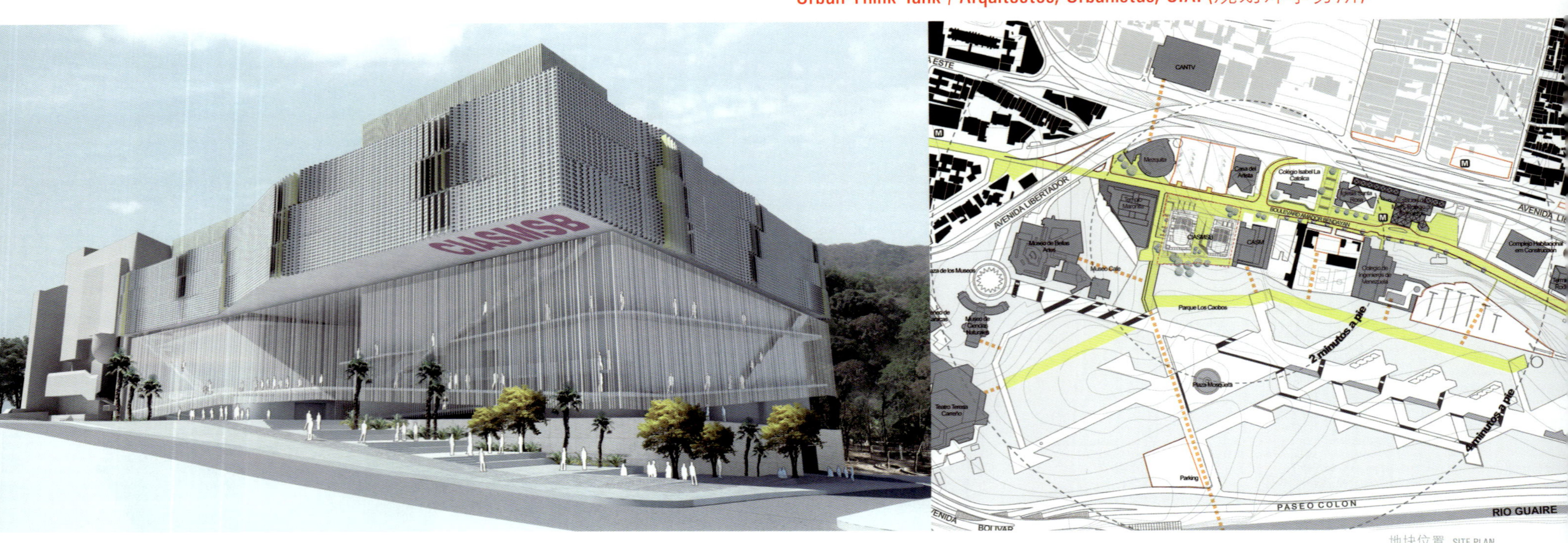


地块位置 SITE PLAN

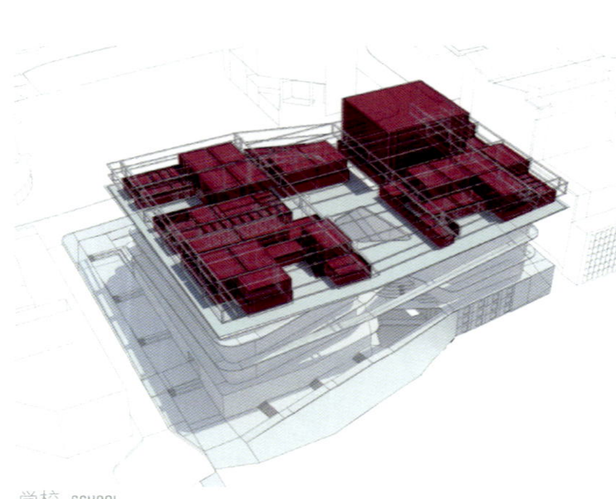

学校 SCHOOL

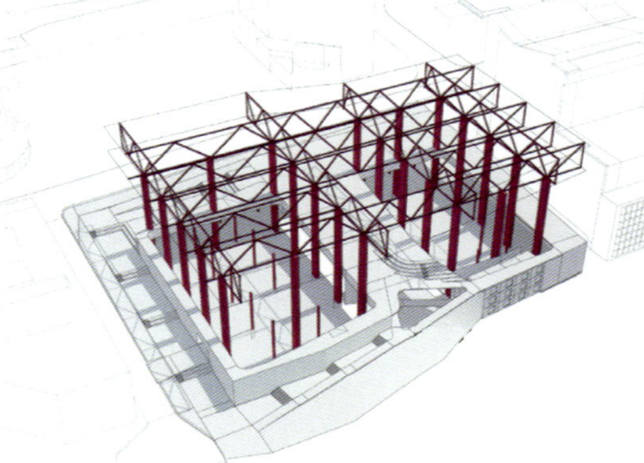

构造 STRUCTURE

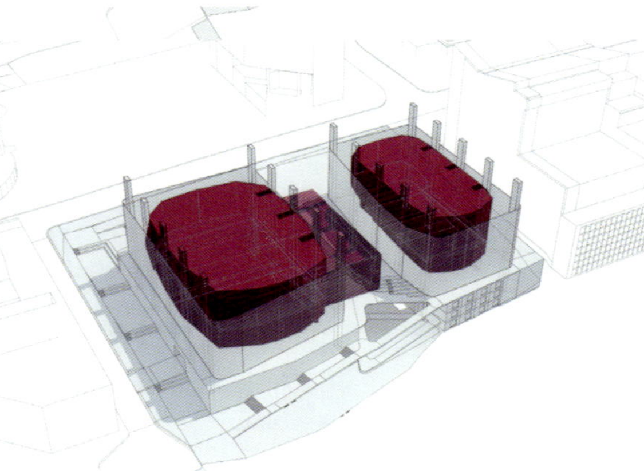

音乐厅 CONCERT HALLS

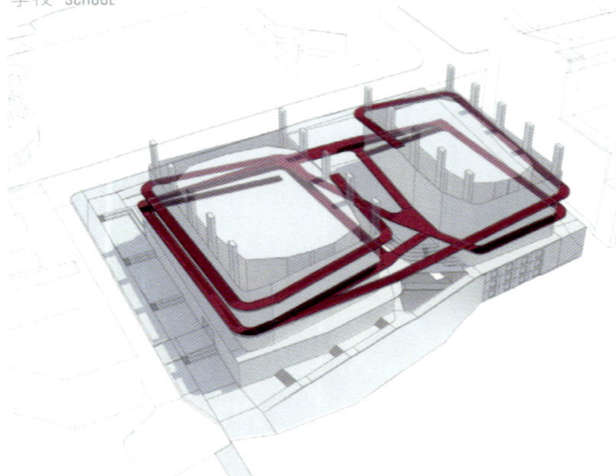

流线 CIRCULATION

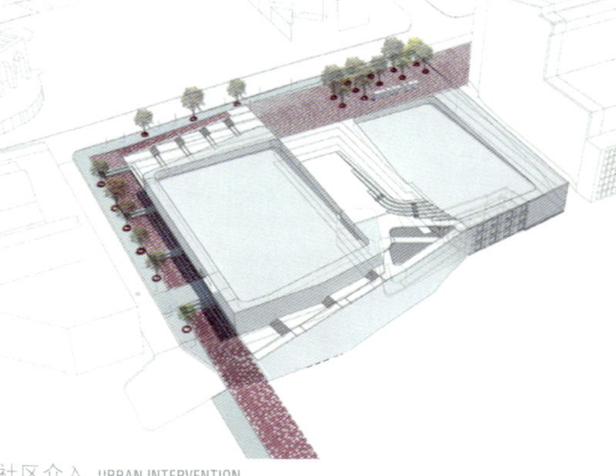

社区介入 URBAN INTERVENTION

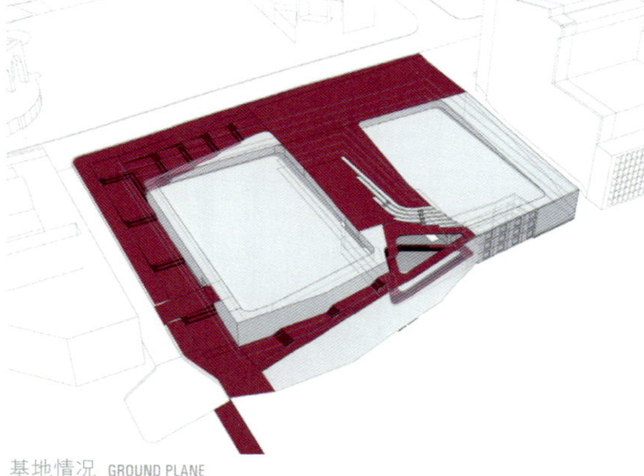

基地情况 GROUND PLANE

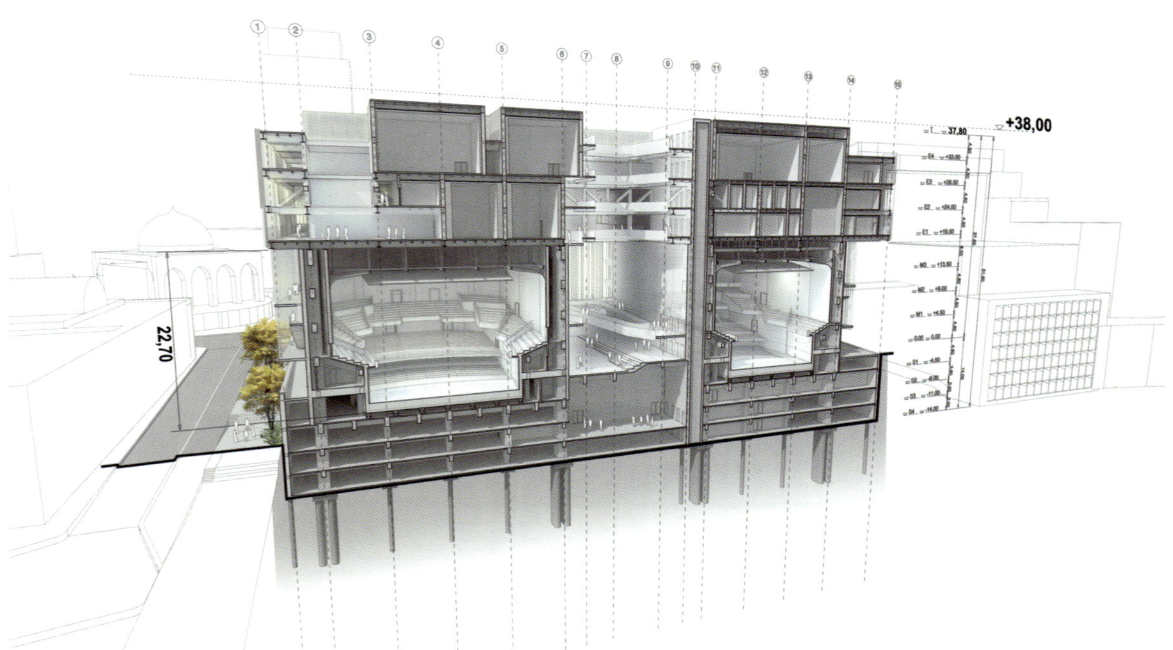


## 白色

该建筑的水平面主要运用斜面作为连结所有建筑区域的主要元素，凸显为一系列不同层次的项目和情景。内部结构的设计主要通过建筑立面的壁板位置来控制和实现，看似布局随意，实则严格根据内部采光、通风和视野观测点的条件而设计。

## WHITE

The horizontality of the building is accentuated as a linear sequence of events and situations throughout the different levels, using the ramp as the main element that connects all the areas of the building. The flow of the internal movement is controlled and produced mainly by the location of the panels of the façade, apparently arbitrary, but that follow strict conditions to generate internal lighting, ventilation and observation points.

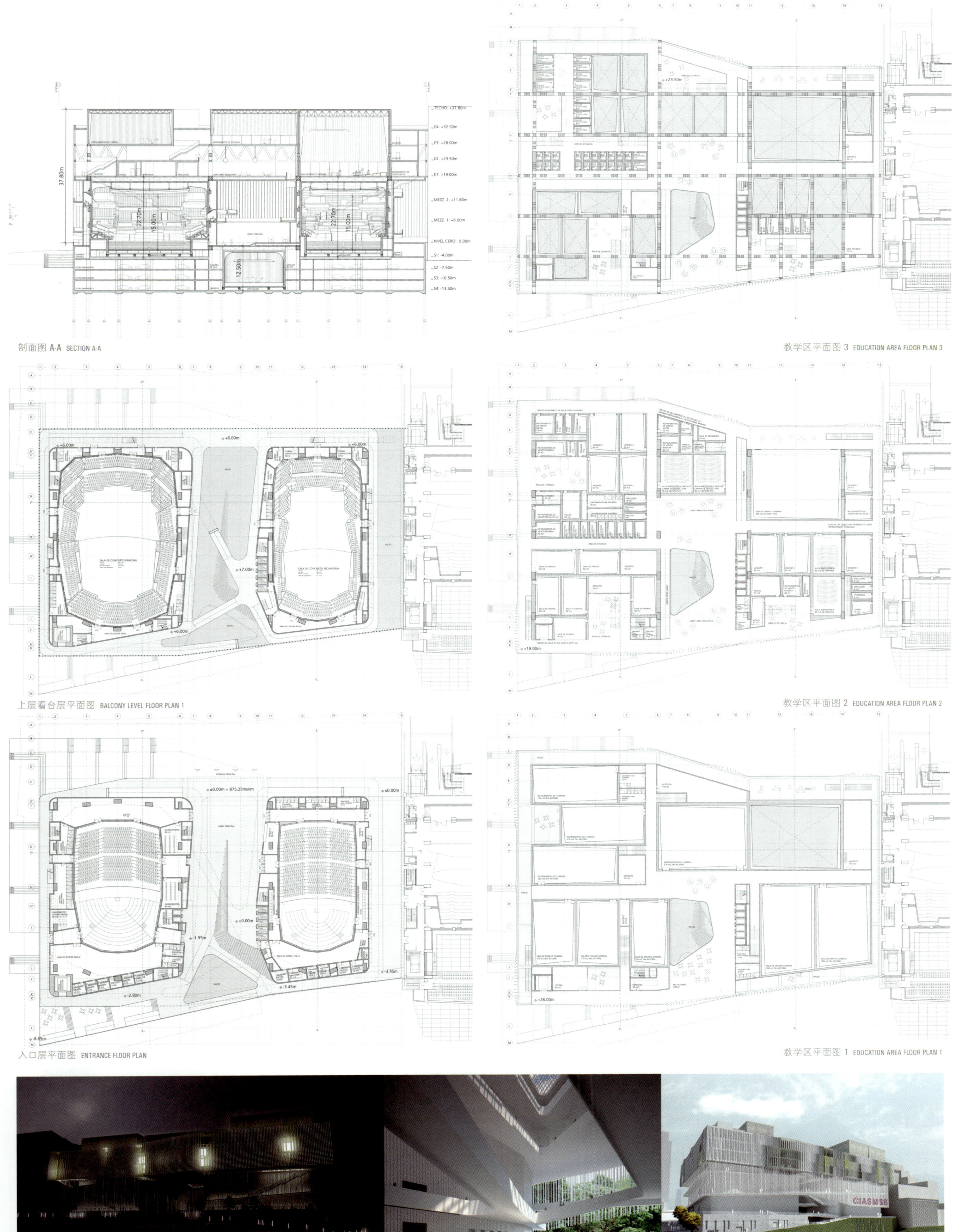

剖面图 A-A SECTION A-A

教学区平面图 3 EDUCATION AREA FLOOR PLAN 3

上层看台层平面图 BALCONY LEVEL FLOOR PLAN 1

教学区平面图 2 EDUCATION AREA FLOOR PLAN 2

入口层平面图 ENTRANCE FLOOR PLAN

教学区平面图 1 EDUCATION AREA FLOOR PLAN 1

国际音乐交流综合体·加拉加斯

International Complex for Social Action through Music · Venezuela

三等奖 · Third Prize

Sullka Lima (建筑师事务所)

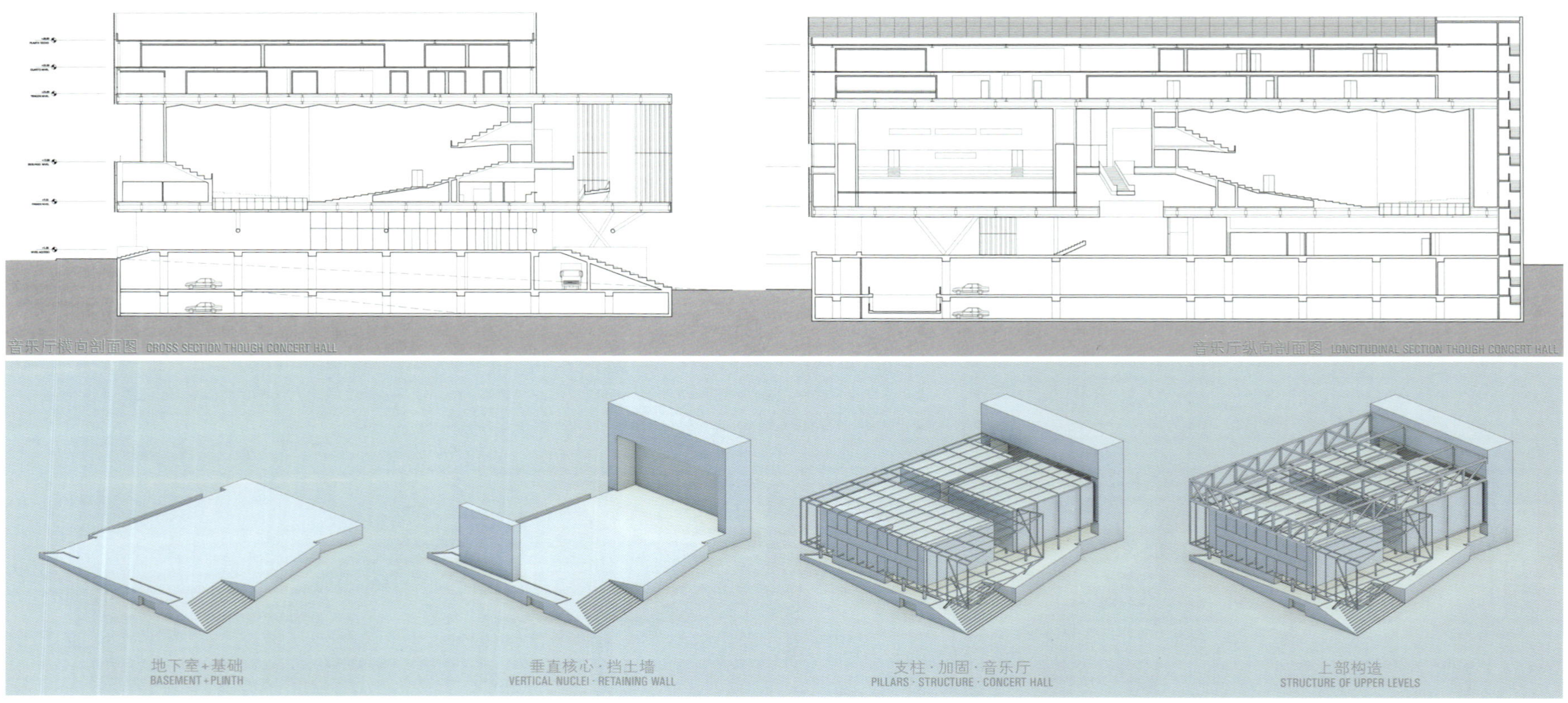

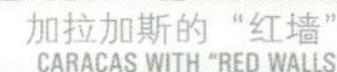
加拉加斯的“红墙”
CARACAS WITH "RED WALLS"

315个像素化单元的重组
RECONSTRUCTION FROM 315 PIXEL CELLS

108个像素化单元的重组=色彩外墙
RECONSTRUCTION FROM 108 PIXEL CELLS=CHROMATIC FAÇADE

# 红墙之城

传统的西班牙红色砖瓦赋予了该城市“红色穹项”的别称。然而，殖民期已过去400多年了，西班牙红砖屋顶现在已几乎不复存在。然而现在仍可在贫民区——所谓“副城”的斑驳的墙上找到这种红黏土。如今，加拉加斯是名副其实的红墙之城。一个用黏土做成的百叶窗盒在阿维拉山和Los Caobos公园的苍翠之间拔地而起。

# THE CITY OF RED WALLS...

The traditional red Spanish tiles, gave the city its "red roofs" pseudonym. But more than 400 years have passed since those colonial times, and the Spanish tiled roofs are almost non-existent. However red clay is still found in its barrios, in the naked walls of the so call informal city. Now Caracas is the city of red walls. With clay as material, a louvered box is built in between the greenness of the Avila Mountain and the Los Caobos park.

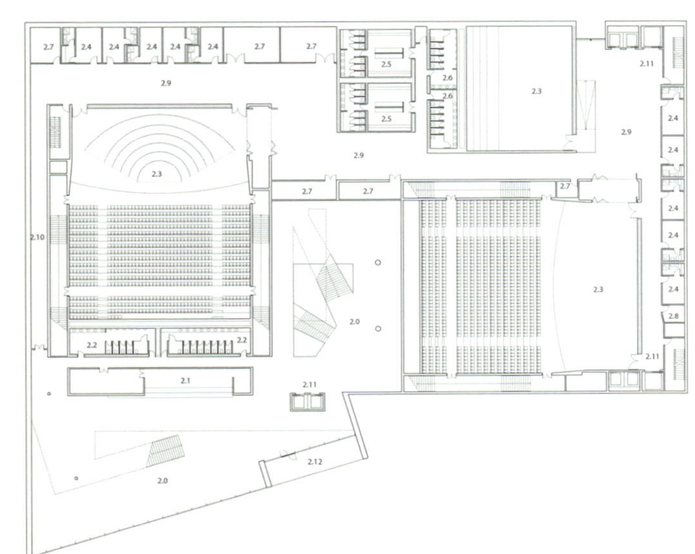
二层平面图 FLOOR PLAN 1

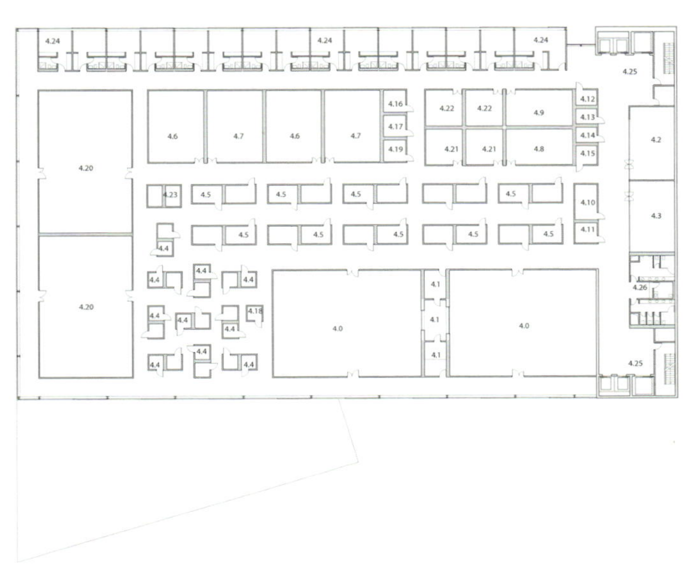
四层平面图 FLOOR PLAN 3

入口层平面图 ENTRANCE FLOOR PLAN

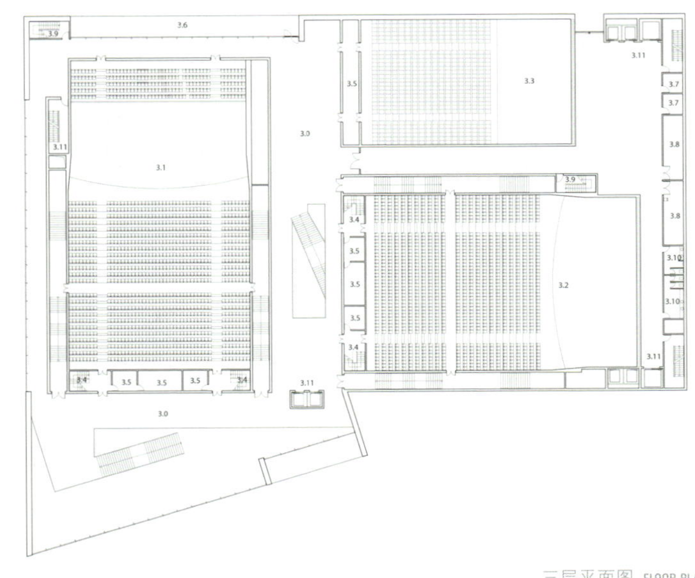
三层平面图 FLOOR PLAN 2

# 一等奖项目展列 Gallery of first prizes

H
herrerosarquitectos

**Juan Herreros**为马德里ETSAM建筑学院建筑项目教授、最高学位项目课堂部主任、纽约哥伦比亚大学实用建筑系教授。他还执教于阿里坎特大学、洛桑理工学院、伦敦建筑协会学院、普林斯顿大学建筑学院、伊利诺伊理工学院等。他目前供职于HerrerosArquitectos公司（www.herrerosarquitectos.com）——一个集专业、教学和研究为一体的协作平台。Juan Herreros还获得英国皇家建筑师学会2008年RIBA国际会士头衔、埃尔·埃斯科里亚皇家修道院2009年美术奖、2009年建筑奖等殊荣，并被提名为美国艺术与文学学院2010年获奖者。

HerrerosArquitectos的作品曾经多次被出版，且屡屡获奖，并在各种私人和集体展览馆展出。最近一次展览为2011年8月至10月在奥斯陆ROM展览馆举办的“盛宴”，展示了其的作品和想法。HerrerosArquitectos最近获得了“2011年西班牙建筑城市规划奖”、“2011年建材展奖”，并入围2009年密斯·凡·德·罗建筑设计奖和2010年FAD奖。

**Juan Herreros** is Professor in Architectural Projects and Director of the Final Degree Project Classroom at the ETSAM School of Architecture in Madrid and Professor in Practice at the University of Columbia, New York. He has also taught at the Universities of Alicante, EPFL-Lausanne, Architectural Association-London, SOA-Princeton, IIT-Chicago, etc. He operates under the name HerrerosArquitectos (www.herrerosarquitectos.com), a collaborative platform that channels its triple activity: professional, teaching and research. Juan Herreros is 2008 RIBA International Fellow by the Royal Institute of British Architects, 2009 Medal of Fine Arts of the City of San Lorenzo de El Escorial, 2009 AD Architecture Award and was nominated for the 2010 Medal of the American Academy of Arts and Letters.

The work of HerrerosArquitectos has been widely published, awarded and exhibited in individual and collective exhibitions. The last is the event “The Banquet” which takes place in ROM Gallery in Oslo between August and October 2011 showing the work and ideas of the office. Among the most recent awards, HerrerosArquitectos has won the Award for Urban Planning of the 2011 Spanish Architecture Biennial, 2011 Construmat Award and was a finalist for 2009 Mies van der Rohe Award and 2010 FAD.

## 中标建成项目 Completed competitions

西班牙 Spain

### Hispasat卫星控制中心大楼
### Building for the satellite control Centre of Hispasat

业主 client: Hispasat
项目负责人 project leader: Diego Barajas
团队 team: Carmen Antón, Angela Ruiz, Ramón Bermúdez, Ricardo Robustini, Verónica Meléndez, Joanna Socha, Jorge Montalván
室内设计 interior design: Paola Simone, HA
外墙 façade: Jens Richter, HA
结构设计 structure: Eduardo Barrón
设备安装 service engineering: Intecsa
外墙顾问 façade consultant: Andrés Rojo, Entorno
监理 surveyor: Arturo Bressel

巴拿马 Republic of Panama

### 巴拿马银行大楼
### Bank of Panama Tower

业主 client: Banco de Panamá
建筑设计 architecture: HerrerosArquitectos+Mallol y Mallol
项目主管 project directors: Jens Richter (HA), Ignacio Mallol T (M&M)
团队 team: Gonzalo Rivas Zinno, Carmen Antón, Joanna Socha (HA), Ruben Taboada (M&M)
可持续性 sustainability: CENER
外墙顾问 façade consultant: Gruppo Entesa

## 中标项目后续设计中 Competitions in progress

西班牙 Spain

### 105户住宅、商铺及车库
### 105 housing units, retail and parking

业主 client: IMPSOL, Barcelona
建筑设计 architecture: Herreros Arquitectos+Benedito/Sanz
项目负责人 project leader: Carmen Antón
团队 team: Jens Richter, Verónica Meléndez, Paola Simone, Filippo Sapienza

挪威 Norway

### 住宅、办公、商业及车库
### Housing, offices, commercial areas and parking

业主 client: HAV Eiendom AS
主案建筑师 principal architect: Juan Herreros
项目主管 project director: Jens Richter
团队 team: Carmen Antón, Margarita Martínez Barbero, Gonzalo Rivas Zinno, Spencer Leaf, Xavier Robledo, David Moreno, Carlos García González
现场建筑师 local architect: Lpo Arkitekter
可持续性 sustainability: Context AS

挪威 Norway

### 蒙克博物馆及斯特内尔森收藏品馆
### Munch Museum and Stenersen Museum Collections

业主 client: Oslo Kommune
主案建筑师 principal architect: Juan Herreros
项目主管 project director: Jens Richter
团队 team: P. Simone, C. Antón, D.Barajas, M. Martínez, G. Rivas, R.Bermúdez, C. Bayod, X. Robledo, S. Leaf, D. Moreno, C. García, P. Vega, R.Robustini, L.Berríos, V. Meléndez, Á.Ruiz, J. Socha
现场建筑师 local architect: Lpo Arkitekter
结构工程 engineering: Kulturplan Bjørvika: Multiconsult - Florian Kosche - Bollinger Grohman, Hjellnes Consult, Brekke og Strand, Civitas
结构工程 (竞标) engineering (competition): IDOM
外墙 façades: ARUP
可持续性 sustainability: Kan Energi
信息和通讯技术 information and communications technology: Rambøll
保安设计 security: COWI
景观设计 (竞标) landscape (competition): Thorbjörn Andersson

哥伦比亚 Colombia

### 波哥达国际会议中心
### International Conventions Centre of Bogota

业主 client: Cámara de comercio de Bogotá (CCB), Corferias.
建筑设计 architecture: Daniel Bermúdez + Herreros Arquitectos
项目主管 project directors: Jens Richter, Ramón Bermúdez
团队 team: Gonzalo Rivas Zinno, Margarita Martínez, Victor Lacima, Iván Guerrero, Ana Torres, Carmen Antón, Verónica Meléndez (HA)
John Oscar Pinzón, Stan van der Maas, Jorge Andrés Pardo, Alejandra Novoa, María Alejandra Carmona (DB)
顾问团 consultants:
规划设计 urbanism: Rafael Obregón, Camilo Santamaría, Daniel Sarabia, Fernando Jiménez (COL)
景观设计 landscape: Diana Wiesner (COL) Bioclimática: Jorge Ramírez, Sandra Varón (COL)
结构设计 structure: BOMA, Xavier Aguiló, Luis Moya, Nicolás Parra, Hernán Sandoval (ESP-COL)
会议中心 conventions centre: Conventional Wisdom Rick Schmidt (USA)
剧院顾问 theatre advisory: Fisher&dachs, Josh Dachs, Bob Campbell (USA)
场景 scenery: Alejandro Luna (MEX)
声学 acoustics: AKUSTIKS Jaume Soler, Chris Blair, Daniel Duplat (USA-COL)
垂直流通 vertical traffic: Rafael Beltrán (COL)
电气安装 electrical services: Jaime Sánchez (COL)
保安及疏散 security and evacuation: Jaime Andrés García (COL)
地板 flooring: Luis Fernando Orozco (COL)
通风及环境质量 ventilation and sanitary: Mauricio Gómez (COL)
概念与特性 concept and identity: Joaquín Gallego, Norberto Chaves (ESP)
厨房功能设计 kitchen use design: Mario Daniel Motta (COL)
规划规范及建造许可 urban regulations and construction license: Juana Sanz, Juan Manuel González, Ignancio Restrepo (COL)
座椅设计顾问 seating advisory: ALIS, Jordi Valverde, Maribel Cedo (ESP)

## 未建成项目 Unbuilt competitions

法国 France

### Iter, 原子能局附属建筑
### Iter, Annex Buildings-CEA

业主 client: C.E.A
项目主管 project directors: Juan Herreros, Jens Richter
团队 team: Verónica Meléndez, Paola Simone
结构和安装设计 structure+services: Intecsa-Inarsa S.A., Ingenierie Studio S.A.S
可持续性 sustainability: CENER

法国 France

### RAPT车库、经济适用房及办公室
### Depots, Social Housing and offices for RAPT

业主 client: Hlm Logis-Transports
项目主管 project director: Juan Herreros
团队 team: Ángela Ruiz, Verónica Meléndez, Jens Richter, Paola Simone

西班牙 Spain

### 生态环保城综合楼
### Hybrid building for the Environmental City

业主 client: Junta Castilla y León
项目主管 project director: Juan Herreros
团队 team: Daniel Marín, Verónica Meléndez, Jens Richter, Paola Simone, Paula Vega, Ángela Ruíz Blanco
结构设计 structure: Julio Martínez Calzón / mc2
模型 model: Juan de Dios Hernández
可持续性 sustainability: Joaquín Guerra / Aster

美国 USA

### Menil收藏馆总体规划
### The Menil Collection Master Plan

业主 client: The Menil Collection
项目主管 project director: Juan Herreros, Jens Richter
团队 team: Paola Simone, Paula Vega, Riccardo Robustini

哥伦比亚 Colombia

### "Yesid Santos"新排球馆
### New Volleyball Coliseum "Yesid Santos"

业主 client: The Menil Collection
项目主管 project directors: Juan Herreros, Jens Richter, Juan Peláez
团队 team: Verónica Meléndez, Paula Vega, Paola Simone

比利时 Belgium

### Heysel社区总体规划
### Heysel Neighbourhood Masterplan

业主 client: Neo Brussels
项目主管 project director: Juan Herreros
项目负责人 project leader: Ramón Bermúdez
团队 team: Gonzalo Rivas Zinno, Hendrick Gruss, Carlos Bayod Lucini, Igor Bragado
结构工程顾问 engineering consultant: IDOM Internacional
景观设计 landscape: TOPOTEK 1, Berlín
能源及功能 energy and program: José Miguel Iribas

西班牙 Spain

### "Tierras del Ebro"新医院
### New Hospital of "Tierras del Ebro"

业主 client: CatSalut (Servei Català de la Salut)
建筑设计 architecture: Herreros Arquitectos+Lahoz López Arquitectos
项目负责人 project leader: Carlos García
团队 team: Margarita Martínez, David Moreno
现场支持 local support: Zaguirre Formatger Arquitectes
结构设计 structure: Xavier Aguiló (BOMA)
设备安装 service engineering: Planho Consultores
可持续性 sustainability: CENER

西班牙 Spain

### 新展览及会议大楼
### New Congress, Exhibitions and Fairs Palace

业主 client: Ayuntamiento de Cuenca
项目负责人 project leader: Verónica Meléndez
团队 team: Carlos Garcia, Gabriel Gutierrez, Igor Bragado, Xavier Robledo, Victor Lacima
结构设计 structure: BOMA
设备安装 service engineering: R. Úrculo Ingenieros Consultores
声学 acoustics: Higini Arau
灯光 lighting: Artec

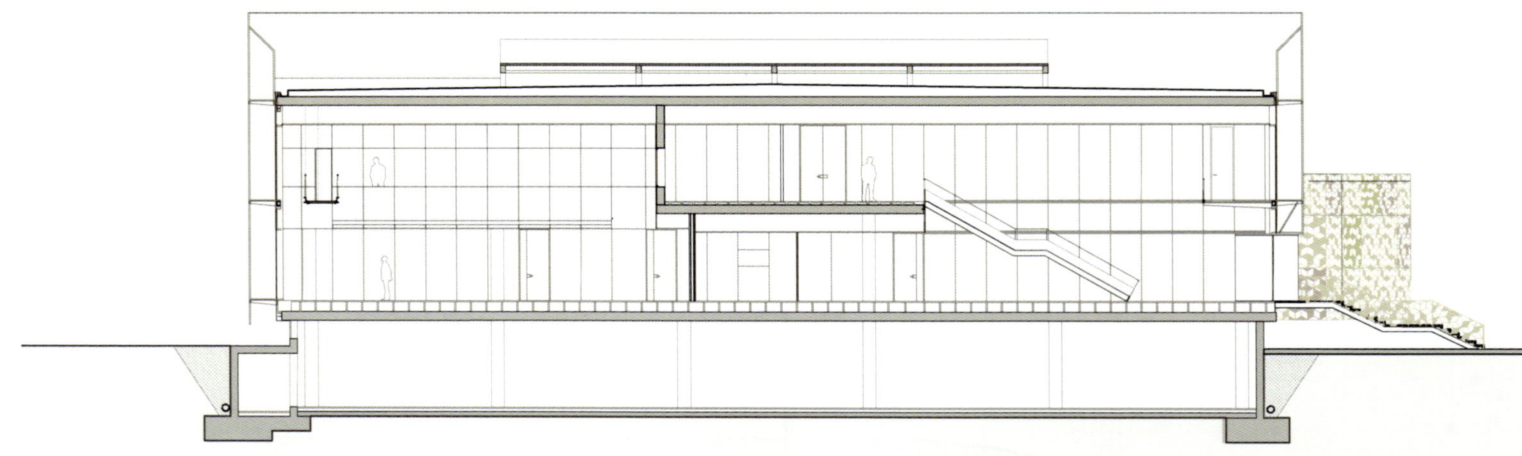
剖面图 SECTION

南立面 SOUTH ELEVATION

北立面 NORTH ELEVATION

新的西班牙卫星（**Hispasat**）总部的设计要求是对既有建筑进行改造，而该建筑唯一还能利用的就是它的结构。该建筑建于20世纪70年代晚期，具有独特的环型平面结构，原先为一块围场，不符合可持续性的要求。

此次改造的目的就是，基于选材、色彩、透明度、可视性以及家具与建筑风格的协调性等，引入当代办公室的理念，将其改造成为更为舒适的建筑体。新建筑体的整体效果呈现出**一个流动空间**，而由于围场与水平面间具有透明性，各层的天花板和楼面看起来连成一体，同时又因中间层具有半透明或不透明性，可为各岗位的工作人员提供足够的隐私空间，从而有助于员工集中注意力。

毫无疑问，该项目的独特之处在于其建筑外观。该建筑外观的设计目的有三：其一，呈现公司的新形象；其二，控制阳光辐射；其三，使其与周围环境相协调，能随着阳光的不同反射呈现出不同的外观，从而带有一种神秘感。

The design of the new headquarters of Hispasat results in the restructuring of an existing building where we only take advantage of the structure. This building, built in the late 1970's, has the distinction of having a circular floor plan and having, in its origin, an enclosure totally divorced from any sustainable criteria.

The commission tries to unfold in its interior, a contemporary office concept under the imprint of comfort based on the choice of materials and their color ranges, the search for transparency and visibility and furniture coherent with architecture. The effect **is a fluid space** in which the ceilings and floors are seen in continuity due to the transparent stripes of the enclosure in contact with the horizontal planes, while the opacity and translucency of the intermediate stripes ensures enough privacy and concentration for the people in their posts.

But undoubtedly, the most special element of the project is its facade. It is designed with the triple objective of providing a new image to the company, controlling of the solar radiation, and harmonizing views, it appears as an enigmatic and changeable surface depending on the reflections of the sun.

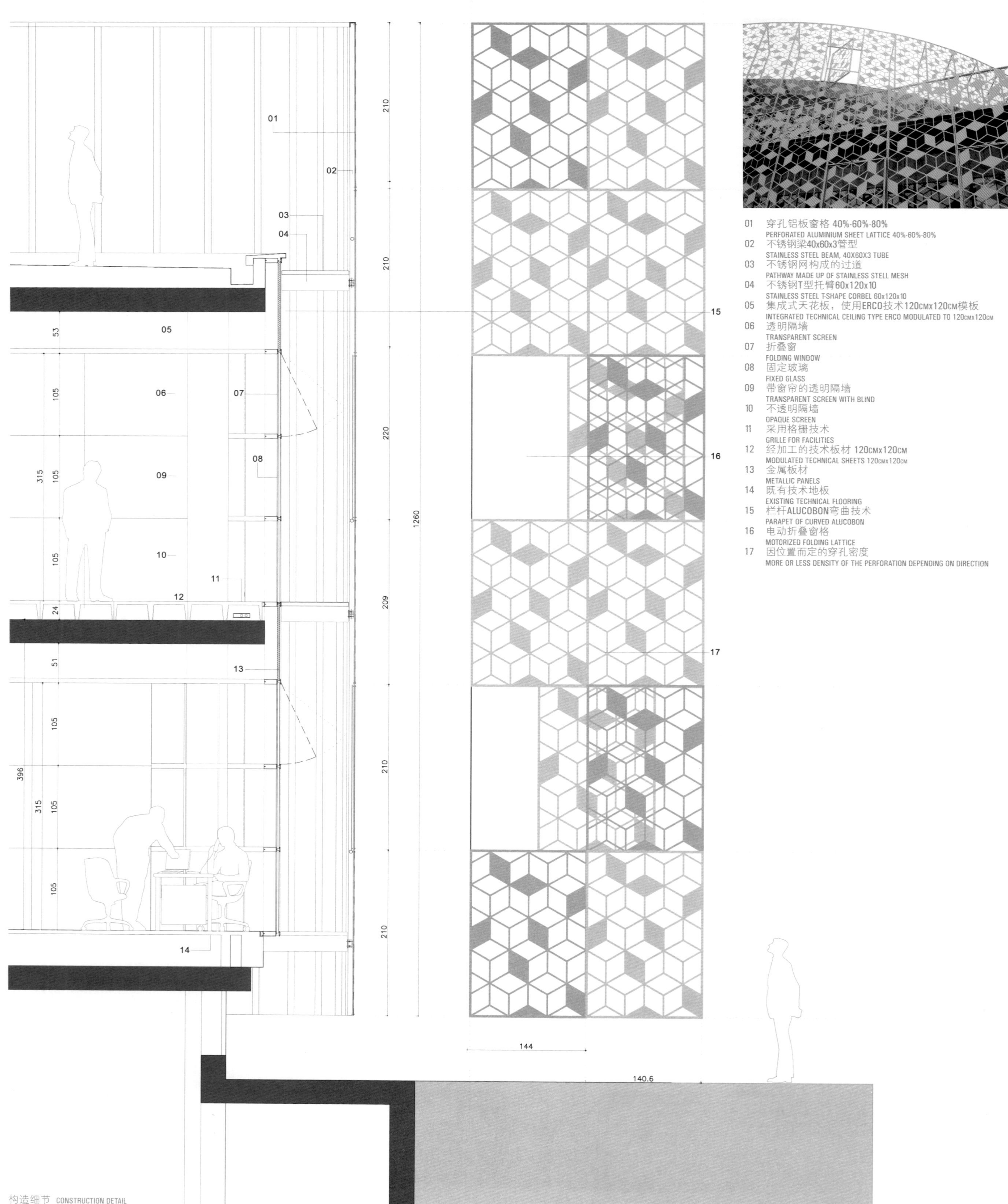


构造细节 CONSTRUCTION DETAIL

巴拿马，巴拿马共和国 Panama, Republic of Panama

## 巴拿马银行大楼
## Bank of Panama Tower

**2008**
邀标项目 Invited competition
一等奖 First Prize

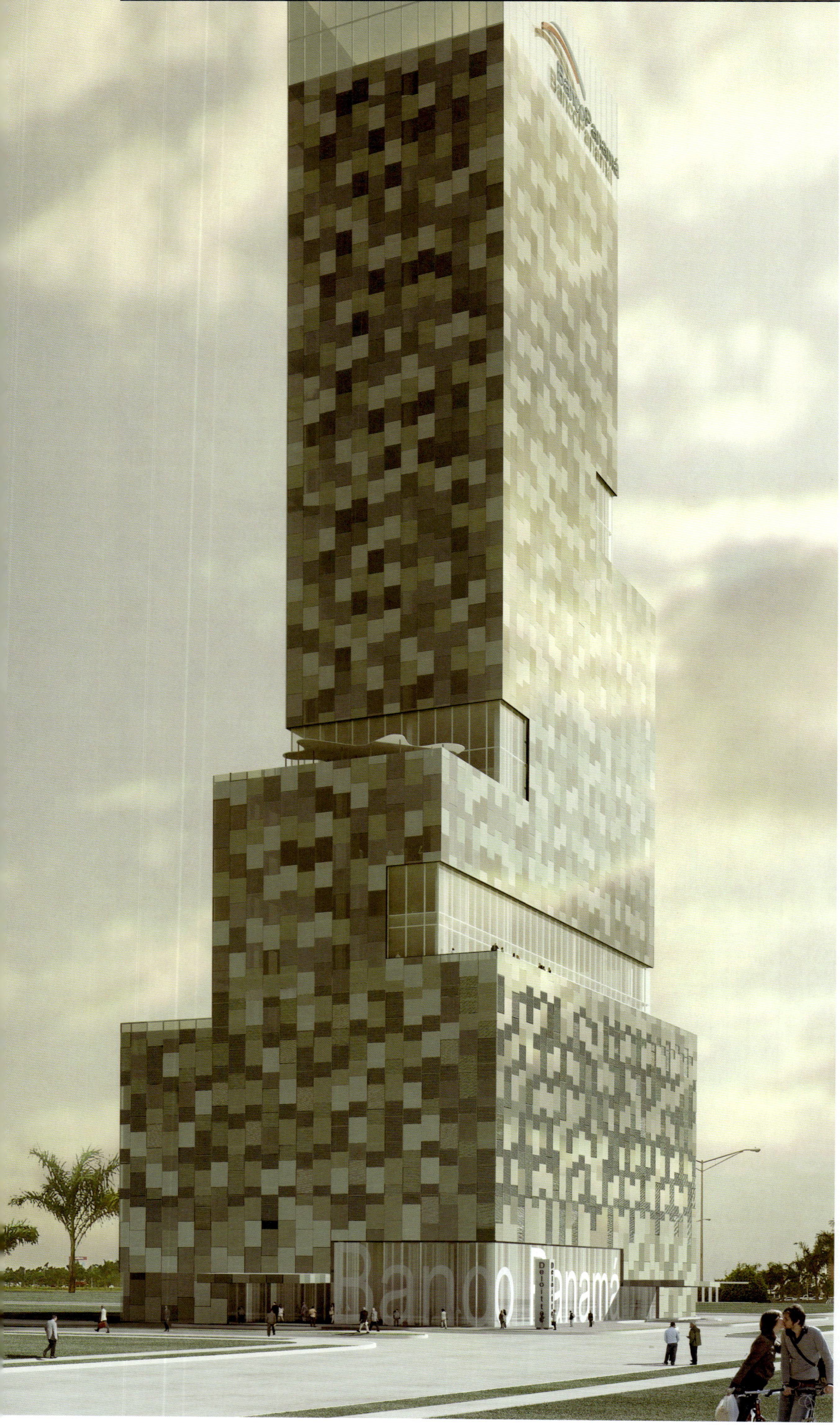


从整体规划来看，塔楼部由**四个分离的建筑物叠加而成**。第一层为入口处， 包括银行办事处，其上为六层式的停车场。其余三部分为体积各异的棱柱型建筑，从下往上依次缩进，形成阶梯结构，可从不同角度观赏旁边的大海和历史中心。

每个棱形建筑可看做一个小型的标准办公楼，配备一个提升的大厅，其设置与所有住户的要求相符合：餐厅、健身房、休息室及俱乐部。最上层有一条具有代表性的过道，可鸟瞰整个城市。

建筑外表采用玻璃幕墙结构，不同的玻璃颜色和透明度赋予其丰富而随机的元素，这不仅源于其半透明和反光效果，而且也源于其典型的模糊效果——一种与建筑水平线不相匹配的效果。

Programmatically, the tower section is organized as **a staggered stack of four separate buildings**. The first welcomes the city on the ground floor housing bank offices on which the six-storey parking stands. The remaining three are prisms of different dimensions which are set back from the previous releasing different terraces which share the different orientations and views over the sea and the historical center.

Each prism can be read as a small canonical office building with a raised lobby which matches with the programs shared by all tenants: a restaurant, a gym and a resting and meeting club. At the top, there is a representative Hall visible from the city.

The envelope is solved with a glass curtain wall with glasses of different colours and transparency which introduce a random and vibrant factor to its reading, not only because of the game between apparent opacity and reflection but also because of the scalar ambiguity of the typical element that does not match with any obvious horizontal line of the building.

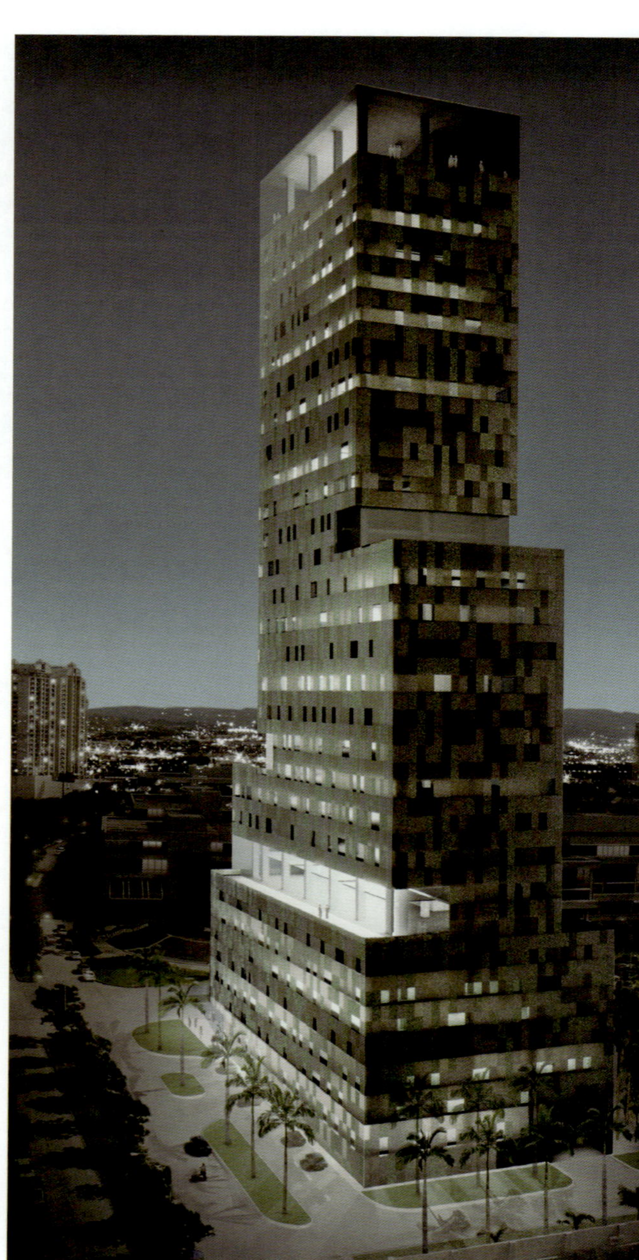

剖面图 SECTION

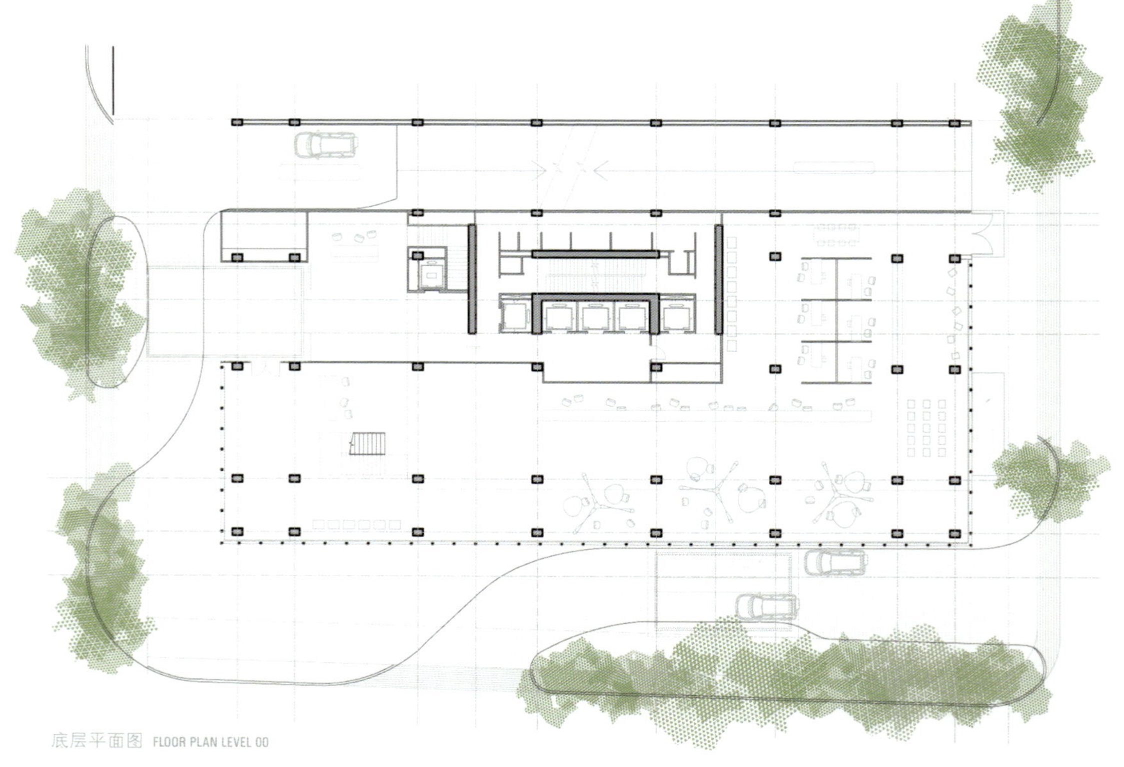

底层平面图 FLOOR PLAN LEVEL 00

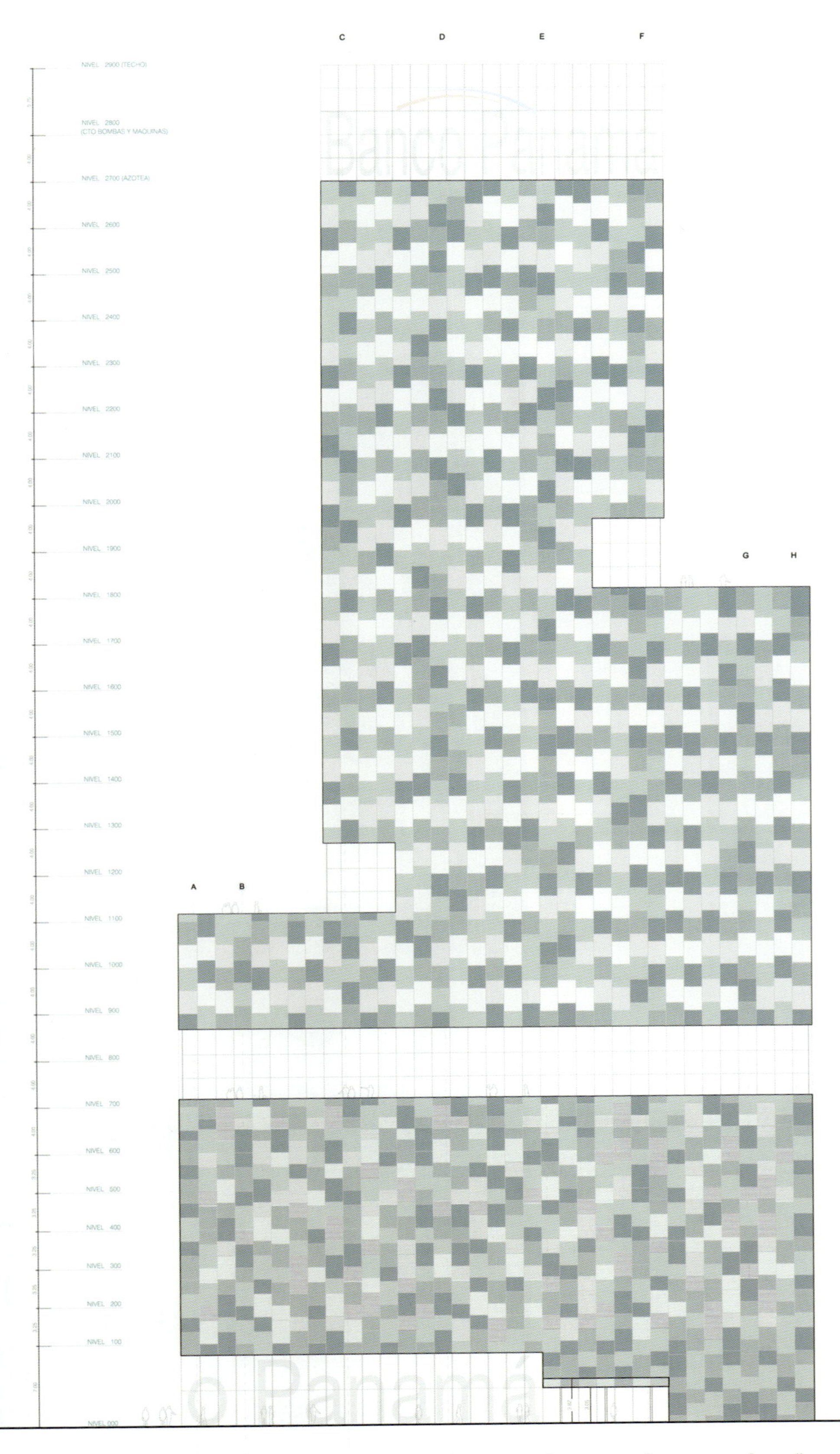

西立面图 WEST ELEVATION

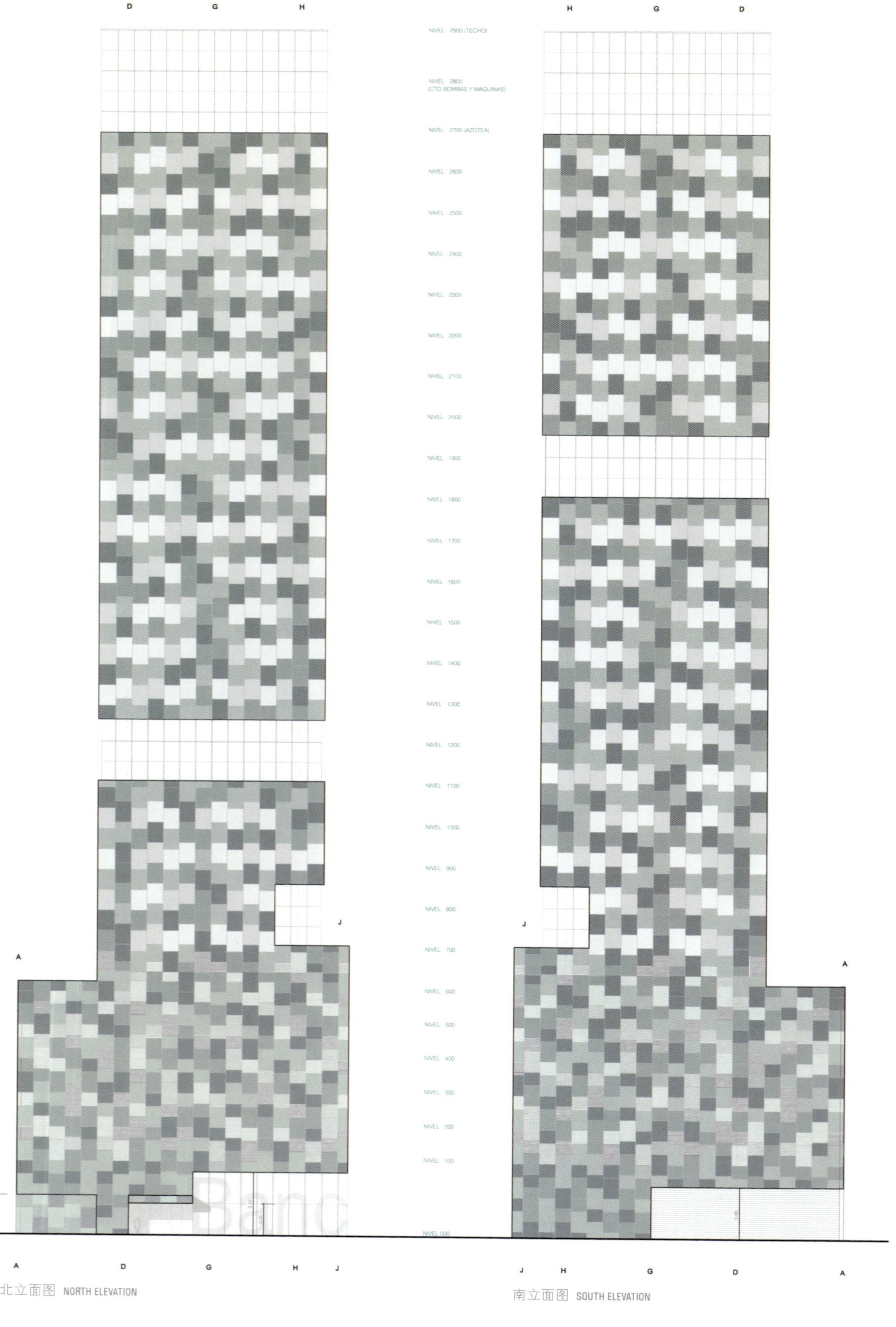

北立面图 NORTH ELEVATION

南立面图 SOUTH ELEVATION

立面细部 FAÇADE DETAIL

巴塞罗那，西班牙 Barcelona, Spain

## 105户住宅、商铺及车库
## 105 housing units, retail and parking

**2009**
资格限制国内竞标 Restricted national competition
一等奖 First Prize

首先，我们要画一张包含16个标准住宅单位的标准层平面图。为了最大化利用空间，我们决定**以建筑中心为主**进行设计，减少7个住宅单位，**以提高体量**，因此可扩大各部分的面积、设计独特的房间造型，使其对流通风，从而具备极佳的环境价值。

We will start by drawing a typical floor plan with 16 ideal housing units. The search for the minimal loss of surface within circulations, lead us to choose to implement **a single core in the centre of the building**. The fulfilment of the maximum surface allows to remove 7 housing units in an operation to **tailor the volume which generates two subtractions** which enriches the section, generate singular housing and provoke a cross ventilation with significant environmental value.

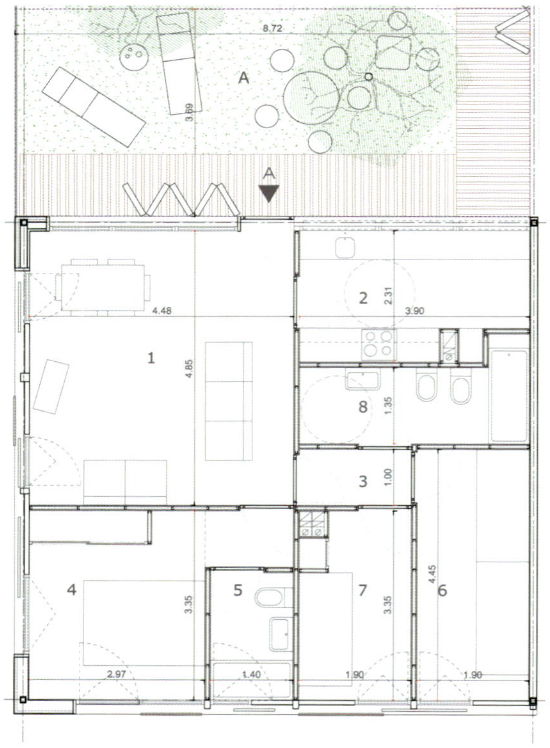

A户型平面图 HOUSING UNIT TYPE A

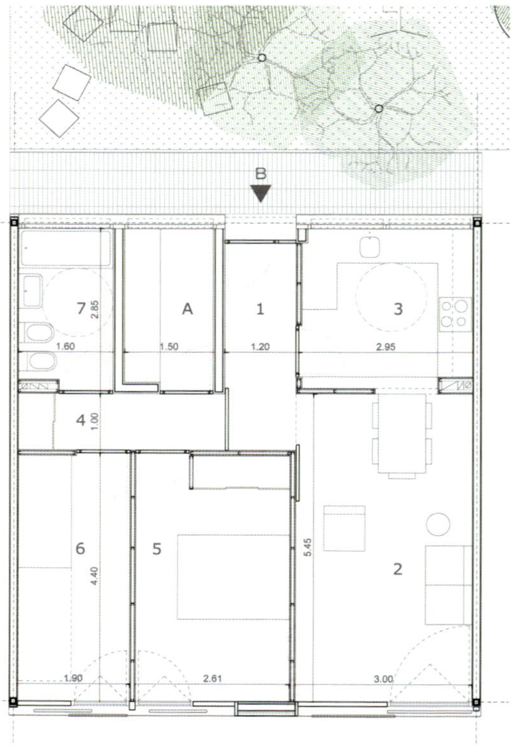

B户型平面图 HOUSING UNIT TYPE B

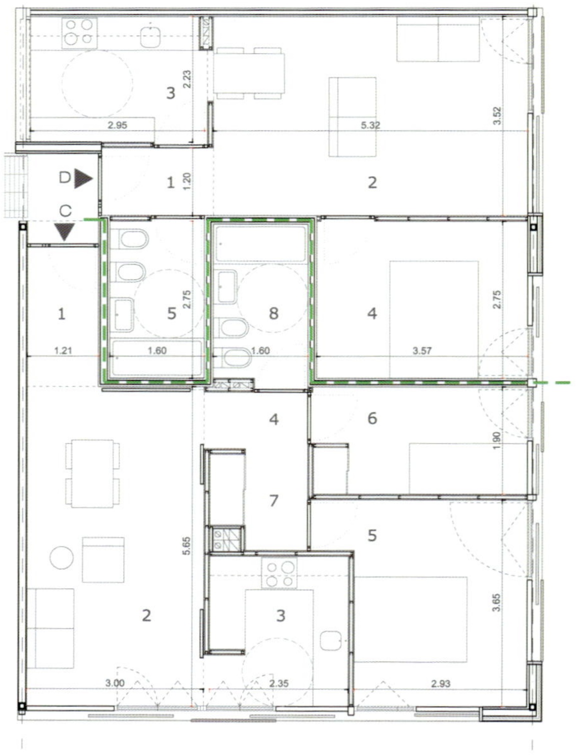

C+D户型平面图 HOUSING UNIT TYPE C+D

东南立面图 SOUTHEAST ELEVATION

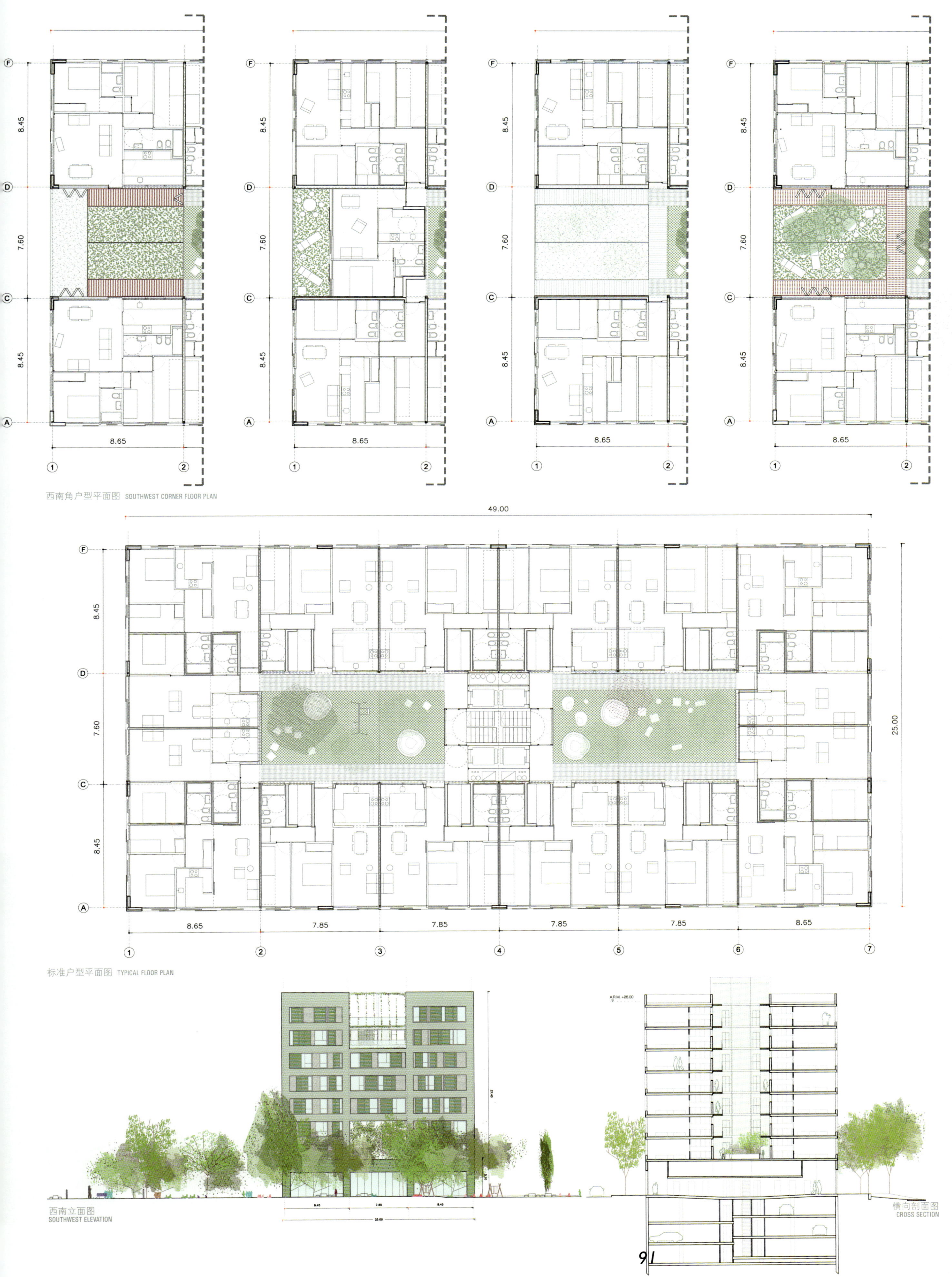

西南角户型平面图 SOUTHWEST CORNER FLOOR PLAN

标准户型平面图 TYPICAL FLOOR PLAN

西南立面图
SOUTHWEST ELEVATION

横向剖面图
CROSS SECTION

## 奥斯陆，挪威 Oslo, Norway

### 住宅、办公、商业及车库
### Housing, offices, commercial areas and parking

**2009**
国际竞标 International competition
一等奖 First Prize

地块位置 SITE PLAN

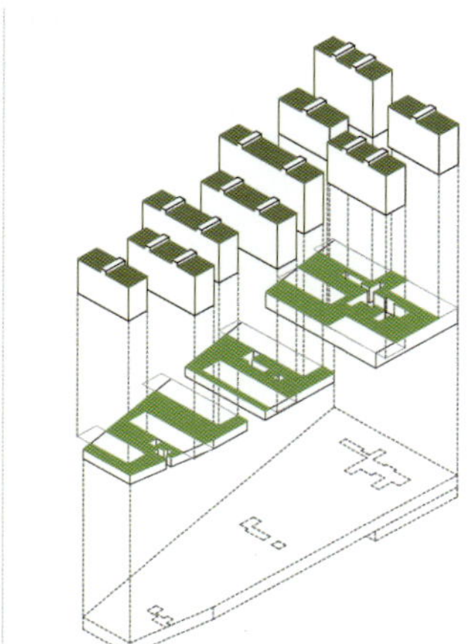
自然资源+屋顶绿化 NATURE+GREEN ROOFS

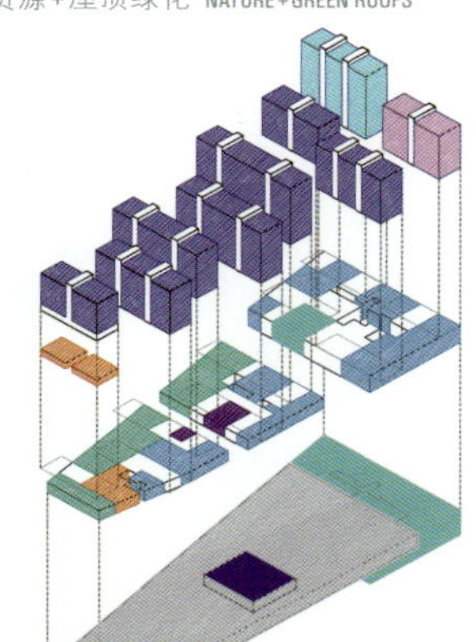

混合功能 HYBRID PROGRAM

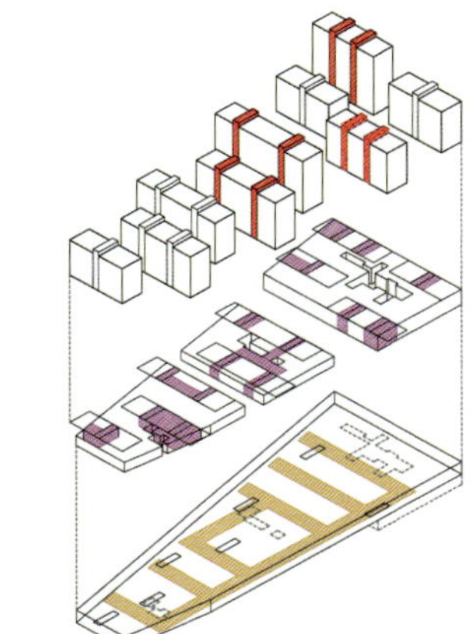
流线·双核心 CIRCULATION·DOUBLE CORES

Paulsenkaia半岛为挪威奥斯陆Bjørvika社区的一部分。这里规划建造Deichman图书馆和蒙克博物馆，将于2015年建成，与大剧院、Barcode项目以及一系列重要的公共项目构成**奥斯陆最具代表性的一块飞地**。

我们希望建筑具有浓厚的城市意蕴，其独特的公共空间由Akerselva河河滨走道和蒙奇广场为主要标志。步行区的规划以大楼为中心，由三个位置紧凑、面积相似的平台构成。

我们希望，公寓楼在朝向方面要大众化，可能的话应两面朝向、视野开阔、采光好、空气流通、东西向透气性强等。通过对不同的结构进行研究后，我们提出了平行大楼的体系：南北向的部分用于采光，东西向的部分采用错叠的方式，以使更多的户型朝向大海。建筑表面拟采用双层结构：内层致密厚实，隔热、隔音、阻挡阳光照射；外层较为轻薄，通风、防水，呈现建筑的象征意义。

该建筑可用于提供商业、小型社区服务及开设餐厅，也可用作办公室。

The Paulsenkaia peninsula is a part of the Bjørvika neighbourhood. Here, The Deichman Library and The Munch Museum will soon rise, which together with the Opera House, the Barcode project and a number of important public spaces will create **the most representative enclave of the City of Oslo in 2015**.

We wish to have a strong urban spirit, with a characteristic public space marked by the presence of the promenade along the Akerselva River and Munch Plaza. The delimitation of the pedestrian space is clearly marked by the footprints of the buildings, consisting of three dense, similarly sized "podiums".

We wish the apartments to have a more democratic form in terms of orientation - double when possible -, distant views, sunlight, air flow, and east-west permeability... After studying different configurations, we propose a system of parallel blocks with a stepping north-south section to receive the sun, and displaced slightly east-west to open fjord views to the greatest number of residents possible. The housing facades are proposed to be constructed in two layers: the interior is a dense and heavy layer that resolves the thermal barrier, the acoustic insulation, sunlight control and contains operable windows and doors; the exterior is a light layer responsible for ventilation, protection from water and creating the emblematic image of the building.

The podiums will house commercial businesses, small neighbourhood services, restaurants and some offices.

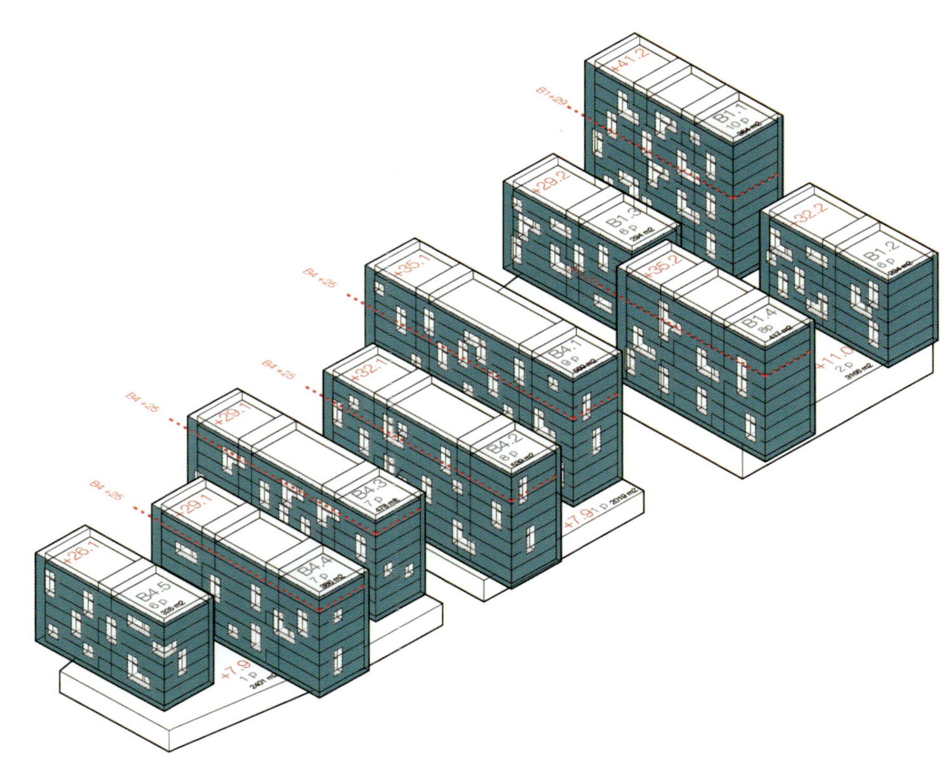

底层平面图 GROUND FLOOR PLAN

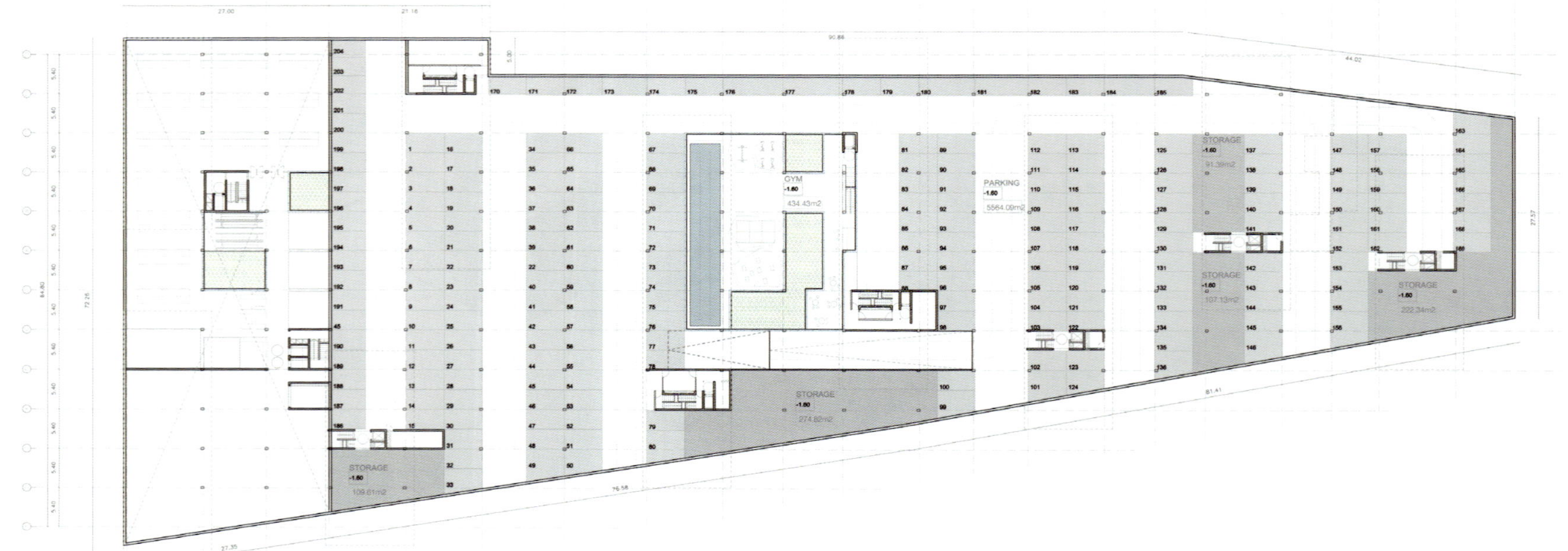

地下一层平面图 FLOOR PLAN -1

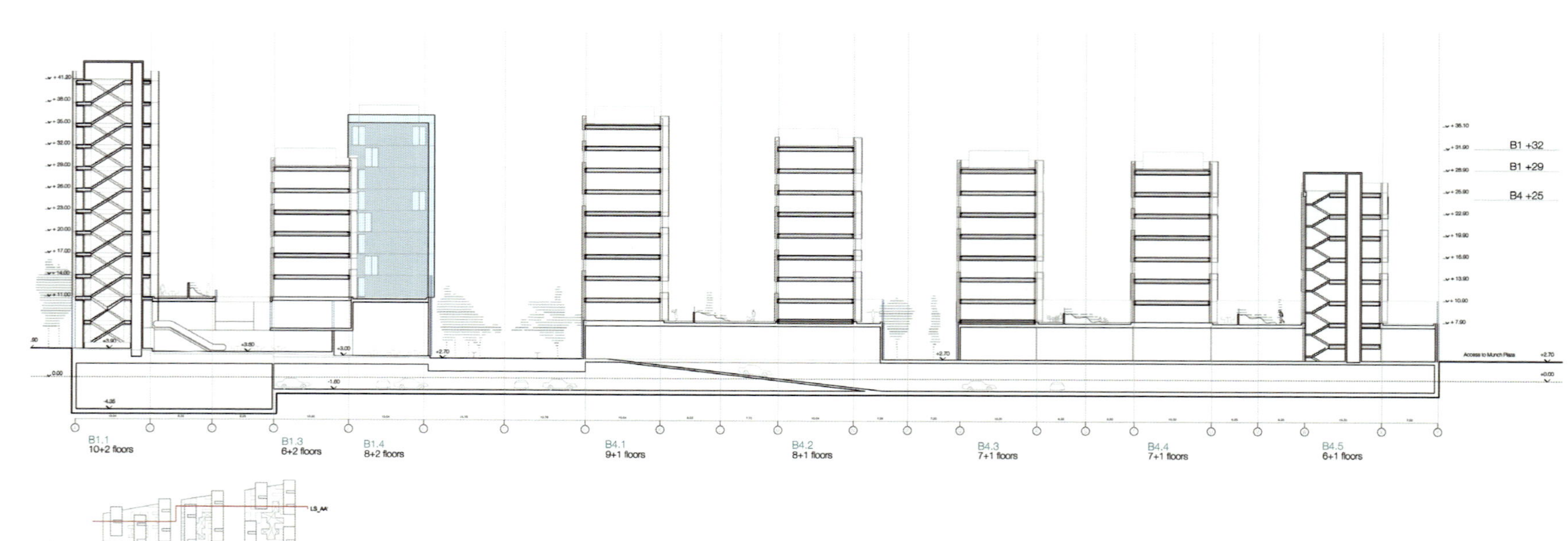


剖面图 AA′ SECTION AA′

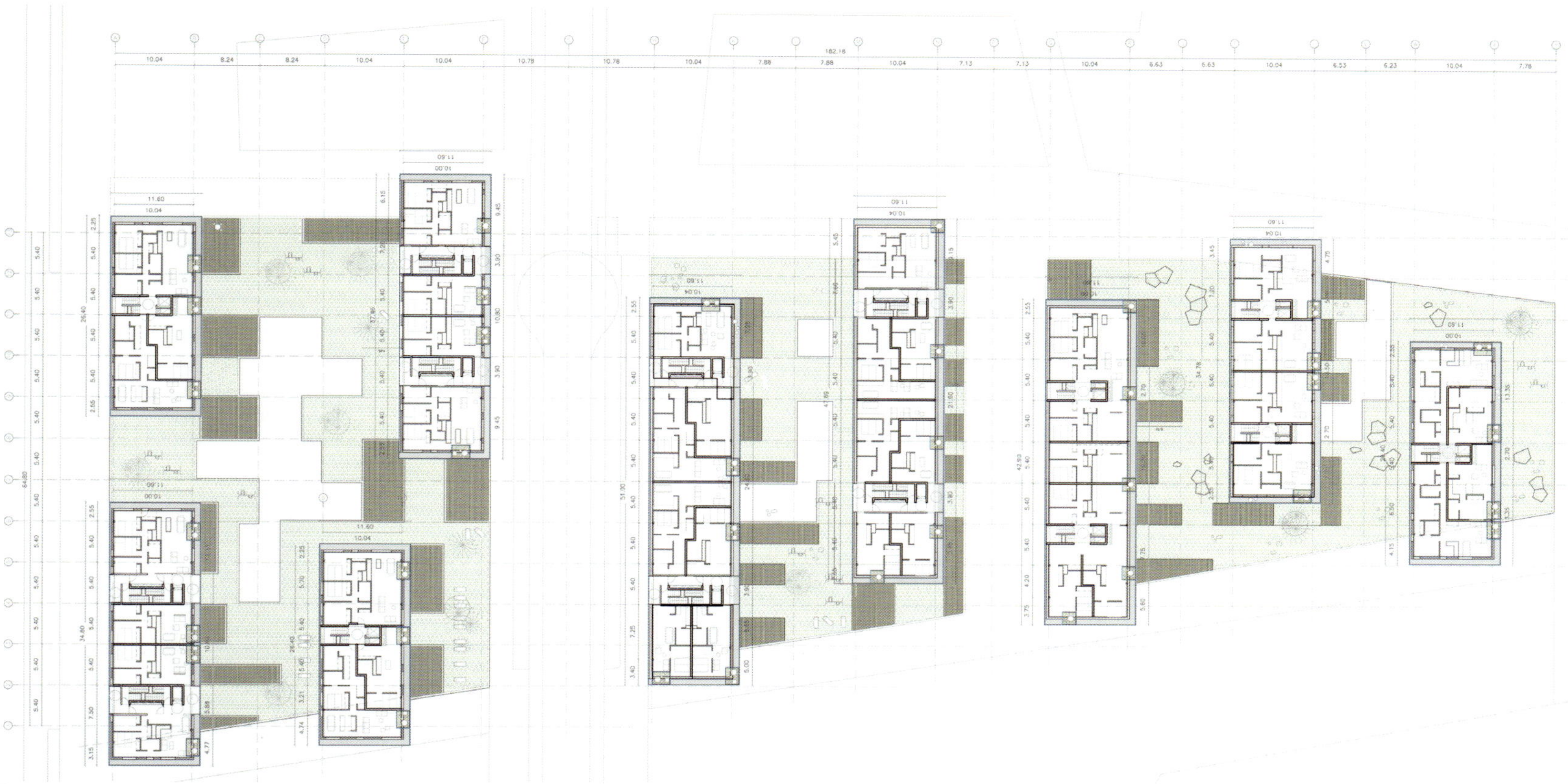

三层平面图 FLOOR PLAN 2

二层平面图 FLOOR PLAN 1

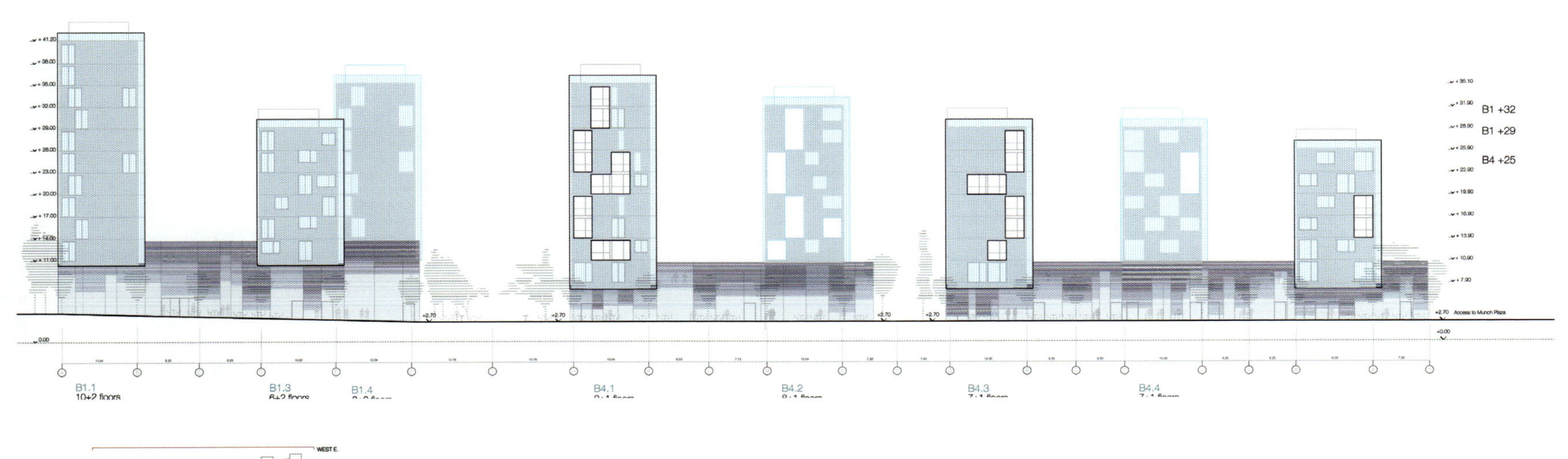


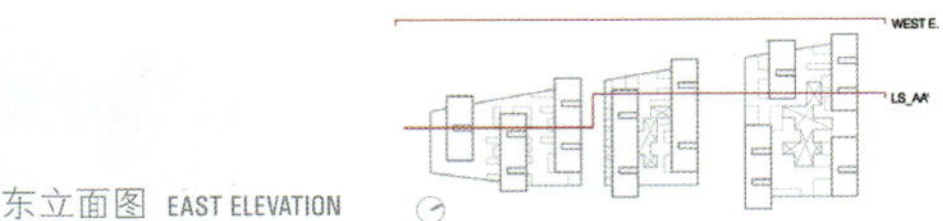

东立面图 EAST ELEVATION

# 蒙克博物馆及斯特内尔森收藏品馆
# Munch Museum and Stenersen Museum Collections

**2009**

国际邀标项目 Invited international competition

一等奖 First Prize

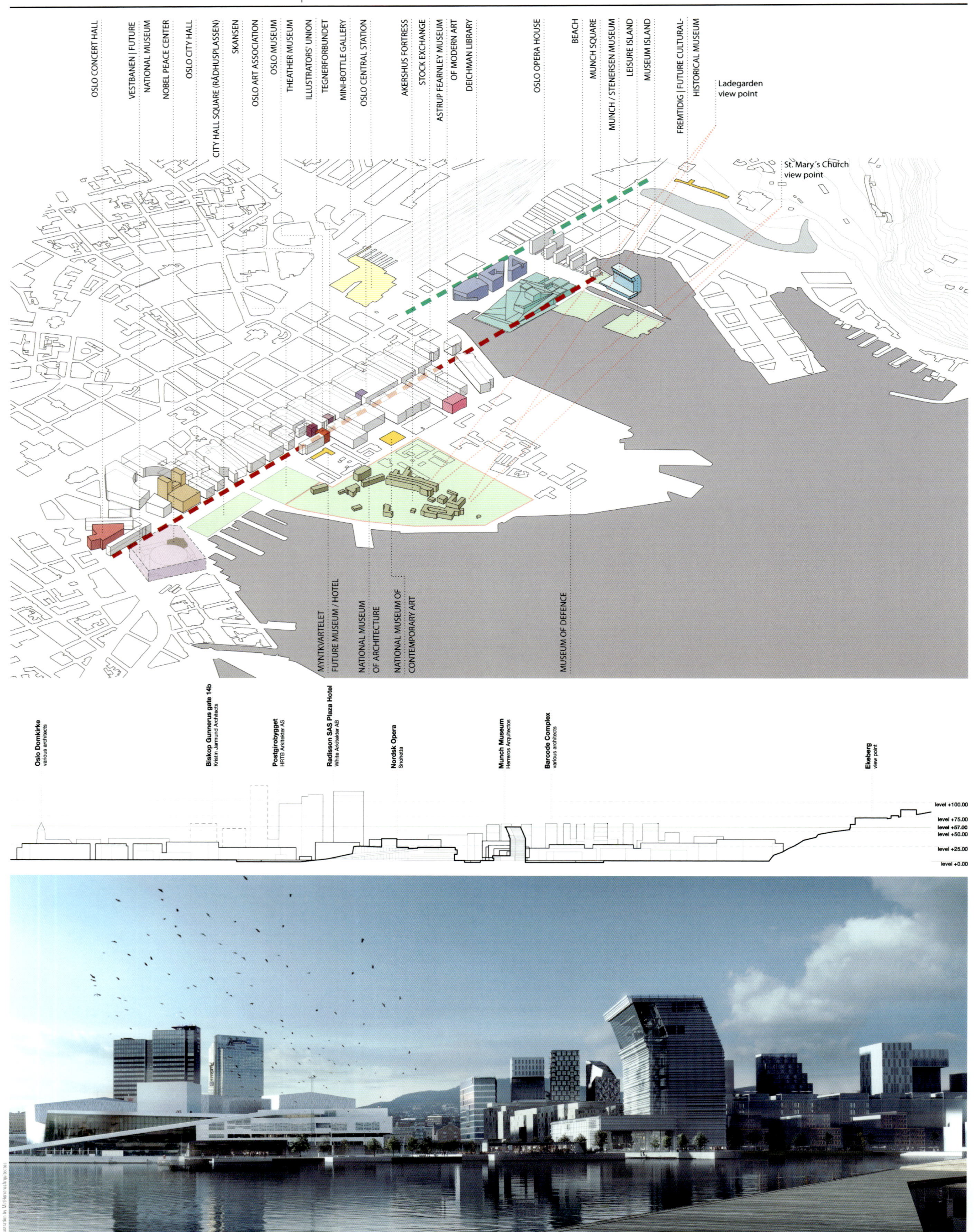

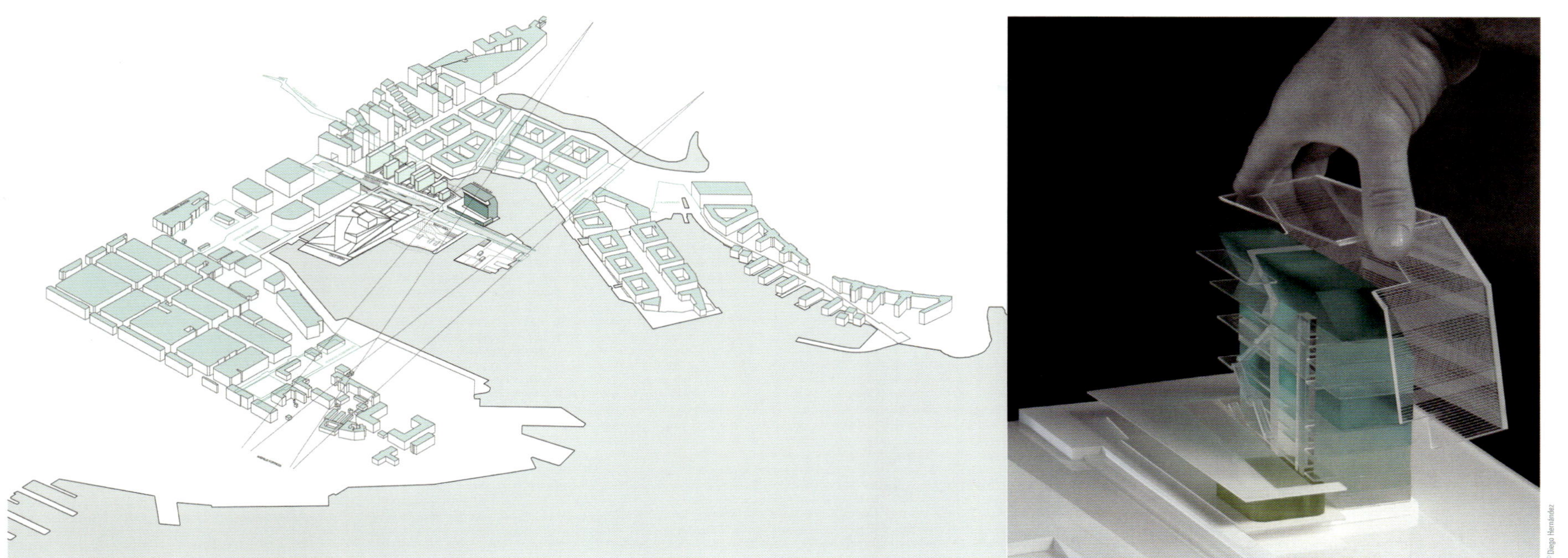

该博物馆呈纵向延伸式，使其大厅变成“室内”广场。建筑主体分为两部分：**静态部分**按高度将展厅分成不同层次，形成管理区、贮存区、档案区；动态部分则为观众聚集区，可从不同角度鸟瞰整个城市。**动态部分**用做纵向延伸的公共区域，在这里可以欣赏到蒙克博物馆的作品，同时可了解艺术家与城市相关联的历史背景。过去、现在和将来，奥斯陆一直在与时俱进。

The museum develops in vertical over that houses the lobby turned into an "in-door" plaza. The main body is divided into two parts: **the static museum** stacks the exhibition halls into different heights mixed with management and conservation spaces and files; the dynamic museum distributes the public and provides successive view-points over the city. **The dynamic museum** functions as a vertical public space which allows discovering the work of Munch and at the same time one can understand the historical urban strata establishing a strong link between the artist and his city; between the past, present and future of Oslo in transformation.

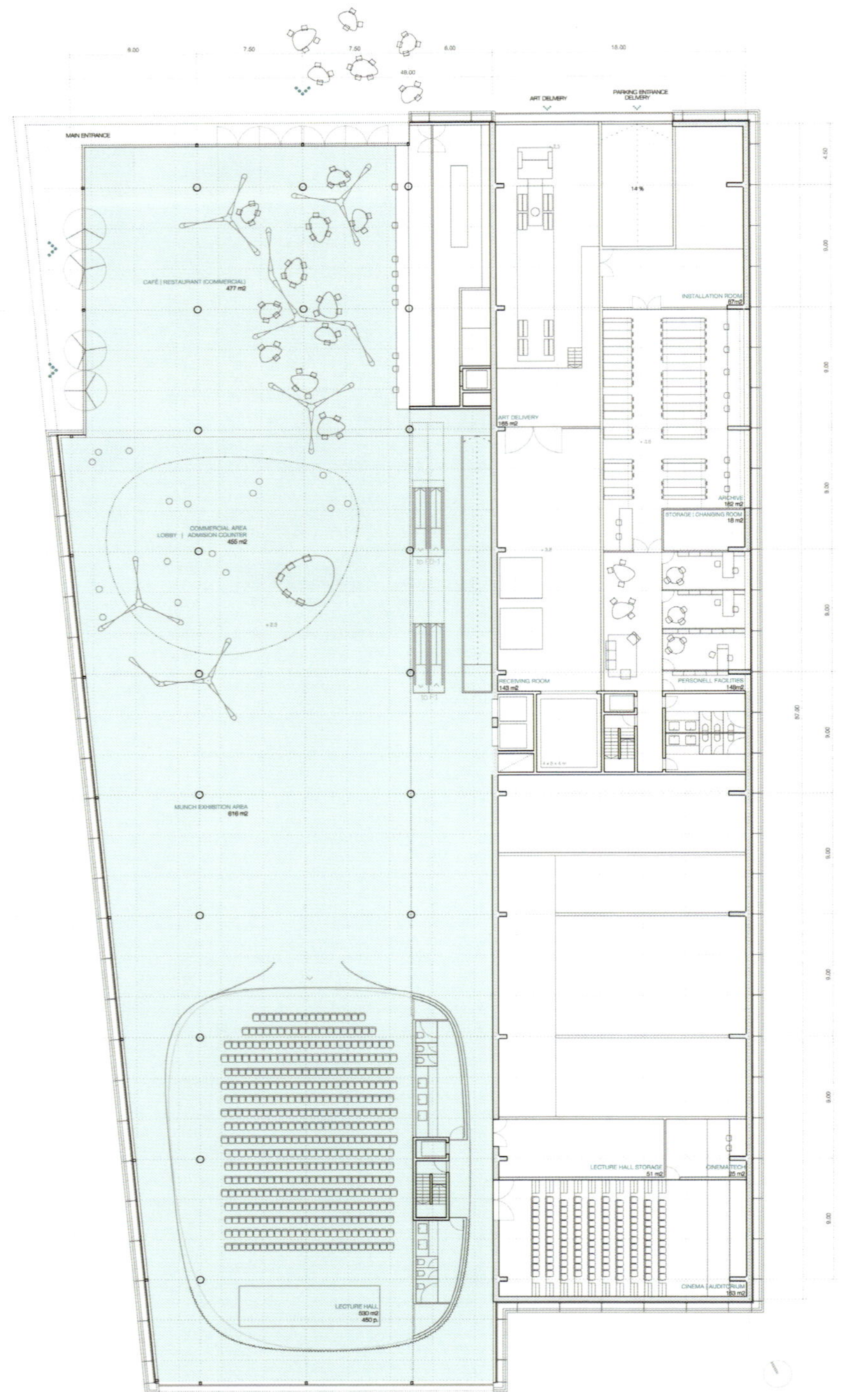

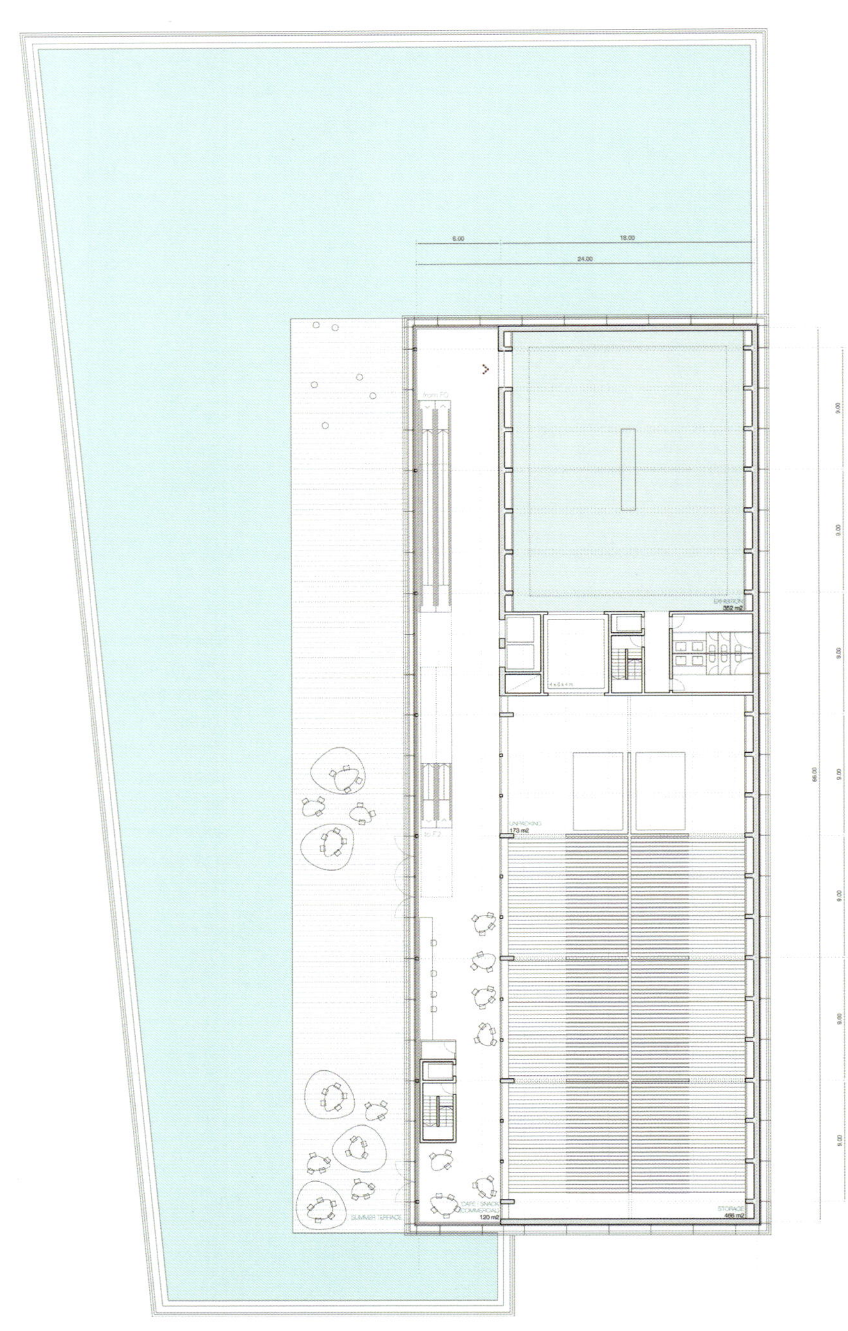

入口层平面图 +2.3/+3.8 ACCESS FLOOR PLAN +2.3/+3.8

平面图 +10.4 FLOOR PLAN +10.4

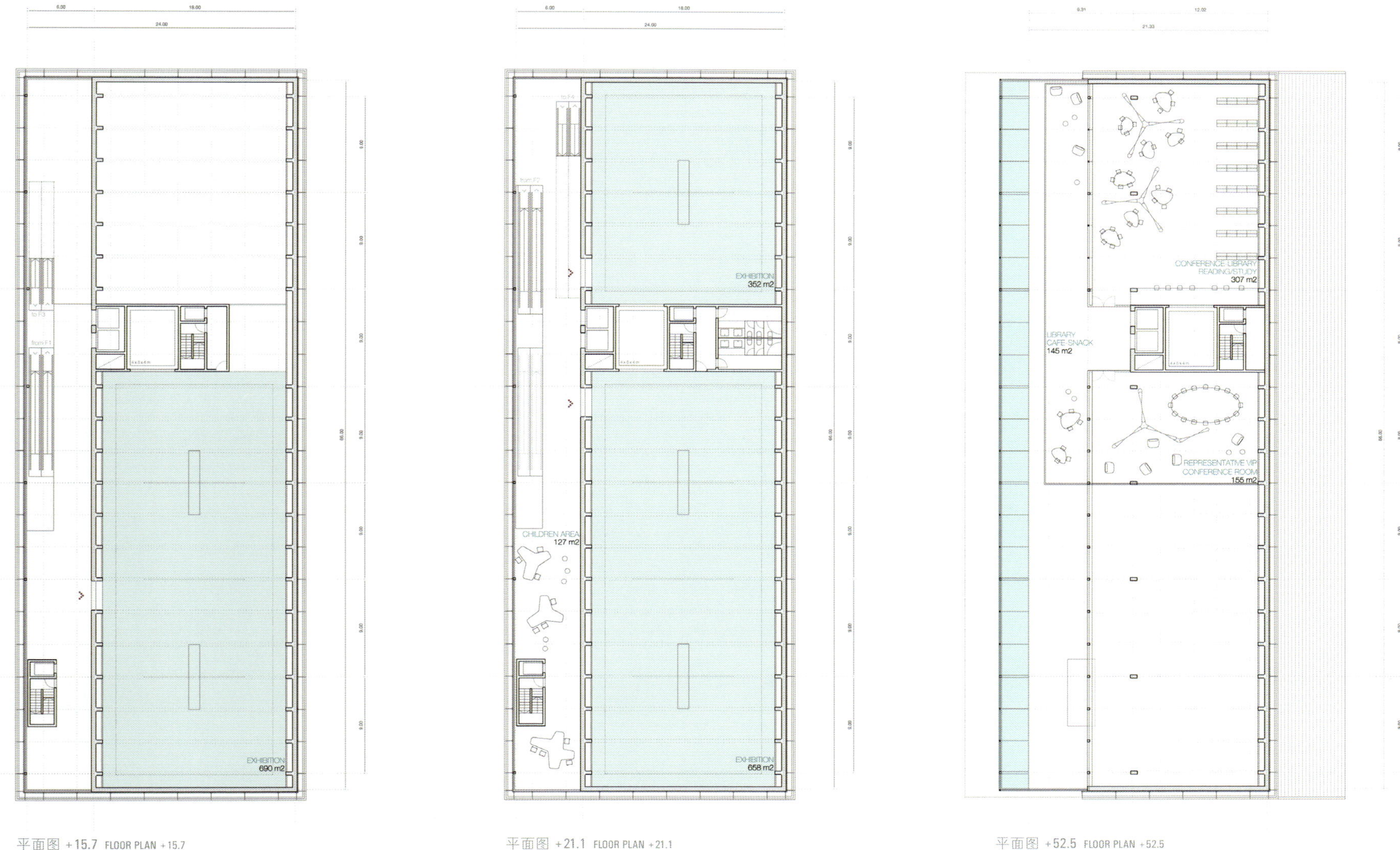


平面图 +15.7 FLOOR PLAN +15.7

平面图 +21.1 FLOOR PLAN +21.1

平面图 +52.5 FLOOR PLAN +52.5

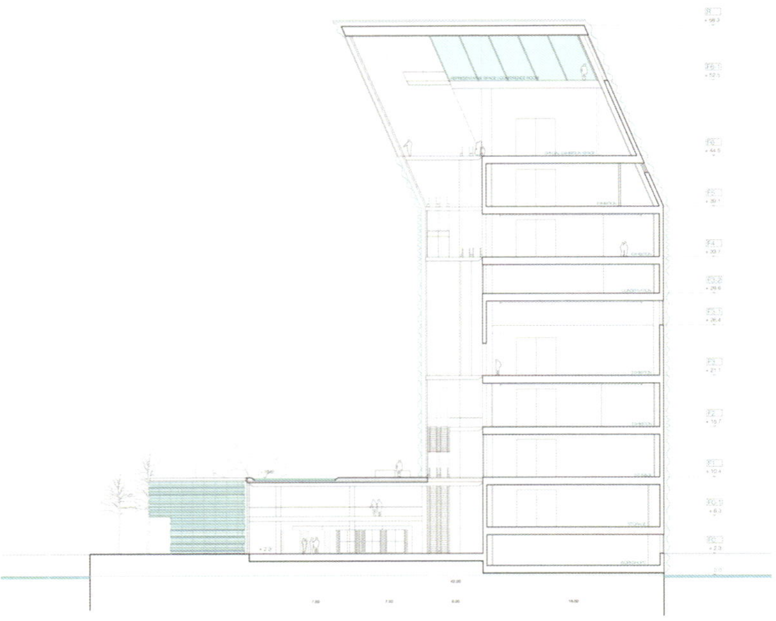

剖面图 1 SECTION 1

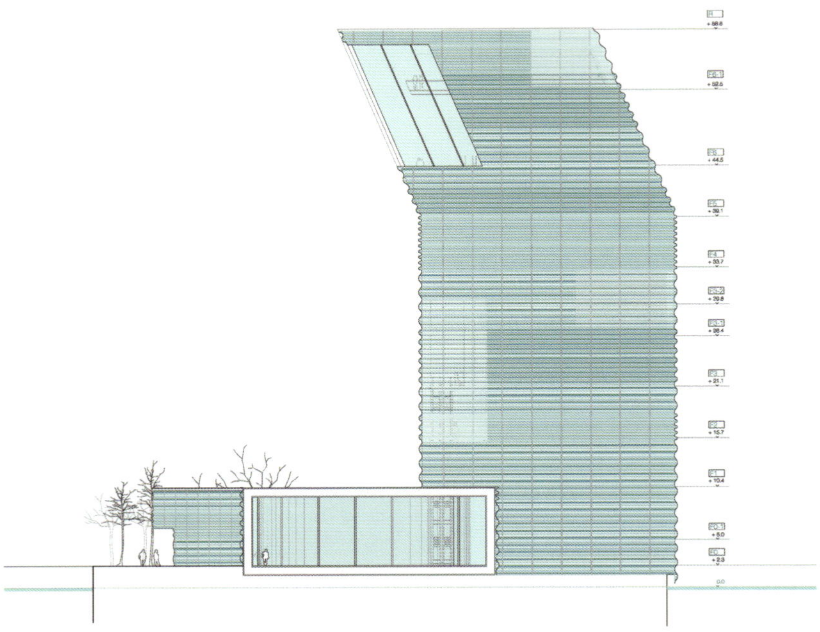

南立面图 SOUTH ELEVATION

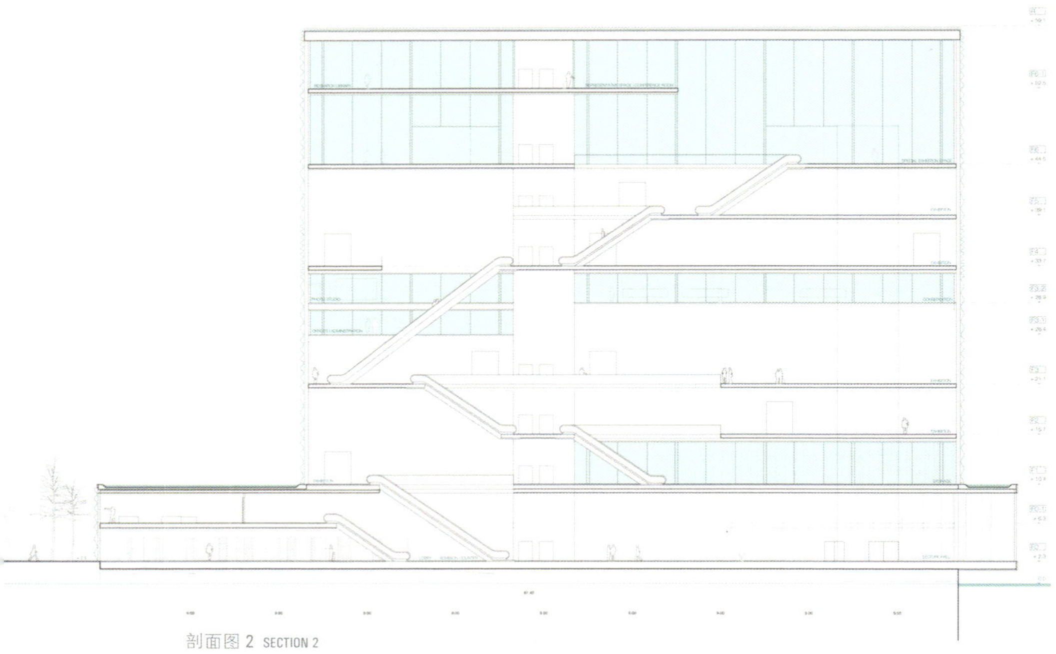

剖面图 2 SECTION 2

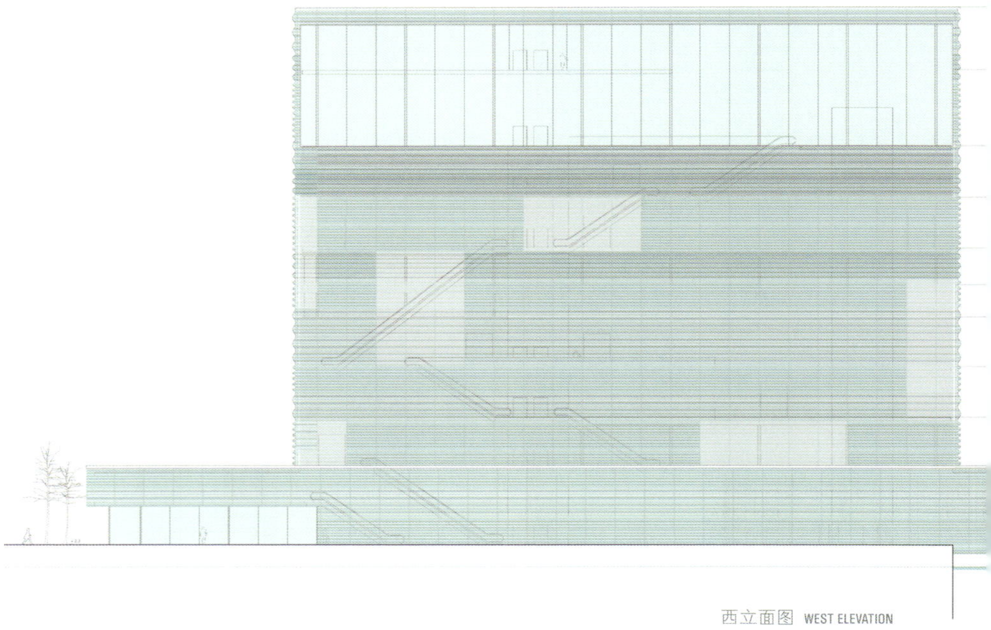

西立面图 WEST ELEVATION

© Illustration by Artefactorylab | Herreros Arquitectos

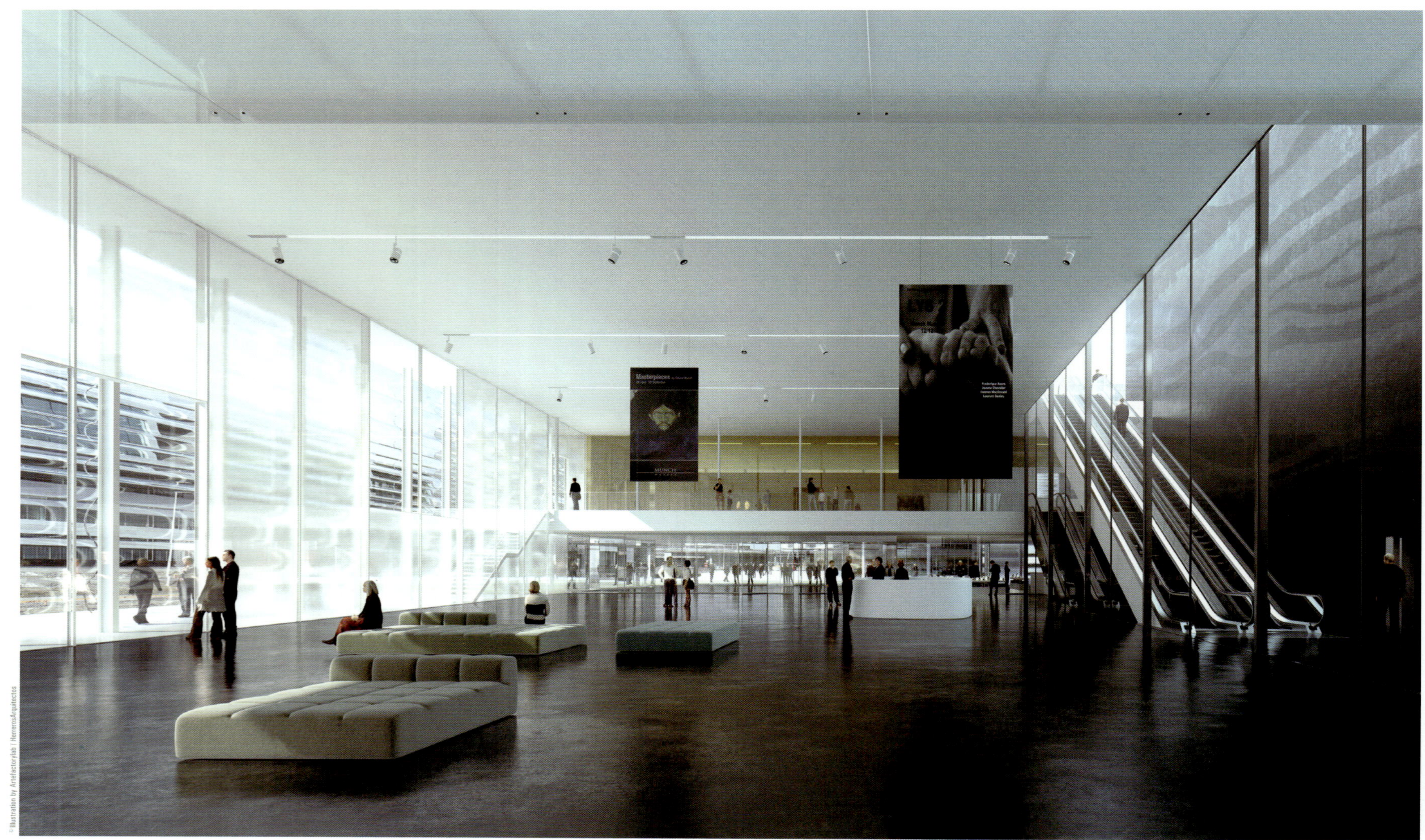

## 波哥达国际会议中心
## International Conventions Centre of Bogota

**2011**

根据履历定向国际邀标 International competition with the selection of portfolios

一等奖 First Prize

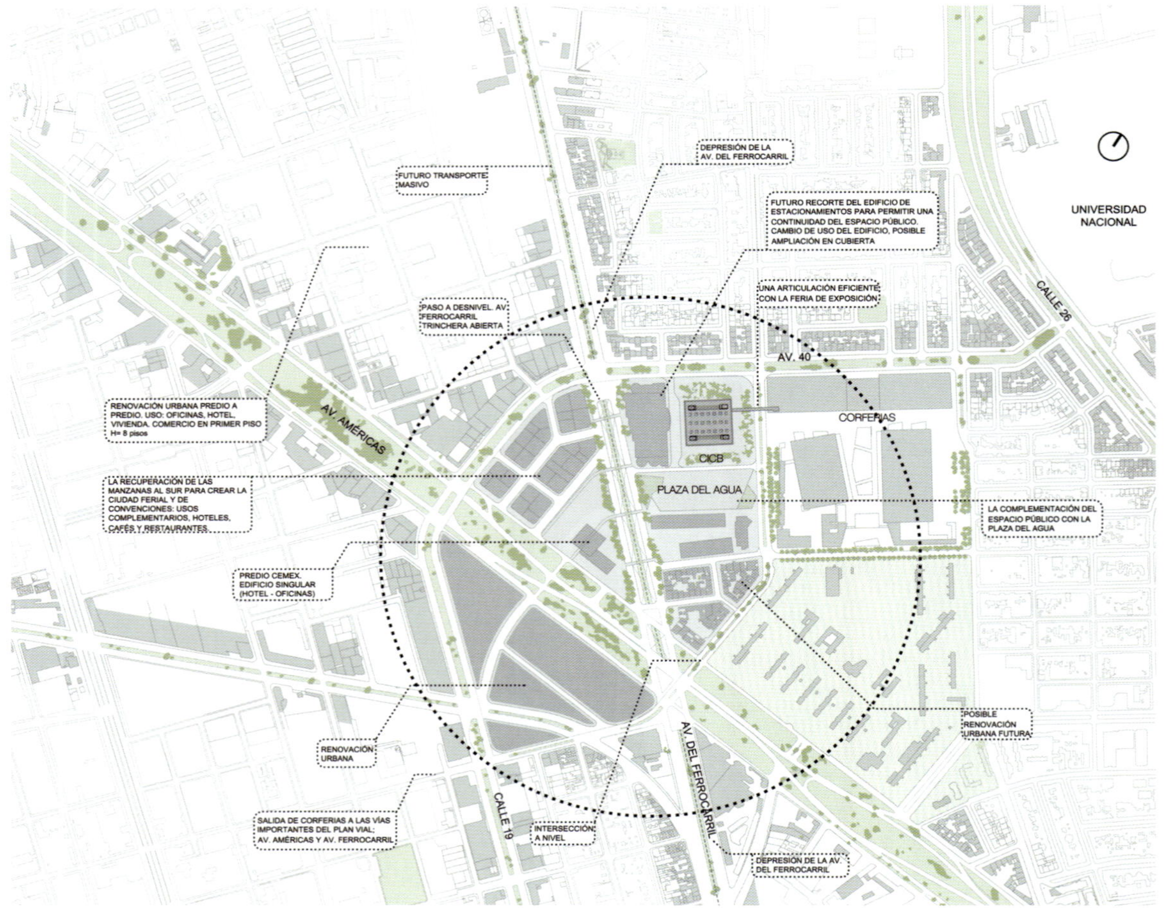


地块位置 SITE PLAN

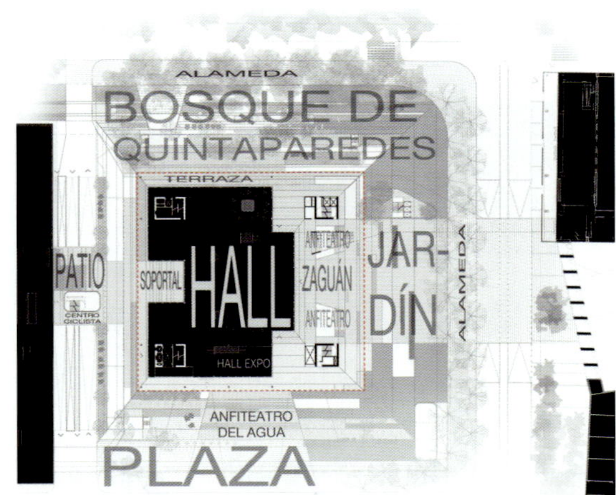


入口层示意图 ACCESS FLOOR PLAN DIAGRAM

该设计凝聚了城市规划的精华，使居民和非居民相互交融，共同分享他们对知识、创新和民间团体作用的看法。该项目组提议，内外层之间要有渗透性，使其具有丰富的公共区域空间和项目的多种选择性。**建筑的空间和技术规格基于四个垂直的沟通空间而形成。**将沟通区和服务区设置于方形结构的四个角落，使得镶嵌于该建筑四个角落的平台得以专门化利用起来。因此，我们有两种平面图：从三个代表性的项目（主大厅、展览厅和礼堂）中即可看出；外围和中心区域将各个核心区域连接起来，为其他项目区域（主要是会议室和公共服务设施）提供更开阔的视阈和充足的自然采光。

The proposal is defined as a condenser of urban experiences in which residents and outsiders will blend sharing their interest in knowledge, innovation and the strength of civil society. The project proposes a determinant permeability between interior and exterior, a remarkable diversity of public spaces and an original wealth of choice of programs. **The spatial and technical order of the building is based on four vertical communication cores.** This arrangement into a square with communication cores and services in the corners allows us to have specialized uses in platforms "embedded" in these "legs." Thus, we have two types of floor plans: clear for the three most representative programs - Main Lobby, Exhibition and Auditorium - and peripheral, releasing the central space, connecting the core and taking advantage of the views and natural lighting for the rest of the program - basically Meeting Rooms and public services.

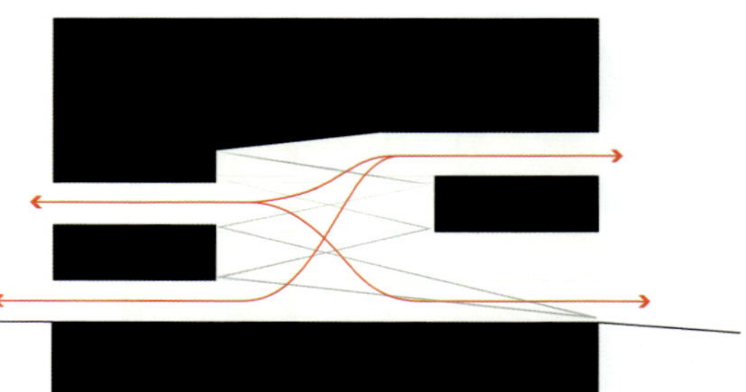

ESQUEMA LLENO-VACÍO SOLID-VOID DIAGRAM

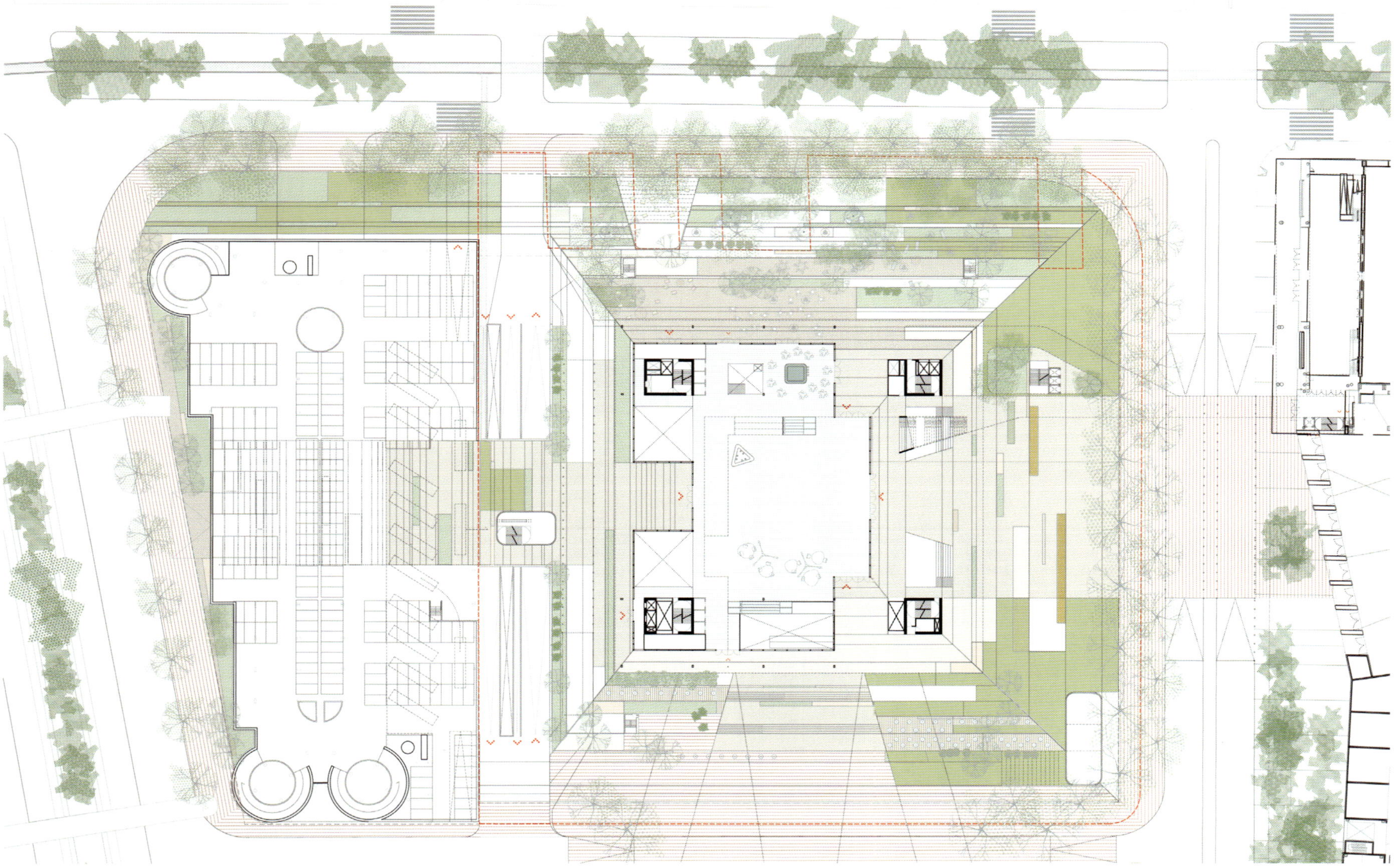

底层平面图 GROUND FLOOR PLAN

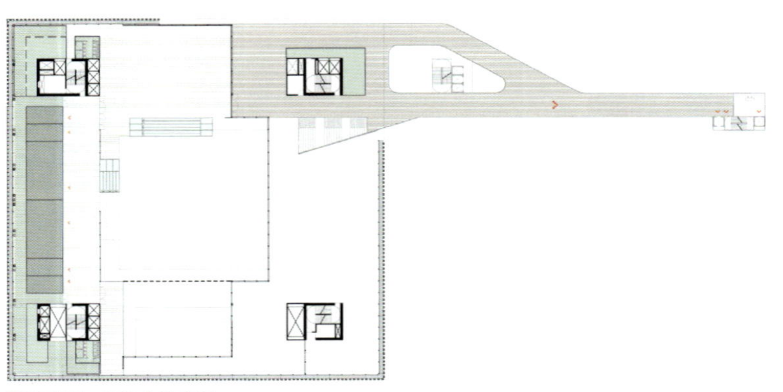

平面图 M2 · 下层会议空间
FLOOR PLAN M2 · LOWER MEETING SPACES

平面图 P2 · 上层会议空间
FLOOR PLAN P2 · UPPER MEETING SPACES

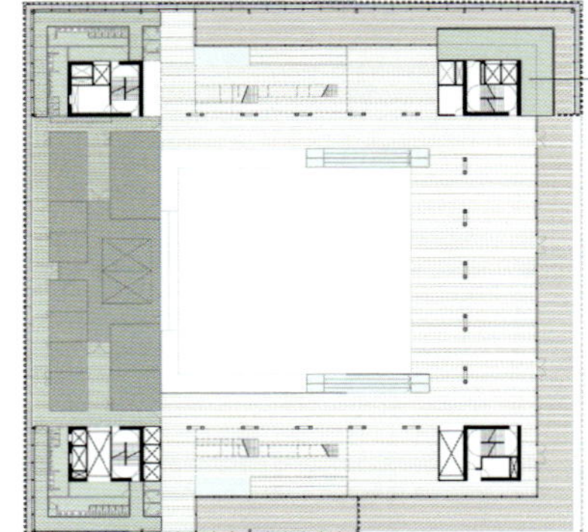

平面图 P3 · 露台
FLOOR PLAN P3 · SKY LOBBY

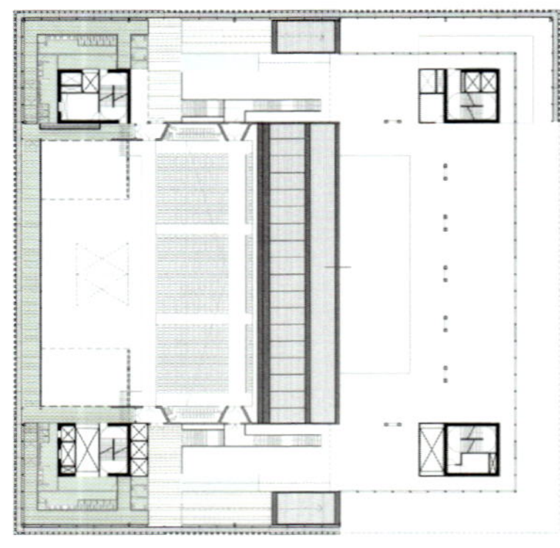

平面图 P4 · 礼堂
FLOOR PLAN P4 · AUDITORIUM

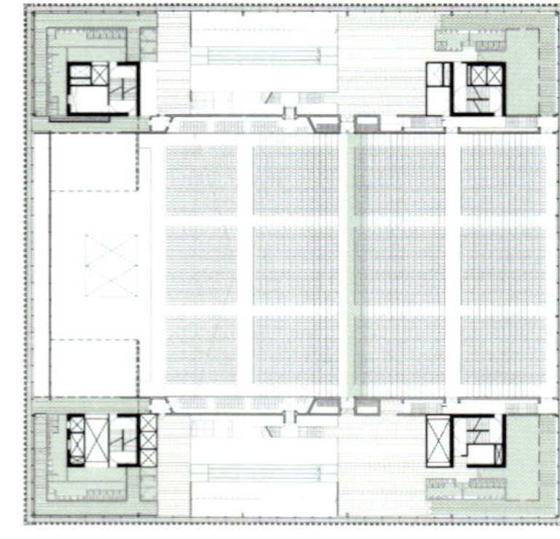

平面图 B1 · 4000人礼堂
FLOOR PLAN B1 · 4000 PEOPLE AUDITORIUM

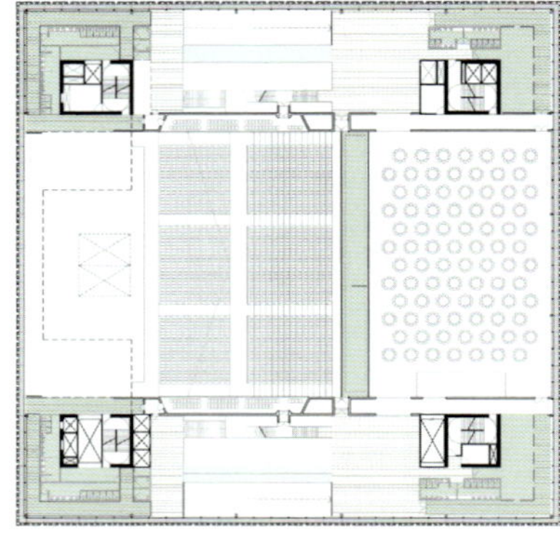

平面图 B1 · 2000人礼堂+2000人大厅
FLOOR PLAN B1 · 2000 PEOPLE AUDITORIUM+2000 PEOPLE HALL

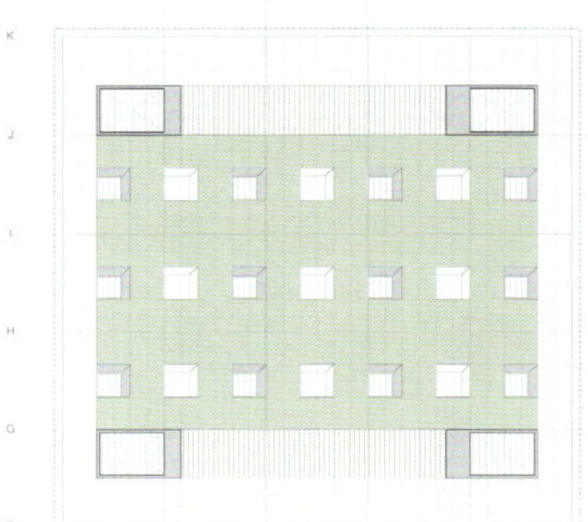

屋顶层平面图
ROOF PLAN

纵向剖面图 LONGITUDINAL SECTION

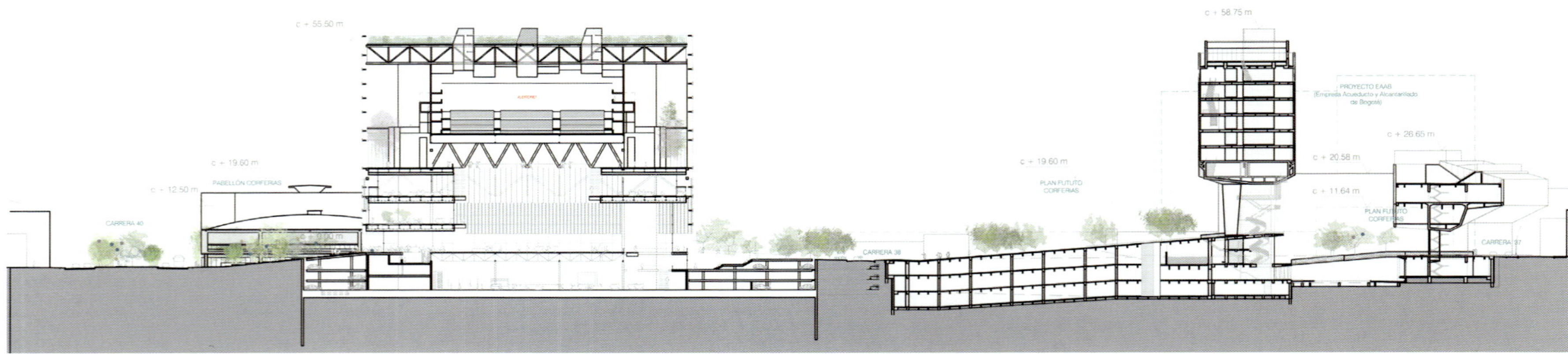
整体横向剖面图 CROSS URBAN SECTION

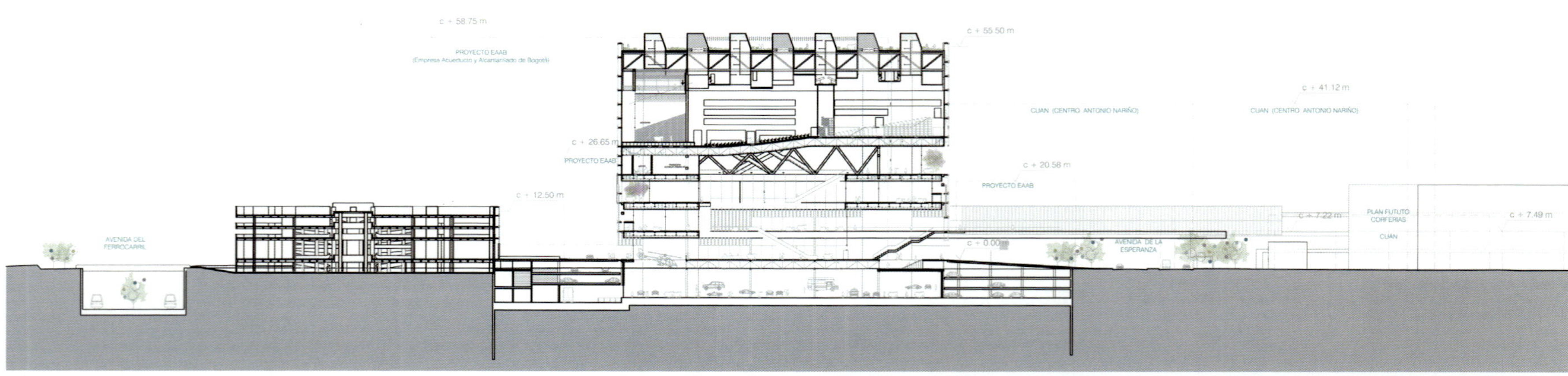
整体纵向剖面图 LONGITUDINAL URBAN SECTION

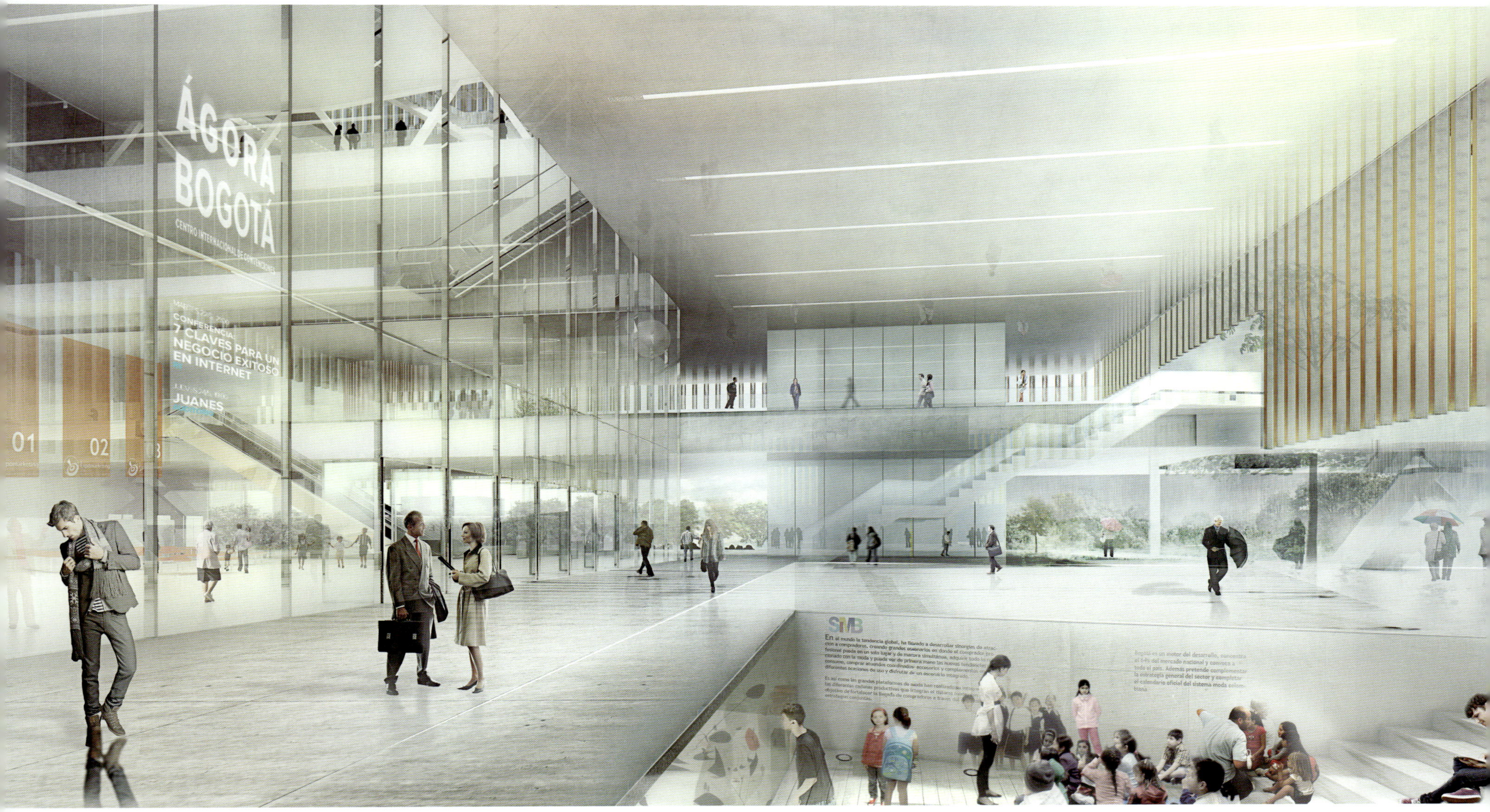
ÁGORA
BOGOTÁ
7 CLAVES PARA UN
NEGOCIO EXITOSO
EN INTERNET
JUANES
01
02

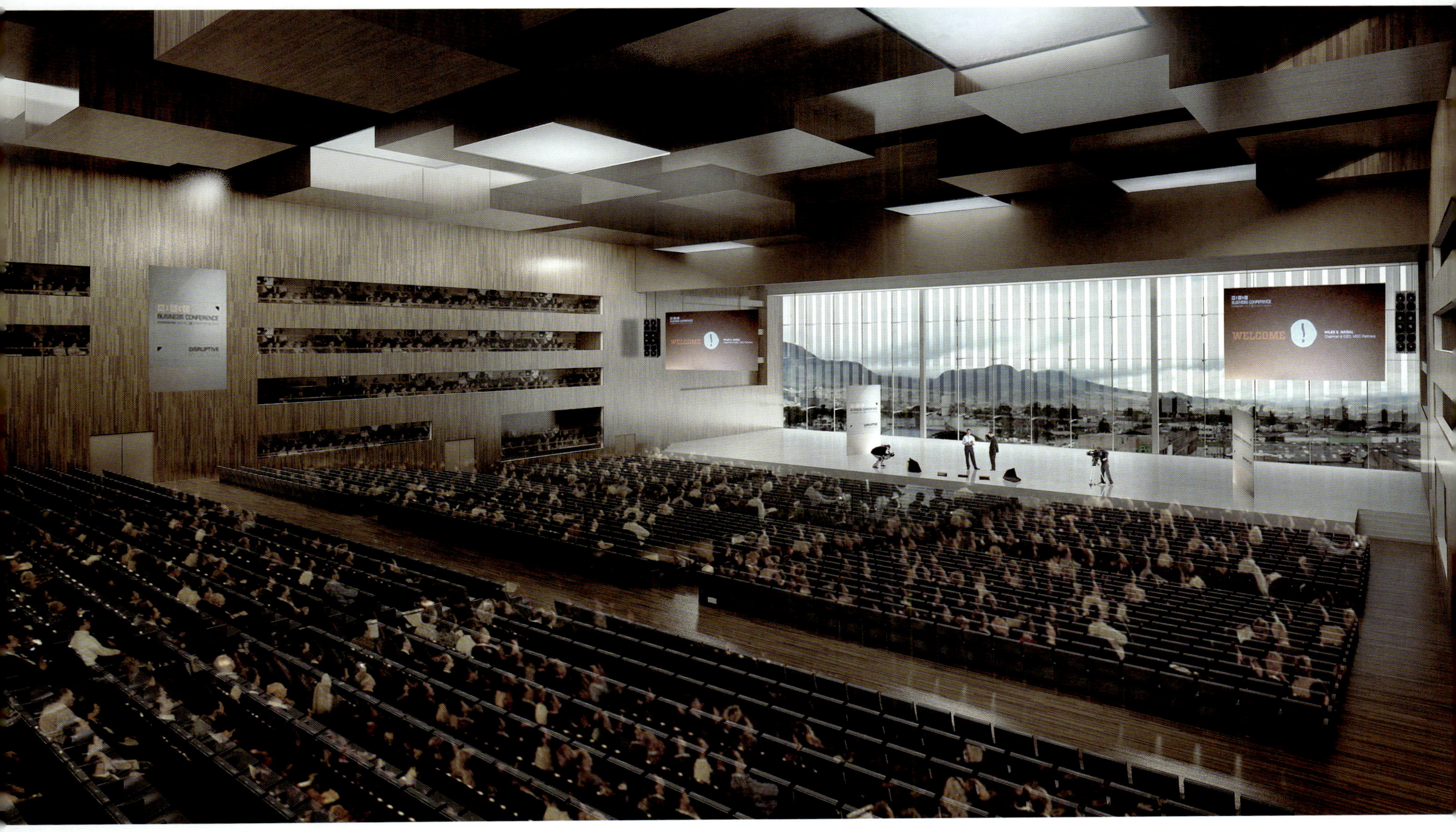
WELCOME

卡达拉奇，法国 Cadarache, France

## Iter, 原子能局附属建筑
## Iter, Annex Buildings-CEA

**2007**
国际邀标项目 Invited international competition
二等奖 Second Prize

该项目选址位于茂密的森林处，由五个元素构成。经过现场勘察，我们得出以下结论：居住于森林之中，要保护和尊重环境，其建筑不得高于树木。**我们不赞同留出空地的做法**，那样会破坏自然循环的连续性；我们建议建造线型建筑，附带等宽的走廊，这样只需砍掉建筑占地面积内的树木，使得其他树木可与建筑各表面保持最大的接触面。

The project consists of five elements fixed in a rich forest with biological activity. From the site visit, we concluded that the best option is to inhabit the forest without exceeding its height making a new kind of architecture that respects and protects it. **We disclaim opening large voids** that break the continuity of natural cycles and we propose to build linear buildings with an optimal and constant width which occupy corridors in which only the necessary trees have been removed to allow the maximum proximity of the remaining trees to the façades.

二层平面图 FLOOR PLAN +1

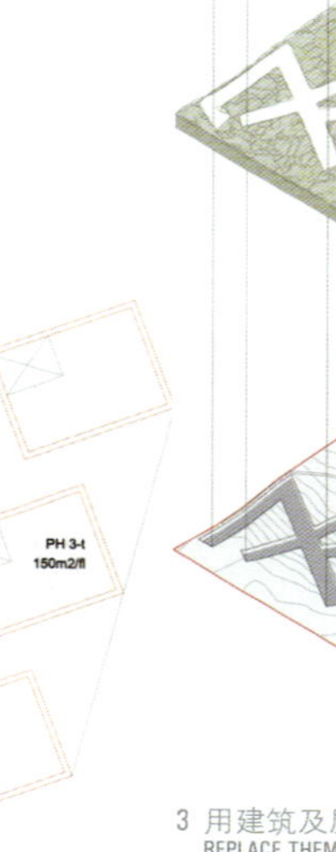

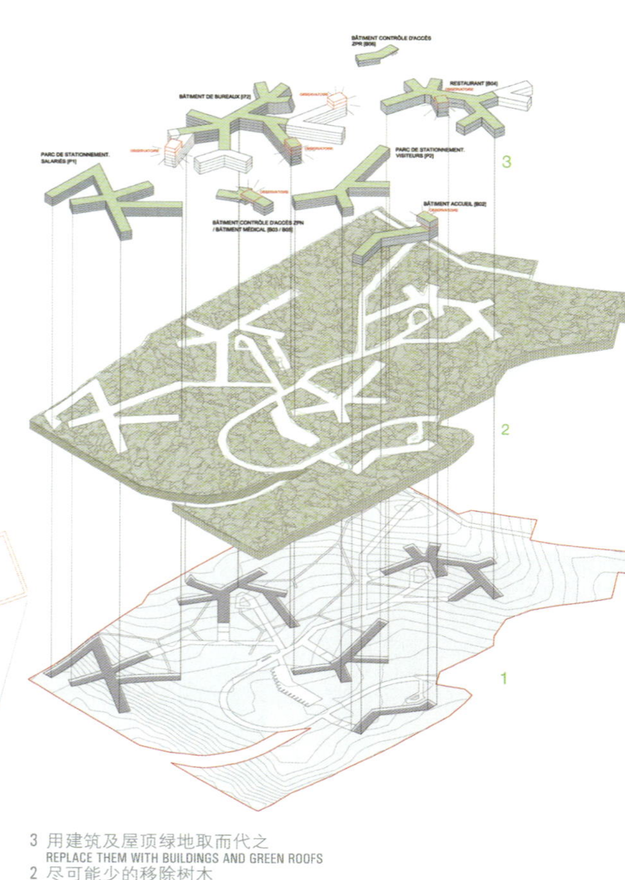

3 用建筑及屋顶绿地取而代之
REPLACE THEM WITH BUILDINGS AND GREEN ROOFS
2 尽可能少的移除树木
REMOVE THE MINIMUM TREES AS POSSIBLE
1 勾画主要线路+划分停车场区域
TRACE THE MAIN ROUTES+SHAPE THE PARKING SPACES

底层平面图 GROUND FLOOR PLAN

## RAPT车库、经济适用房及办公室
## Depots, Social Housing and offices for RAPT

**2007**
国际邀标项目 Invited international competition
二等奖 Second Prize

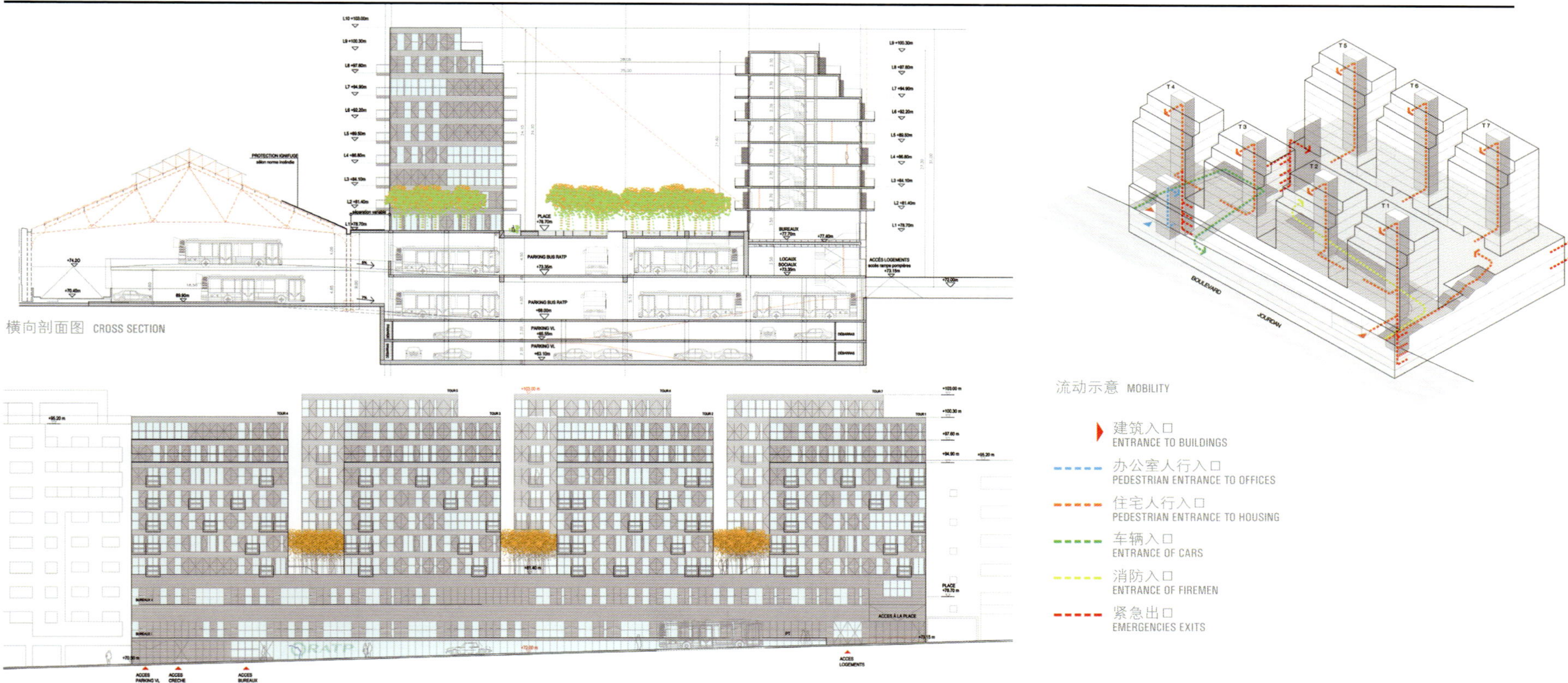


横向剖面图 CROSS SECTION

流动示意 MOBILITY

南立面图 SOUTH ELEVATION

总平面图 OVERALL PLAN

该居住项目旨在一块提升的公园地块建造一个拥有7座楼房的小型社区，建成后与幼儿园一并成为最具吸引力的公共项目。该建筑结构简单而又有重复。一些材料和施工原则以表面和小块单元为基础。**彩色树脂板具有丰富的色彩**，再加上居民的活动以及各家各异的窗户，与公园的绿化环境相融于一体，将构成一幅变幻多彩、令人向往的图景。

The residential program offers a small community of seven towers built in a raised garden over the street which constitute, together with the nursery, the most attractive communal program. The construction is simple and repetitive. A few materials and construction principles resolve on facades and subdivisions. The **chromatic vibration introduced by the coloured resin panels** and by the different movements of people and windows will create a diverse and optimistic state when it is mixed with the vegetation of the garden.

## 生态环保城综合楼
## Hybrid building for the Environmental City

**2007**
国际邀标项目 Invited international competition
入围 Finalist

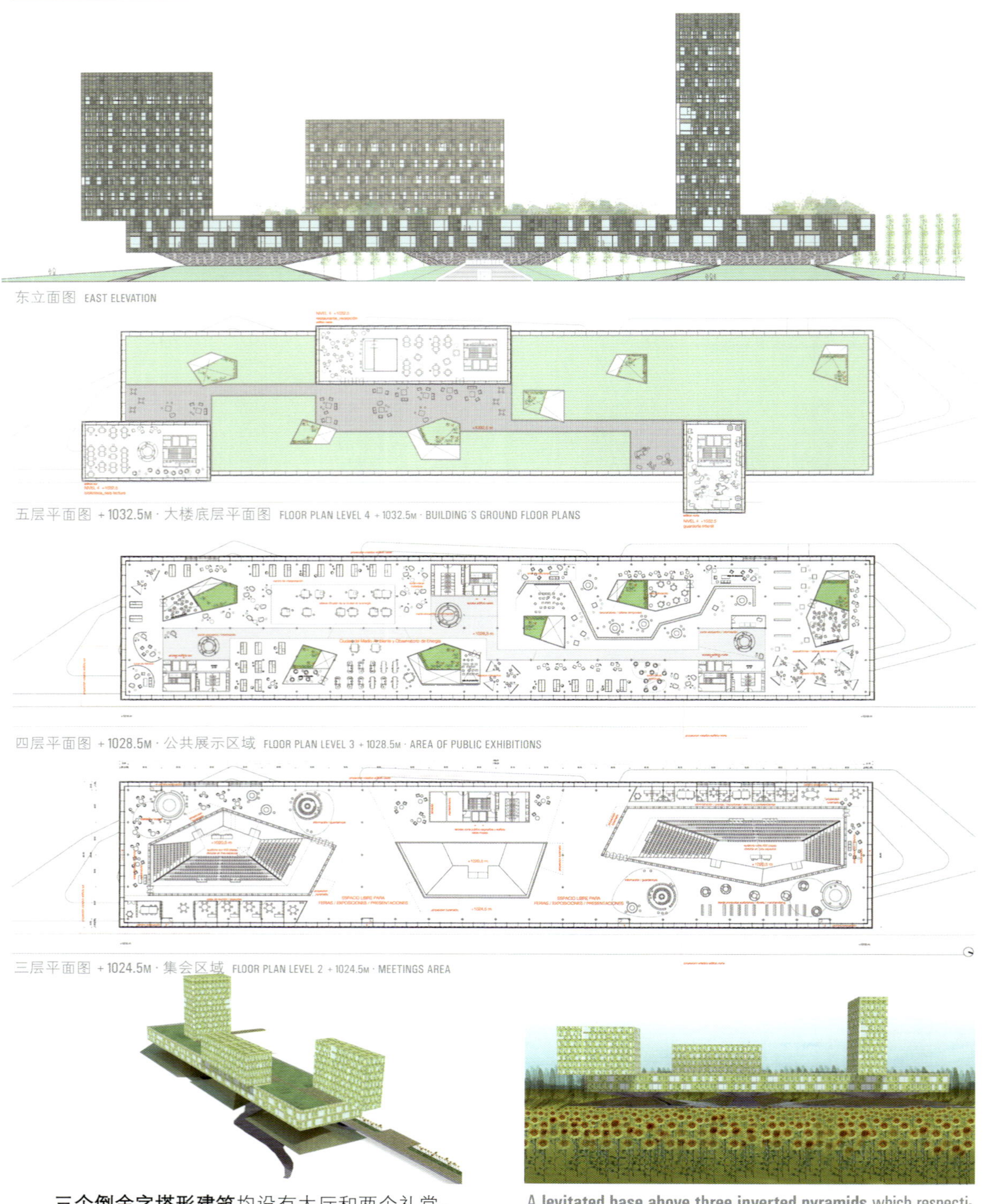

东立面图 EAST ELEVATION

五层平面图 +1032.5M · 大楼底层平面图 FLOOR PLAN LEVEL 4 +1032.5M · BUILDING'S GROUND FLOOR PLANS

四层平面图 +1028.5M · 公共展示区域 FLOOR PLAN LEVEL 3 +1028.5M · AREA OF PUBLIC EXHIBITIONS

三层平面图 +1024.5M · 集会区域 FLOOR PLAN LEVEL 2 +1024.5M · MEETINGS AREA

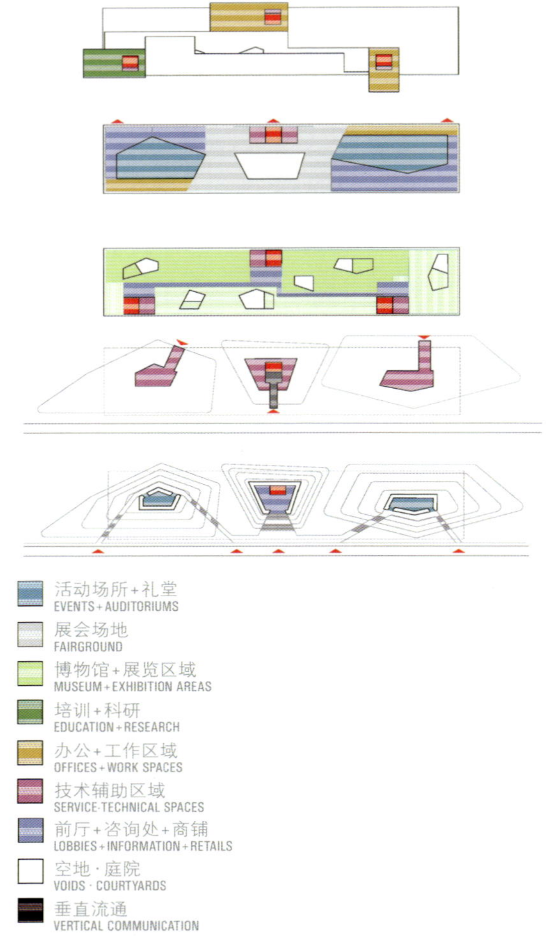


**三个倒金字塔形建筑**均设有大厅和两个礼堂，**其上方为提升型的基座**，包括展览中心、博物馆和日常公共活动项目。新基座上方的**三个区域**可鸟瞰四周，囊括了余下的所有项目。平面图上设有三个独特的公共项目（图书馆、幼儿园和餐厅），可作为两个体系间的过渡点。从整体规划图来看，其他区域均为模糊区。

A **levitated base above three inverted pyramids** which respectively house the lobby and two auditoriums, contains the exhibition center, museum and everyday public programs. **Three volumes** arising from the new benchmark emerge over the landscape and contain the rest of the programs. There are three communal programs of special character (library, nursery, restaurant) located on the floor plan which acts as contact between the two systems . The rest is programmatically neutral and undefined spaces.

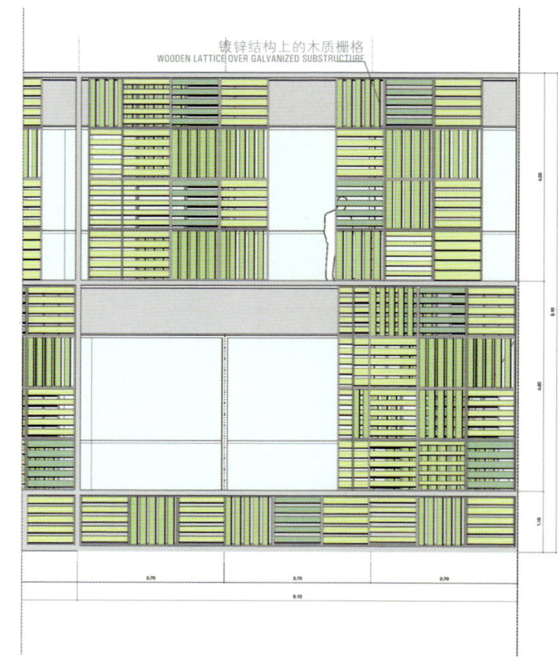


外墙细部 FAÇADE DETAIL

## 休斯敦，美国 Houston, USA

### Menil收藏馆总体规划
### The Menil Collection Master Plan

**2008**

国际邀标项目 Invited international competition

入围 Finalist

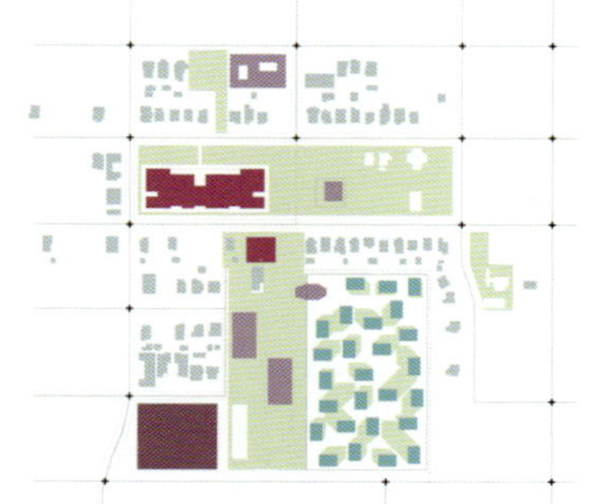

MENIL收藏馆
MENIL COLLECTION

原有附属建筑
EXISTING ANNEX

MENIL住宅
HOUSING MENIL

补充功能设施
COMPLEMENTARY USE

新附属建筑・画廊
NEW ANNEX · GALLERIES

新住宅・RICHMOND
NEW HOUSING · RICHMOND

新补充功能设施
NEW COMPLEMENTARY USE

原有绿地空间
EXISTING GREEN SPACE

新绿地空间
NEW GREEN SPACE

公共建筑
PUBLIC BUILDINGS

大街
STREET

私人空间
PRIVATE SPACE

停车场
PARKING

停车场 RICHMOND
RICHMOND APARTMENTS

公共空间
PUBLIC SPACE

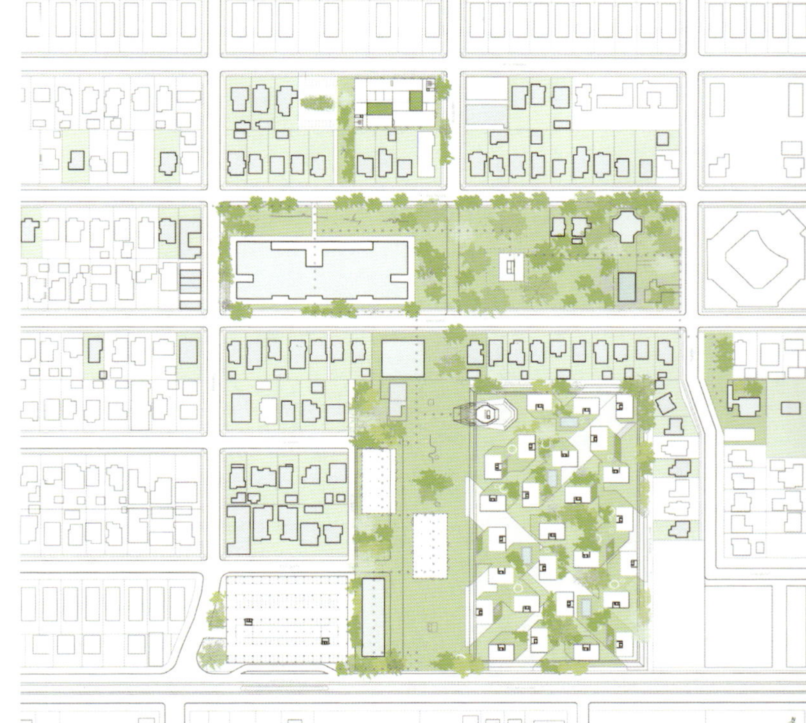

总分布图 OVERALL PLAN

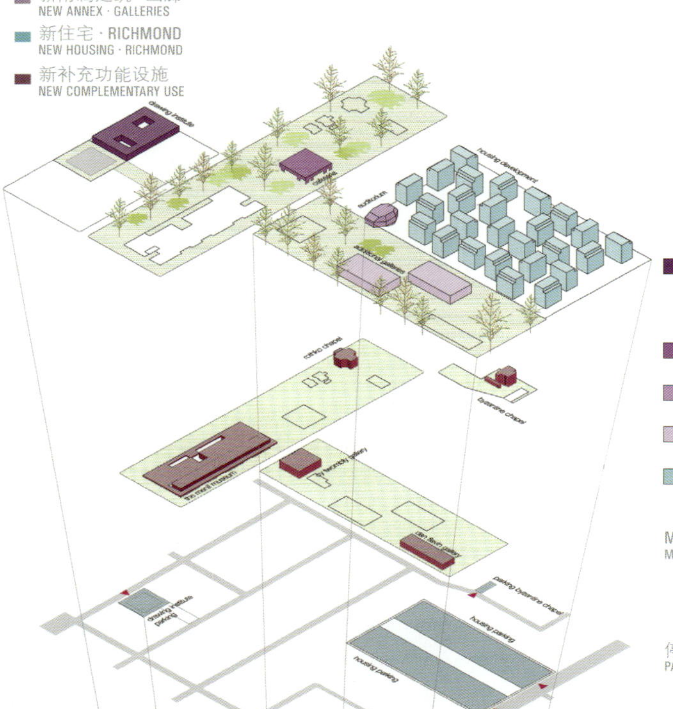

MENIL所有的建筑
MENIL PROPERTY BUILDINGS

总体规划已将Menil校园安谧的环境考虑在内，我们计划建造一个档案馆、礼堂、咖啡厅、图书馆和公寓。各个区域绿草如茵，一直从校园延伸至城市的主要街区，与公共区域、公园以及新项目融为一体，成为其有机的组成部分。文化、商业、社区和环境成为新Menil校园的代名词，现在Menil校园已然转变成一个适合人居的地方。

The sensitive and quiet spirit of Menil Campus pervades the Master Plan that wants to enrich its program with a drawing file, an auditorium, a cafe, a library and a residential complex. Everything inhabits a green carpet that extends the boundaries of the Campus to the main streets of the city looking for a hybrid image between public space, communal garden and new programs which colonize it as if it were a new species. Culture, commerce, society and environment are the equations of the new Menil, transformed now into a place to live.

住宅总分布图 OVERALL HOUSING PLAN

**"Yesid Santos"新排球馆**
New Volleyball Coliseum "Yesid Santos"

**2008**
国际公开竞标 Open international competition
二等奖 Second Prize

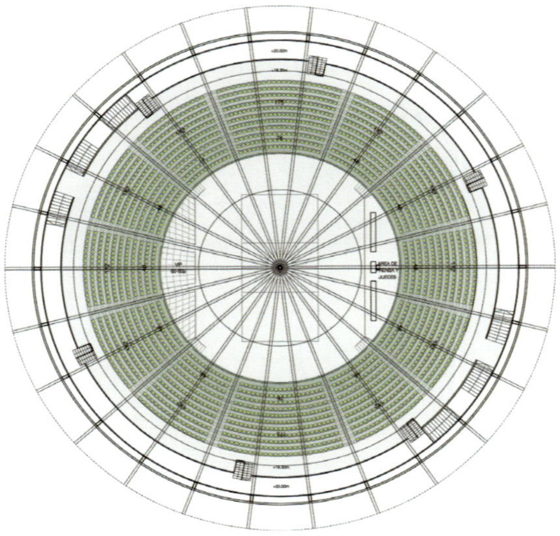

平面图标高 +20.55 FLOOR PLAN LEVEL +20.55

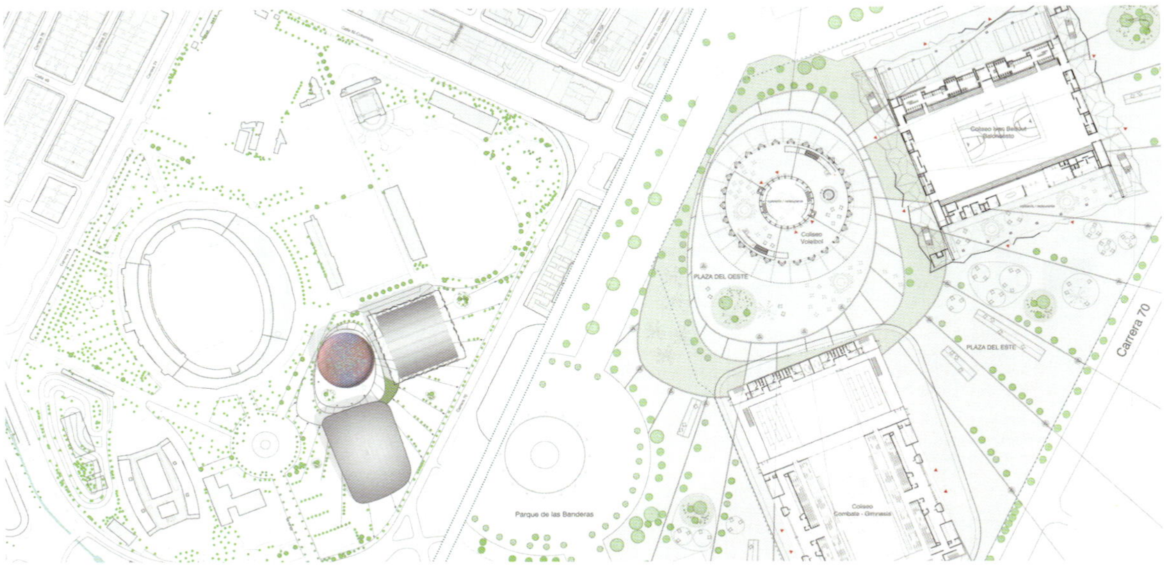

地块位置 SITE PLAN

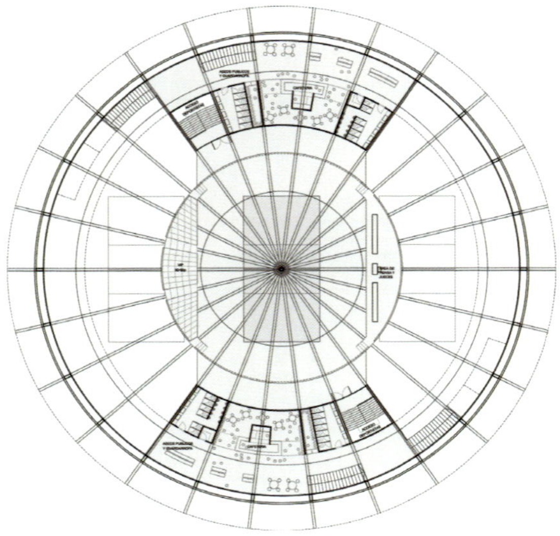

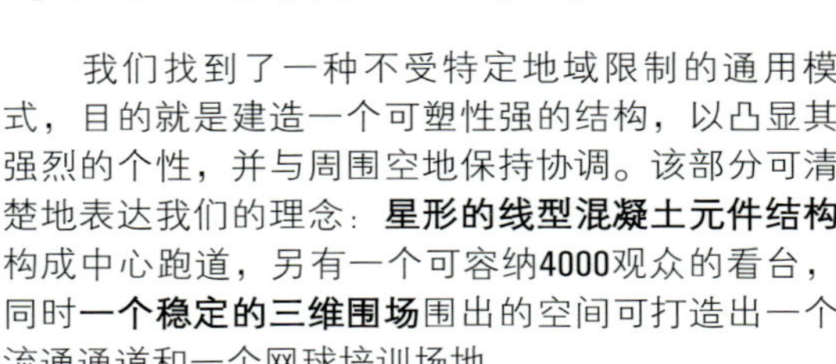

我们找到了一种不受特定地域限制的通用模式，目的就是建造一个可塑性强的结构，以凸显其强烈的个性，并与周围空地保持协调。该部分可清楚地表达我们的理念：**星形的线型混凝土元件结构**构成中心跑道，另有一个可容纳4000观众的看台，同时**一个稳定的三维围场**围出的空间可打造出一个流通通道和一个网球培训场地。

剖面图 SECTION

We have looked for a generic form not committed to the place where it is settled. The aim is to build a plastic object which shows its monumental character providing scale to the surrounding esplanades. The section clearly explains our concept: **a radial structure of linear concrete elements** shape the central circular track and the stands for 4,000 spectators, while **a resistant three-dimensional enclosure** creates the space for circulations and tennis training courts.

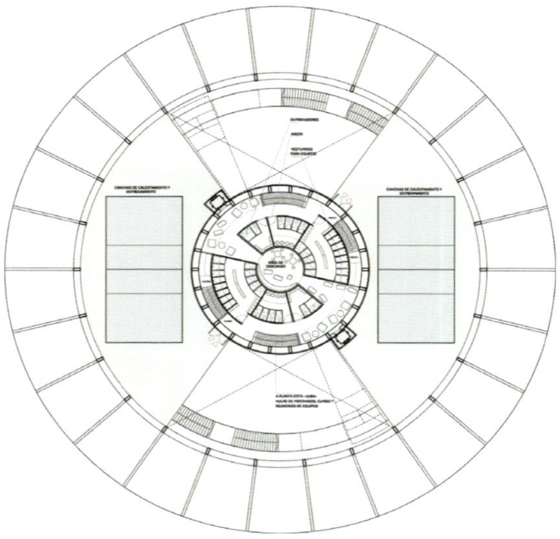

平面图标高 +12.30 FLOOR PLAN LEVEL +12.30

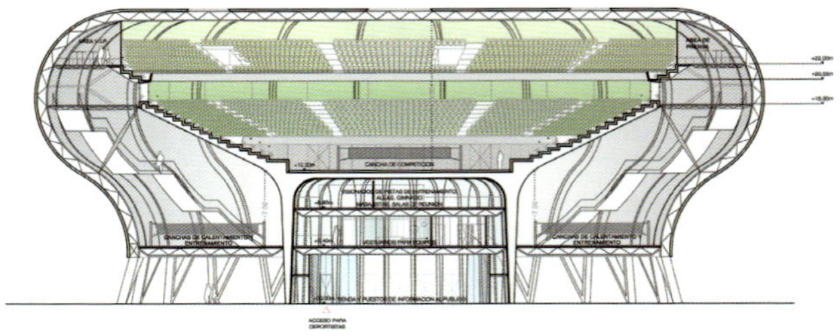

立面图 ELEVATION

平面图标高 +5.40 FLOOR PLAN LEVEL +5.40

屋顶层平面图 ROOF PLAN

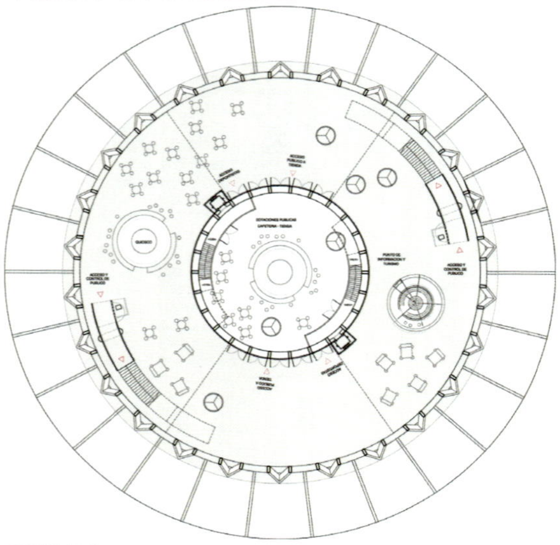

平面图标高 +0.00 FLOOR PLAN LEVEL +0.00

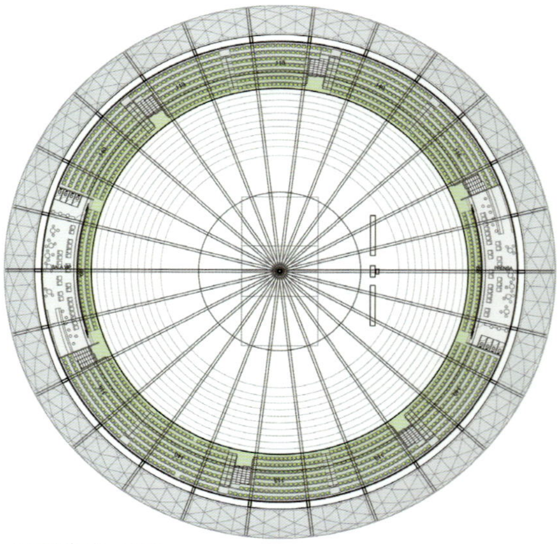

平面图标高 +22.00 FLOOR PLAN LEVEL +22.00

## Heysel社区总体规划
## Heysel Neighbourhood Masterplan

**2010**

国际邀标项目 Invited international competition

地块位置 SITE PLAN

公共空间系统 SYSTEM OF PUBLIC SPACES

自由公共空间 OPEN PUBLIC SPACES

社区单元的主要结构 MAIN STRUCTURE OF URBAN CELLS

剖面图 1 SECTION 1

剖面图 2 SECTION 2

- 商业 COMMERCIAL
- 会议 CONVENTIONS
- 商务公园 BUSINESS PARK
- 演出 SHOWS
- 办公室 OFFICES
- 住宅 HOUSING
- 娱乐场点 ENTERTAINMENT POLE
- 酒店 HOTEL
- 绿地 GREEN
- 停车场 PARKING
- 体育场 STADIUM

目前Heysel地区是一边缘社区。该设计方案为其连接整个城市提供了一个极好的机会，其住宅、办公、商业项目将成为城市结构中最具活力的元素，而困难之处在于测量建筑数量，以达到城市“密度”项目和“主”项目平衡共处的状态：3500席位的会议中心、1.5万席位的剧院、城市级规模的运动场和10万平方米的商业地块等。

Heysel district is currently perceived as peripheral neighbourhood. The approach offers a great opportunity to generate a complete fragment of city in which residential, office and commercial programs are the real dynamic substance of its urban structure. The difficulty is then to measure the "amount of architecture" required achieving a coexistence that balances urban "dense" programs with the "main" programs requested: A Conference Centre of at least 3,500 seats, a Theater for 15,000 spectators , a Sports Pole of urban scale, a Commercial Pole up to 100,000 m², etc.

## 塔拉哥纳，西班牙 Tarragona, Spain

**"Tierras del Ebro"新医院**
New Hospital of "Tierras del Ebro"

**2010**
公开概念竞标 Open ideas competition

我们的目标是建造一个呈矩形阵列结构的建筑，用其的简易性打造所含项目的多样性，并为未来的建筑改造提供机会。矩阵内的所有区域可以在不影响整体结构的情况下进行修缮，但最重要的是布局清晰明朗。清晰的布局便于使用，且益于表达建筑的内涵。该方案设计是基于一系列的循环拱肋，并在各循环拱肋之间打造出一个个纵向延伸的空间，**将由中庭隔开的分散建筑连接起来。**

The goal is based on building a matrix which uses its simplicity to support the variety of programs and opportunities for future changes. All areas of this matrix should have the possibility to be refurbished without affecting the whole complex. But most important will always be the clarity of its layout. A clarity will result in ease of use and interpretation of the building by all types of users. This clarity is based on a series of circulation ribs and longitudinal facilities that host in between, **programmatic strips shaping autonomous buildings separated by courtyards.**

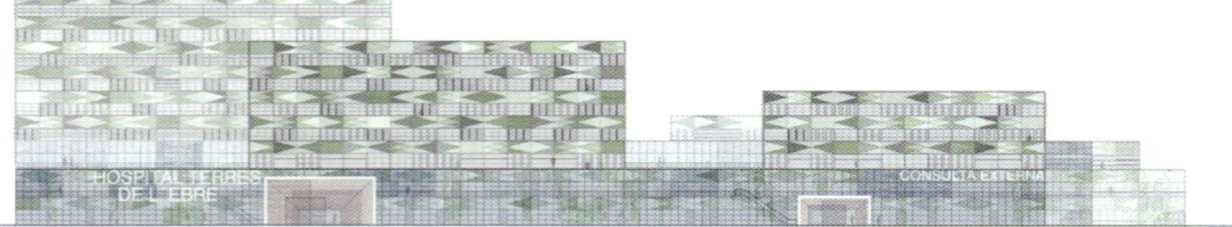

立面图 ELEVATION

四层平面图 FLOOR PLAN LEVEL 3

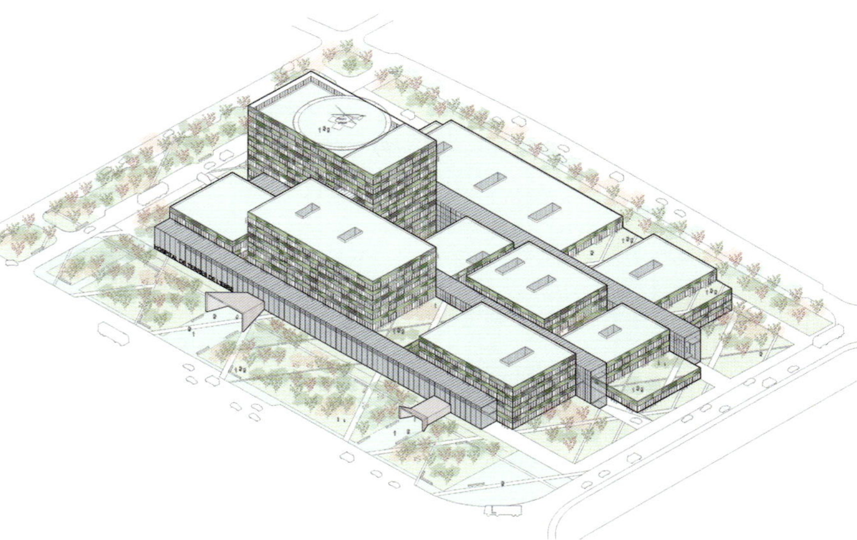

地块位置 SITE PLAN

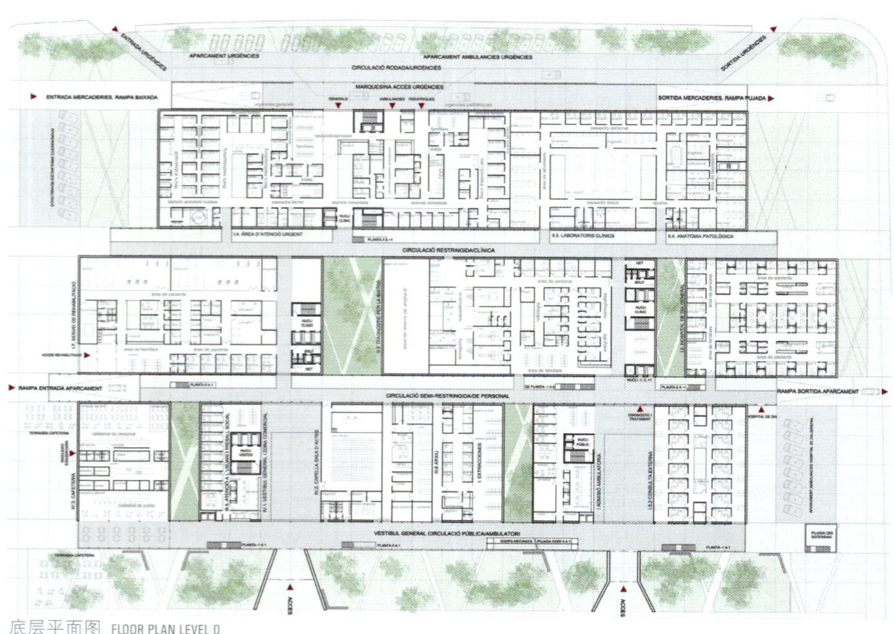

底层平面图 FLOOR PLAN LEVEL 0

剖面图 SECTIONS

## 新展览及会议大楼
## New Congress, Exhibitions and Fairs Palace

**2010**
国内公开竞标 Open national competition

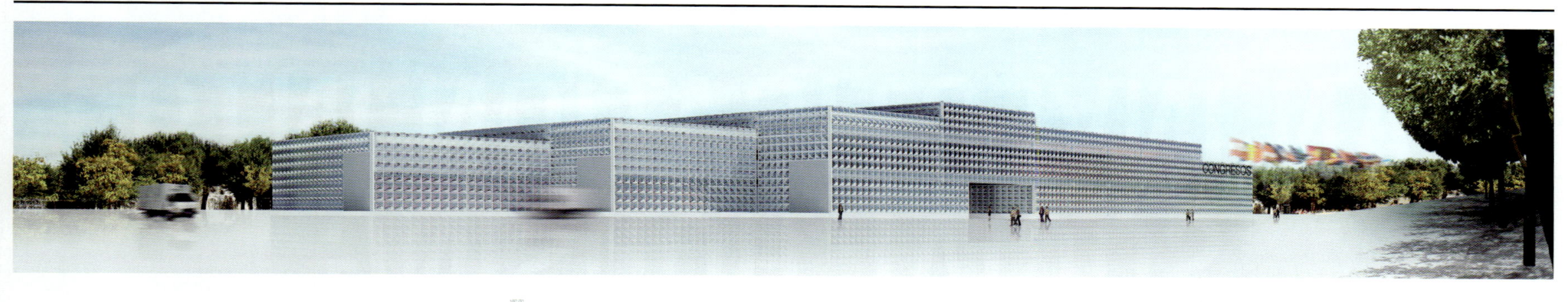

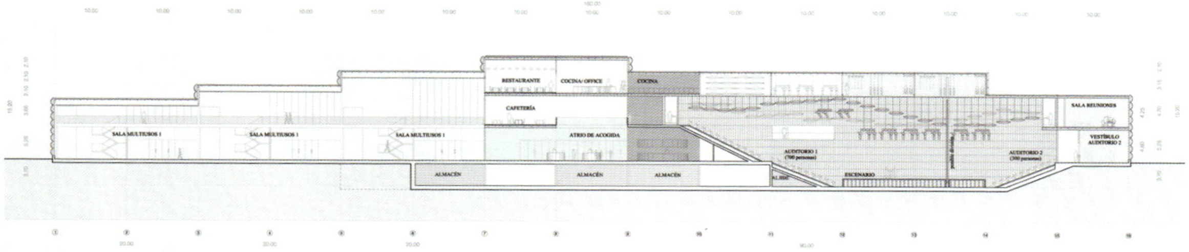

纵向剖面图 LONGITUDINAL SECTION

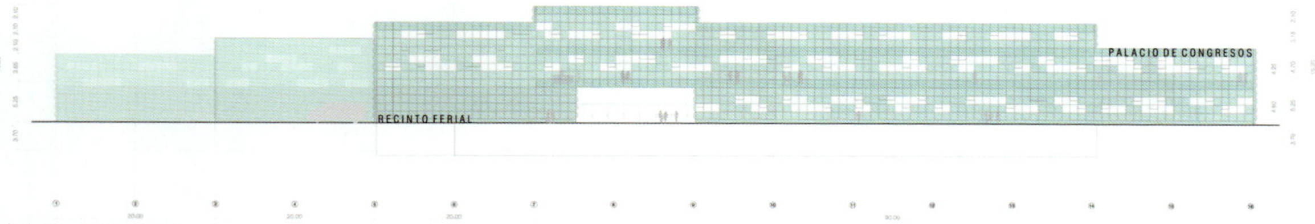

南立面 SOUTH ELEVATION

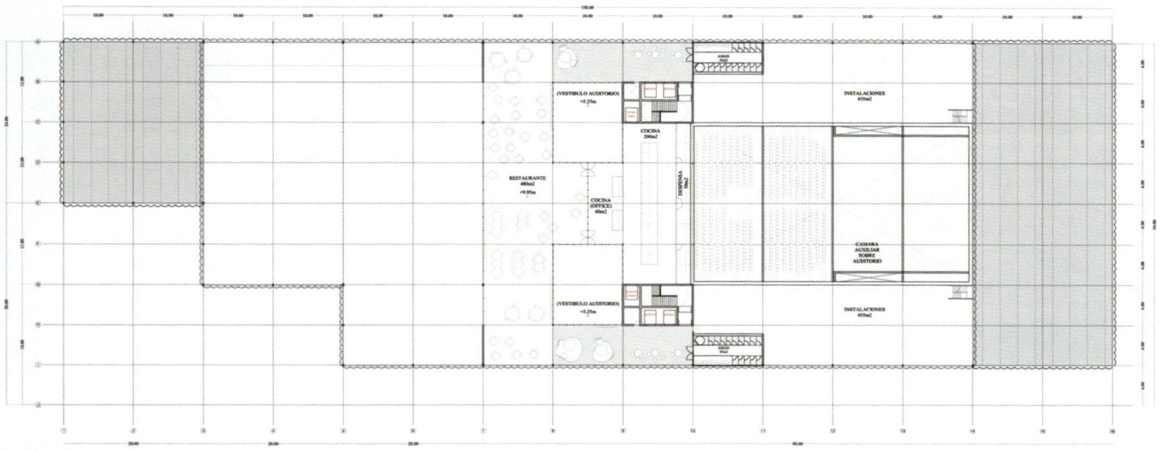

三层平面图 FLOOR PLAN 2

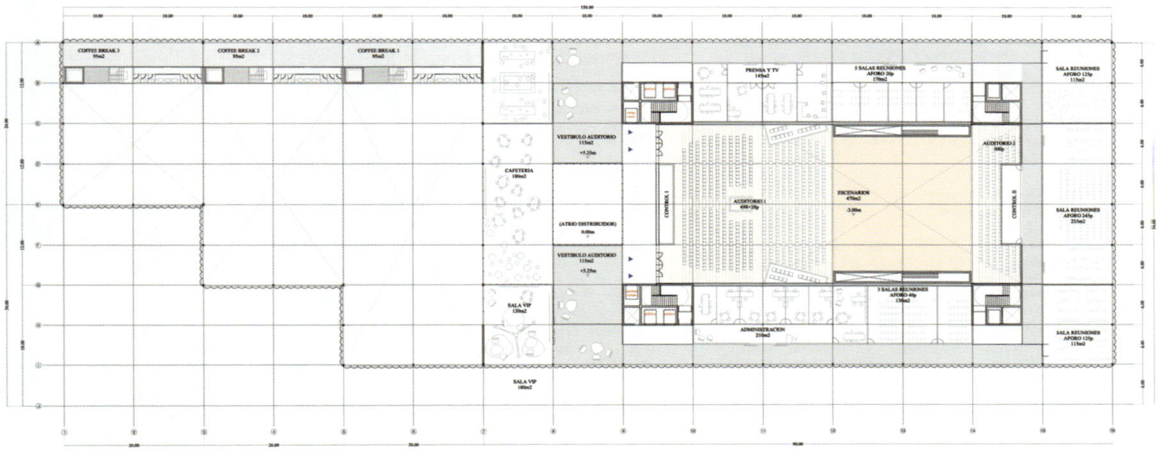

二层平面图 FLOOR PLAN 1

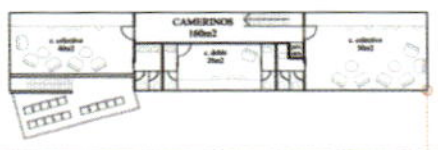

夹层平面图 MEZZANINE

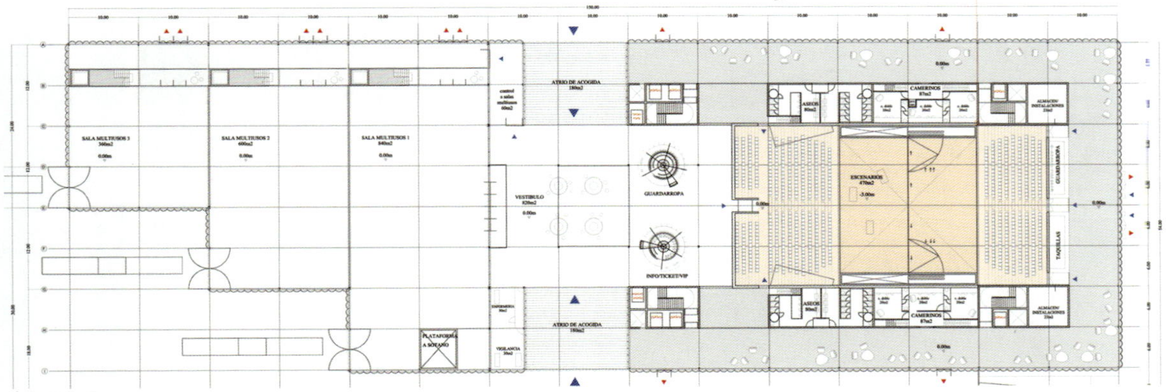

底层平面图 GROUND FLOOR PLAN

从地理位置来看，不可能完全利用整个地块;从经济角度来看，更适合建造独幢建筑，因此，我们建议**向东西方向扩展，使其更加紧凑**。你可能会发现，我们在建筑上方打造了几何图形的建构，使用了柔和的材料，因为这样更符合自然光的要求，并着力让其显得雅致、协调而不突兀，看起来就像一幢“既成建筑物”。

The shape of the plot, the impossibility of a total occupation and the interpretation that the economy will negotiate better with a single building, leads us to propose **the building extended at east-west direction and compact**. You will notice that some geometric resources that alter its top, the use of kind materials which react to natural light and the search for an effect of a "laid building", try to naturalize its presence making it friendly and elegant.

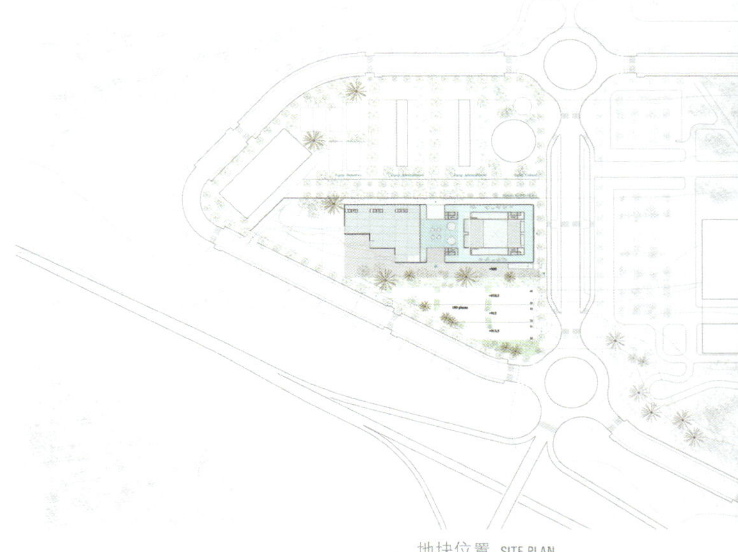

地块位置 SITE PLAN

# 经济适用房考察
# AN INVESTIGATION INTO SOCIAL HOUSING

## 11个竞标项目
## 11 COMPETITIONS

马德里正在着手建造迄今为止世界上最大的社会保障住房项目”

“Madrid is developing one of the most ambitious social housing programs that exist nowadays worldwide”

**建筑与新社区**

对于大多数城市而言，建造经济适用房让人想到的只是大量的普通房子，并未考虑居住者的实际需求和环境因素。但在马德里，却是个例外。目前马德里正在组织公开竞标，以期与国内外顶尖的建筑师共同建造社会保障性住房的“公共画廊”。马德里土地住房局拥有“城市创意实验室”，其推出的创新想法，有助于建造集美学、可持续性、功能性于一体的社会保障性住房。

**ARCHITECTURE AND NEW NEIGHBOURHOODS**

For most cities, the need to create affordable housing would mean erecting mass quantities of mediocre housing stock without much regard to the real needs of the occupants or the environment. Not so for Madrid, has been organizing open competitions to work with some of the best national and international architects to create what can only be described as an open gallery of social architecture. The "Municipal Land and Housing Authority" of the City of Madrid has a Laboratory for Urban Ideas, from where it launches innovative ways to build social housing buildings where aesthetics, sustainability, and functionality are top priorities.

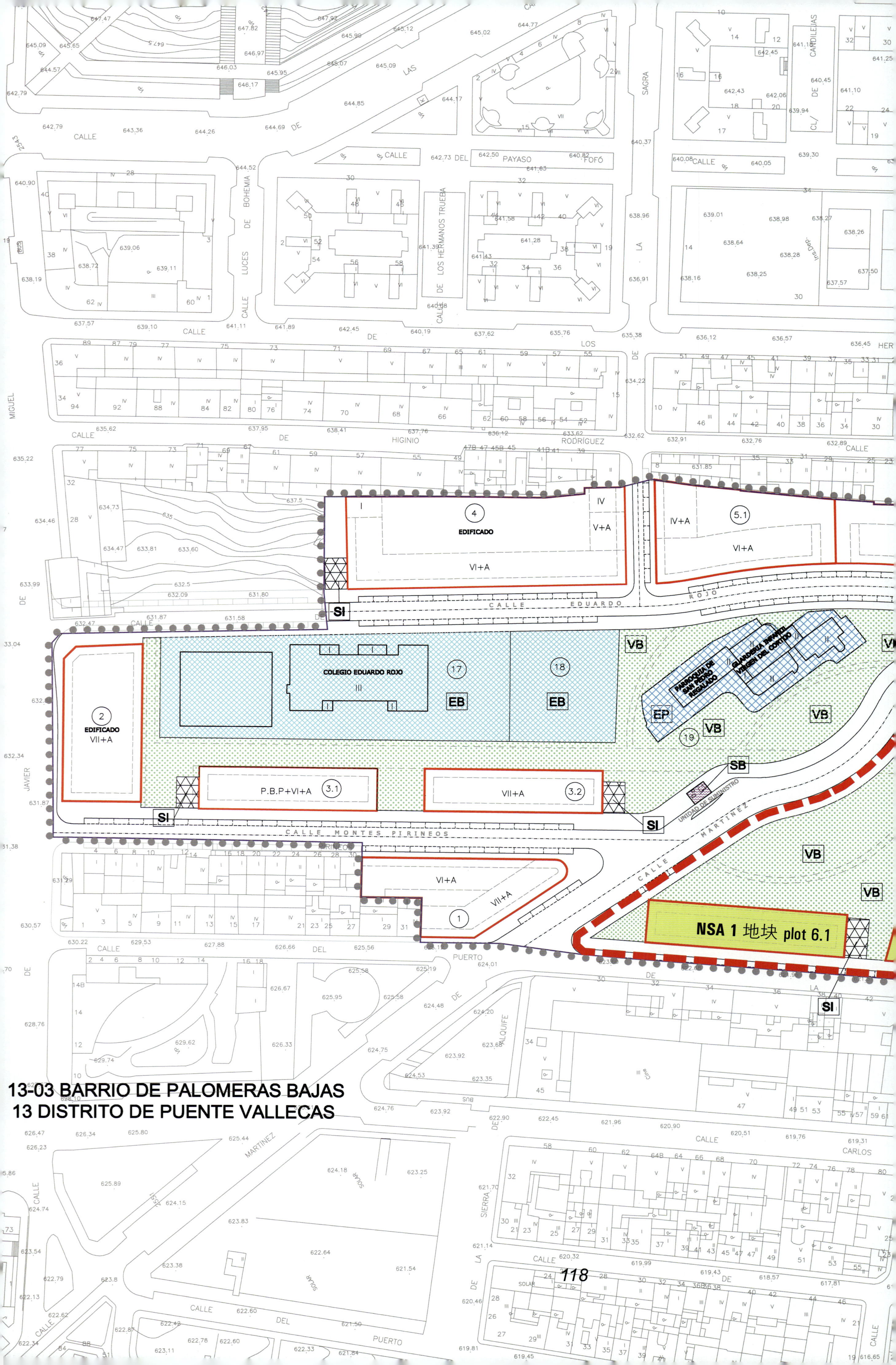
13-03 BARRIO DE PALOMERAS BAJAS
13 DISTRITO DE PUENTE VALLECAS
NSA 1 地块 plot 6.1
COLEGIO EDUARDO ROJO
EDIFICADO
CALLE EDUARDO ROJO
CALLE MONTES PIRINEOS
CALLE MARTINEZ
PARROQUIA DE SAN PEDRO REGALADO
GUARDERIA INFANTIL VIRGEN DEL CORTIJO
UNIDAD DE SUMINISTRO

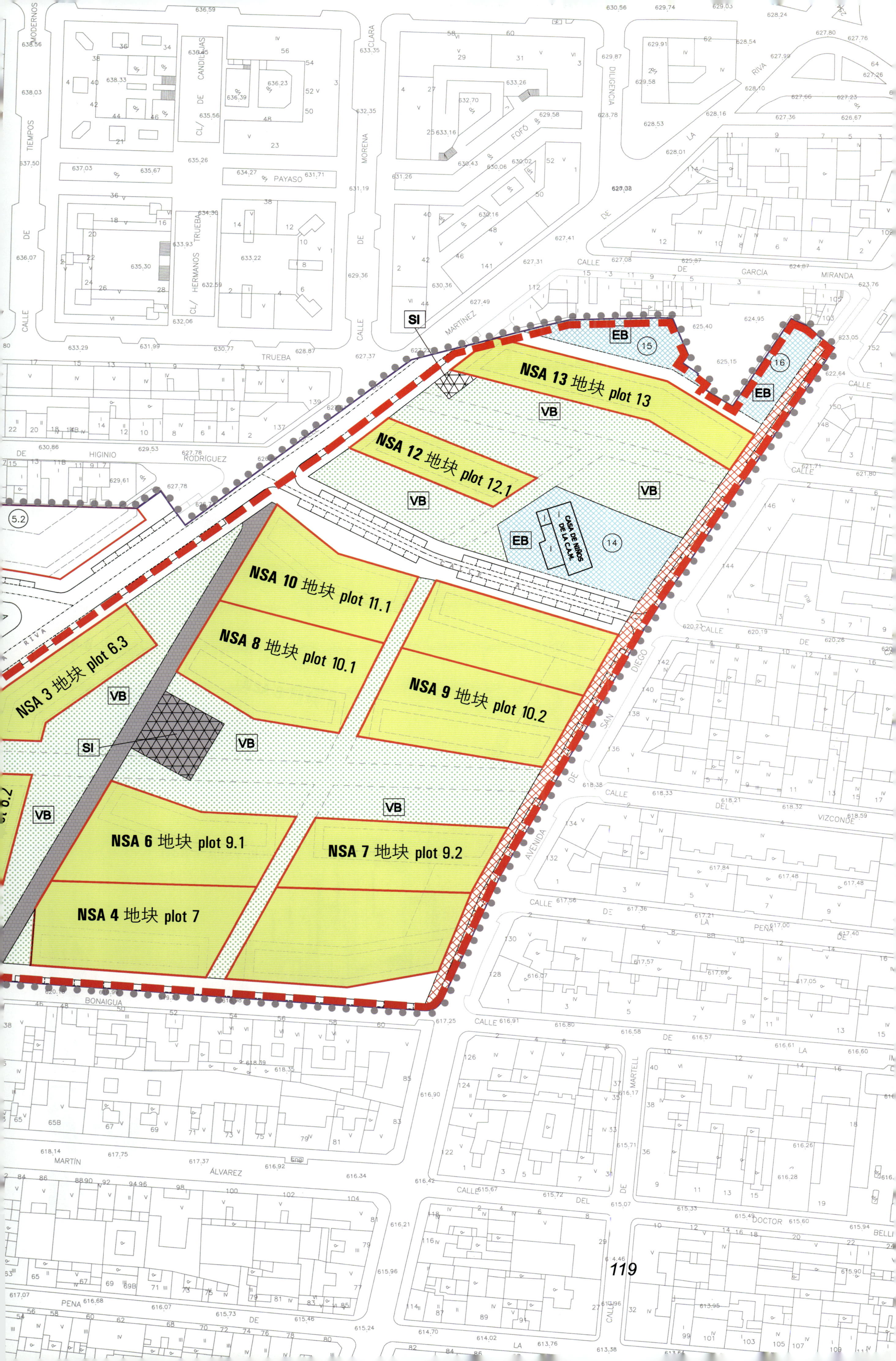
NSA 13 地块 plot 13
NSA 12 地块 plot 12.1
NSA 10 地块 plot 11.1
NSA 8 地块 plot 10.1
NSA 9 地块 plot 10.2
NSA 3 地块 plot 6.3
NSA 6 地块 plot 9.1
NSA 7 地块 plot 9.2
NSA 4 地块 plot 7
VB
EB
SI
CASA DE NIÑOS DE LA C.A.M.
CALLE DE GARCÍA MIRANDA
AVENIDA DE SAN DIEGO
CALLE DEL VIZCONDE
CALLE DE LA PEÑA
MARTÍN ÁLVAREZ
CALLE DEL DOCTOR
CALLE DE HIGINIO RODRÍGUEZ
TRUEBA
CL/ HERMANOS TRUEBA
CL/ DE CANDILEJAS
PAYASO
FOFÓ
MARTÍNEZ
CALLE DE MORENA
CLARA
DILIGENCIA
RIVA
BONAIGUA
MARTELL
CALLE DE TIEMPOS MODERNOS

## camacho-macia arquitectos (Javier Camacho) (建筑师)

合作 (c) María Eugenia Maciá · María Navascués · Alvaro Rivera · Juan Santana

中标 winner

正立面 MAIN ELEVATION

# 双层结构

为了使外墙更具表现力，我们决定在双层之间打造窗式阳台，作为房子的外观。

–内层表面：在天花板和楼板、不透明金属之间使用玻璃隔断。

–外层表面：使用铝板型折叠式百叶窗，具有更灵活的结构，使临街的外层达到所期望的动感。

# DOUBLE SKIN

In order to enhance the expressiveness of the new front, we choose to generate windowed balconies between two skins which generate the façade:

-Interior skin: glazed walls from floor to ceiling and opaque metal.

-External skin: folding shutters made by aluminium plates, with a more flexible layout, generating a desired vibration of the street façade.

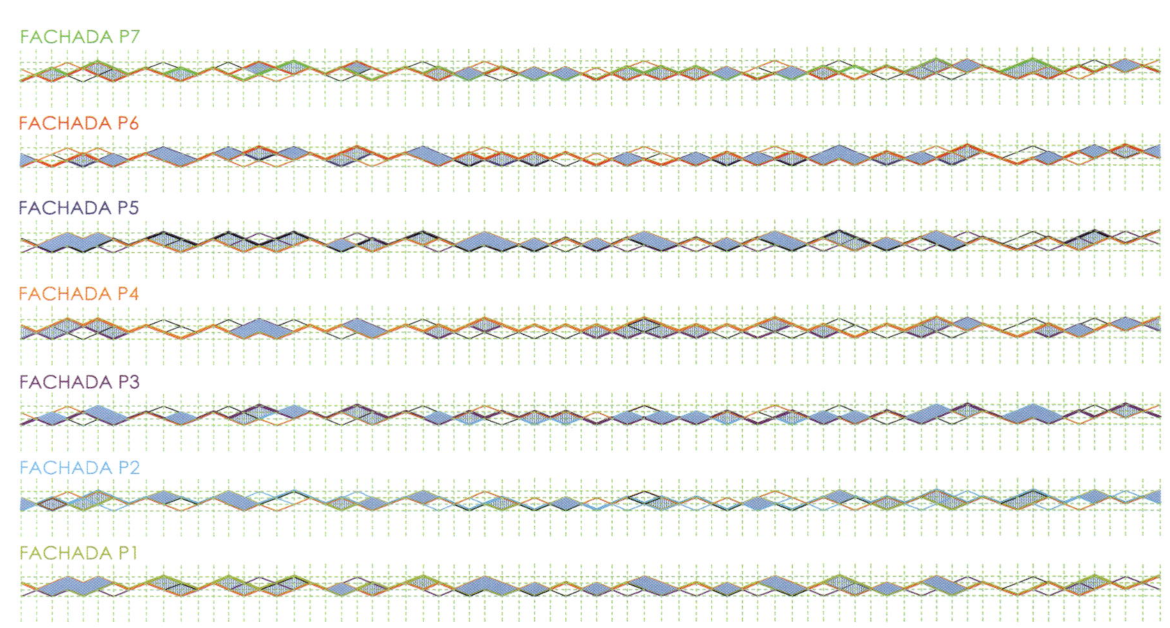


外立面组成 FORMATION OF THE FAÇADE

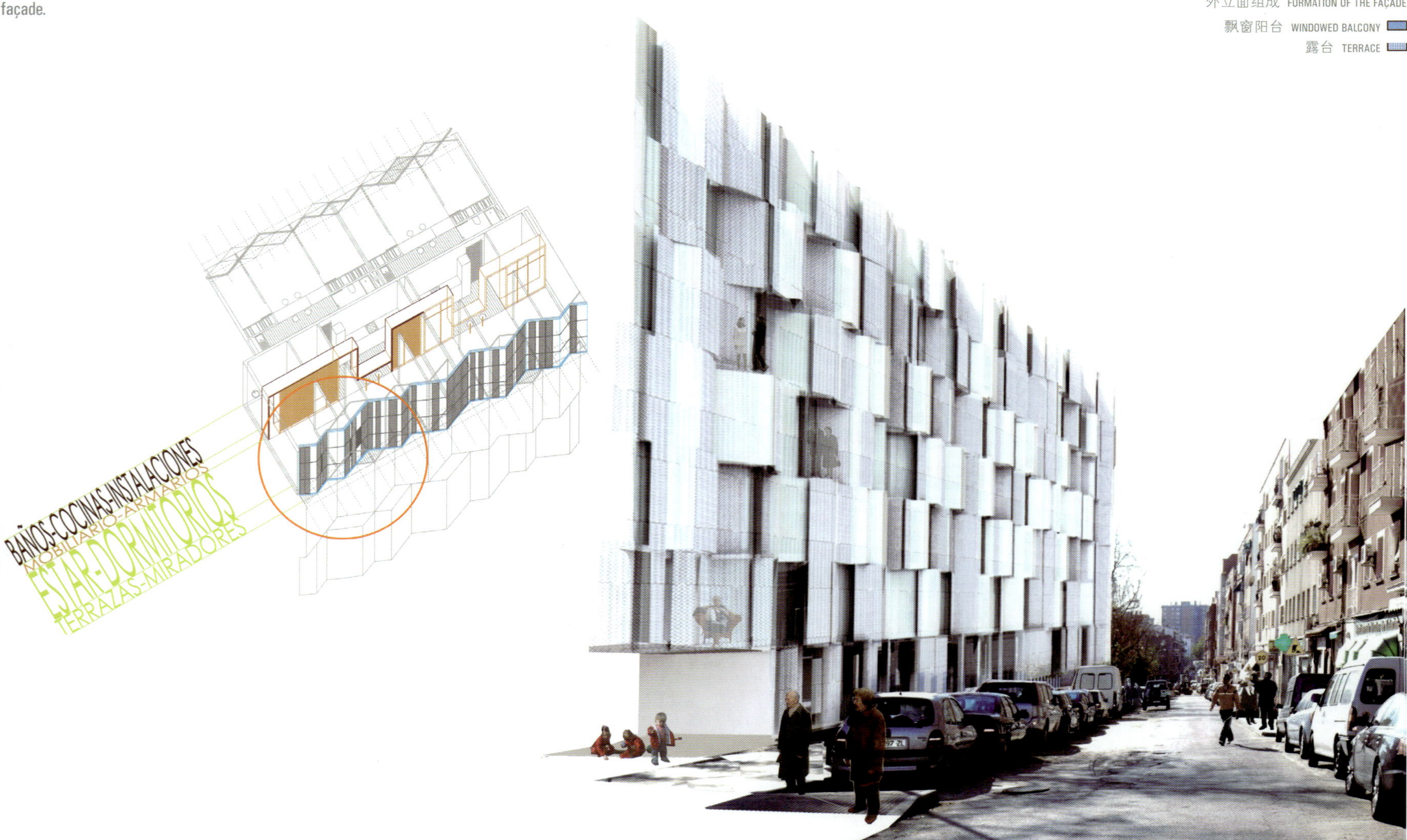

阁楼 PENTHOUSE
6
5
4
3
2
1
0
-1
-2
阁楼层平面图 PENTHOUSE FLOOR PLAN
1-ACCESO VIVIENDAS. 2-PORTALES-BUZONES. 3-SOPORTALES ABIERTOS. 4- PATIO INTERIOR. 5-COMUNICACIONES VERTICALES. 6- ACCESO GARAJES. 7-RAMPAS DE ACCESO A GARAJE. 8-PORTERÍA.
PLANTA ENTORNO. E: 1/300
四层平面图 FLOOR PLAN 3
底层平面图 GROUND FLOOR PLAN
飘窗阳台 WINDOWED BALCONY
露台 TERRACE
流通核心 COMMUNICATION CORE
家具摆设 STRIP OF FURNITURE
潮湿空间带 STRIP OF HUMID SPACES
3卧室单元房 HOUSING UNITS 3 BEDROOMS
2卧室单元房 HOUSING UNITS 2 BEDROOMS
1卧室单元房 HOUSING UNITS 1 BEDROOM

# FRPO Rodriguez & Oriol + COMA Arquitectura (建筑师事务所)

Pablo Oriol · Fernando Rodríguez · Constanza Temboury · Mara Sanchez Llorens (建筑师)

第一提名奖 first mention

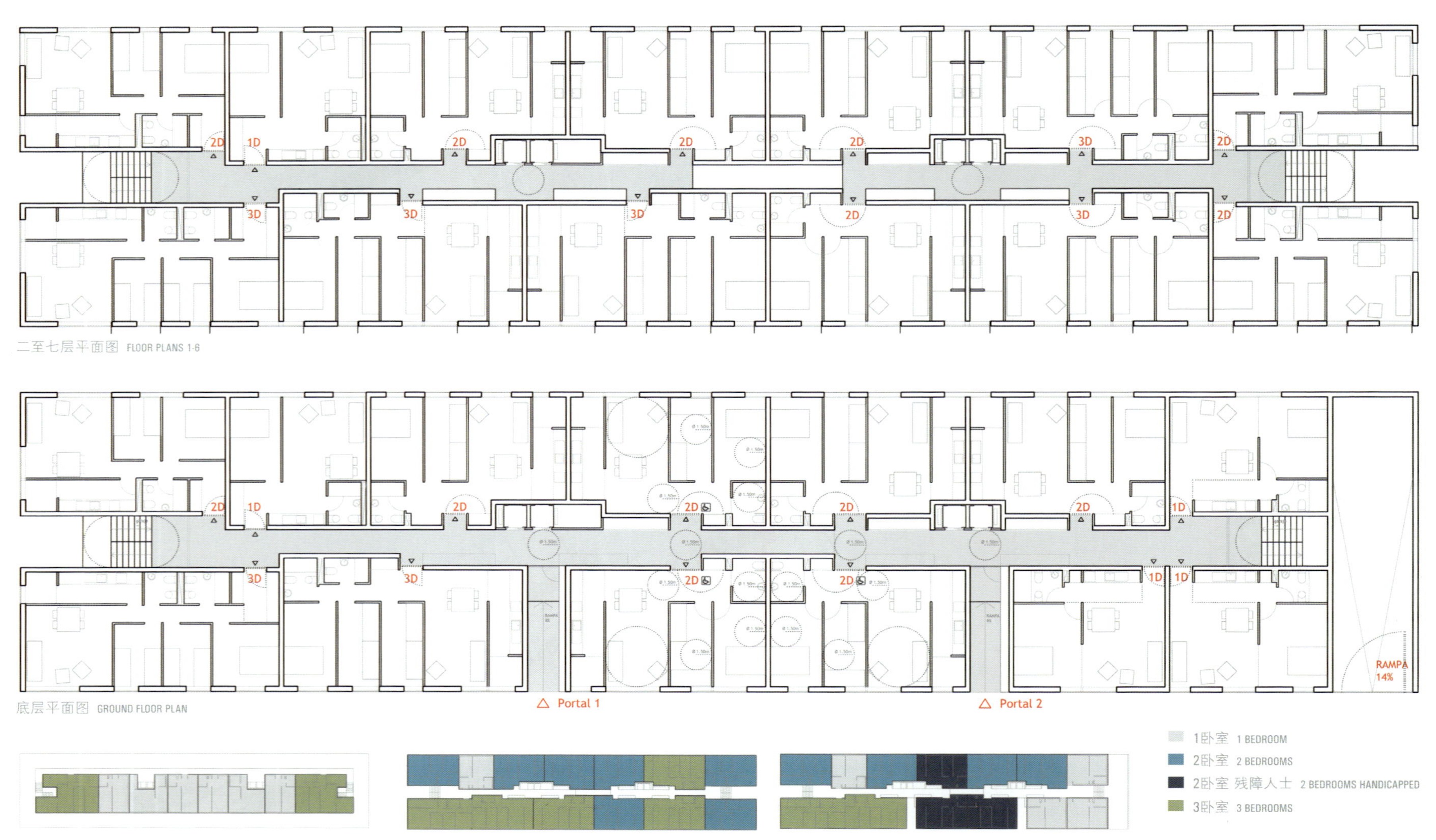


二至七层平面图 FLOOR PLANS 1-6

底层平面图 GROUND FLOOR PLAN

1卧室 1 BEDROOM
2卧室 2 BEDROOMS
2卧室 残障人士 2 BEDROOMS HANDICAPPED
3卧室 3 BEDROOMS

阁楼层 PENTHOUSE

标准层 TYPICAL FLOOR PLAN

底层 GROUND FLOOR PLAN

## 两地块及其间的空隙

我们将原来15米的地块切割成两块纵向区域（宽6.5米），以减少侧墙的面积，同时有助于室内单元和其他区域形成更好的布局，随外表面呈流线型延伸开去。两部分之间形成空隙，可作为各个单元之间的衔接流通区域。

## TWO PIECES AND A GAP

We transform the original 15-metre block into two longitudinal pieces of 6.5 meters width, reducing the scale of the side walls and allowing a clear organization of housing units and all its spaces abroad, following a linear configuration of spaces throughout the façade. The shift of both parts creates a gap which houses the circulations, communications and access to the different housing units.

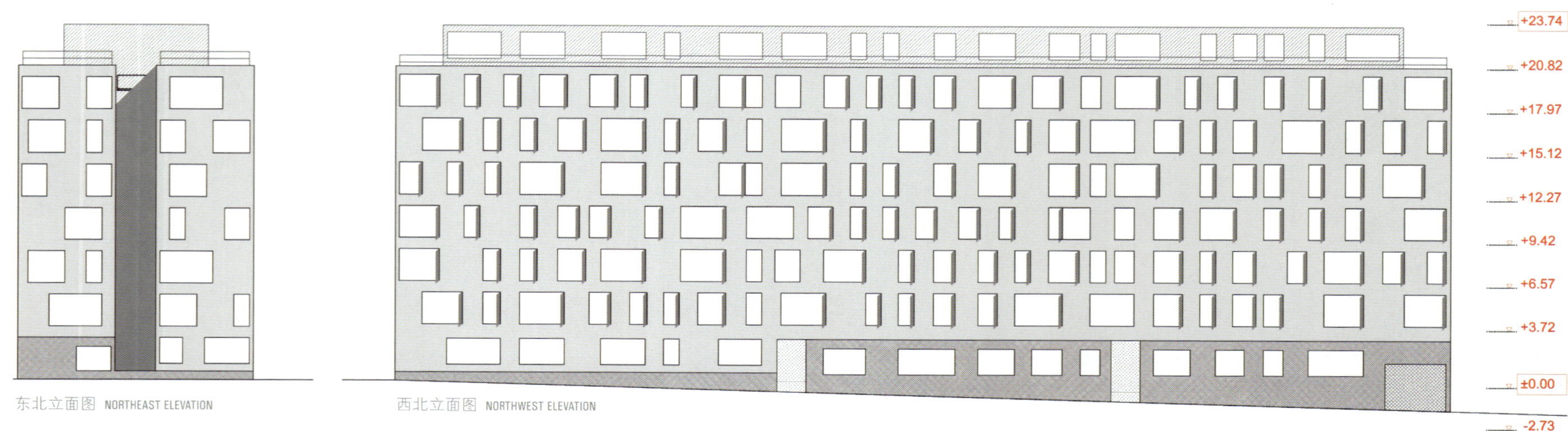


东北立面图 NORTHEAST ELEVATION

西北立面图 NORTHWEST ELEVATION

透视 PERSPECTIVE 1

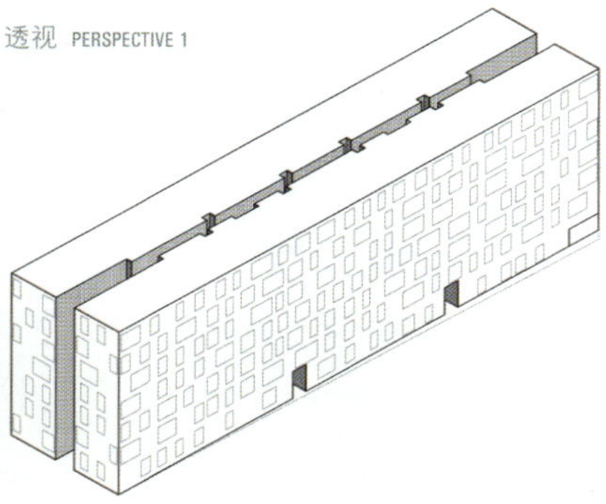

透视 PERSPECTIVE 2

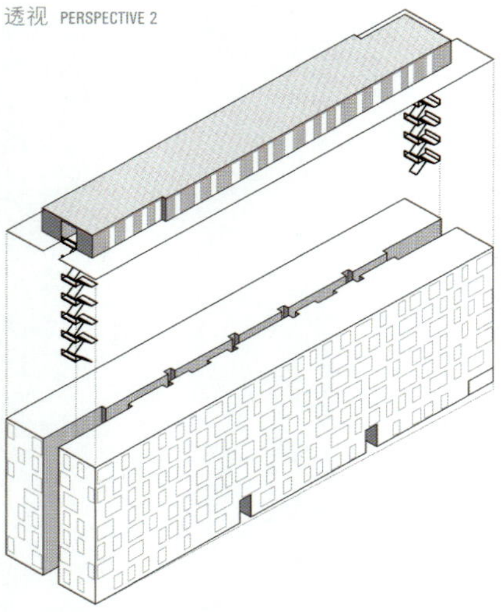

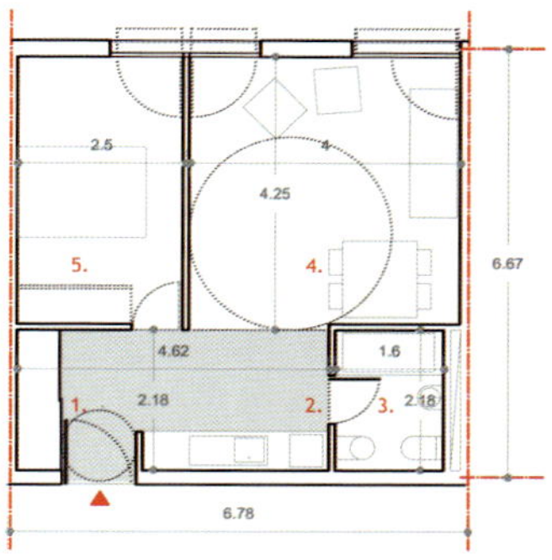

1卧室单元房 HOUSING UNIT 1 BEDROOM 40.65M2

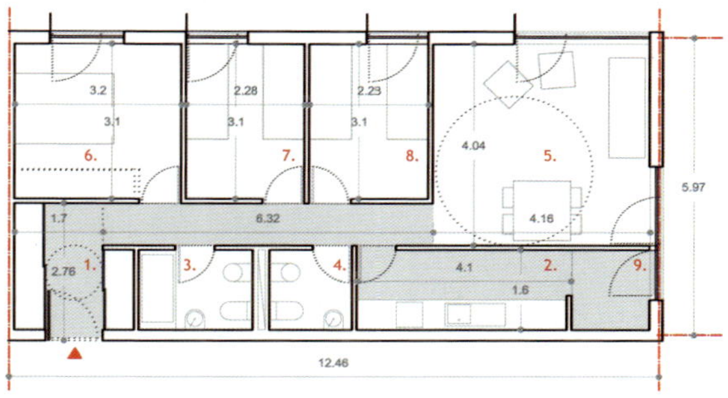

3卧室单元房 HOUSING UNIT 3 BEDROOMS 64.44M2

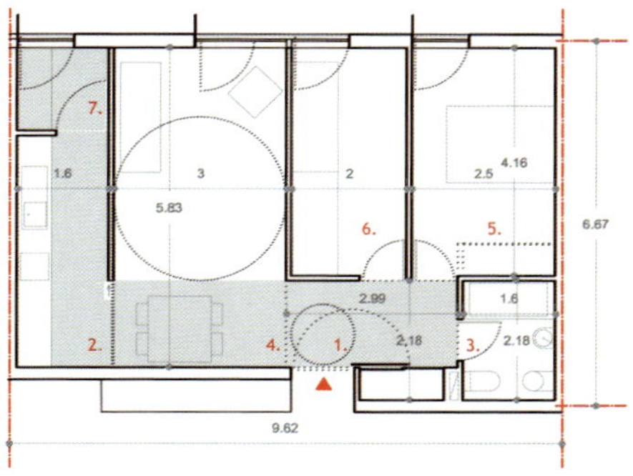

2卧室单元房 HOUSING UNIT 2 BEDROOMS 52.35M2

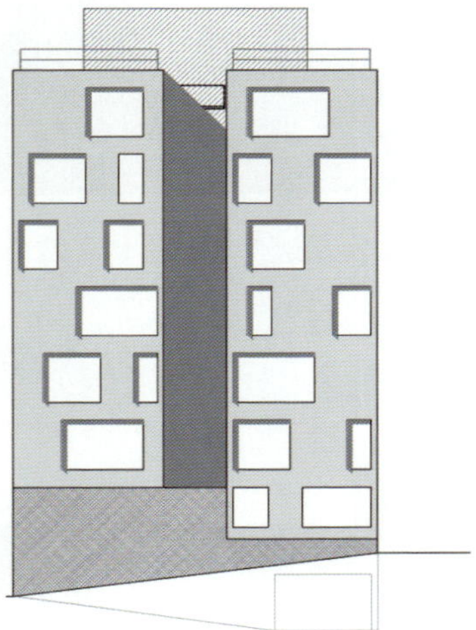

+23.74
+20.82
+17.97
+15.12
+12.27
+9.42
+6.57
+3.72
±0.00
-2.73

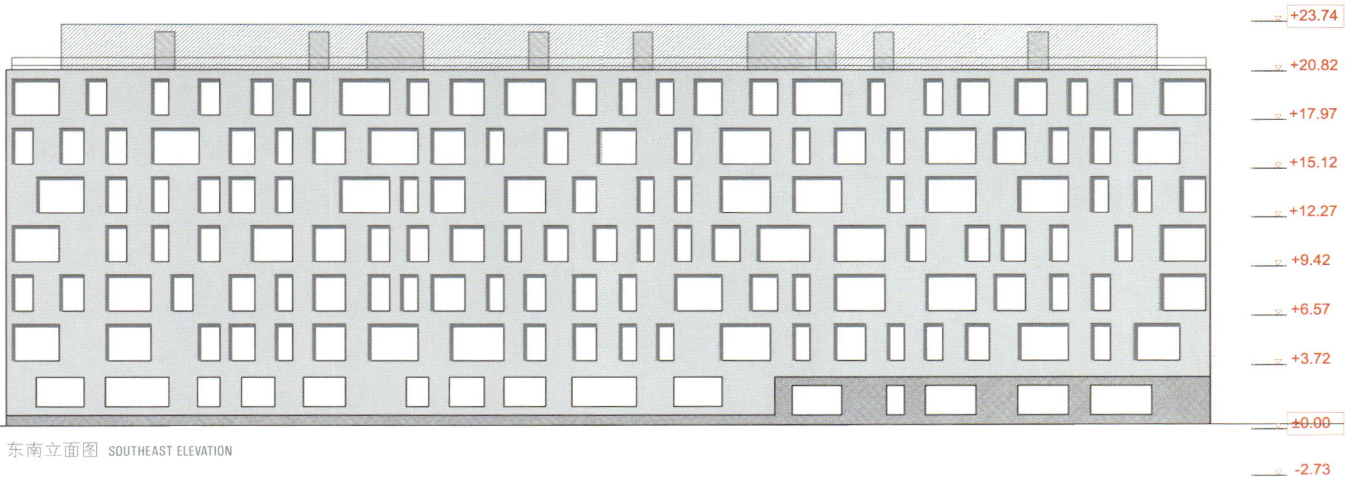

东南立面图 SOUTHEAST ELEVATION

西南立面图 SOUTHWEST ELEVATION

VERMREV

# Fernando Araujo Fuster · Ana Dolado Cosín (建筑师)

合作 (c) Nuria Heras Donoso

第二提名奖 second mention

## 室内的“大自然”

地面层采用透明结构，使其与主干道及周围的绿地形成通透结构。连续性的表层结构依地块而建，向后折叠，形成两个庭院。同时，将边界最大化，以取得最佳的通风和采光效果，保证住房项目气候方面的舒适性。

## NATURE INSIDE THE BUILDING

We release the ground floor to create a visual permeability between the main street and green spaces across the plot. The continuous skin-façade adjusts to the edge of the plot, folds back on itself, creating two courtyards. At the same time, it multiplies its edges looking for the best ventilation and sunlight, to ensure climatic comfort on the strict housing program.

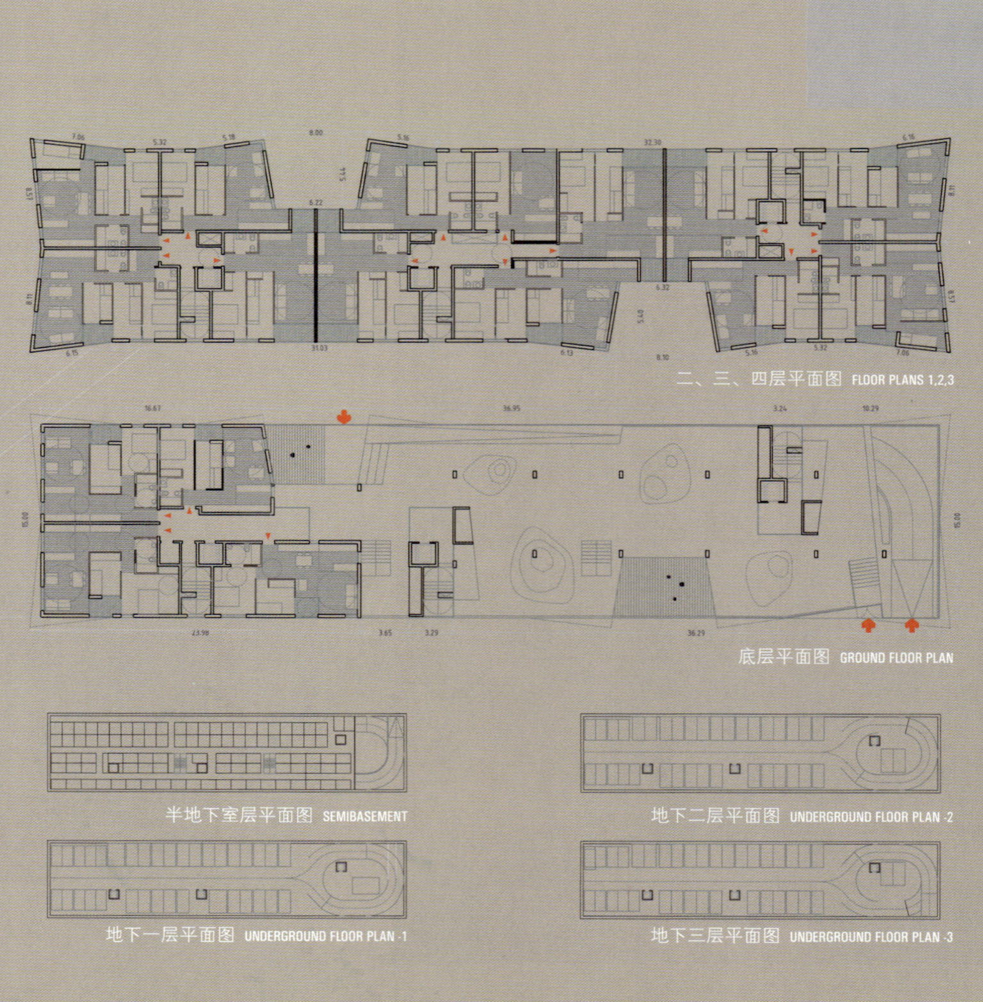

二、三、四层平面图 FLOOR PLANS 1,2,3

底层平面图 GROUND FLOOR PLAN

半地下室层平面图 SEMIBASEMENT

地下二层平面图 UNDERGROUND FLOOR PLAN -2

地下一层平面图 UNDERGROUND FLOOR PLAN -1

地下三层平面图 UNDERGROUND FLOOR PLAN -3

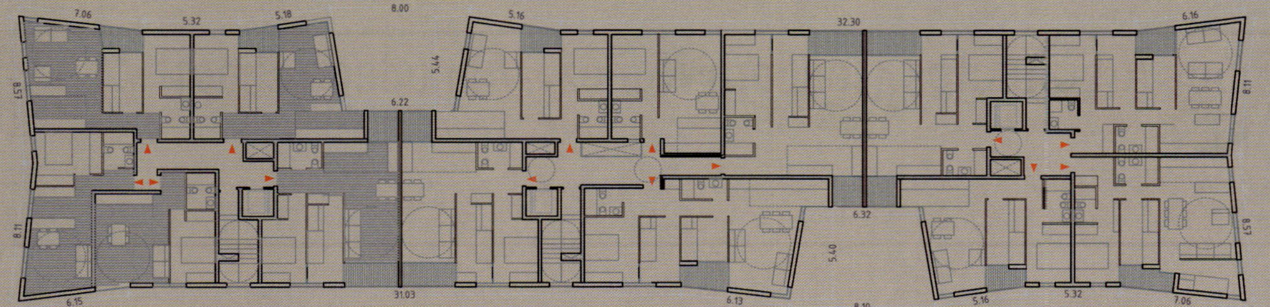

五至八层平面图 FLOOR PLANS 4-7

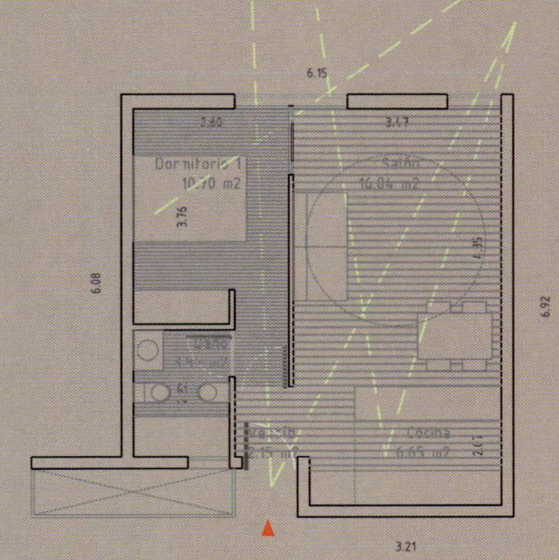

conexiones visuales

1卧室单元房 HOUSING UNIT 1 BEDROOM

conexiones visuales

2卧室单元房 HOUSING UNIT 2 BEDROOMS

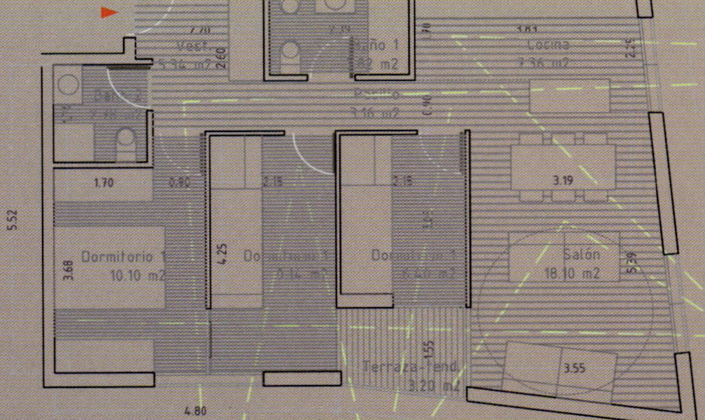

3卧室单元房 HOUSING UNIT 3 BEDROOMS

东南立面图 SOUTHEAST ELEVATION

2VAA450

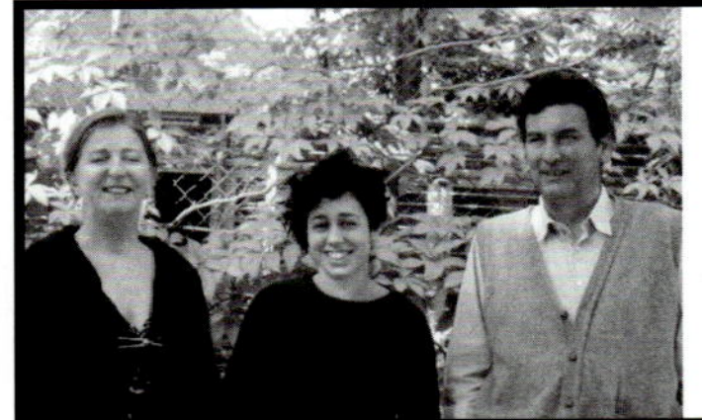

# Nuñez Ribot Arquitectos (Teodoro Nuñez · Almudena Ribot) (建筑师)

合作 (c) Lucía Acedo · Elena Cuerda

并列第三提名奖 third mention ex-aequo

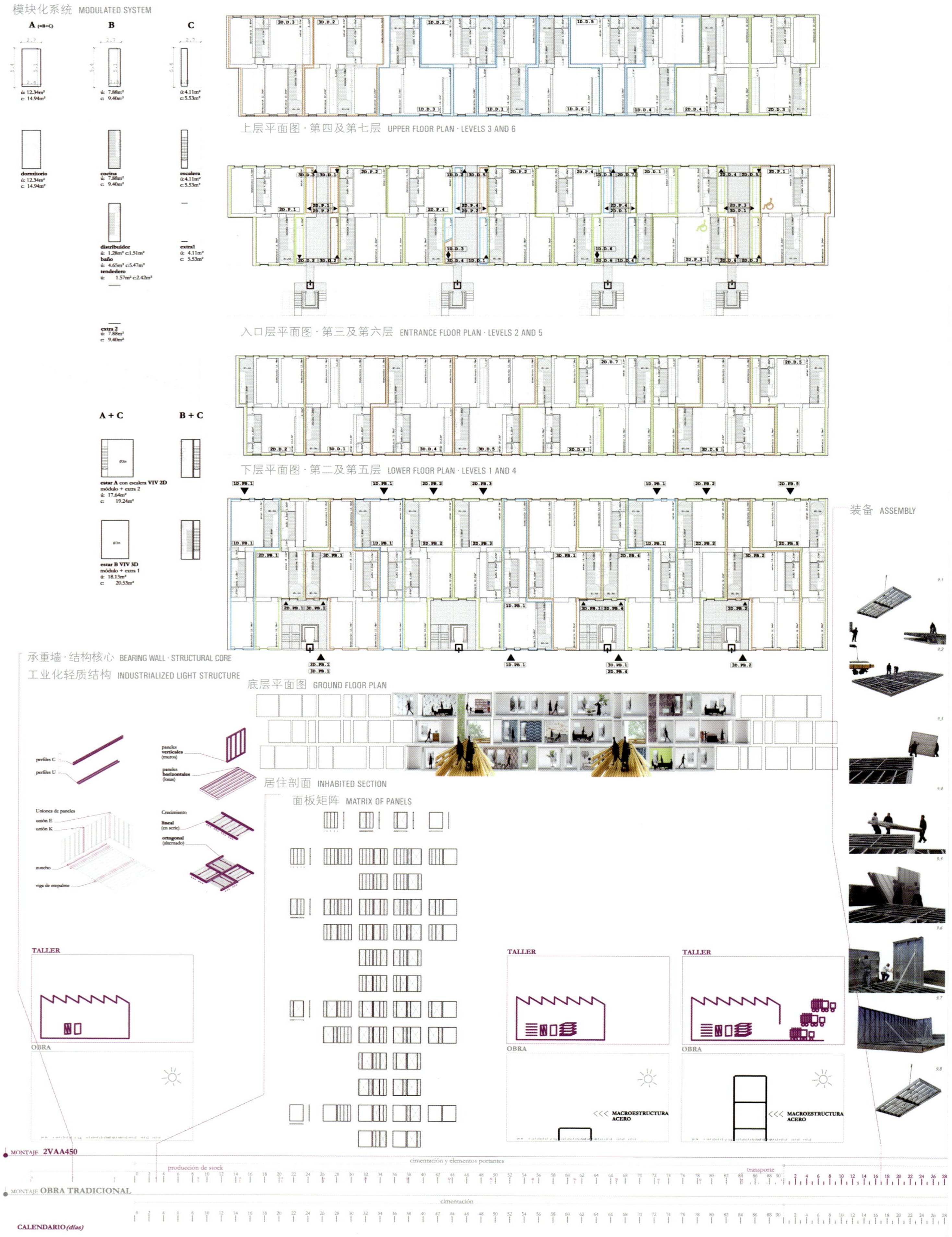

模块化系统 MODULATED SYSTEM
A (=B+C)
B
C
dormitorio
cocina
escalera
distribuidor
baño
tendedero
extra1
extra 2
A + C
B + C
estar A con escalera VIV 2D
estar B VIV 3D
上层平面图 · 第四及第七层 UPPER FLOOR PLAN · LEVELS 3 AND 6
入口层平面图 · 第三及第六层 ENTRANCE FLOOR PLAN · LEVELS 2 AND 5
下层平面图 · 第二及第五层 LOWER FLOOR PLAN · LEVELS 1 AND 4
装备 ASSEMBLY
承重墙 · 结构核心 BEARING WALL · STRUCTURAL CORE
工业化轻质结构 INDUSTRIALIZED LIGHT STRUCTURE
底层平面图 GROUND FLOOR PLAN
居住剖面 INHABITED SECTION
面板矩阵 MATRIX OF PANELS
perfiles C
perfiles U
paneles verticales
paneles horizontales
Uniones de paneles
Crecimiento
lineal
ortogonal
TALLER
OBRA
<<< MACROESTRUCTURA ACERO
MONTAJE 2VAA450
producción de stock
cimentación y elementos portantes
transporte
MONTAJE OBRA TRADICIONAL
cimentación
CALENDARIO (días)

# 结构矩阵

# STRUCTURAL MATRIX

我们提出的构造体系来自于组装型结构，使用的是一种叫做“钢架结构”的商业结构系统，这是美国经典“轻捷型构架”的镀锌钢版本，以利用该种材料轻便、强度高的特性。

We propose a construction system that starts from the assembly. We use a commercial structure system called "steel-frame", galvanized steel version of the classic American "ballon-frame", taking advantage of the qualities of lightness and strength typical of this material.

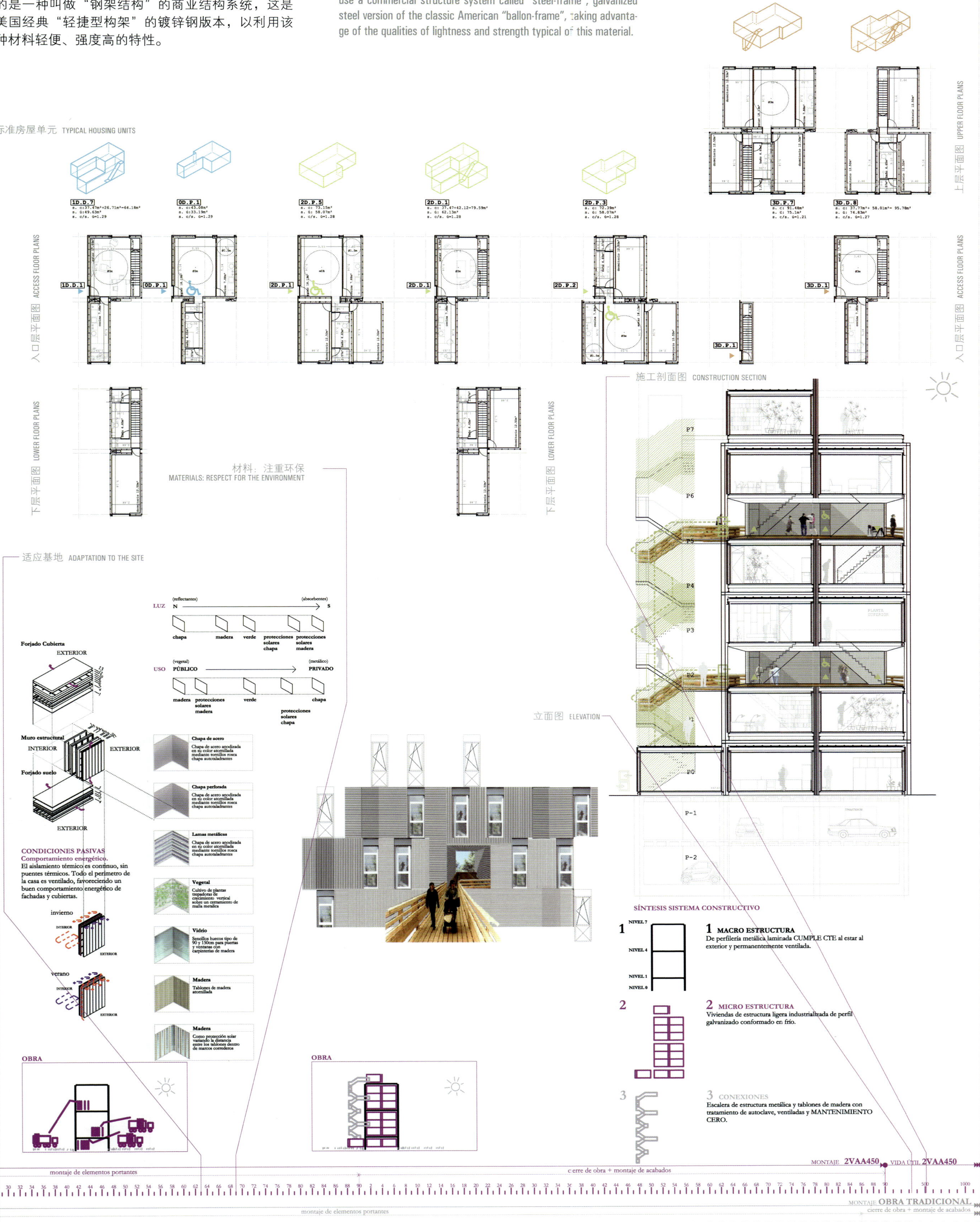

ASOMA2S

# AL PUNTO ARQUITECTURA (建筑师事务所)

Hugo Sebastián de Erice · Ricardo Sánchez · Sergi Artola (建筑师)

并列第三提名奖 third mention ex-aequo

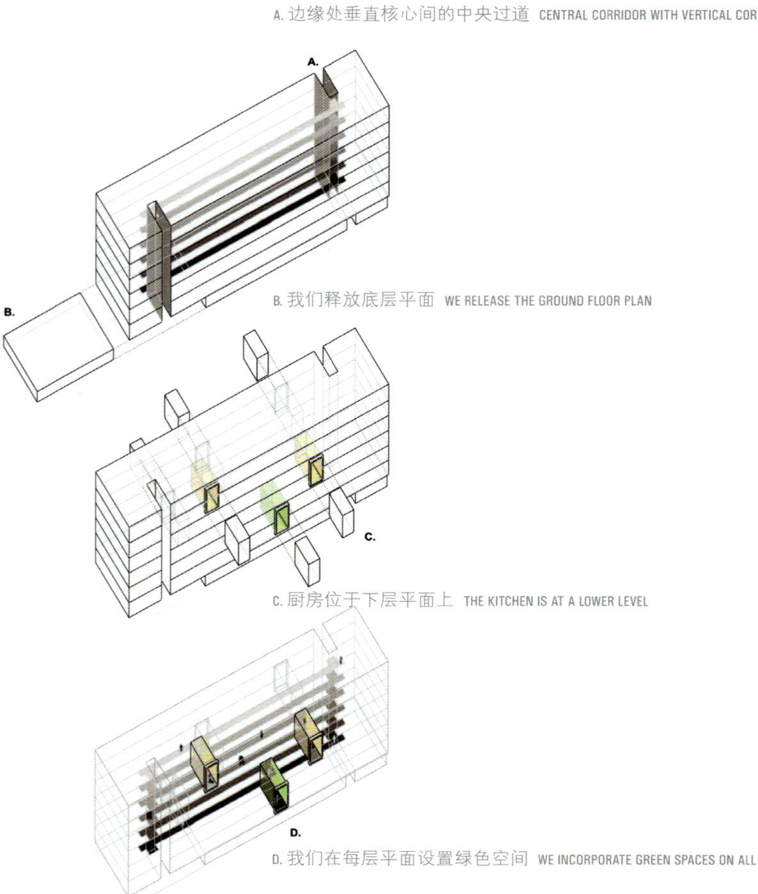


## 上下层

我们知道，该项目为优化项目。客厅位于上方，面对餐厅，形成双层的感觉，使客厅成为整套房的中心。

## LOWER AND UPPER LEVEL

We understand the project as the optimization of the section. The living room looks out over the dining room, creating a sense of double height. The living room becomes the center of the apartment.

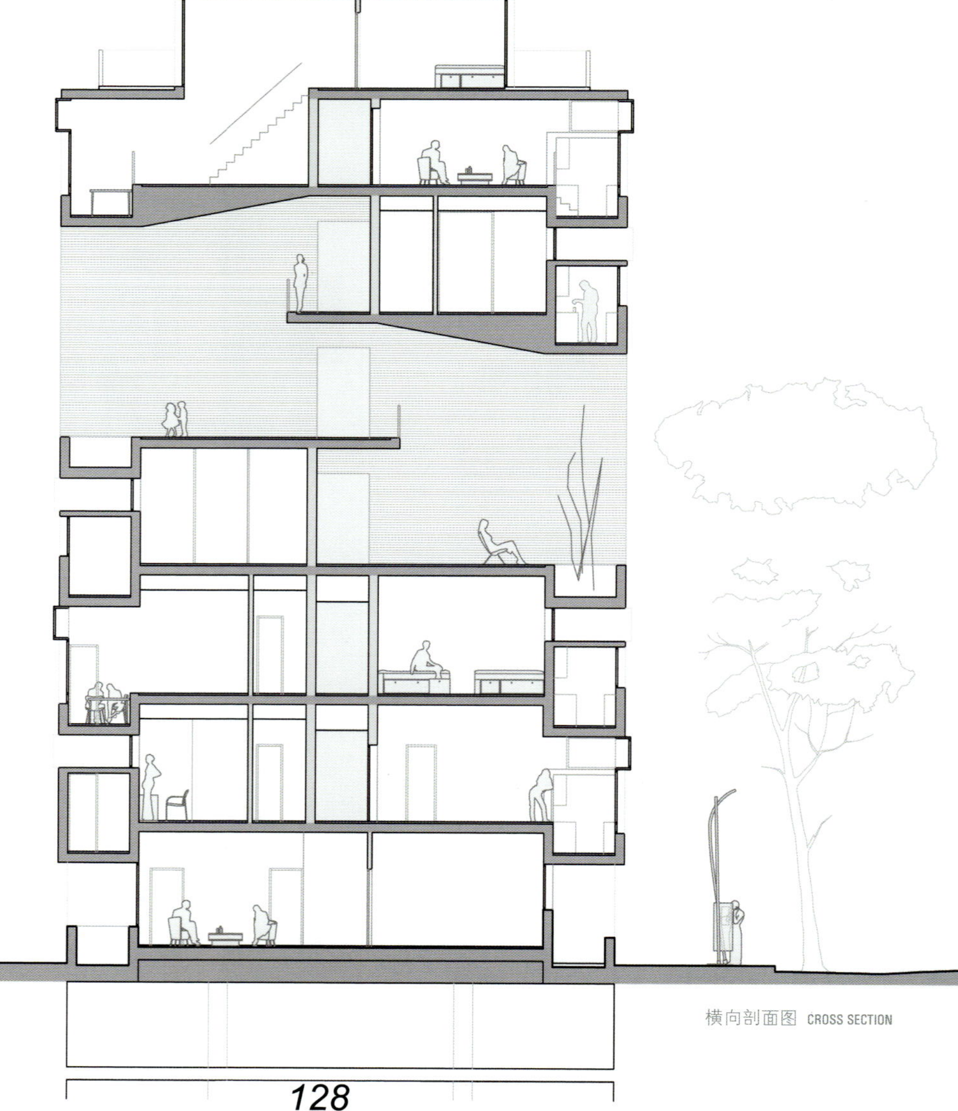

横向剖面图 CROSS SECTION

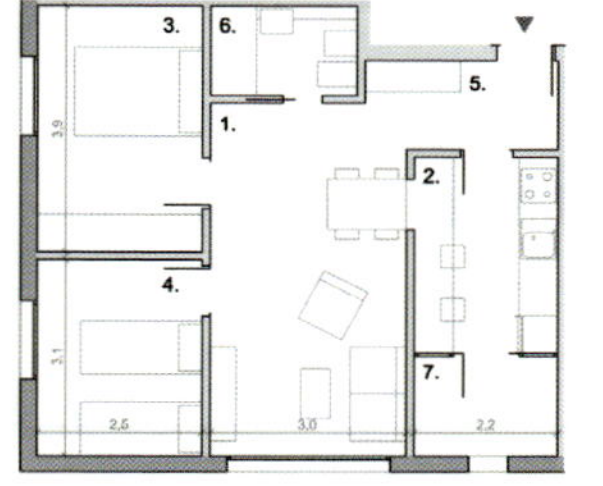

标准房屋单元 TYPICAL HOUSING UNIT 2B. 50.60M2

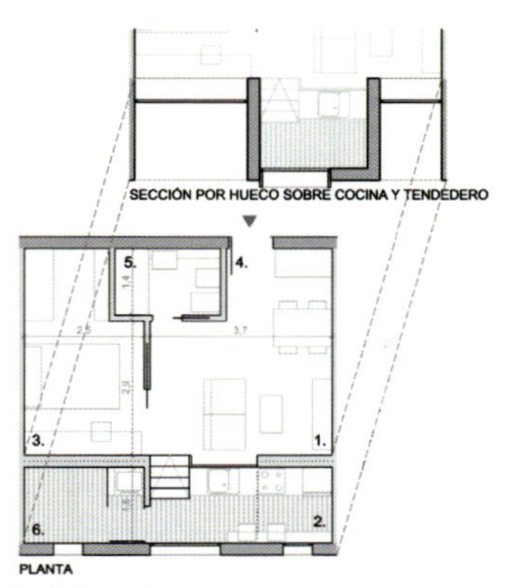


标准房屋单元 TYPICAL HOUSING UNIT 1A. 35.70M2

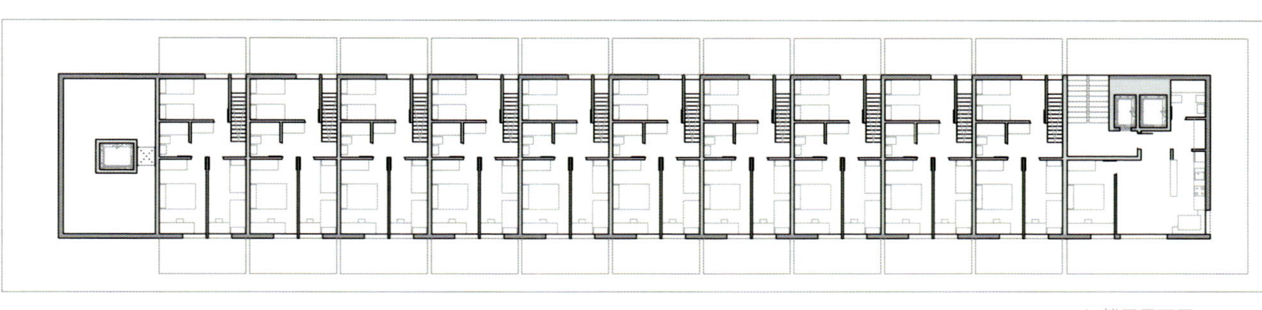

阁楼层平面图 PENTHOUSE

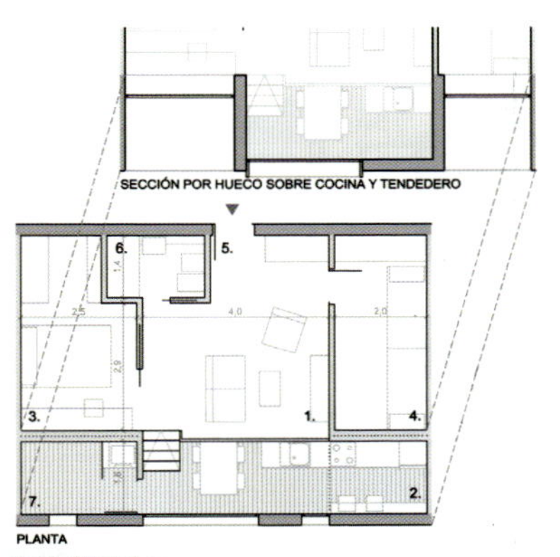


标准房屋单元 TYPICAL HOUSING UNIT 2A. 50.70M2

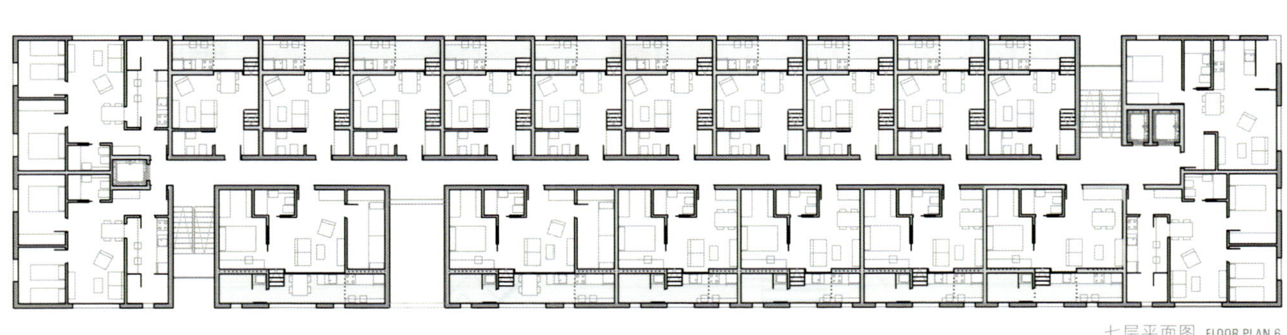

七层平面图 FLOOR PLAN 6

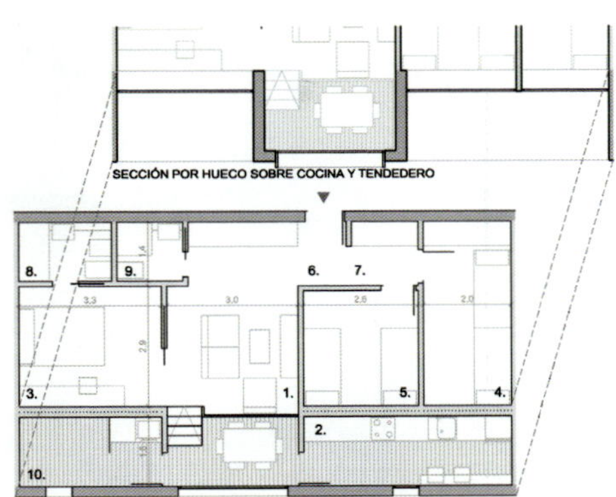


标准层平面图 三至六层 TYPICAL FLOOR PLAN 2-5

标准房屋单元 TYPICAL HOUSING UNIT 3A. 63.60M2

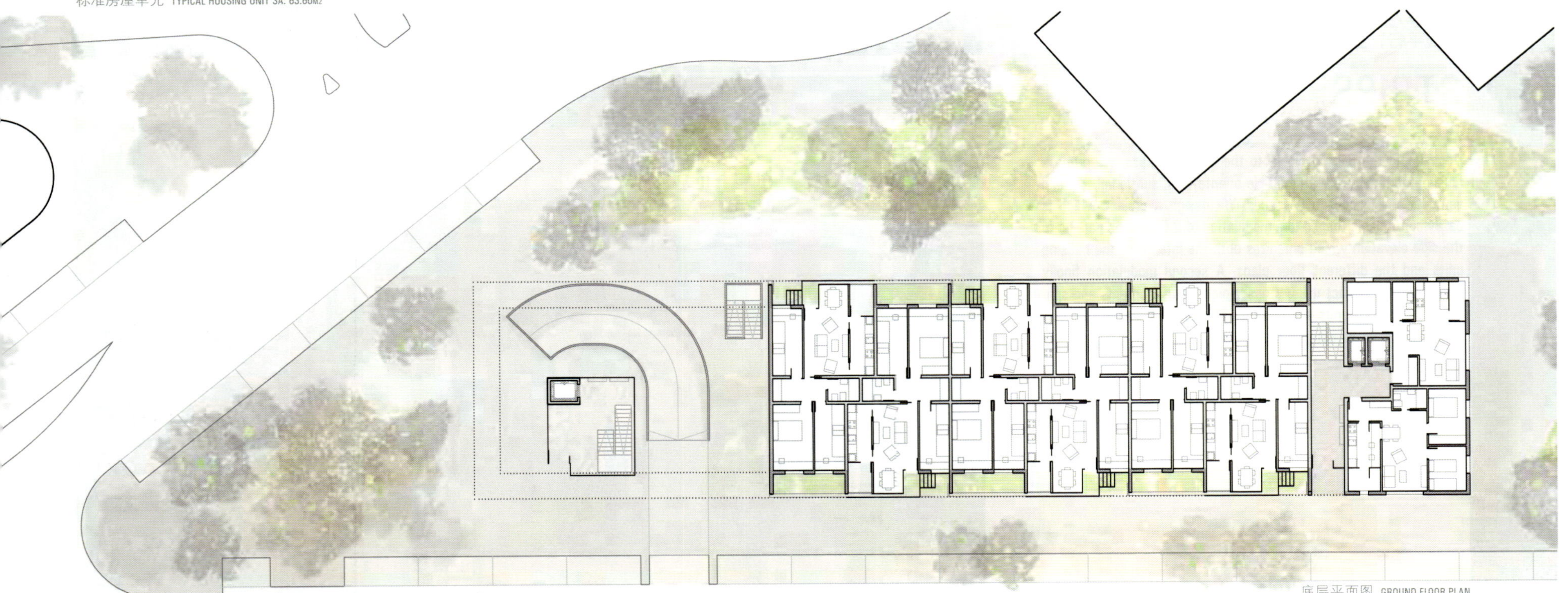

底层平面图 GROUND FLOOR PLAN

# Fernando Maniá project director (项目主管)

团队 team: Fernando Maniá · Santiago Cifuentes · Juan Manuel Palacios · José Blázquez Martina Sibona

第一提名奖 first mention

## 悬浮空间

房子中间降起，在底层形成连续性空间。楼层平面图的设计中未包含中央走廊，目的有二：一是避免直接看到塔楼，二是为所有室内单元提供最佳的通风和采光效果。该建筑以“悬浮”的形式修建。

## SUSPENDED VOLUME

The building rises in the central area to produce a continuous space on the ground floor. The distribution of the floor plans has avoided the central corridor with a double objective: to avoid direct unique views towards the towers, and obtain cross ventilation for all housing units opening to the best sun exposure. The building is developed as a "suspended" volume.

遮檐 · 防止南向日晒 VISOR · SOLAR PROTECTION FROM SOUTH

栅栏 · 防止西向日晒 RIBS · SOLAR PROTECTION FROM WEST

露天观景台 · 气候垫 TERRACE VIEWPOINT · CLIMATIC CUSHION

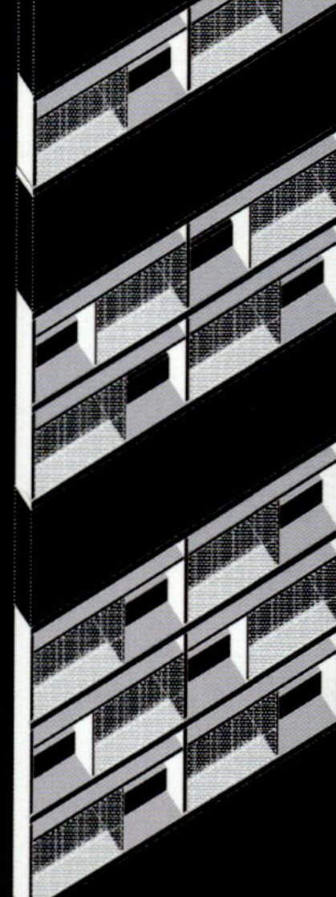

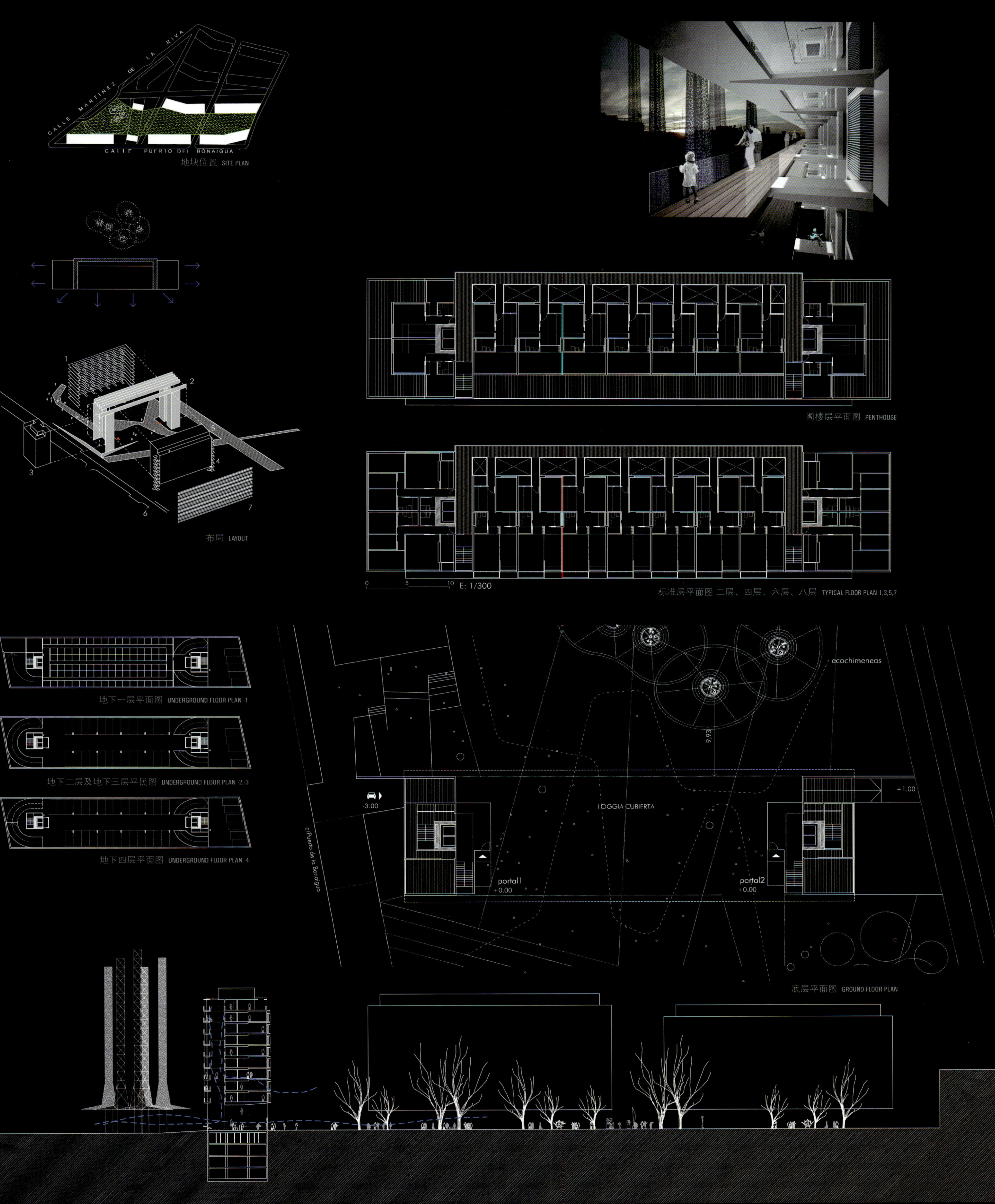
CALLE MARTINEZ DE LA RIVA
CALLE PUERTO DEL BONAIGUA
地块位置 SITE PLAN
布局 LAYOUT
阁楼层平面图 PENTHOUSE
E: 1/300
标准层平面图 二层、四层、六层、八层 TYPICAL FLOOR PLAN 1,3,5,7
地下一层平面图 UNDERGROUND FLOOR PLAN -1
地下二层及地下三层平民图 UNDERGROUND FLOOR PLAN -2,3
地下四层平面图 UNDERGROUND FLOOR PLAN -4
ecochimeneas
LOGGIA CUBIERTA
c/Puerto de la Bonaigua
portal1
portal2
底层平面图 GROUND FLOOR PLAN

3TORRES

# 1004arquitectos (建筑师事务所)

Coral Álvarez de Miguel · Jaime Lamúa Chueca · Pedro López Quintas · Sergio Soria Soria (建筑师)

第二提名奖 second mention

地块位置 SITE PLAN

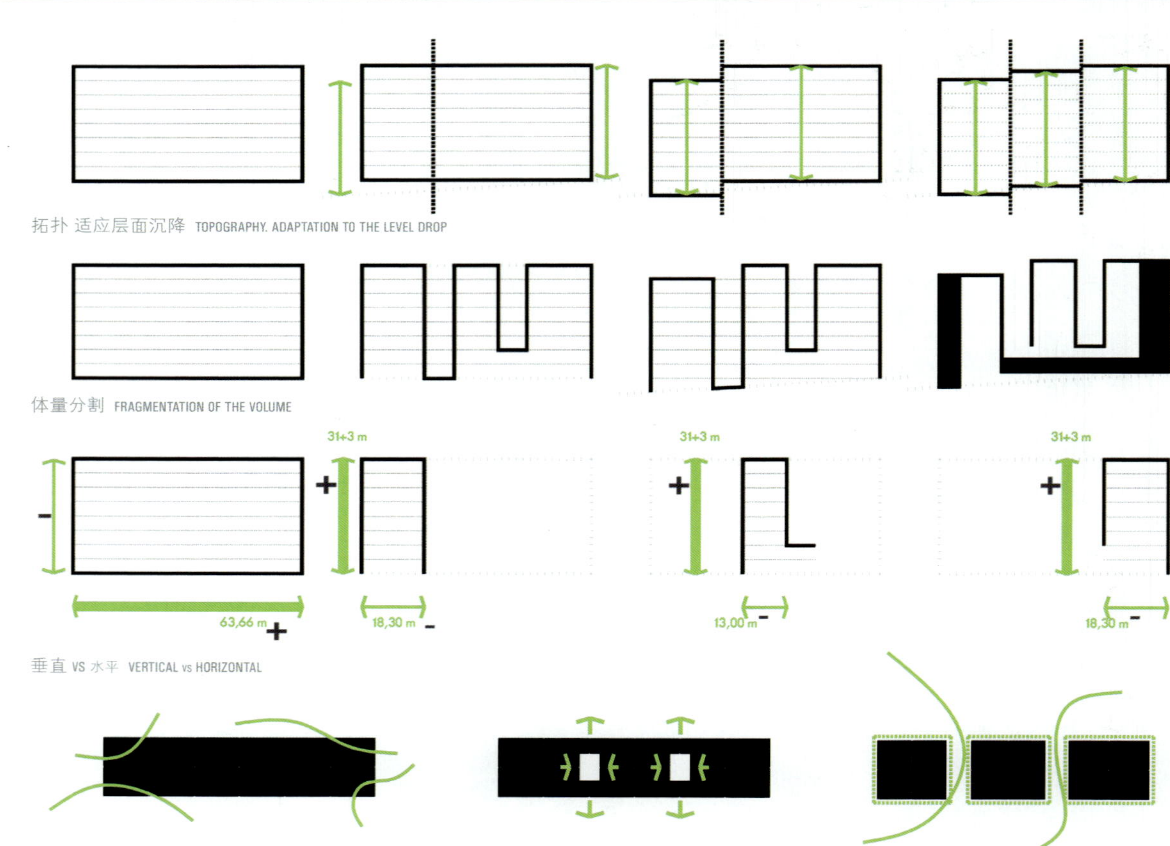


## 三个直线元素

该项目产生一透明型空间，不会破坏视觉的连贯性。房子（大型）以连续性的折叠结构为基础，在其间设置功能性较强的室内单元（小型）。

## 3 VERTICAL ELEMENTS

The project generates a permeable volume that does not block the visual continuity of the set. A continuous folded piece, defines a container (large scale) where we introduce the functional packages of the housing units (small scale).

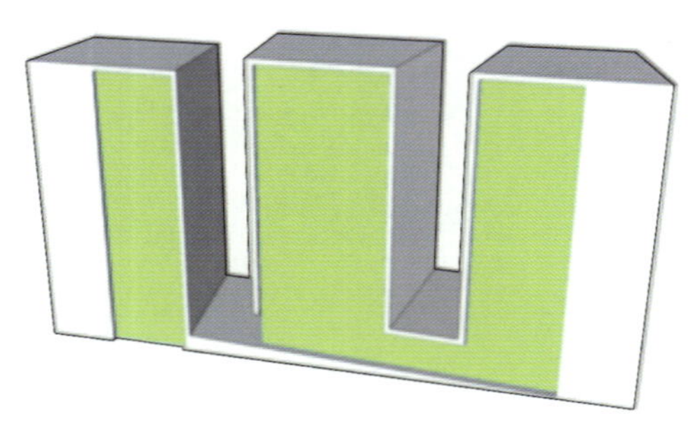

体块折叠 FOLDED CONTAINER

悬挑+飘窗阳台 CANTILEVERS+WINDOWED BALCONIES

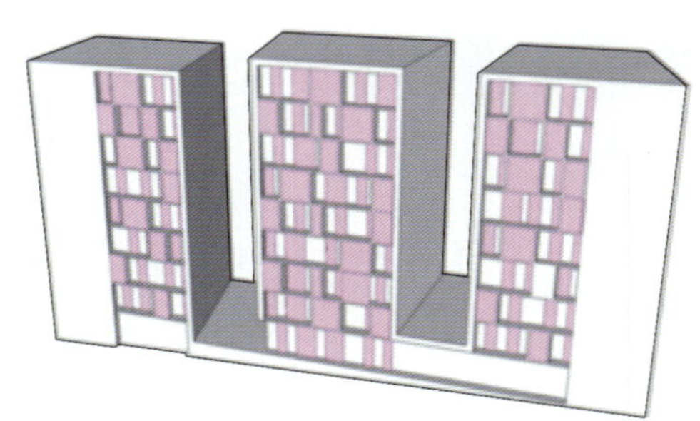

透明+不透明 TRANSPARENT+OPAQUE

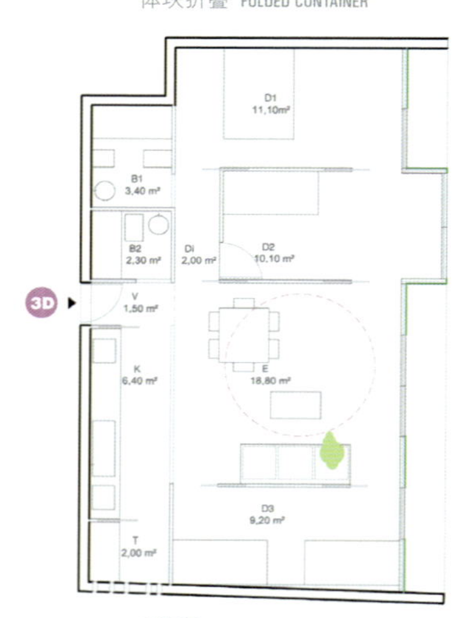

3卧室 3 BEDROOMS. 66.80M2

2卧室 2 BEDROOMS. 51.80M2

1卧室 1 BEDROOM. 44.20M2

建议体量 PROPOSED VOLUME

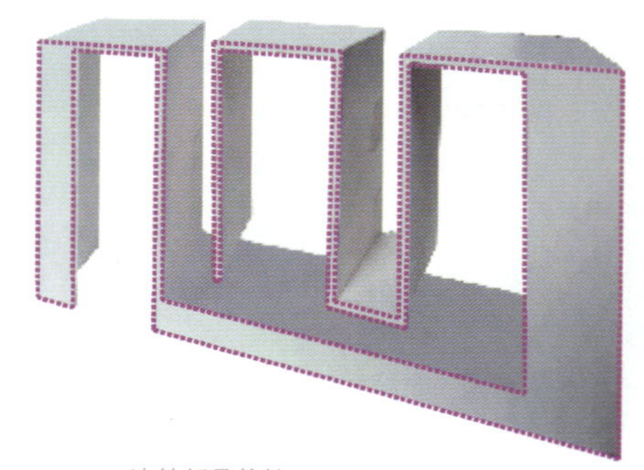

连续折叠体块 CONTINUOUS FOLDED PIECE

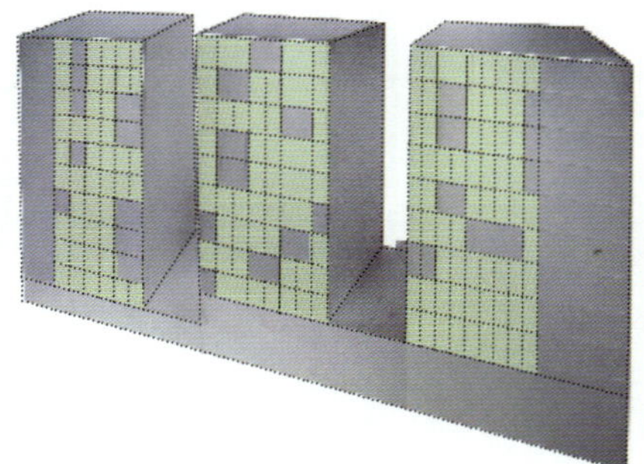

功能体量·房屋单元 FUNCTIONAL VOLUME · HOUSING UNITS

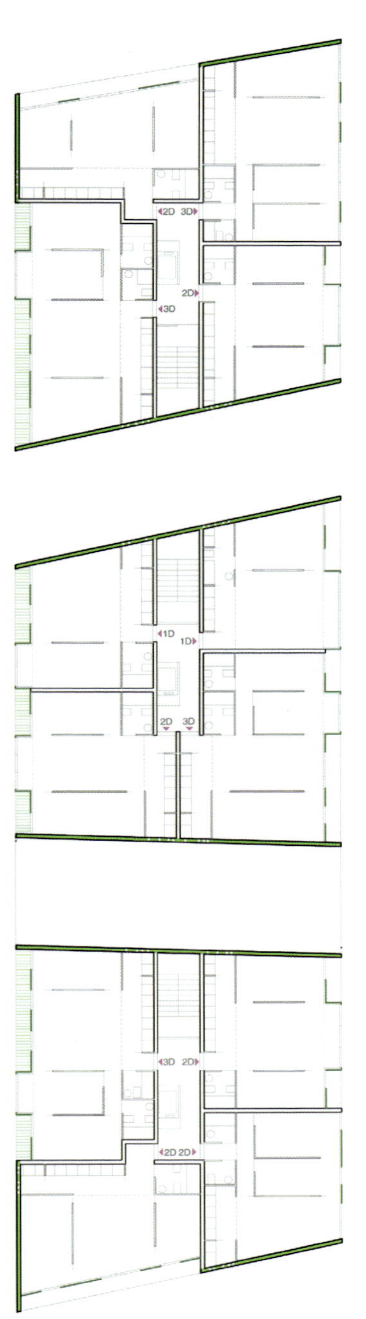

标准层平面图 TYPICAL FLOOR PLAN

底层平面图 GROUND FLOOR PLAN

阁楼 PENTHOUSE

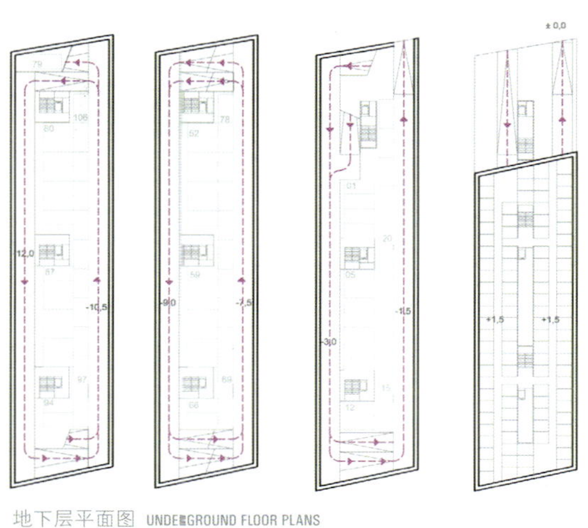

地下层平面图 UNDERGROUND FLOOR PLANS

09CALLES

# Mixuro (建筑师事务所)

María Oliver Sanz · Javier Alejandro Molinero Domingo · José Rafael Mira Albero · Javier Matoses Merino (建筑师)

第三提名奖 third mention

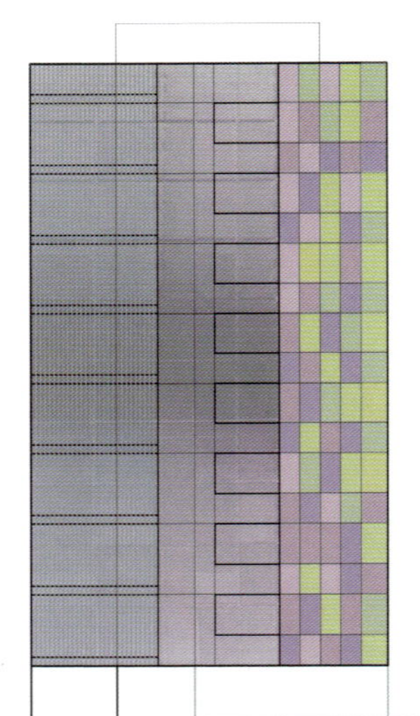

东立面 EAST ELEVATION

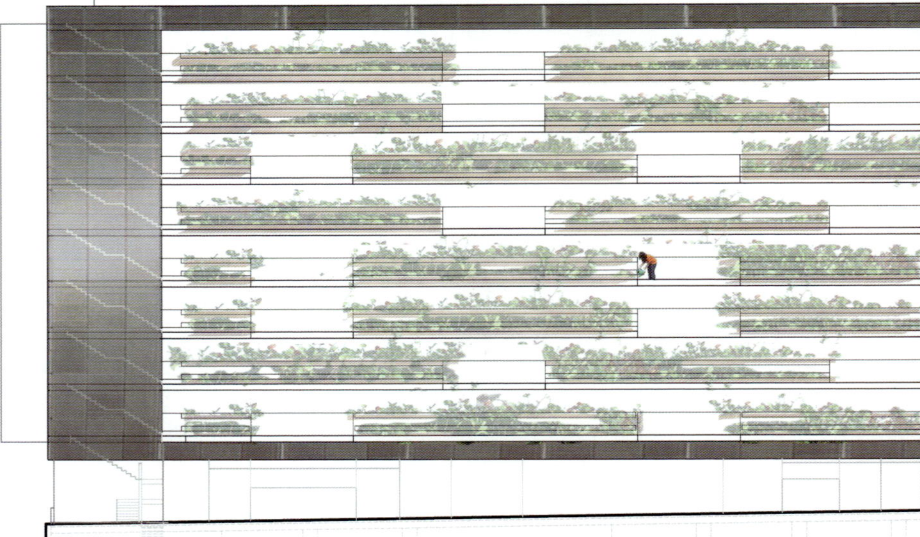

北立面 NORTH ELEVATION

2卧室 2 BEDROOMS. 62.45M2

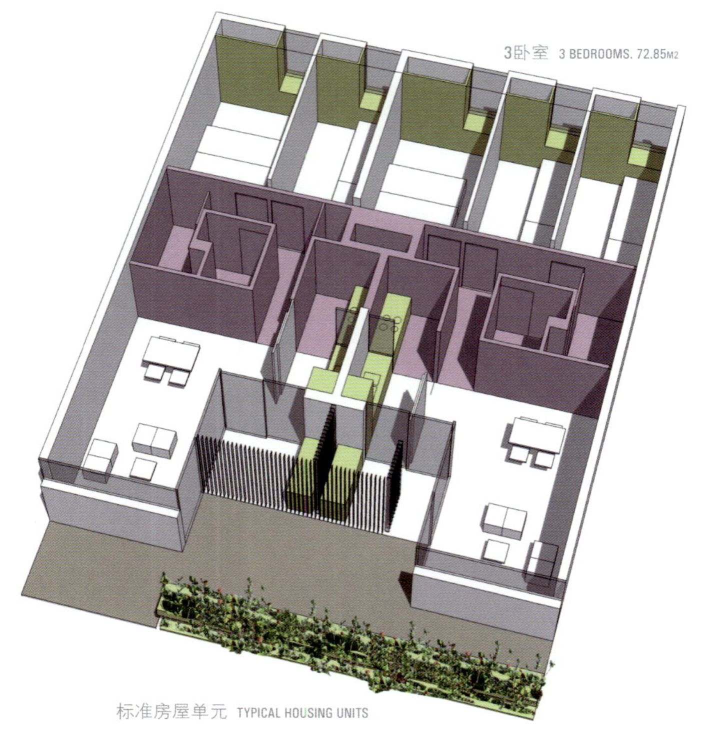


标准房屋单元 TYPICAL HOUSING UNITS

## 植物绿化

人们可通过位于两端的两个垂直楼道穿行于九条提升的街道。拓宽的街道提供了更多的住宅或公共空间。从外面看，街道犹如大型的阳台，我们将其用作各个建筑的绿化平台。

## VEGETAL TAPESTRY

There are nine raised streets which each one can have accessed through two vertical communication cores located at the ends. The streets are widened creating relative spaces associated with housing or communal use. The streets are seen from the outside as large balconies where we promote cultivation platforms allocated to each dwelling.

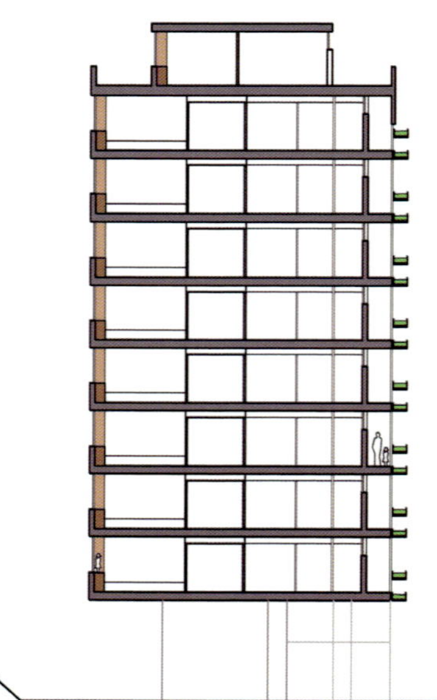

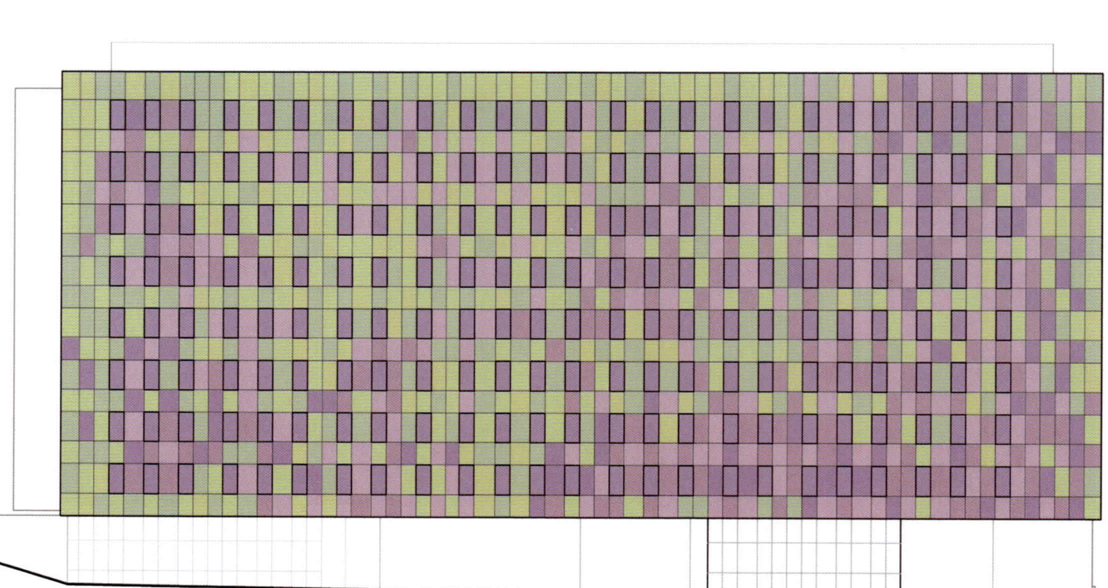

横向剖面图 + 南立面图 CROSS SECTION + SOUTH ELEVATION

5,45
5,80
5,90
5,80
5,90
5,70
6,00
5,80
5,90
5,90
4,13
4,60
5,25
4,60
阁楼层平面图 PENTHOUSE
标准层 TYPICAL FLOOR PLAN
底层平面图 GROUND FLOOR PLAN
地下一层平面图 UNDERGROUND FLOOR PLAN -1

# AL PUNTO ARQUITECTURA (建筑师事务所)

Hugo Sebastián de Erice · Ricardo Sánchez · Sergi Artola (建筑师)

中标 winner

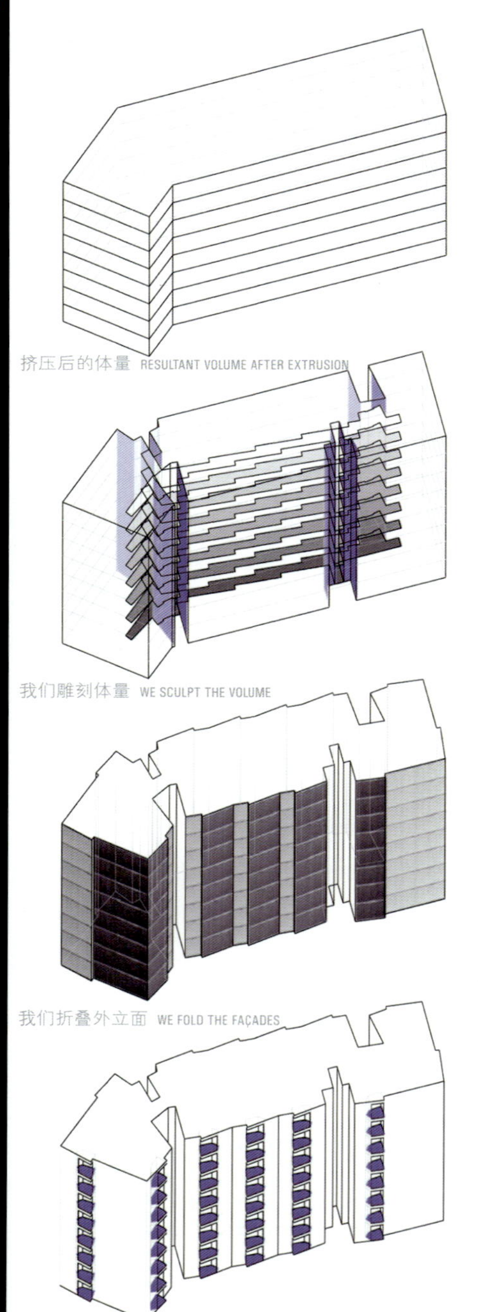
挤压后的体量 RESULTANT VOLUME AFTER EXTRUSION

我们雕刻体量 WE SCULPT THE VOLUME

我们折叠外立面 WE FOLD THE FAÇADES

每个住户都有一个露台 · 新体量 ONE TERRACE PER DWELLING · NEW VOLUME

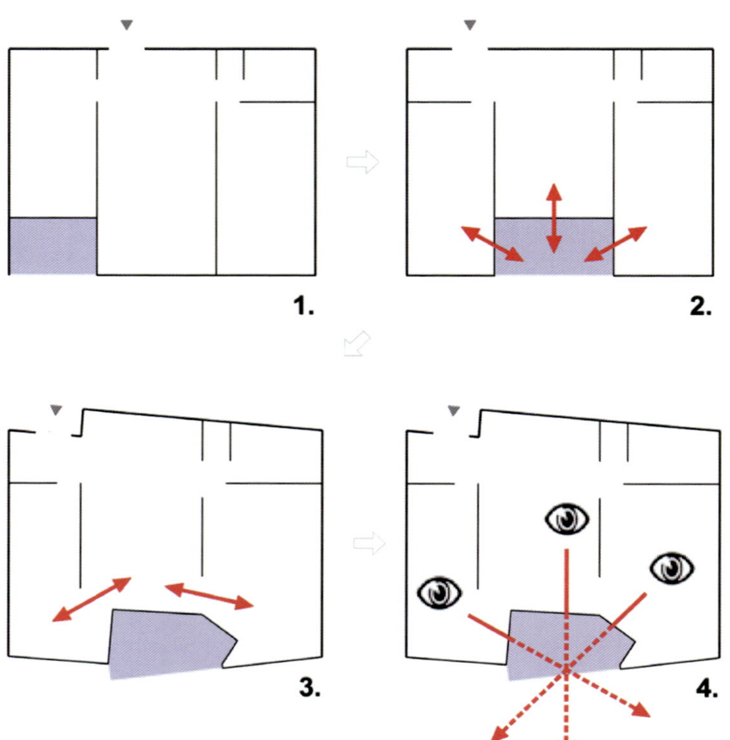

房屋党员概念 HOUSING UNITS CONCEPT

## 倾斜效果的折叠外形

每个住宅单元都是庭院的缩微体，连起来在各个楼层可以组成一个小型村庄。住宅单元的布局，如厨房、客厅和主卧室与室外相连，营造一种亲密同时又具有私密性的氛围。我们刻意使用折叠的外形以柔化房子的外观及改善房子的采光效果。

## A FOLDED PROFILE WITH OBLIQUE VIEWS

Each housing unit is a small scale version of the house-courtyard, and we can say that its grouping reproduces a small village on each floor. The layout of the primary uses of the housing units, such as the kitchen, the living room and master bedroom, around an outdoor space, creates an intimate and private atmosphere. We intentionally looked for folding profile of the building, softening and lightening the volume.

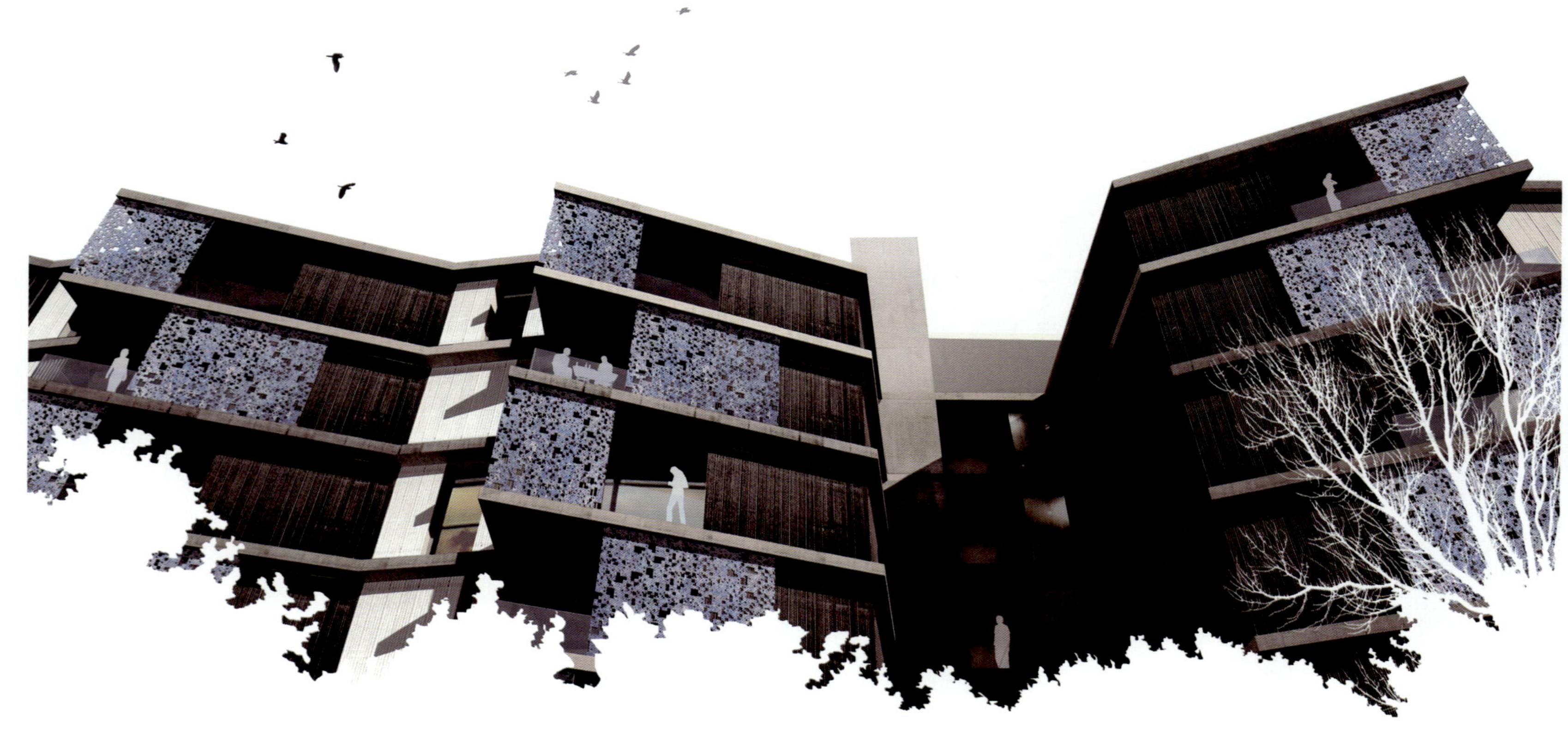

阁楼层平面图 PENTHOUSE
标准层平面图 TYPICAL FLOOR PLAN
底层平面图 GROUND FLOOR PLAN
房屋单元 HOUSING UNIT 1A
房屋单元 HOUSING UNIT 1D
房屋单元 HOUSING UNIT 2C
房屋单元 HOUSING UNIT 2E

F288GTO

# Oscar Teijeiro Castro (建筑师)

第二提名奖 second mention

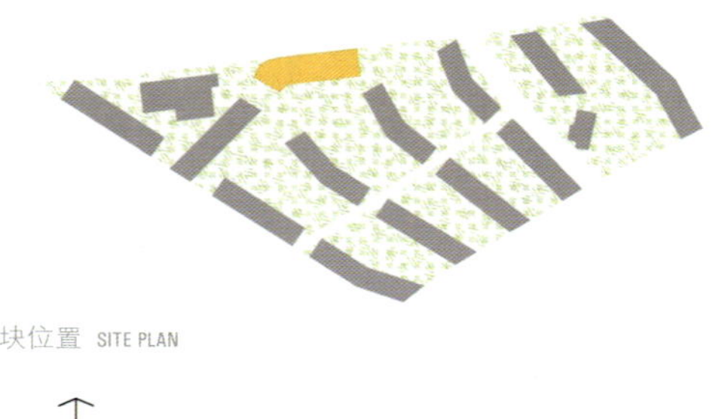

地块位置 SITE PLAN

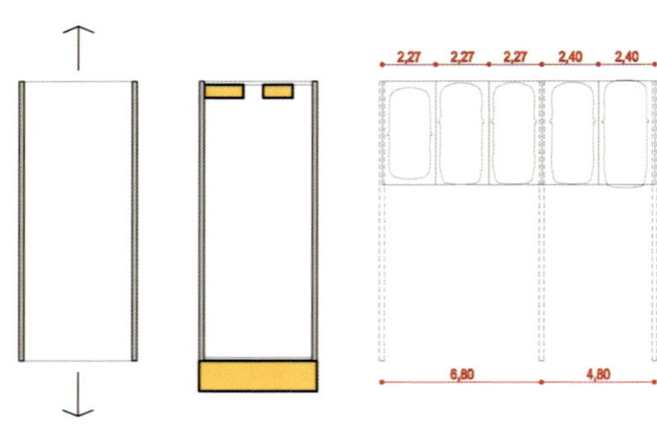

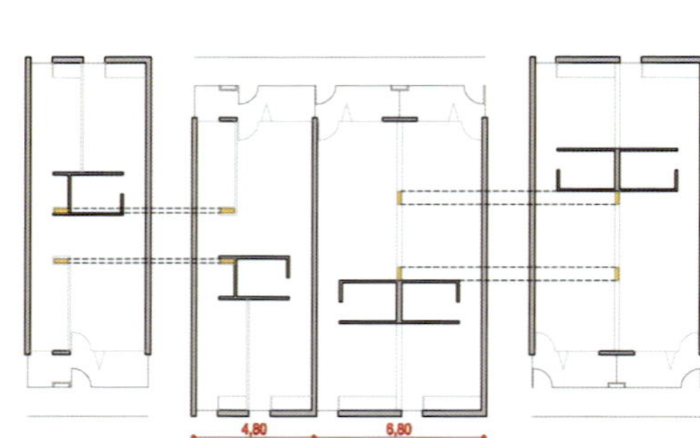

概念 CONCEPTS

## 人行道

建筑前后两面外观的双重朝向，可保证建筑内具有更高的能源效率，中间区域（走廊和储藏室）可起隔热作用，并提高绝缘效果。公寓建于两面支撑墙之间，中心区域（卫生间）将其分隔成两个空间，分别为白天和夜晚所用。

## WALKWAYS

The dual orientation in opposite façades ensures better energy efficiency in the building, the intermediate spaces (corridor and closets) act as thermal cushion insulation and reinforce the insulation of the enclosure. The apartments are developed between two supporting walls. A central element (bathroom) divides the space into day and night areas.

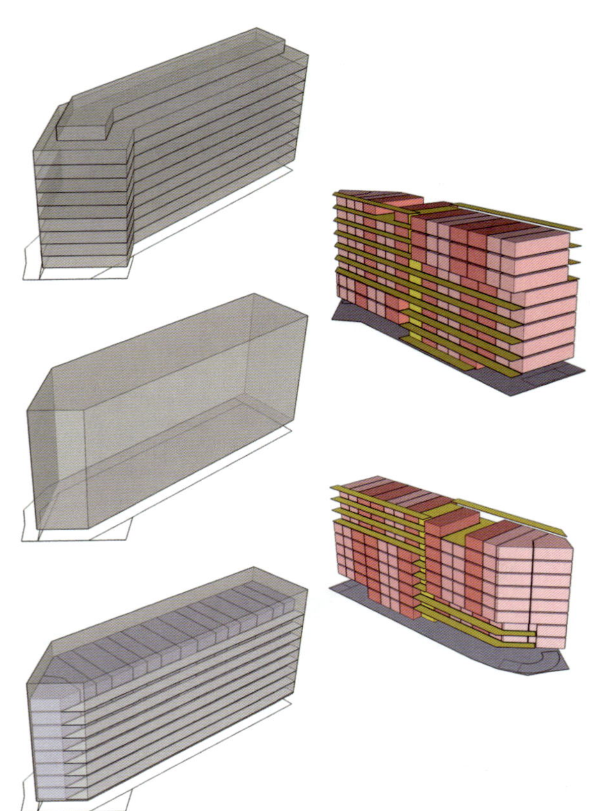

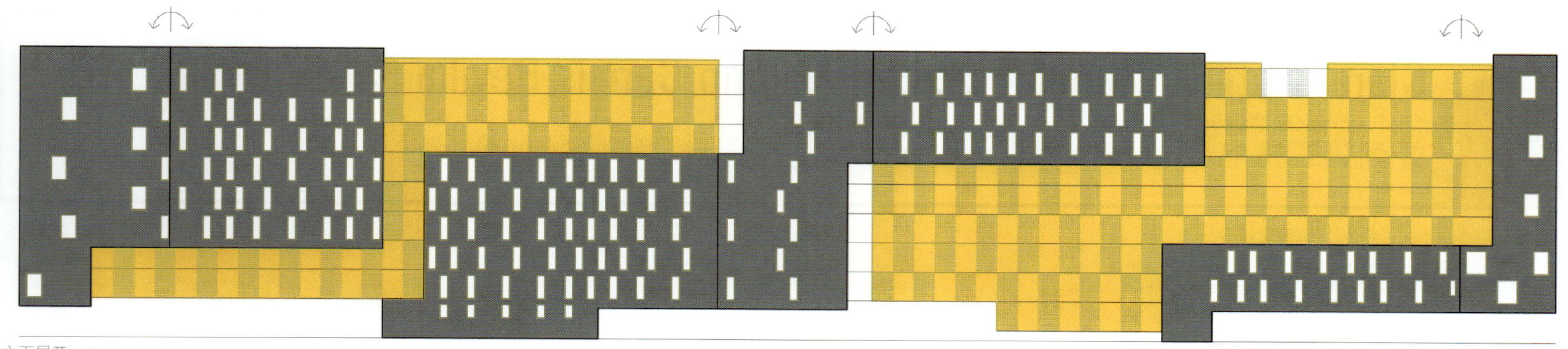
立面展开 UNFOLDED ELEVATION

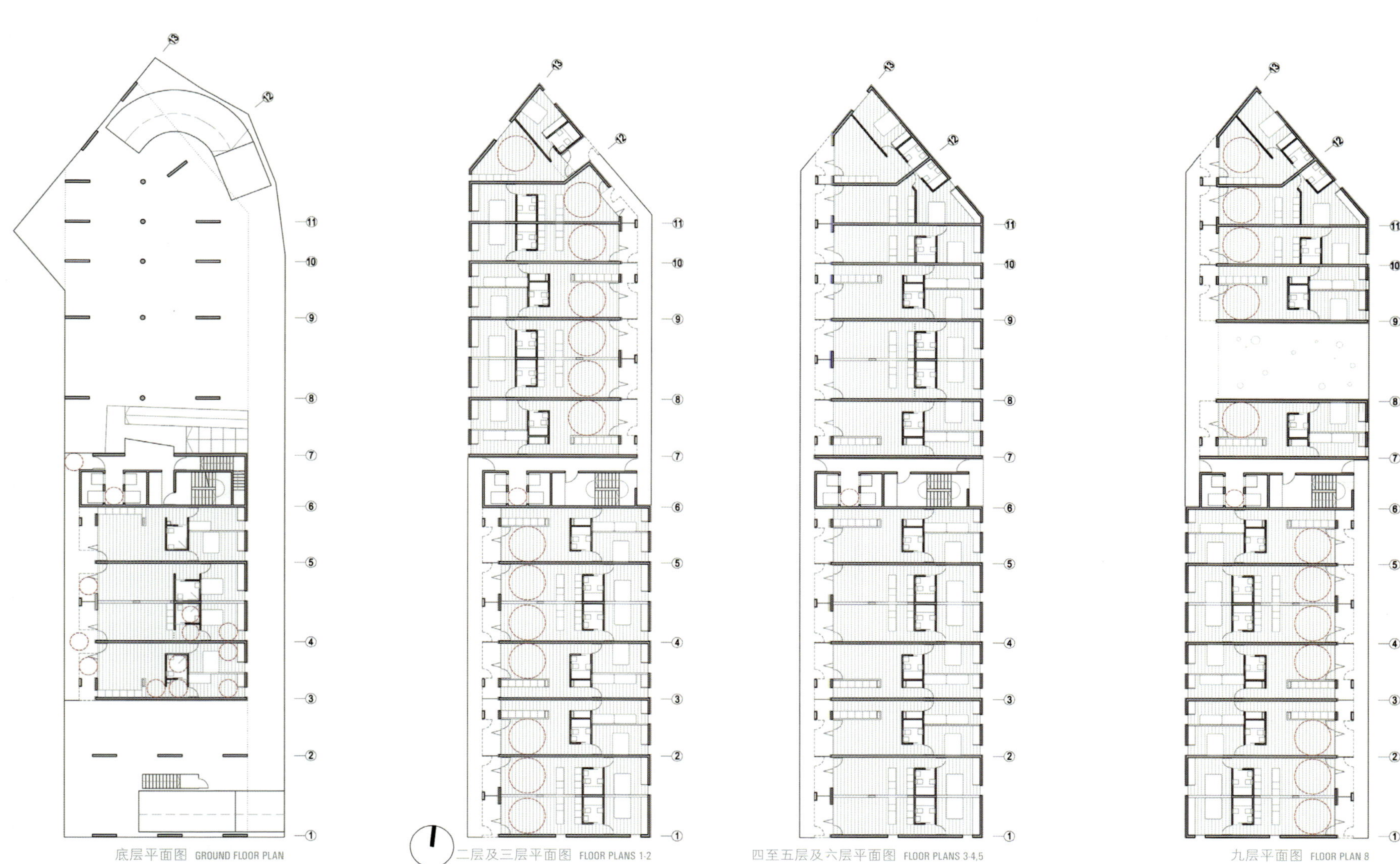
底层平面图 GROUND FLOOR PLAN

二层及三层平面图 FLOOR PLANS 1-2

四至五层及六层平面图 FLOOR PLANS 3-4,5

九层平面图 FLOOR PLAN 8

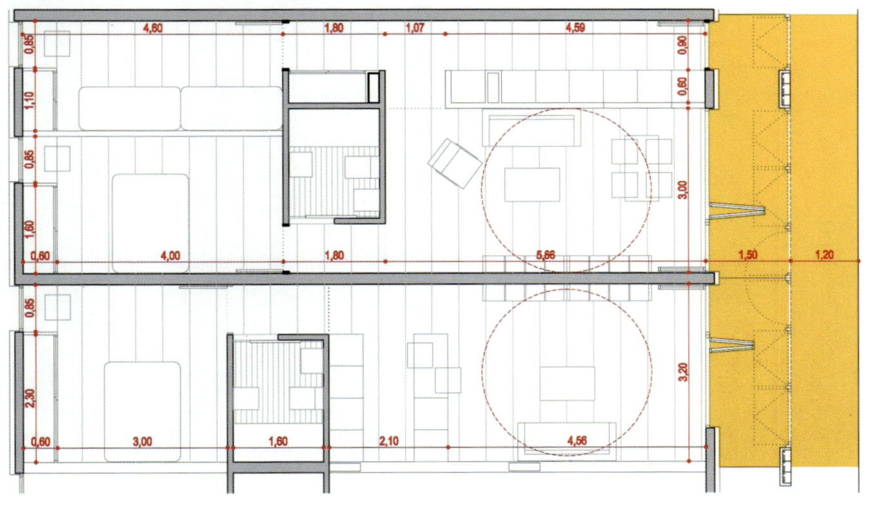
平面图 FLOOR PLAN

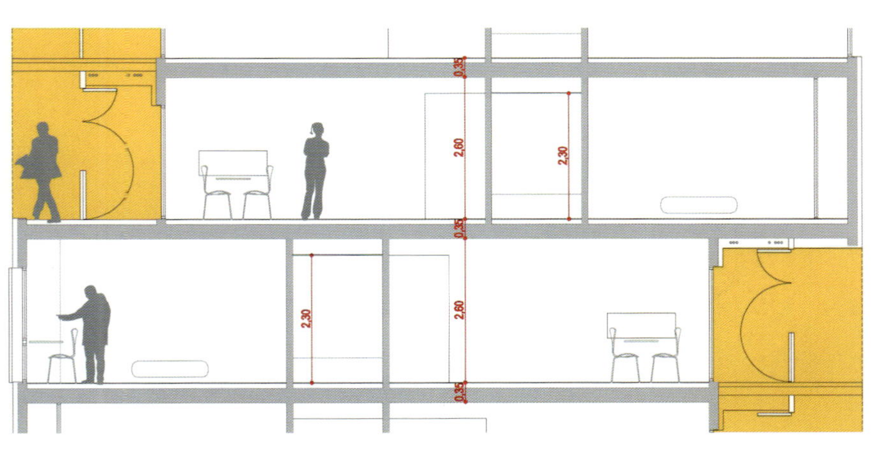
剖面图 SECTION

标准房屋单元 TYPICAL HOUSING UNITS

## Clara Matilde Moneo Feduchi · Valerio Canals Revilla (建筑师)

合作 (c) Lucía Martínez Martínez

并列第三提名奖 third mention ex-aequo

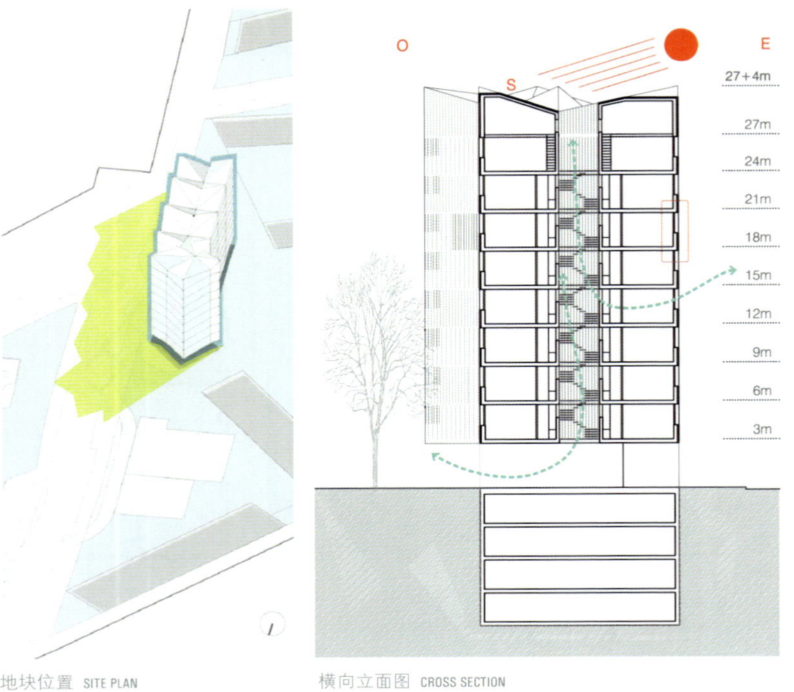


地块位置 SITE PLAN　　横向立面图 CROSS SECTION

大街立面 STREET ELEVATION

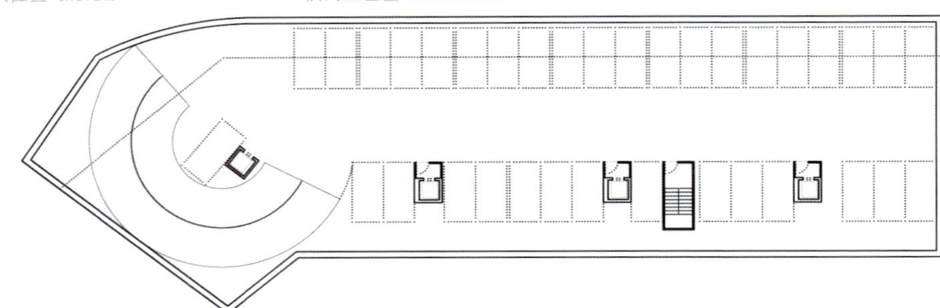

地下二、三、四层平面图 UNDERGROUND FLOOR PLANS -2,-3,-4

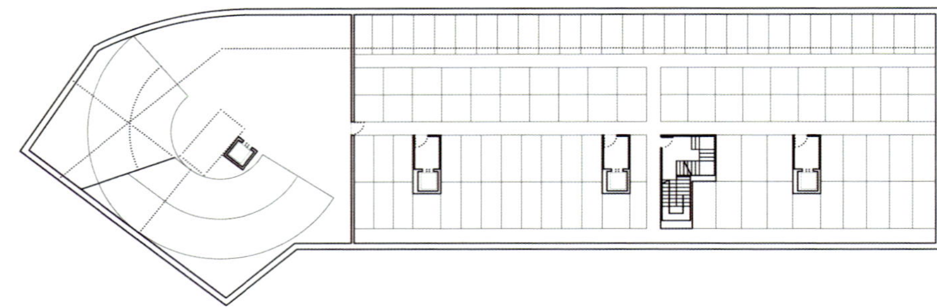

地下一层平面图 UNDERGROUND FLOOR PLAN -1

底层平面图 GROUND FLOOR PLAN

# 旋转

该项目是通过对空间进行分隔，实现良好的采光、通风，以及在功能上灵活性较强的目的，以满足城市空间的要求。单元依街而建，并以直线型方式组合；第四单元则依据地块的地形建造，正面朝向公园，因此从街道看其背面也颇具视觉美感。

# ROTATION

We propose a project which allows, by the fragmentation of the volume, to give an adequate urban answer with sunlight, cross ventilation and functional flexibility. The units are linearly grouped following the direction of the street and the fourth unit rotates following the alignment of the plot, opening towards the park while it becomes a visual frame from the street.

共九层+夹层
8 STOREYS+MEZZANINE
单元归类：
房屋单元含自然通风
GROUPING UNITS:
HOUSING UNITS WITH CROSS VENTILATION
根据地块外形调整形状
DEFORMATION ADJUSTED TO THE SHAPE OF THE PLOT
房屋单元 HOUSING UNITS
标准层平面图 TYPICAL FLOOR PLAN
九层平面图 FLOOR PLAN 8
九层上部夹层平面图 MEZZANINE ABOVE FLOOR PLAN 8

# Esther Collado Quirantes · David Samuel Ares Esteve (建筑师)

并列第三提名奖 third mention ex-aequo

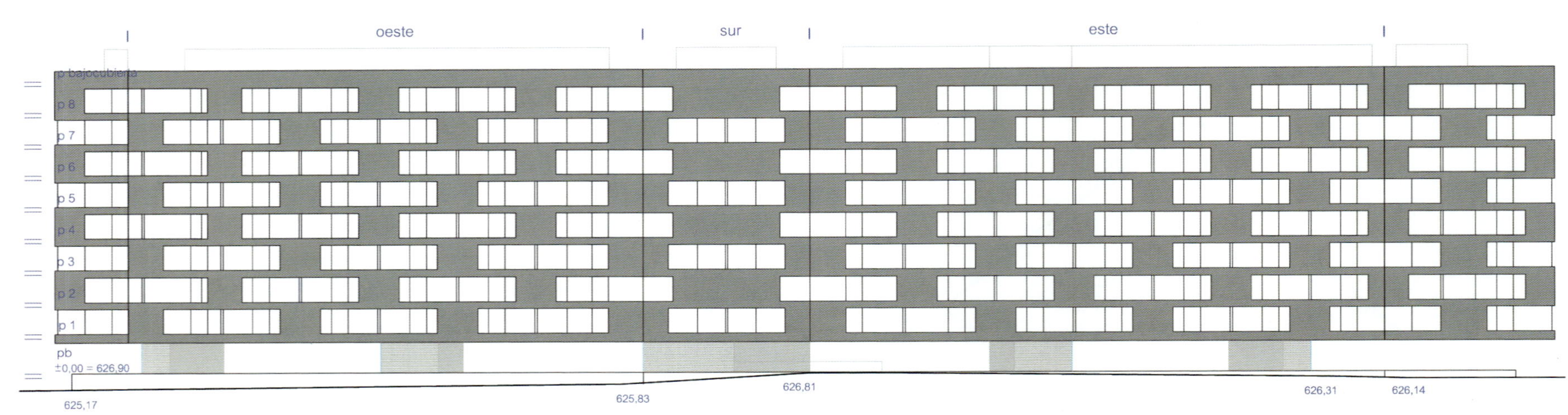


立面展开 UNFOLDED ELEVATION

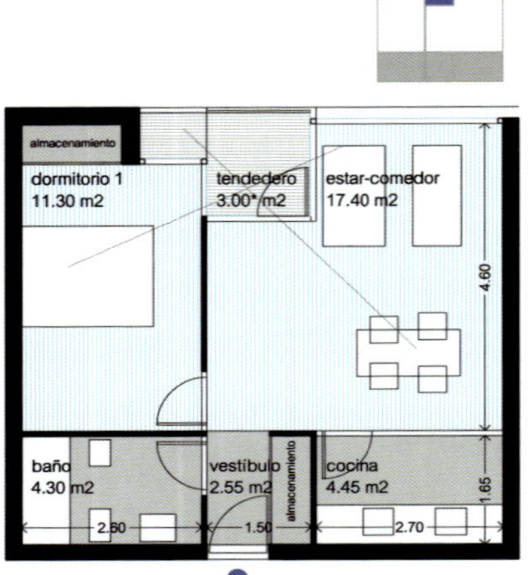


1卧室房屋单元 HOUSING UNIT 1 BEDROOM 41.50M2

横向剖面图 CROSS SECTION

## 开放式结构

该项目通过使用“丝带”结构来解决狭窄的空间问题：折叠后融入地块地形，然后将不同的住宅单元改造成具有一个或两个卧室的房间。从体量上看，该项目契合了“丝带”包围空间的想法。建筑周围新的绿化空间也支持使用大型敞开式窗户的决定，使住户能充分享用这个地块的城市特质。

## OPEN WEBBING

The project solves a tight program from the generation of a "ribbon" that folds and fits inside the plot and subsequently modulates the different housing units to one and two bedrooms. Volumetrically, the project also remains true to the idea of evenly wrapping the volume formed by the "ribbon". The presence of new green spaces surrounding the building, justifies the decision to open large windows that allow users to enjoy the urban qualities of the site.

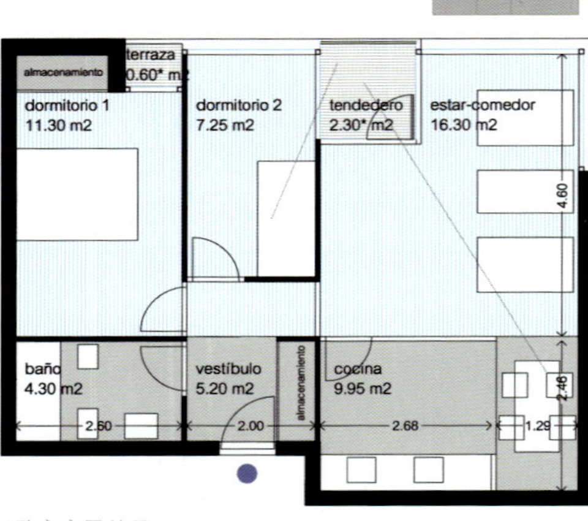


2卧室房屋单元 HOUSING UNIT 2 BEDROOMS 55.70M2

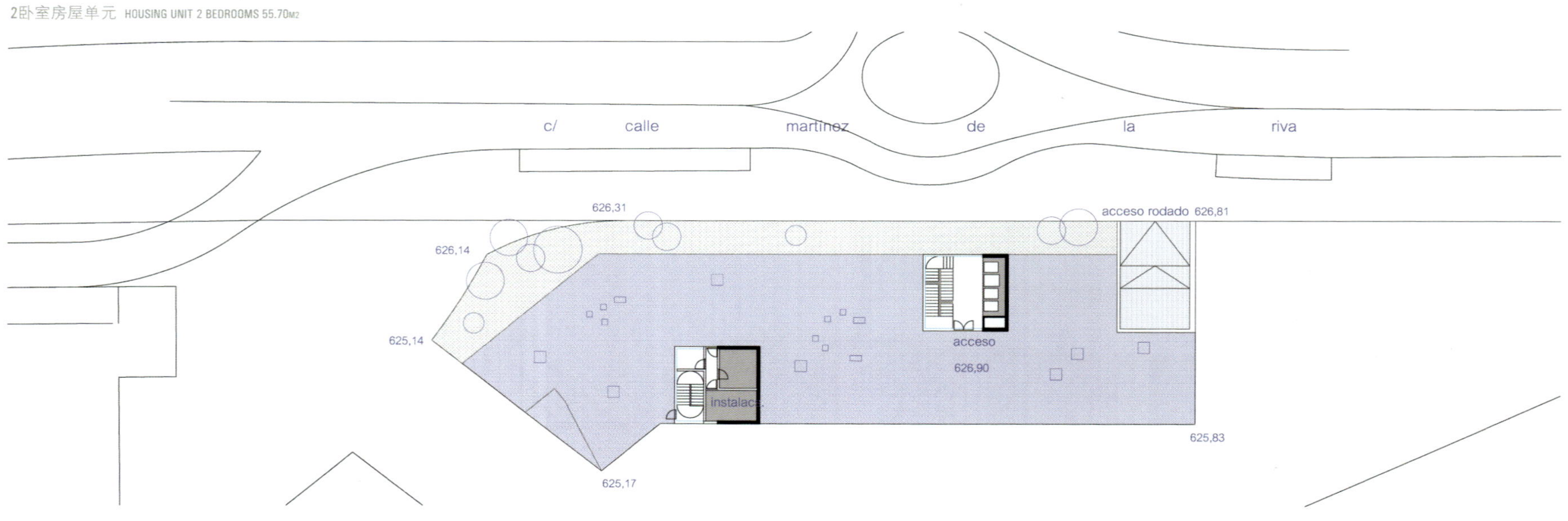


底层平面图 GROUND FLOOR PLAN

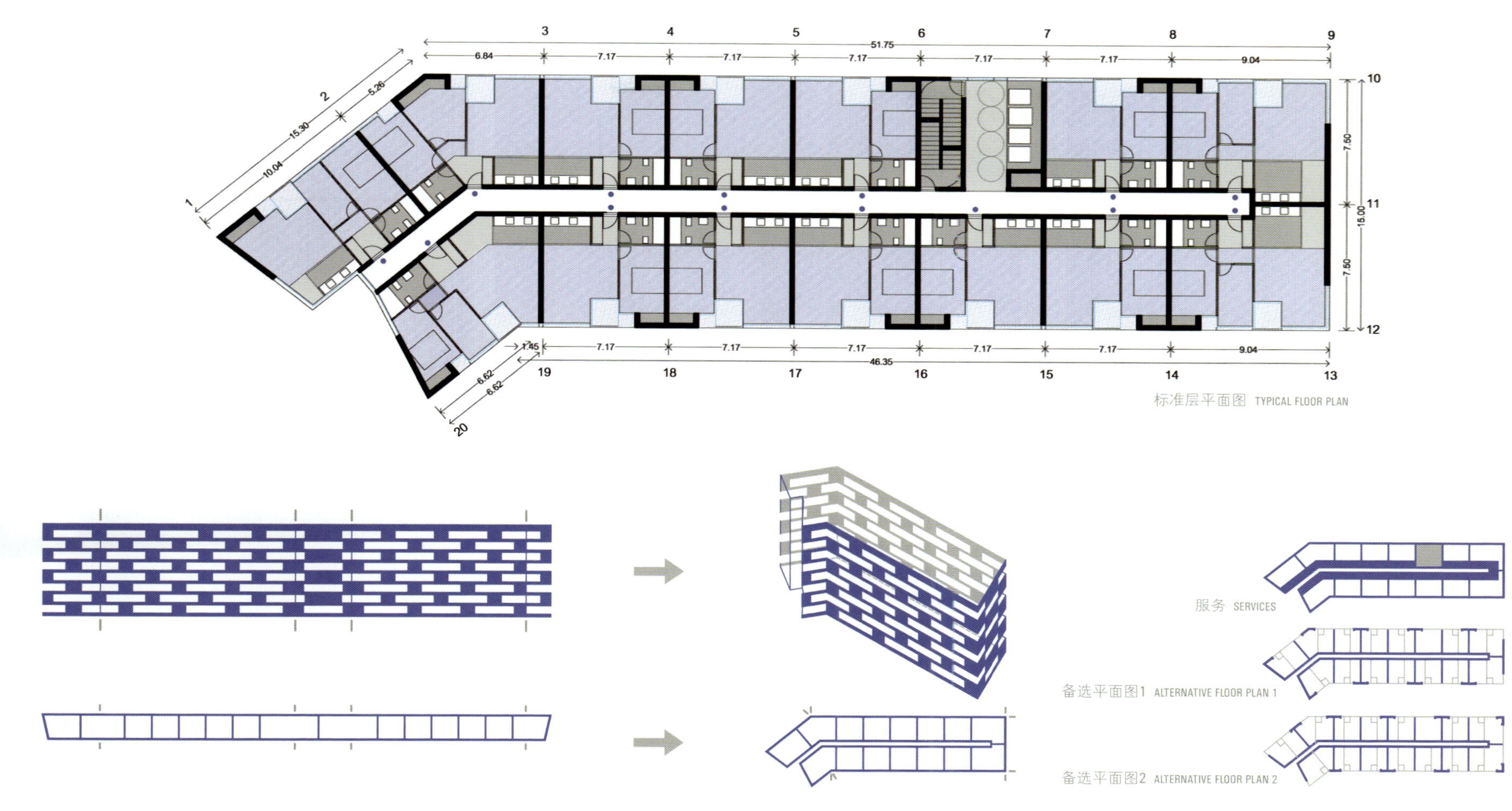

51.75
6.84
7.17
9.04
15.30
10.04
5.26
7.50
15.00
46.35
1.45
6.62
标准层平面图 TYPICAL FLOOR PLAN
服务 SERVICES
备选平面图1 ALTERNATIVE FLOOR PLAN 1
备选平面图2 ALTERNATIVE FLOOR PLAN 2

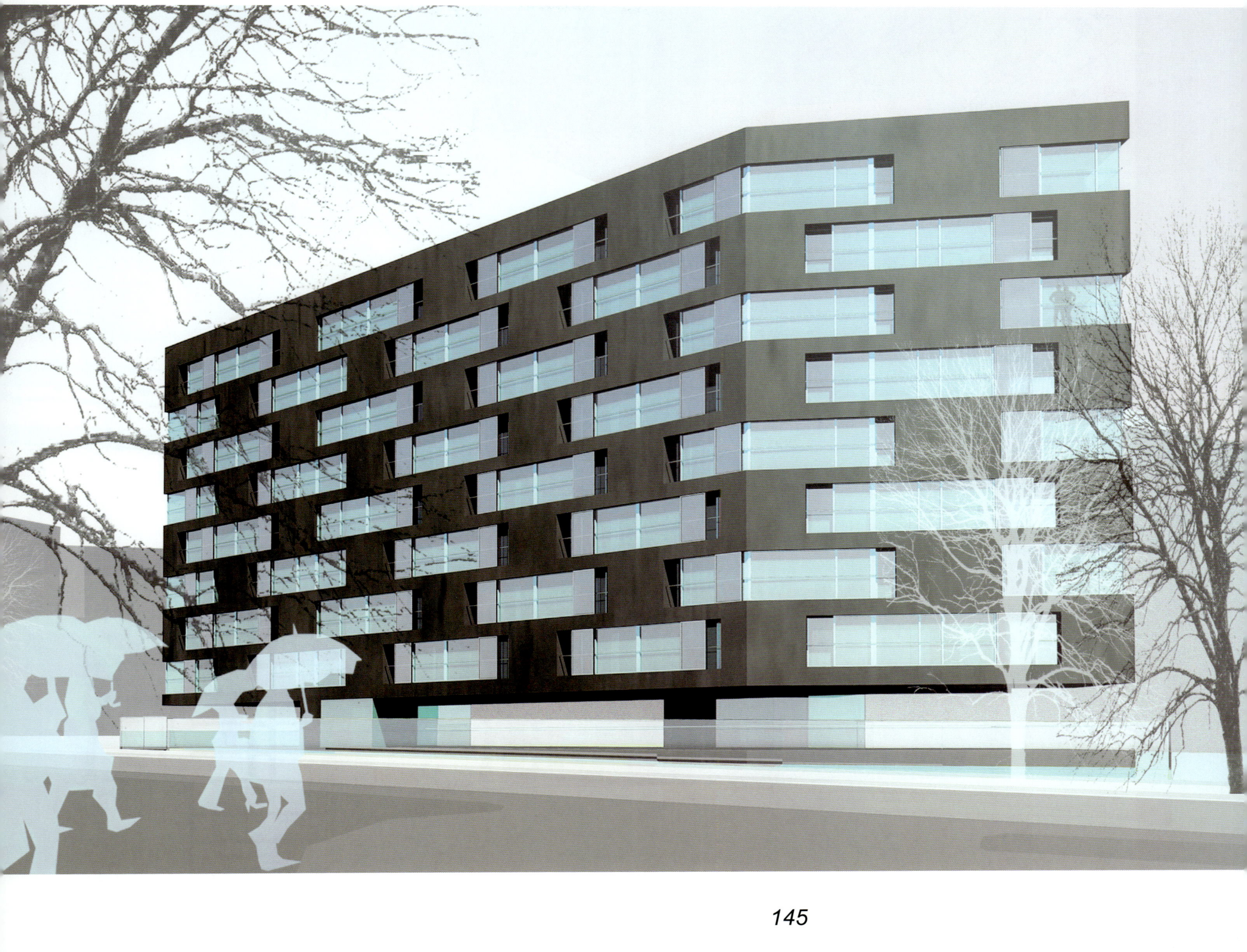

# DJarquitectura (Diego Jimenez López · Juana Sanchez Gómez)(建筑师)

合作 (c) Jorge Salguero Ropero

中标 winner

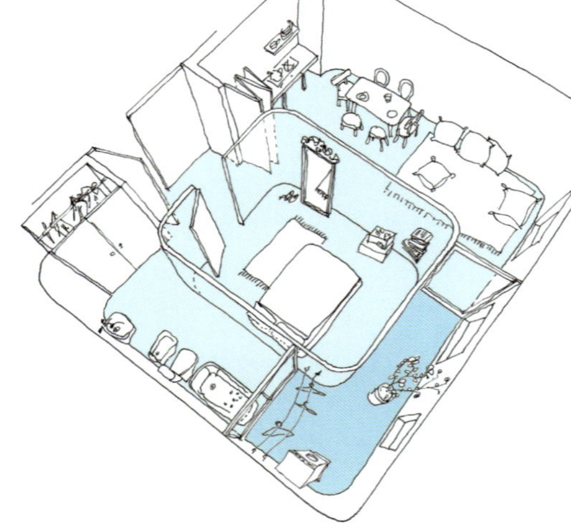

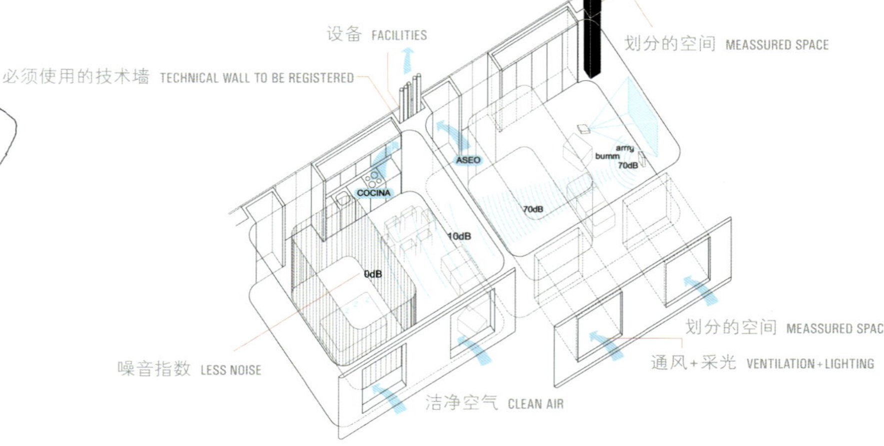


## 不间断空间

所有公寓设计均带有“一个卧室”，这表明公寓适合单身或人员少的家庭居住。因此我们认为建筑周围应设计为热闹地带，且生活便利，以使住户生活更舒适。该公寓形状犹如巢室，中间区域由其他设施所包围，功能更完整。

## A CONTINUOUS SPACE

The fact that all apartments should have "one bedroom" means that they will be inhabited by a single person or a very small family unit. We imagine this resident surrounded by all actions and surrounded by everything you need to make it more comfortable. And his apartment is organized like a cell, where the core is surrounded by other bodies to complete its functions.

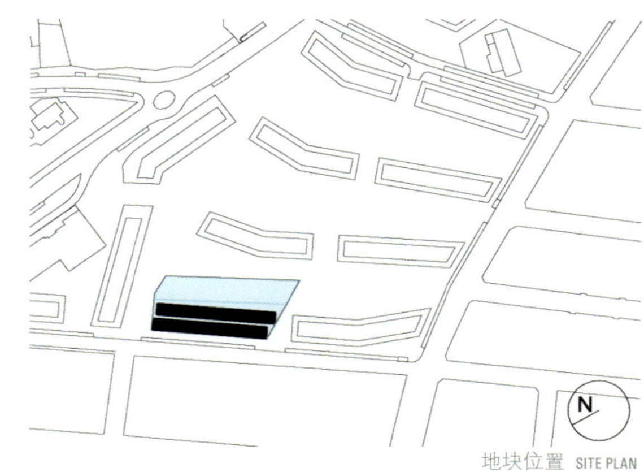


地块位置 SITE PLAN

立面 ELEVATION

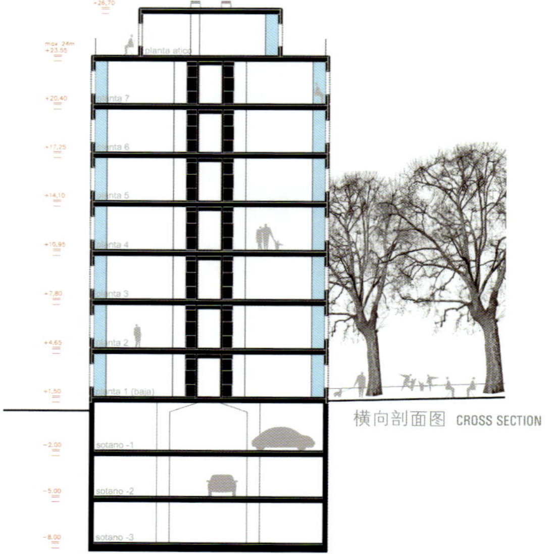

横向剖面图 CROSS SECTION

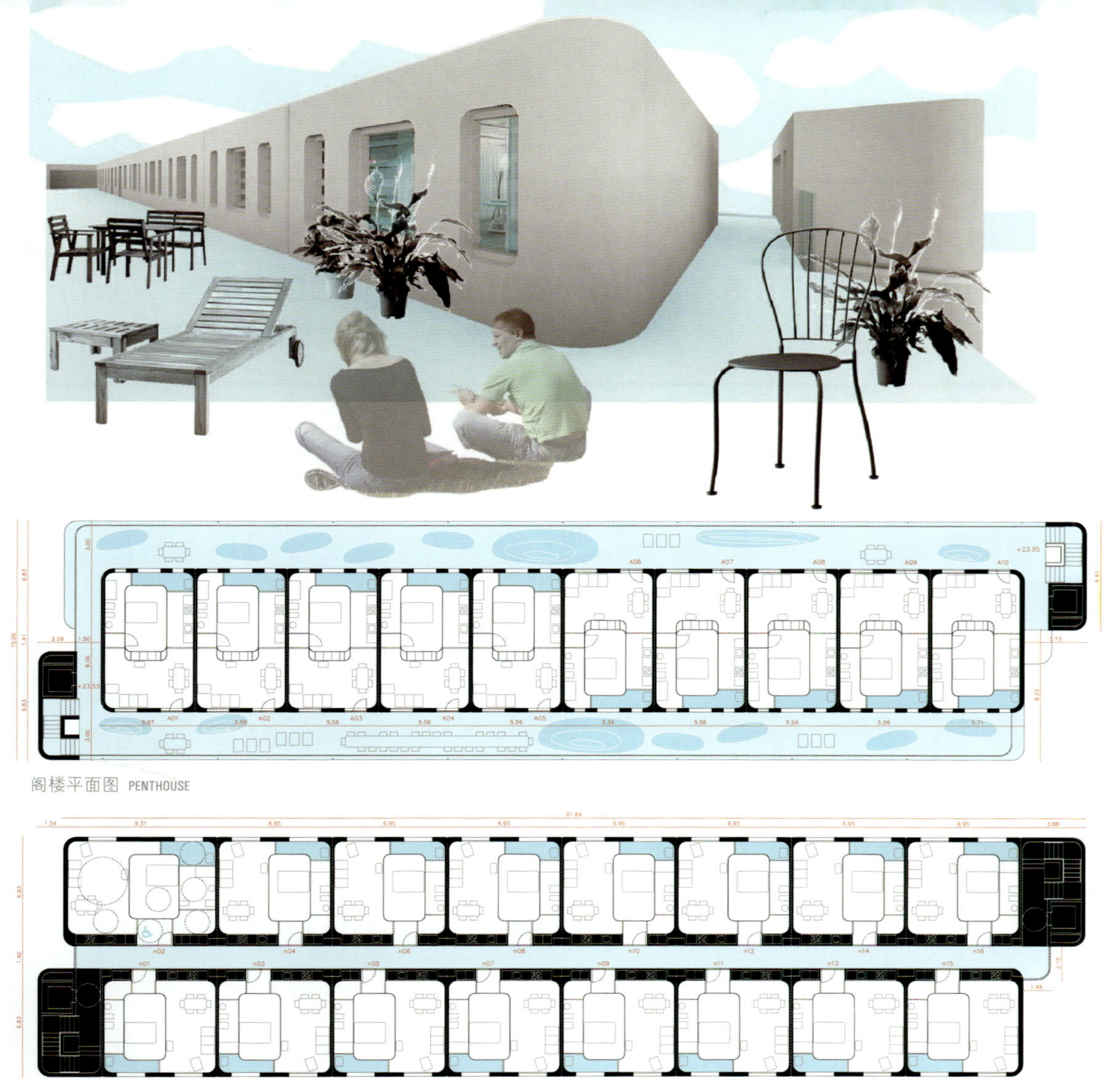

阁楼平面图 PENTHOUSE

标准层平面图 TYPICAL FLOOR PLAN

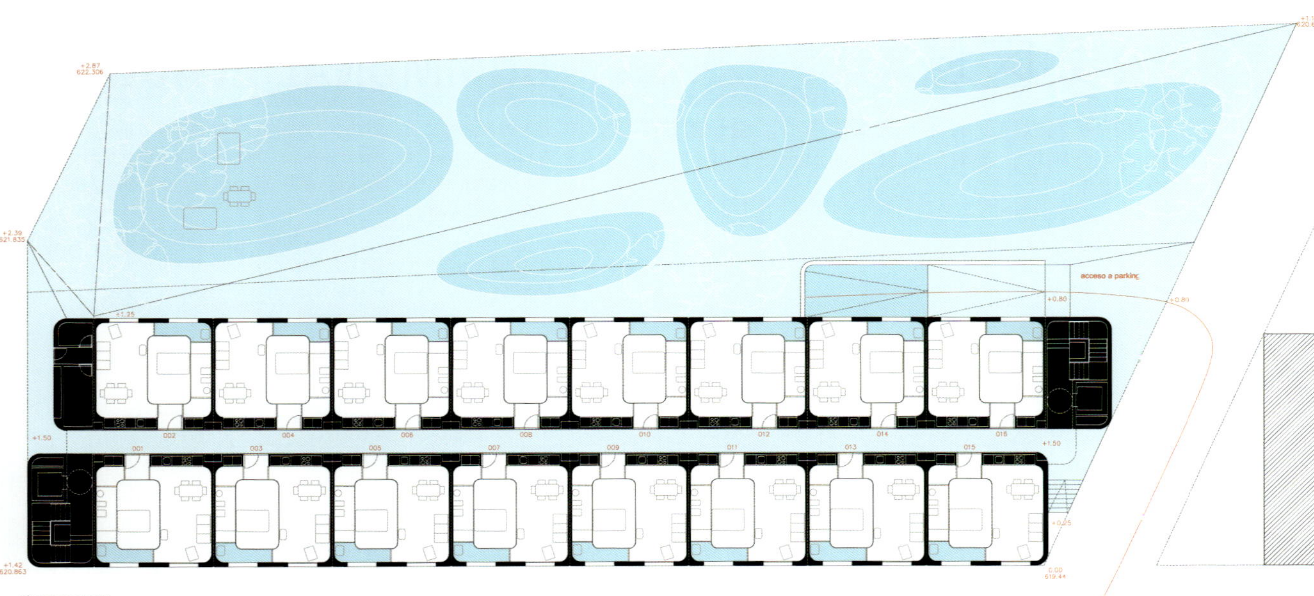

底层平面图 GROUND FLOOR PLAN

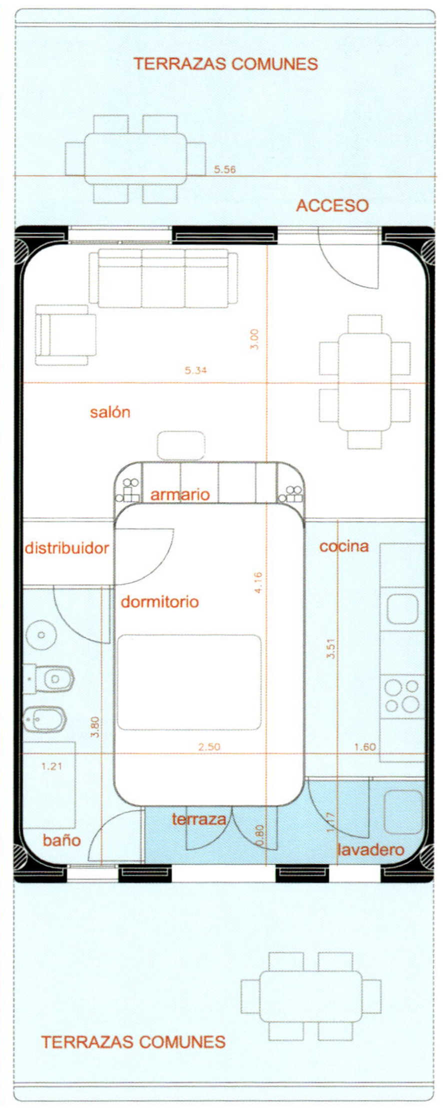


标准阁楼房屋单元 PENTHOUSE TYPICAL HOUSING UNIT

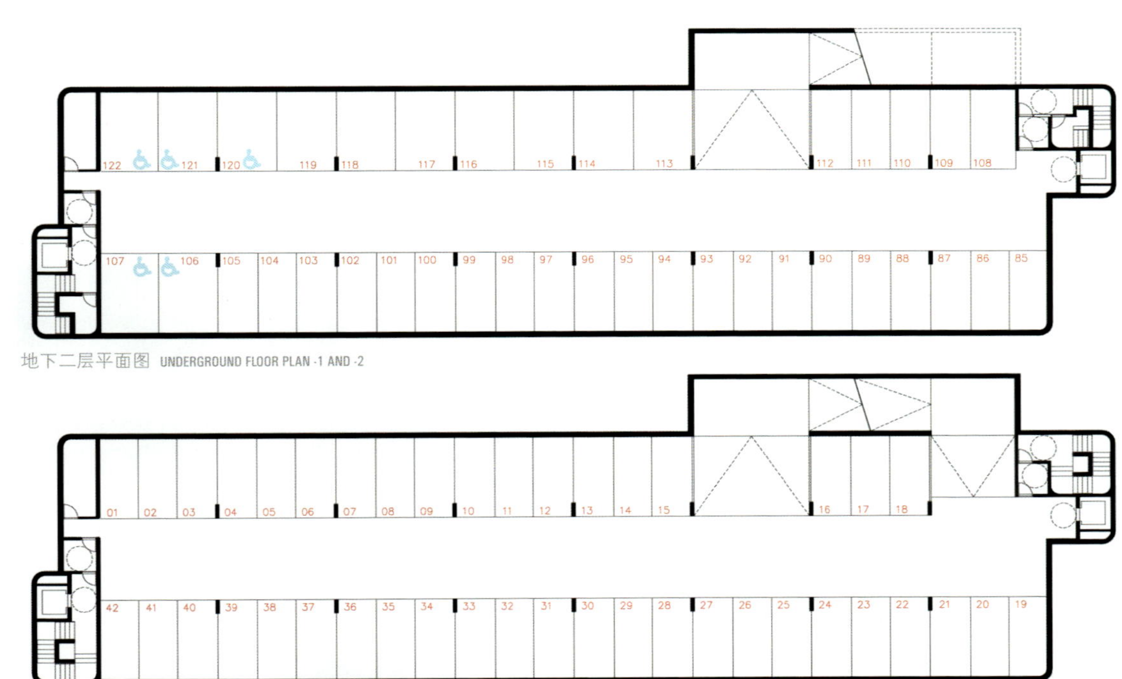

地下二层平面图 UNDERGROUND FLOOR PLAN -1 AND -2

地下三层平面图 UNDERGROUND FLOOR PLAN -3

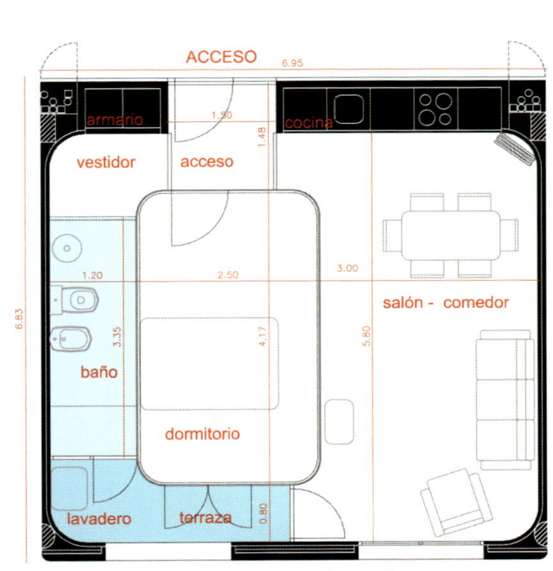


标准房屋单元 TYPICAL HOUSING UNIT

MIRADOR

# estudio km.0 (建筑师事务所)
## (Jose Blazquez · Santiago Cifuentes · Juan Manuel Palacios) (建筑师)
第一提名奖 first mention

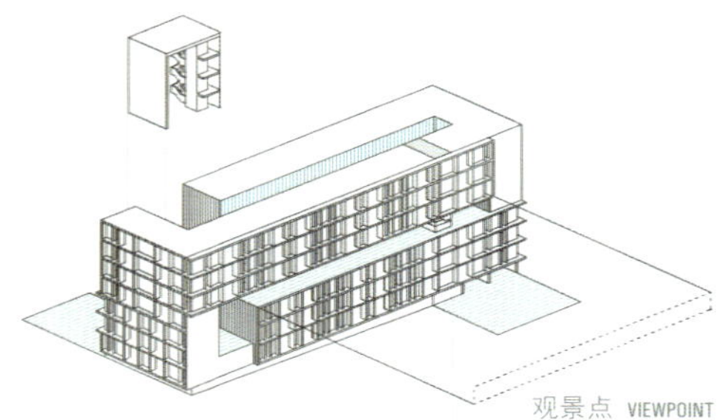


# 新型的公共区域

公寓的布局以内部的环形走廊为中心。由于建筑空间的可分割性，走廊可扩展，通风良好，并可鸟瞰整个城市。

# A NEW COMMUNAL SPACE

The apartments are arranged around an interior circulation gallery. A ventilated gallery which extends thanks to the decomposition of the volume of the building, looking out towards the panoramic views of the city.

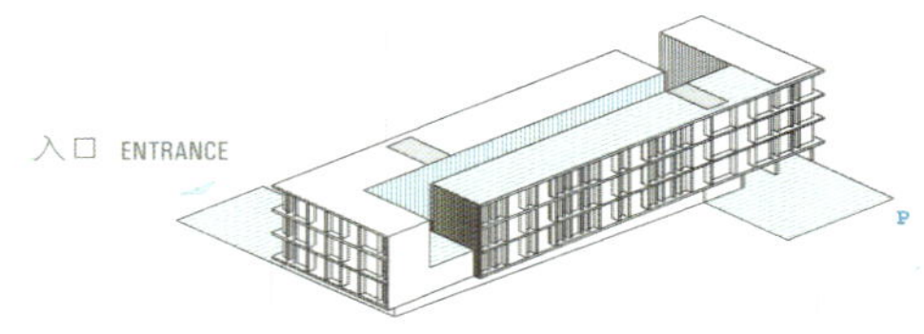


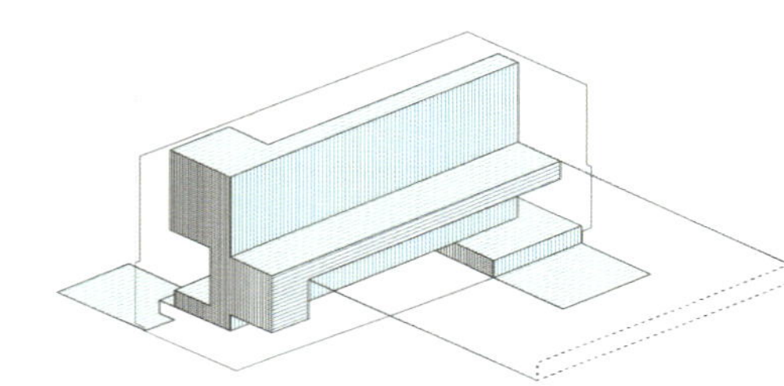


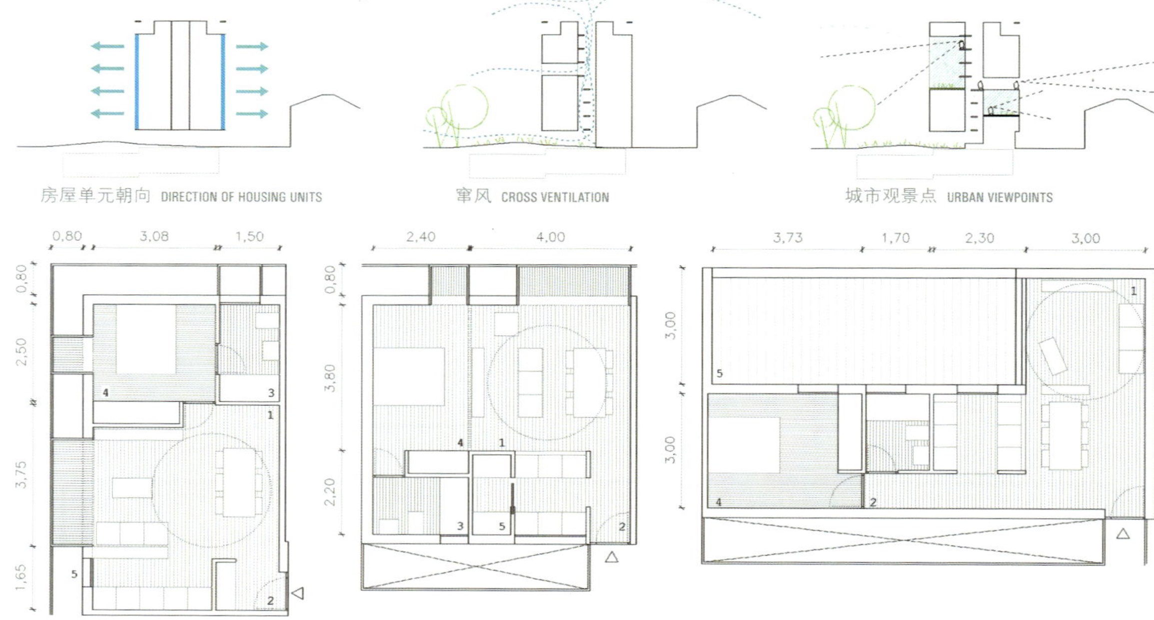


标准房屋单元 TYPICAL HOUSING UNITS

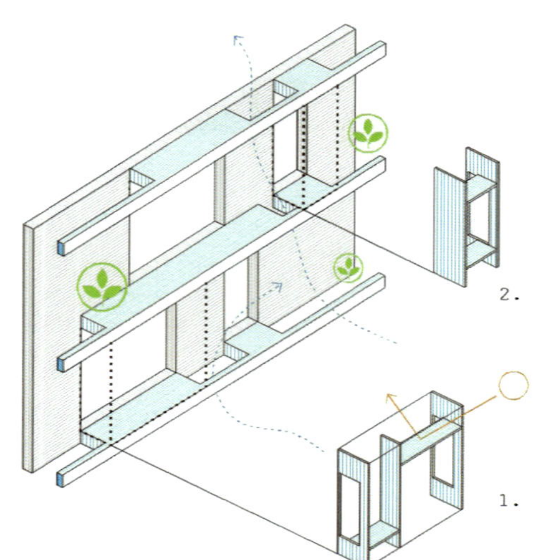


住户外墙 · 宽度80厘米 INHABITED FAÇADE · 80cm WIDTH
1. 起居室模型 LIVING ROOM MODULE
2. 卧室模型 BEDROOM MODULE

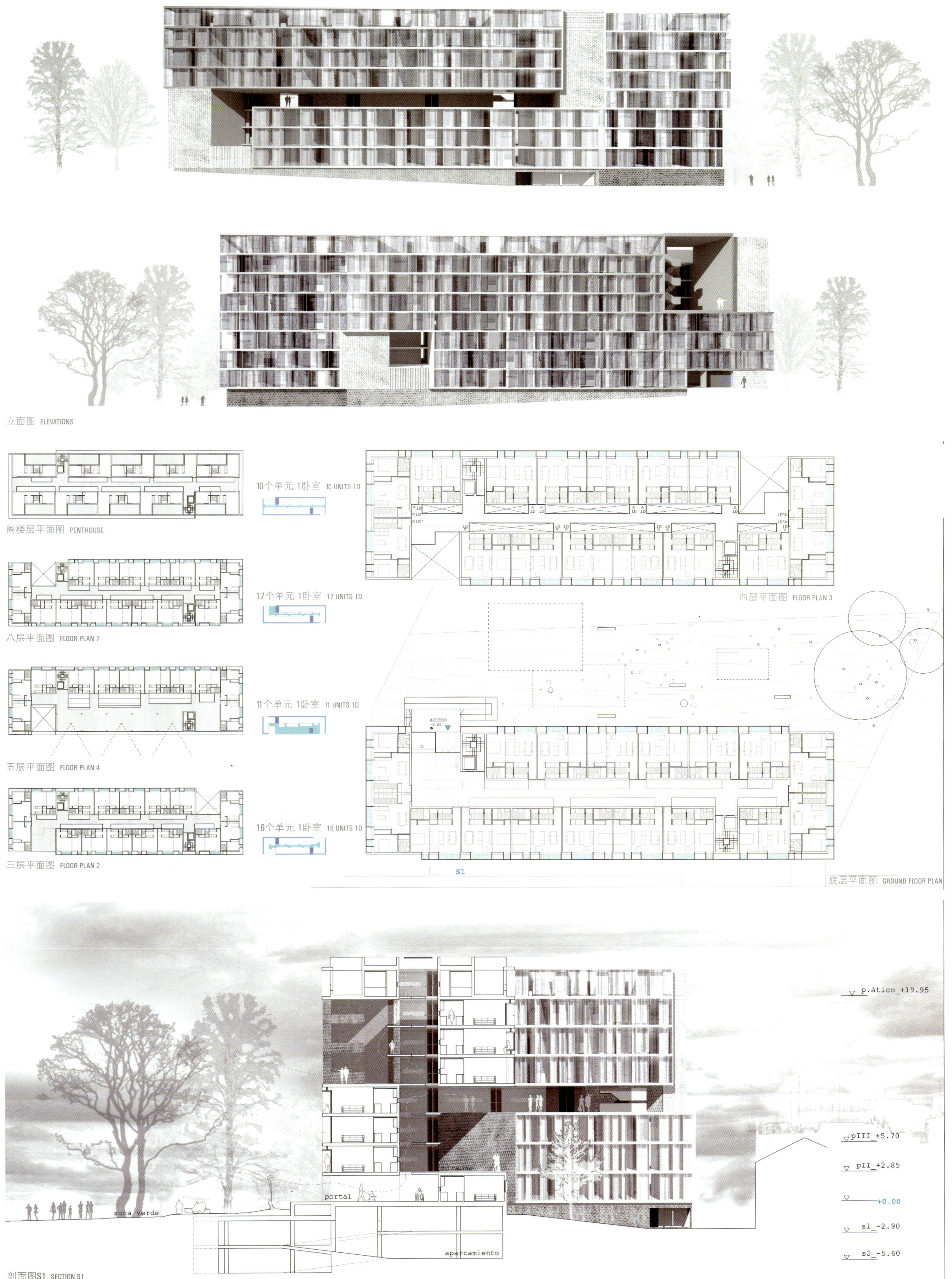
立面图 ELEVATIONS
10个单元 1卧室 10 UNITS 1D
阁楼层平面图 PENTHOUSE
17个单元 1卧室 17 UNITS 1D
八层平面图 FLOOR PLAN 7
11个单元 1卧室 11 UNITS 1D
五层平面图 FLOOR PLAN 4
16个单元 1卧室 16 UNITS 1D
三层平面图 FLOOR PLAN 2
四层平面图 FLOOR PLAN 3
ACCESO
S1
底层平面图 GROUND FLOOR PLAN
p.ático +19.95
pIII +5.70
pII +2.85
+0.00
s1 -2.90
s2 -5.60
portal
mirador
zona verde
aparcamiento
剖面图S1 SECTION S1

REVERSE

# Ramon Gónzalez · Fernando Pancorbo · Mathias Schuette (建筑师)

第二提名奖 second mention

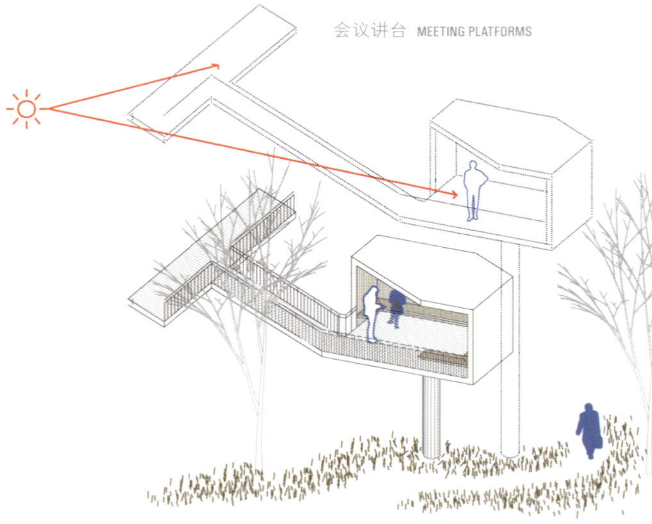

互动辅助设施 SUPPORTS FOR INTERACTION

运动区 SPORTS AREA

公共区 COMMUNAL SPACE

会议讲台 MEETING PLATFORMS

走道 CORRIDORS
+
对角楼梯 DIAGONAL STAIRCASE
+
会议讲台 MEETING PLATFORMS

grada

杨树林 FOREST OF POPLARS

拉尔夫·厄斯金，1980年斯德哥尔摩大学图书馆
RALPH ERSKINE, LIBRARY OF THE UNIVERSITY OF STOCKHOLM 1980

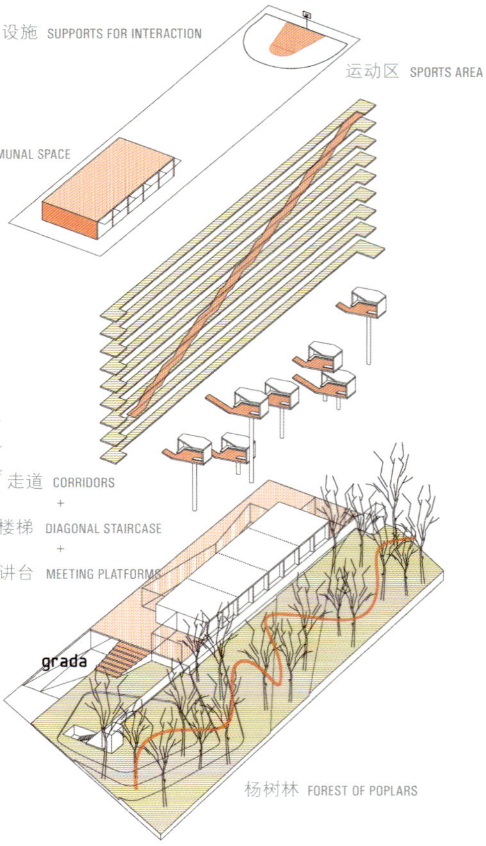

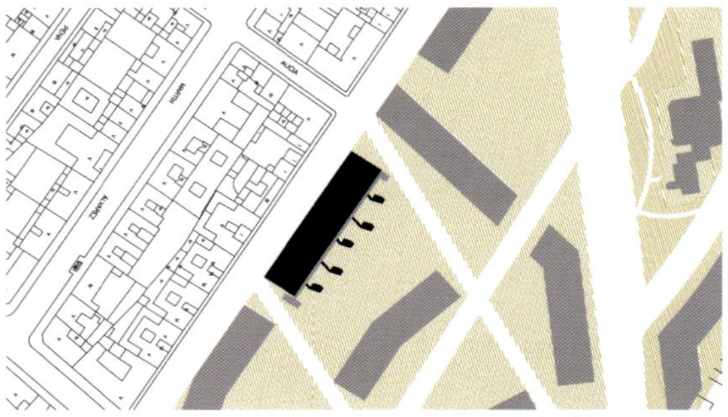

地块位置 SITE PLAN

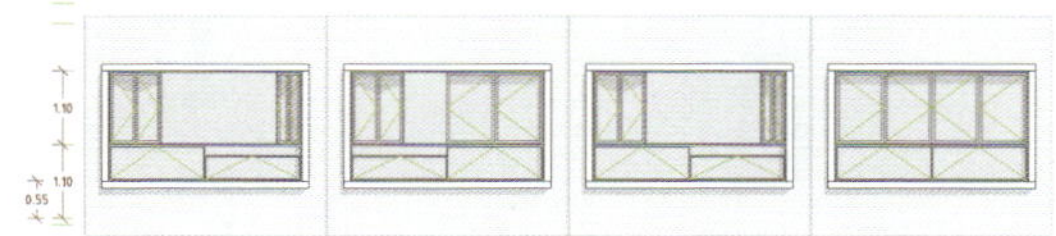

东北立面图 NORTHEAST ELEVATION

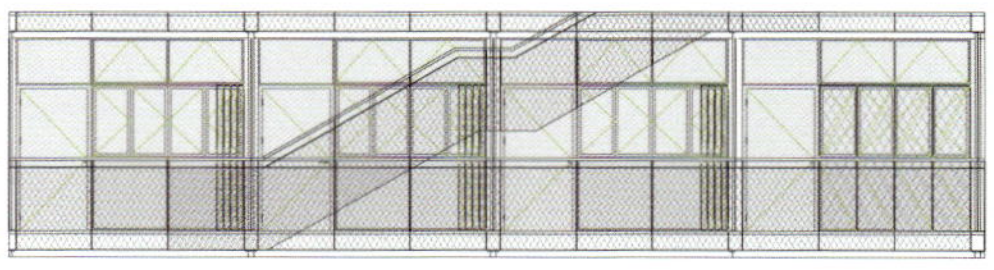

东南立面图 SOUTHEAST ELEVATION

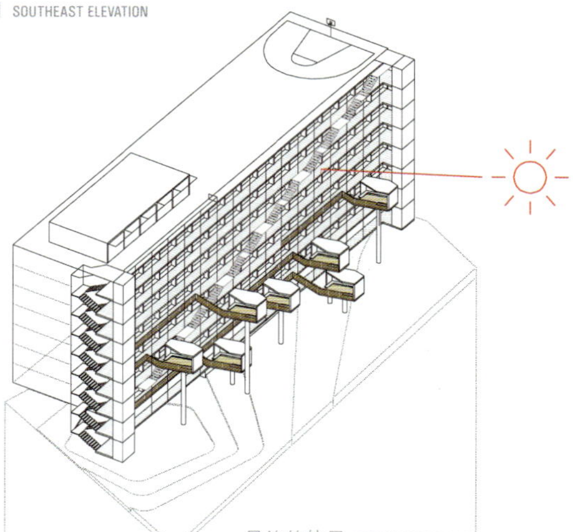

最终的体量 RESULTANT VOLUME

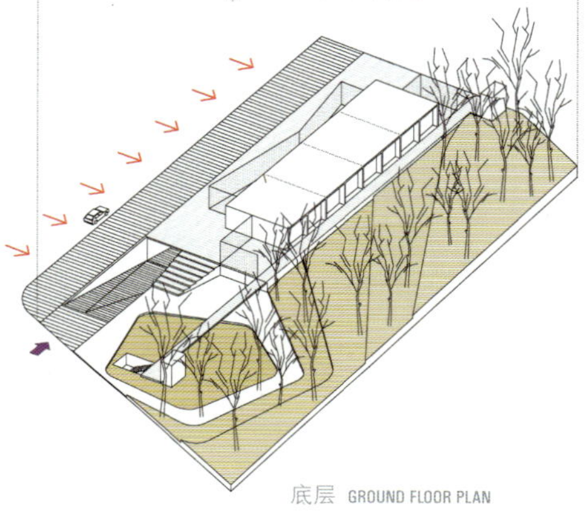

底层 GROUND FLOOR PLAN

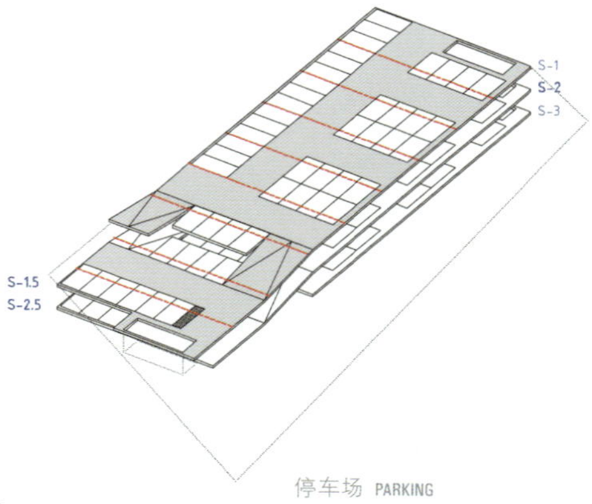

S-1
S-2
S-3
S-1.5
S-2.5

停车场 PARKING

底层平面图 GROUND FLOOR PLAN

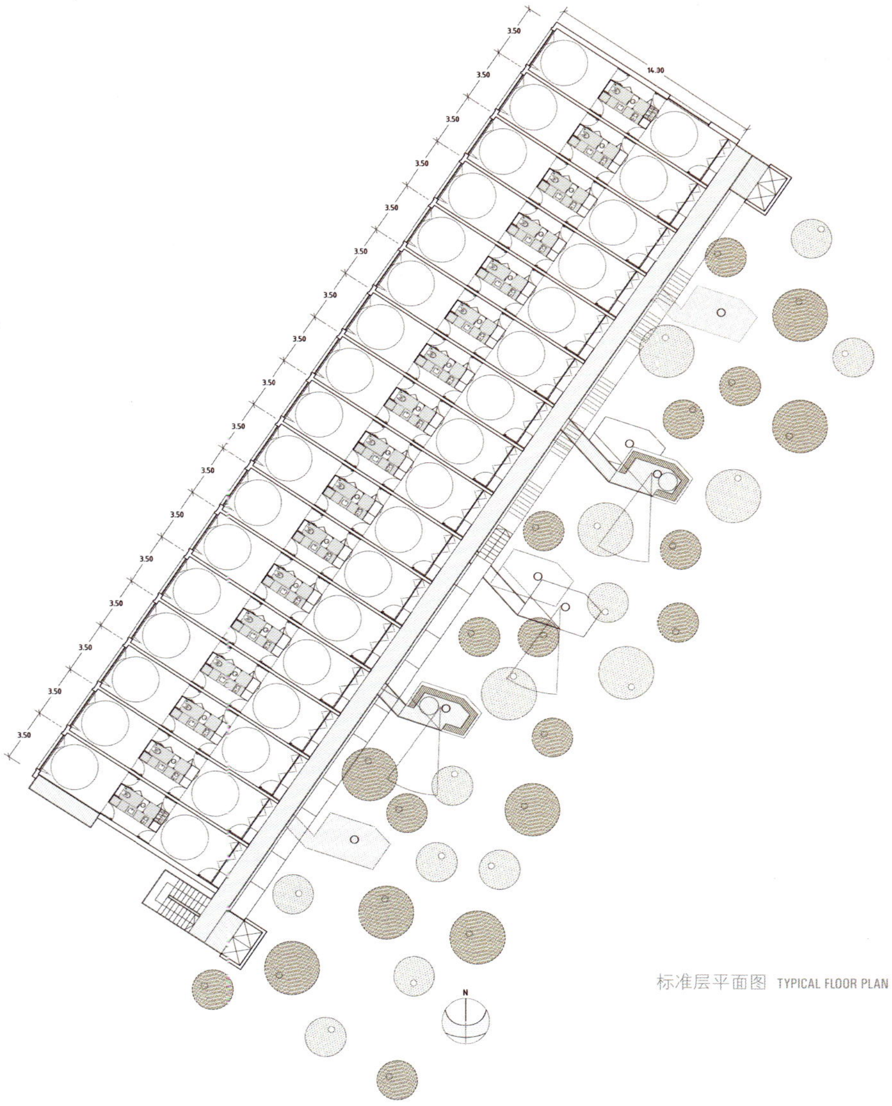

标准层平面图 TYPICAL FLOOR PLAN

# 两种不同的建筑外观

# TWO DIFFERENT FAÇADES

该地块体量小、空间紧凑，所释放的工作区域应考虑该地块各建筑之间的关系：城市－街坊－社区－住宅。该地块的绿地涵盖大片的绿色空间以及白杨树林。顶部可打造成大型公共区域、公用房间、运动场及花园等。

It is a dense and compact volume which releases the working spaces which must set the relationships inside the building: city-neighborhood-community-dwelling. A large green space, a forest of tall poplars, continues the green plot of the site. The roof is a great communal space, a large communal room, a sports area, a garden.

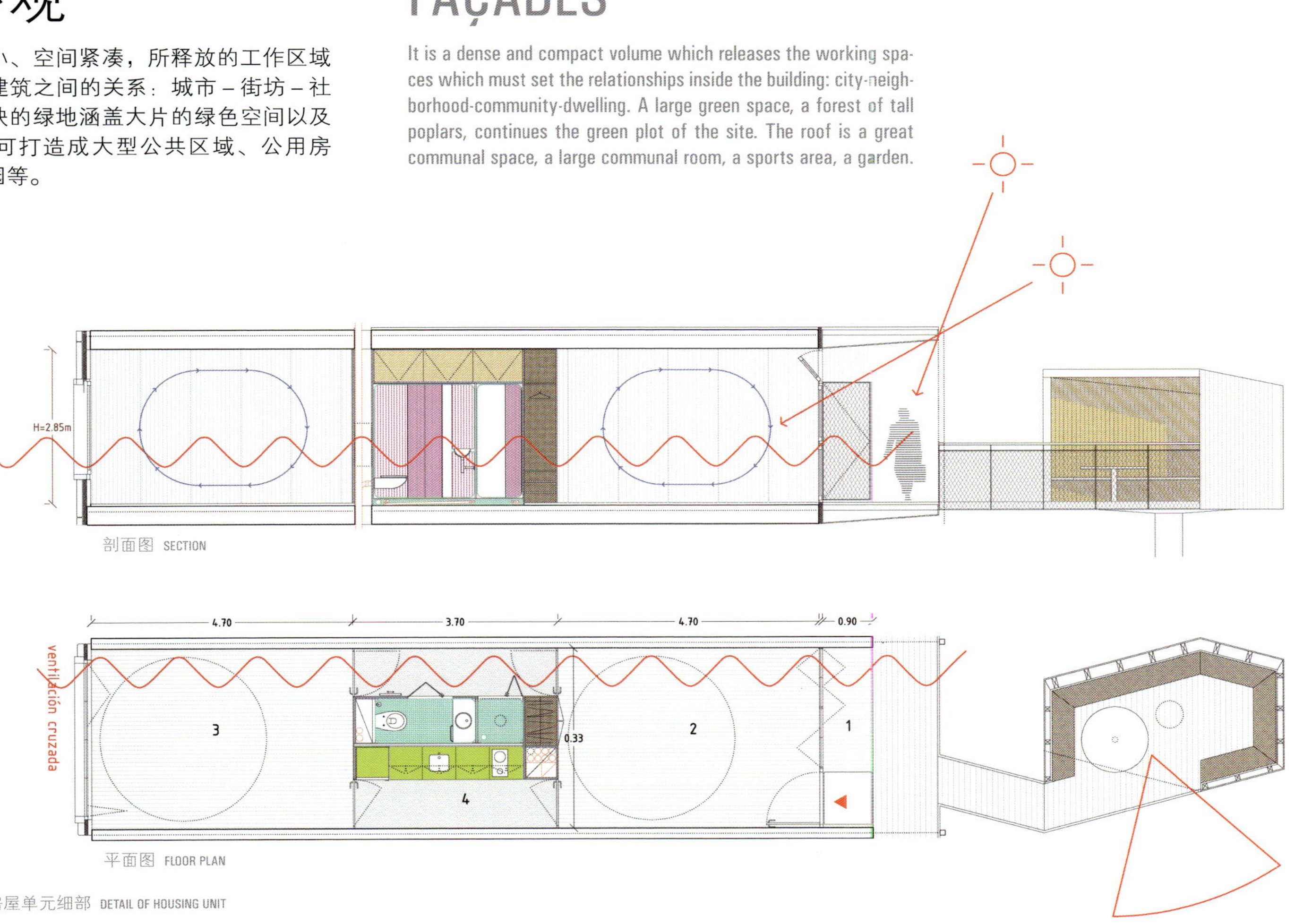


剖面图 SECTION

平面图 FLOOR PLAN

房屋单元细部 DETAIL OF HOUSING UNIT

## Borja Peña (beta.0 architecture & landscape) (建筑师事务所)

合作 (c) Carla Arranz · Almudena Cano · Lucía Espinosa de Los Monteros · Gonzalo Peña

第三提名奖 third mention

# 可塑性模式

# MOULD PLASTIC

该项目基于三种战略而建：城市与环境一体化、能源效率及成本削减。客厅设计灵活，可适应各种变化。住宅单元内部无支柱，也无安装各种家具设施的孔洞，而是将所有的结构和设施近于建筑正面而设计。

The project proposes development around 3 strategic lines: Urban and environmental integration, energy efficiency, and reduction of costs. The living space is flexibly designed to simply adapt to the changes over time. The housing unit is entirely free, in its interior, from pillars and facilities holes designing the structure and facilities close to the façade.

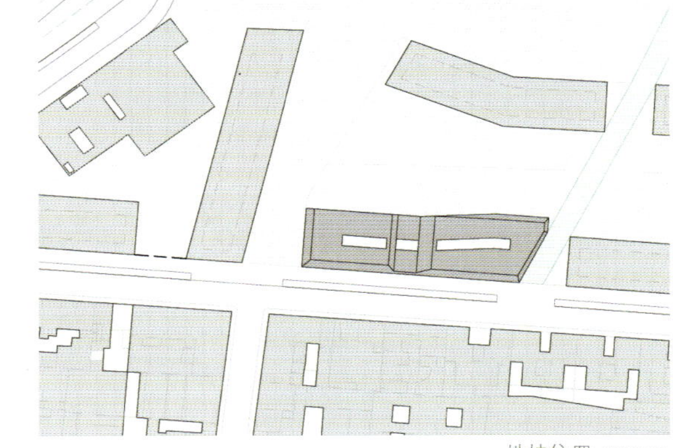

地块位置 SITE PLAN

基于规则地块上的实体 SOLID BASED ON REGULATIONS

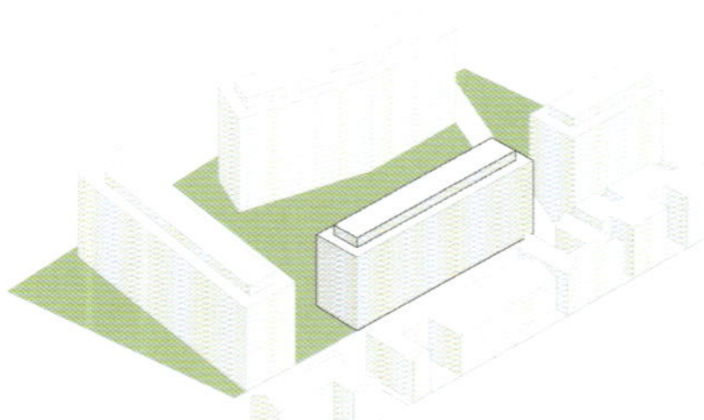

最大占用和分隔 MAXIMUM OCCUPATION-SEPARATION

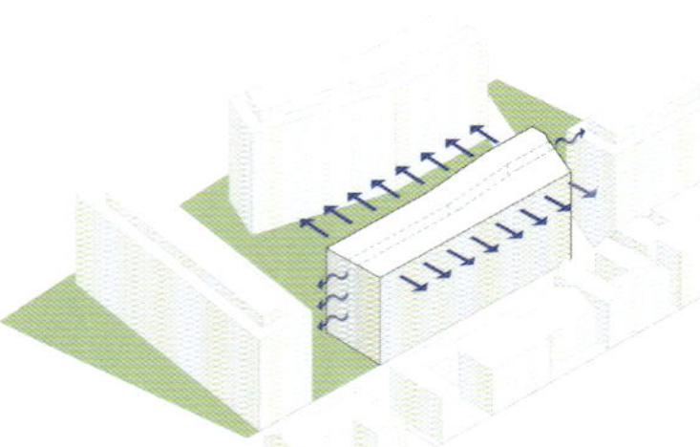

调整外形 SHAPING THE ENVELOPE

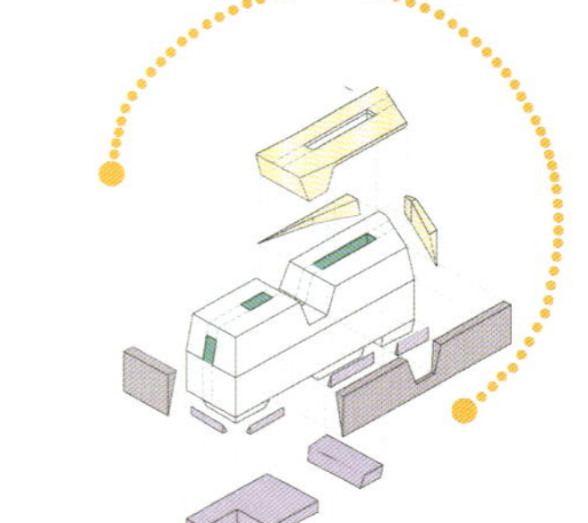

植入地块 INSERTING INTO THE URBAN PLOT

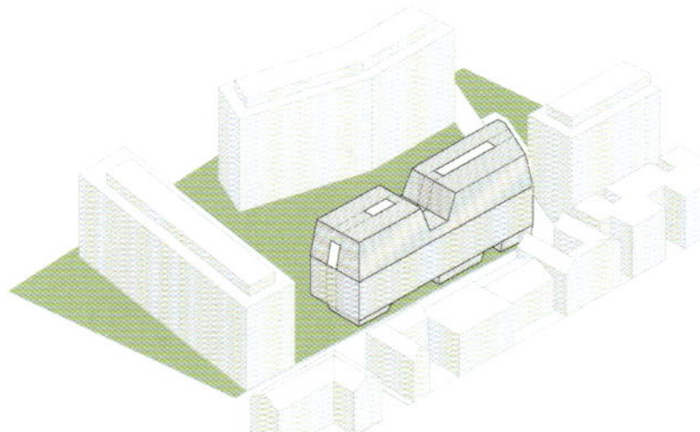

生成过程 CONFIGURATION PROCESS

八层平面图(阁楼) FLOOR PLAN 7 (PENTHOUSE)

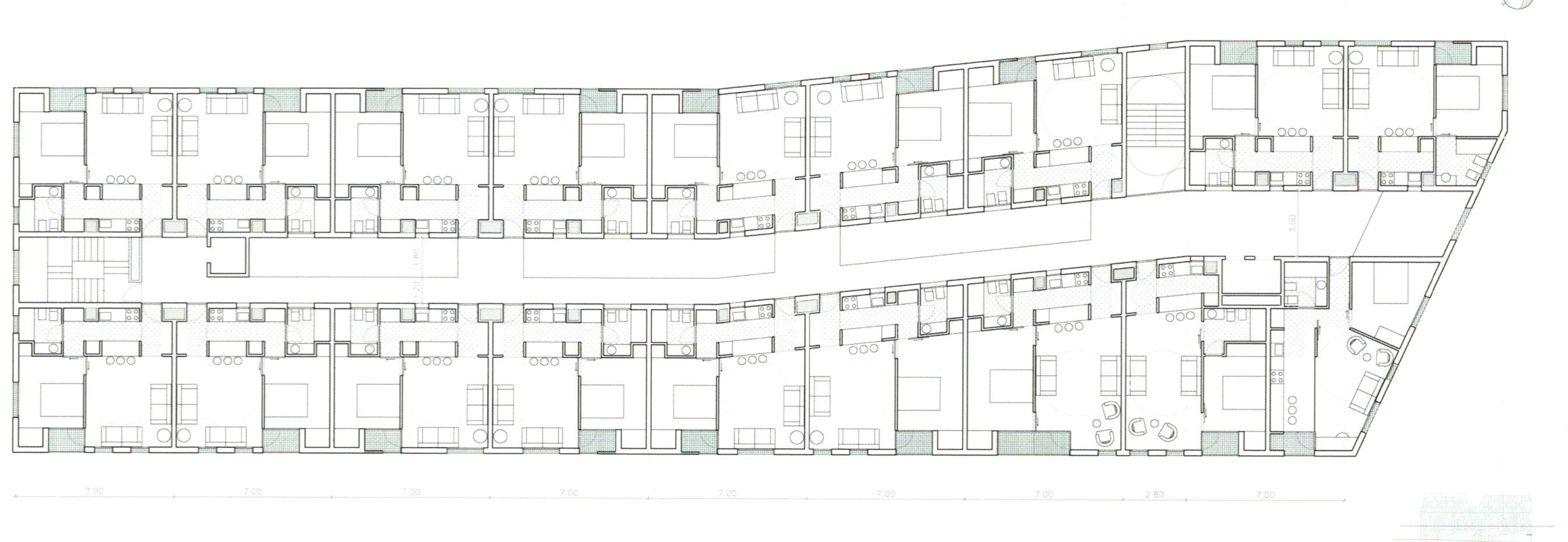

标准层平面图 三、四、五、六层 TYPICAL FLOOR PLANS 2, 3, 4, 5

北立面 NORTH ELEVATION

西立面 WEST ELEVATION

南立面 SOUTH ELEVATION

东立面 EAST ELEVATION

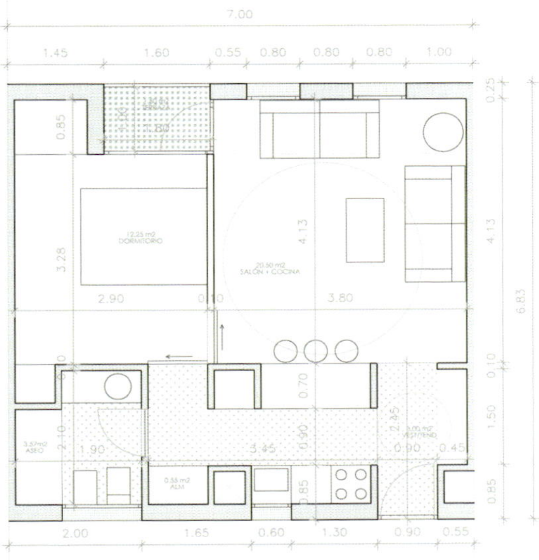

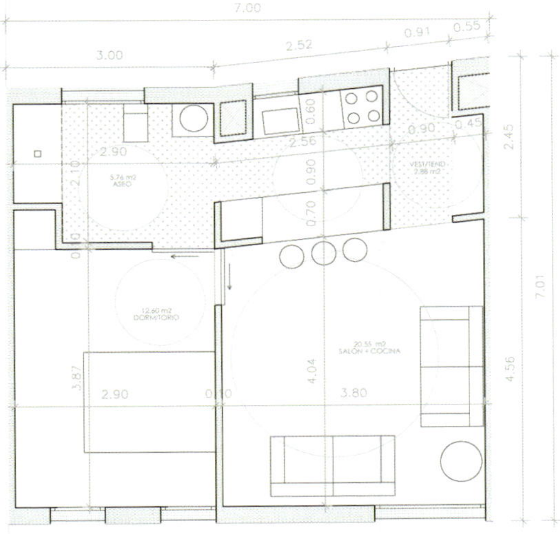

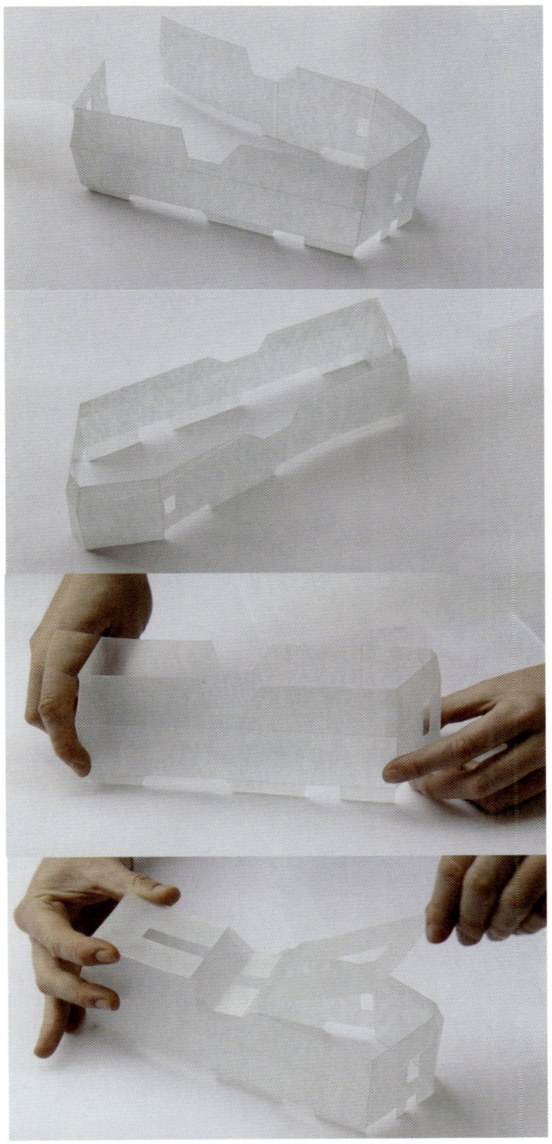

七层平面图 FLOOR PLAN 6

六层平面图 FLOOR PLAN 5

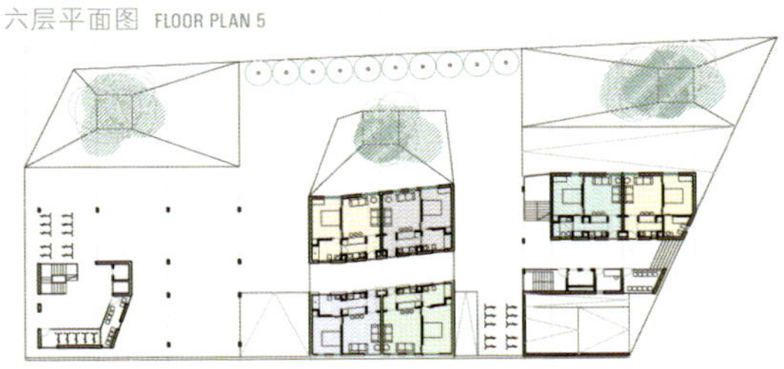

底层平面图 GROUND FLOOR PLAN

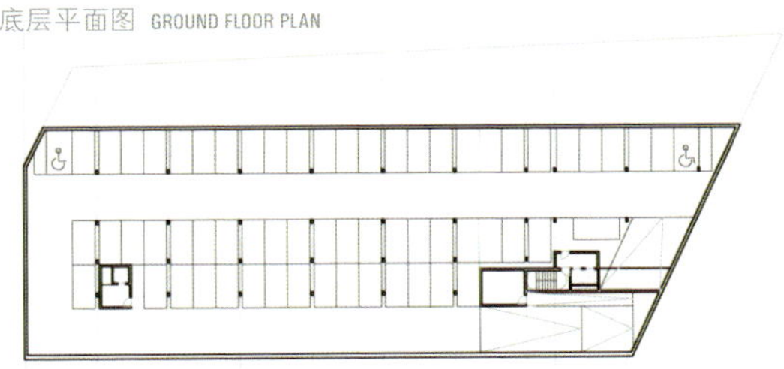

地下一层平面图 UNDERGROUND FLOOR PLAN -1

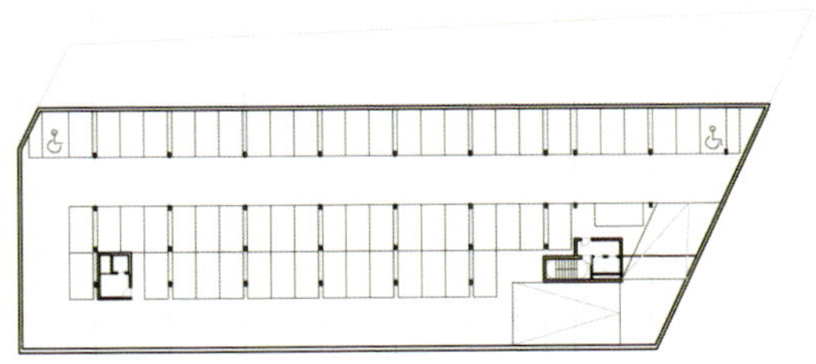

地下二层平面图 UNDERGROUND FLOOR PLAN -2

房屋单元细部 DETAIL OF HOUSING UNIT

ENERGIA

# Eva Sanchez Moya · Pablo Roel Herranz (建筑师)

第四提名奖 fourth mention

## 人字形

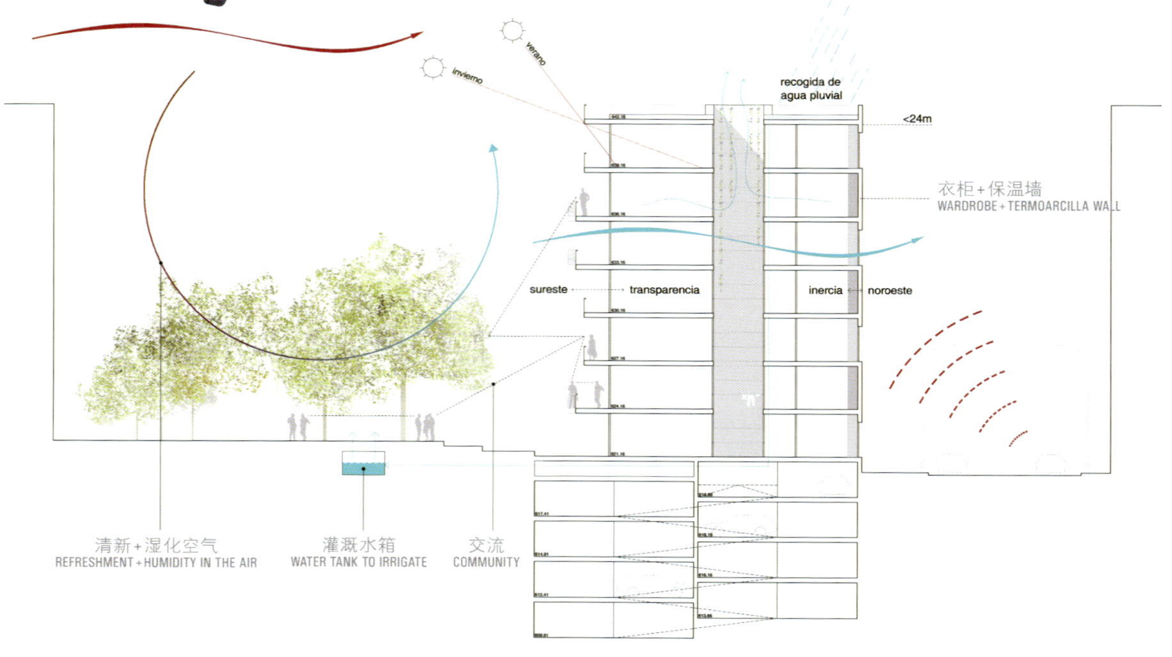


该项目的设计基础在于，所有户型均对流通风，内部的花园和街道的外部空间均能为所有住户享用。建筑的布局以及各个单元的位置均强调了其双面朝向：正面是朝向花园的走道，作为出入各单元的通道；背面朝向街道，使建筑更具隐密性。

## HERRINGBONE

The project involves the assumption that all households have cross ventilation and receive benefit from the interior garden of the plot and the outer space of the street. The organization of the building and the location of housing units emphasize the duality of both sides. On one side the walkways which open towards the garden solve the access to the housing units. On the other side of the street, the building becomes more closed and introverted.

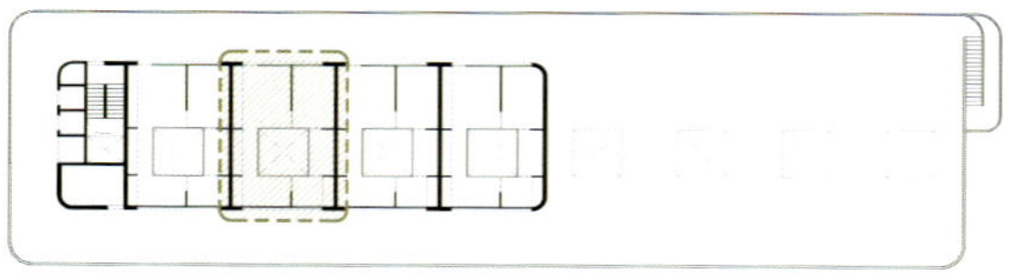

阁楼层平面图 PENTHOUSE

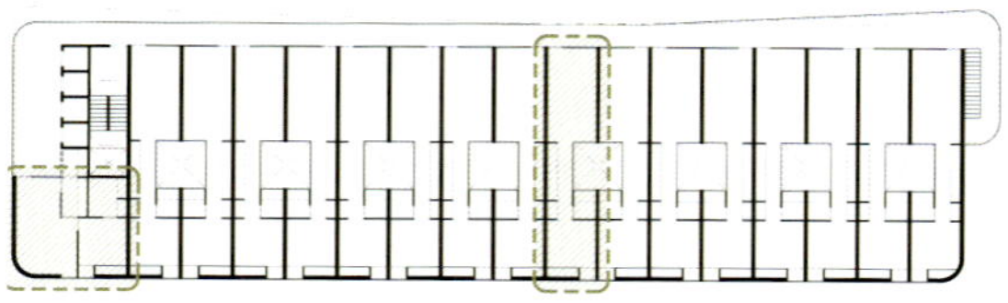

标准层平面图 TYPICAL FLOOR PLAN

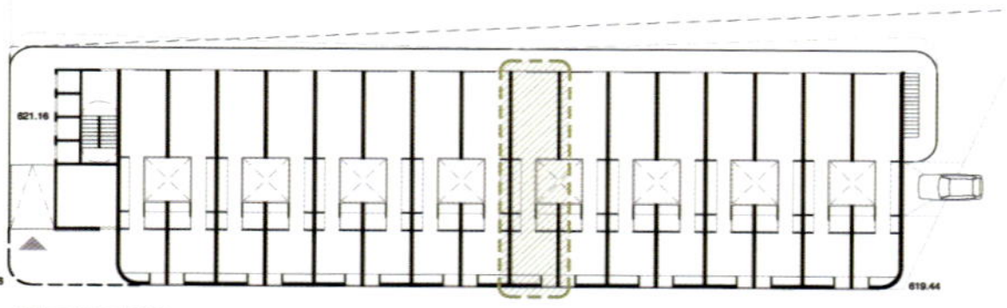

底层平面图 GROUND FLOOR PLAN

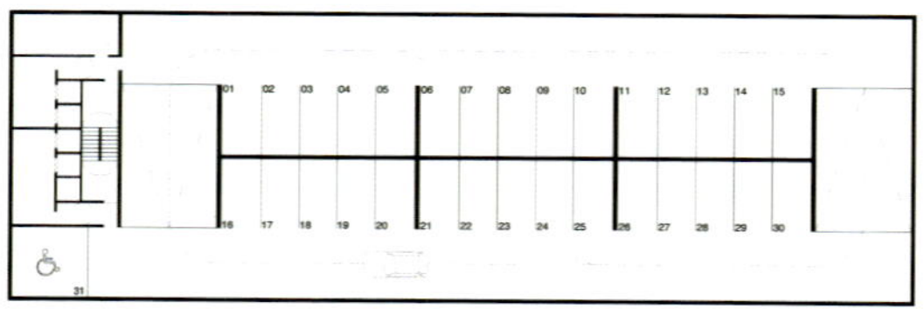

地下层平面图 UNDERGROUND FLOOR PLAN

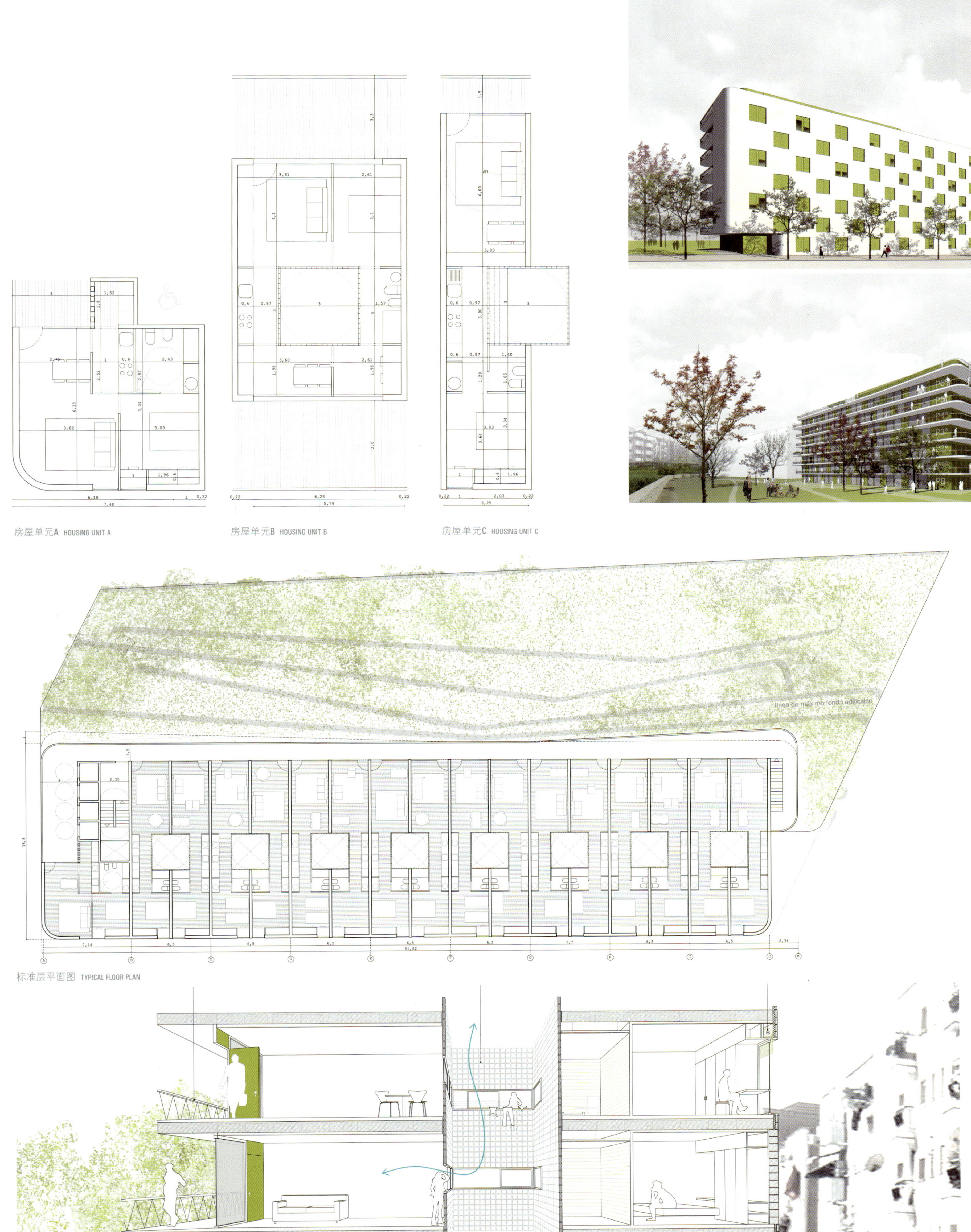

房屋单元A HOUSING UNIT A

房屋单元B HOUSING UNIT B

房屋单元C HOUSING UNIT C

标准层平面图 TYPICAL FLOOR PLAN

B型房屋单元 纵向剖面 HOUSING UNIT TYPE B. LONGITUDINAL SECTION

UPNDOWN

# Espegel-Fisac Arquitectos (Carmen Espegel · Concha Fisac)
# (建筑师事务所)

合作 (c) Laia Lafuente · Marcela Aragüez · Miriam Llamazares · Teresa Serrano

第五提名奖 fifth mention

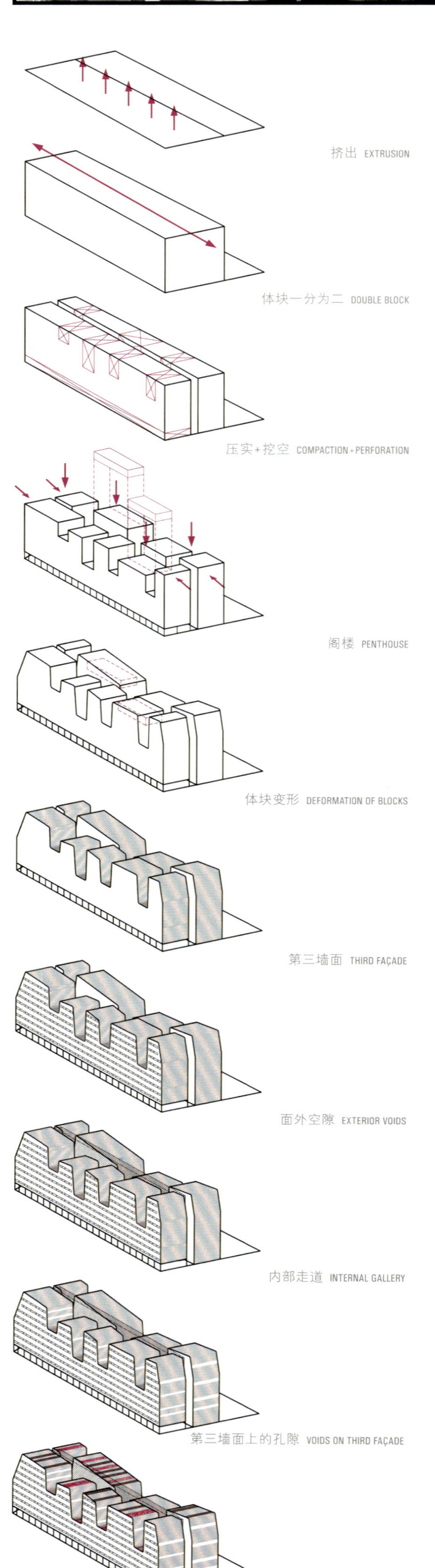

起初条件 STARTING CONDITIONS

max.24m

提议方案 PROPOSAL

max.24m

## 运动

住宅单元如此布局，可释放地面空间，形成自由、透明、开放、灵活的公共区域，为整个建筑提供一个小型的城市空间。另外，特意设计的不规则形状使得露台可释放更多的室外空间。

## MOVEMENT

The layout of the housing units releases the ground floor, which is free and transparent, open and flexible, designed for public use, providing the complex with a small-scale urban space. The intentional fluffy block creates new outdoor spaces on the terraces.

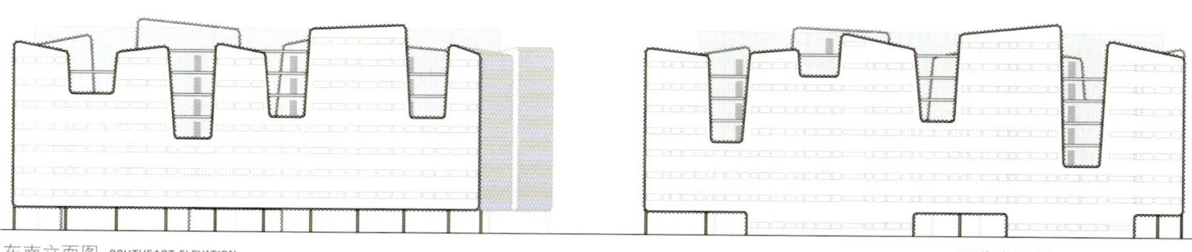

东南立面图 SOUTHEAST ELEVATION

西北立面图 NORTHWEST ELEVATION

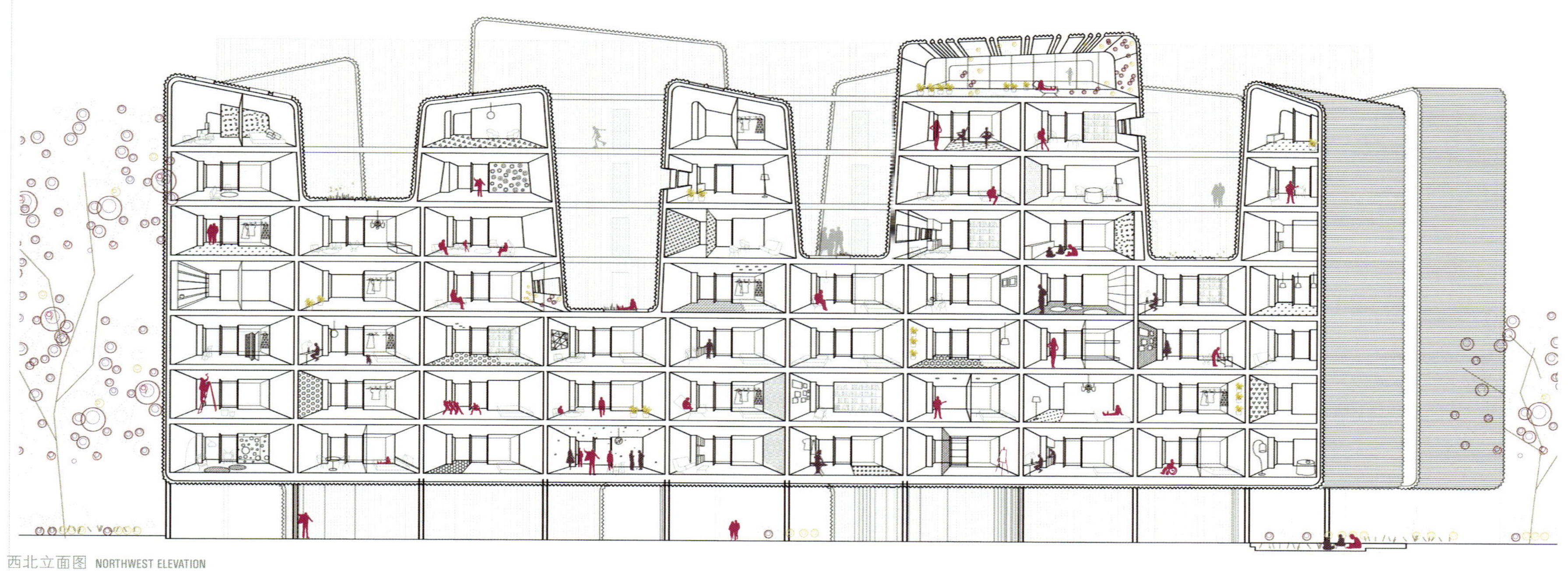

西北立面图 NORTHWEST ELEVATION

房屋单元 可能的类型 HOUSING UNIT POSSIBILITIES

六层平面图 FLOOR PLAN 5

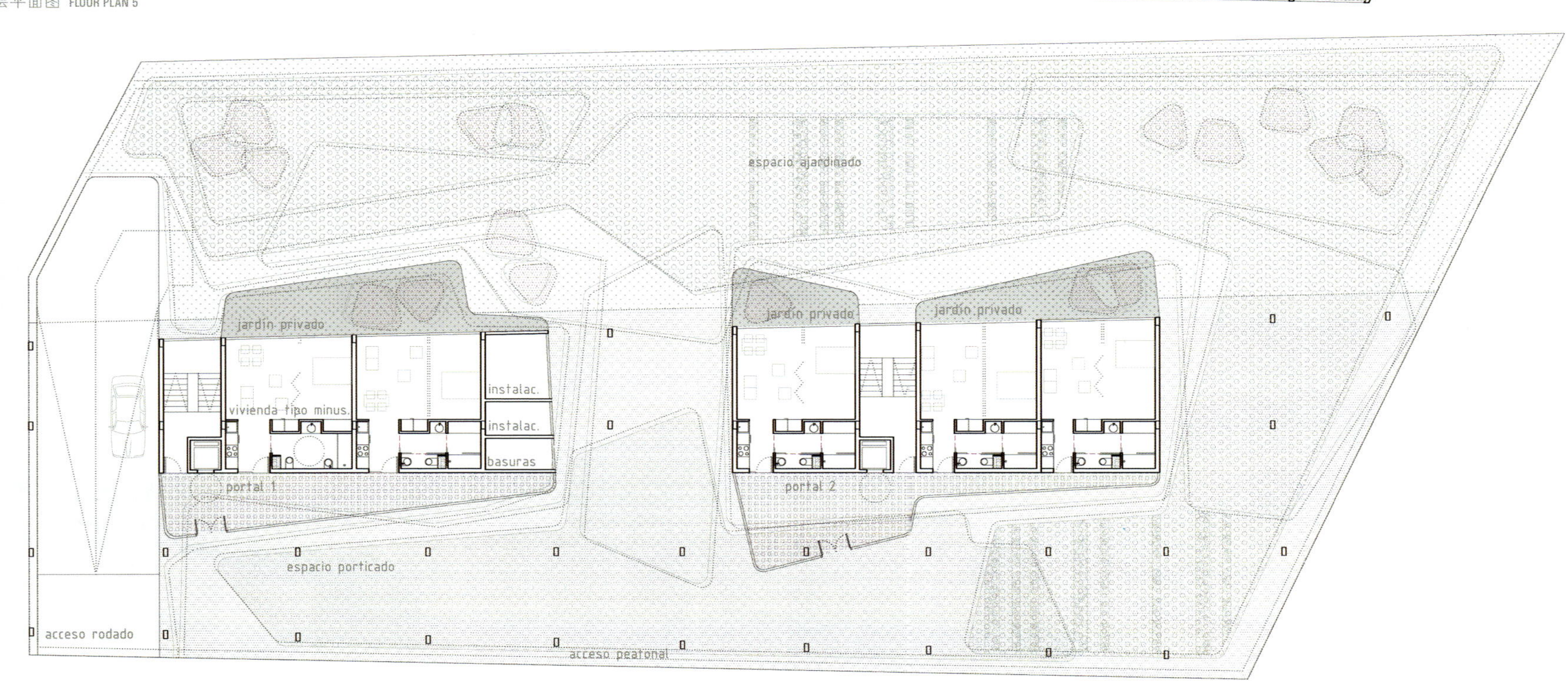


底层平面图 GROUND FLOOR PLAN

# Estudio UNTERCIO (建筑师事务所)

Marina del Mármol · Daniel Bergman · Mauro Bravo · Miguel Herraiz (建筑师)

中标 winner

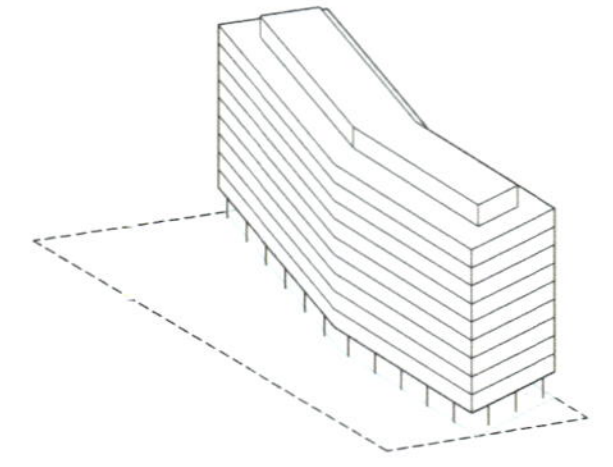
区块内部走道 INTERIOR BLOCK-CORRIDOR

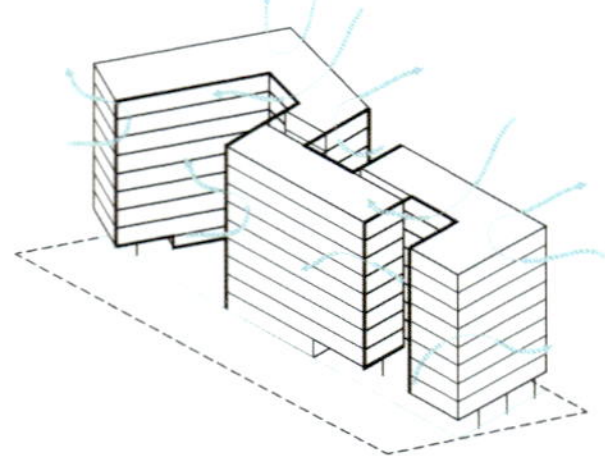
3个敞开的庭院 3 OPEN COURTYARDS

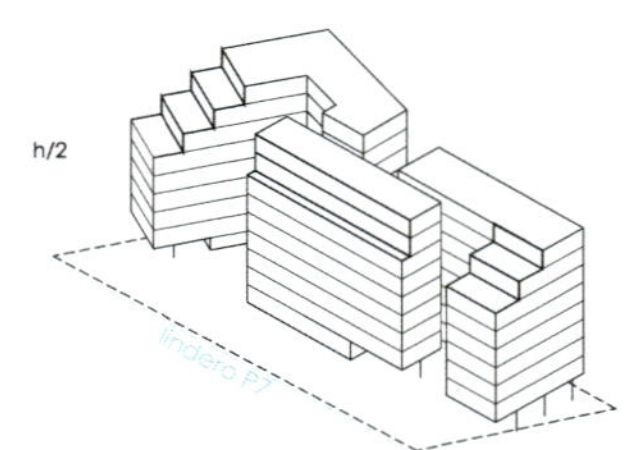

减小尺度 REDUCTION OF THE SCALE

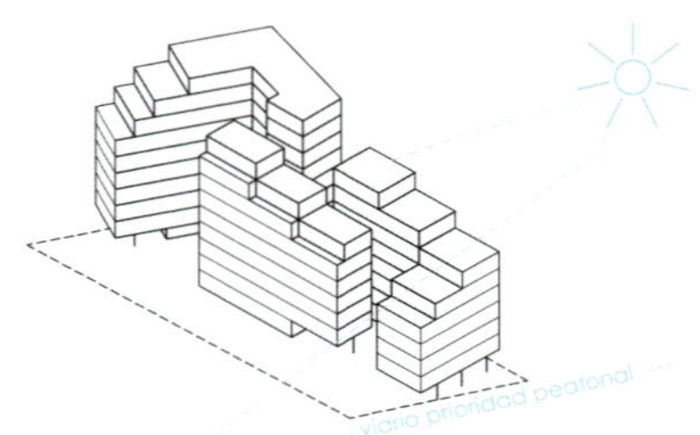
调整尺度和功能 ADJUSTMENT OF THE SCALE AND PROGRAM

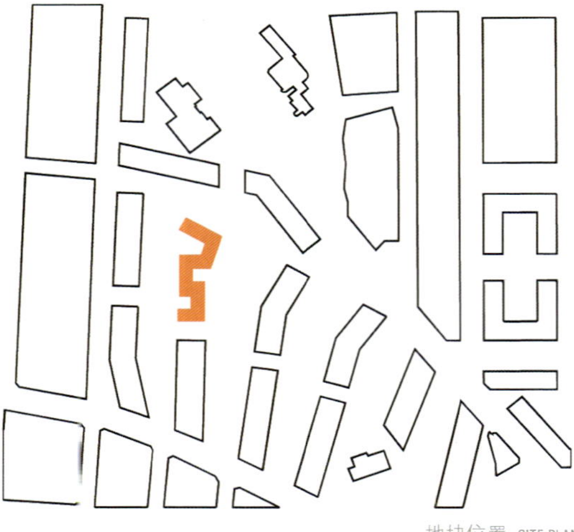
地块位置 SITE PLAN

## 城乡一体化

与直线型的解决方案相比，该项目作为一个面向城市的解决方案，旨在减小建筑规模，有人提出通过降低建筑高度、采用空间分隔等方法予以解决。该解决方案还意味着要提高住宅单元的室内条件和通风质量。另外，打造后退型结构和变更上层建筑材料有助于缩减建筑规模。

## URBAN INTEGRATION

As an urban solution, compared to a linear solution, the project proposes to reduce the building scale which someone can perceive through height reduction and fragmentation of volumes. This decision also means to improve indoor and ventilation of the housing units. The setback and the change of the material of the upper floors help reduce the scale.

底层平面图 GROUND FLOOR PLAN

二至五层平面图 FLOOR PLANS 1-4

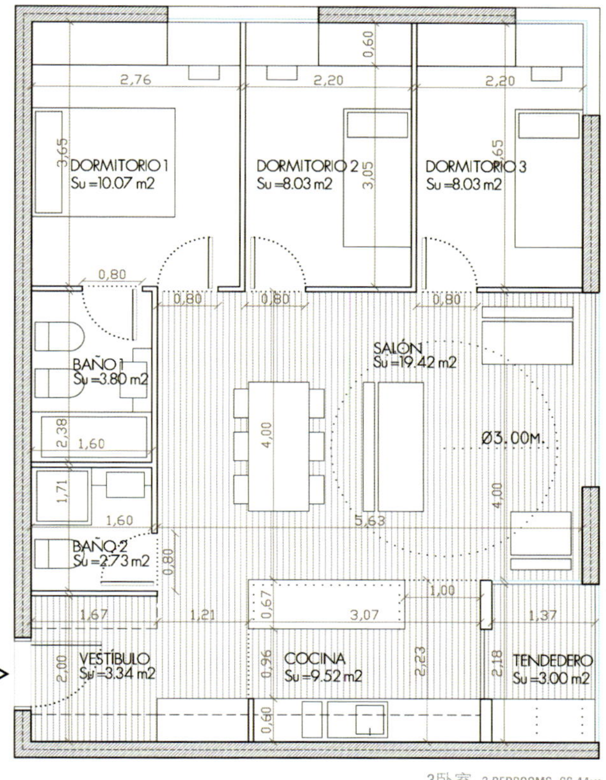


3卧室 3 BEDROOMS 66.44M2

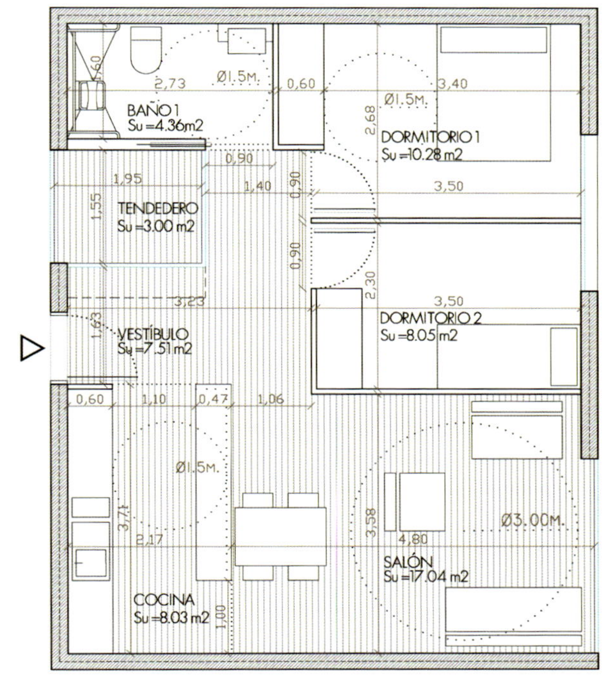


2卧室 2 BEDROOMS ADAPTED 56.77M2

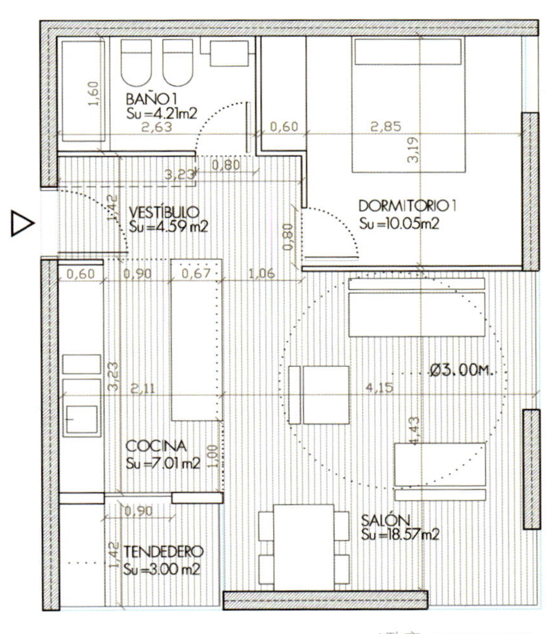


1卧室 1 BEDROOM 45.93M2

标准房屋单元 TYPICAL HOUSING UNITS

105TREE

## Agantangelo Soler Montellano · Luz Sempere Sanchez (建筑师)

第一提名奖 first mention

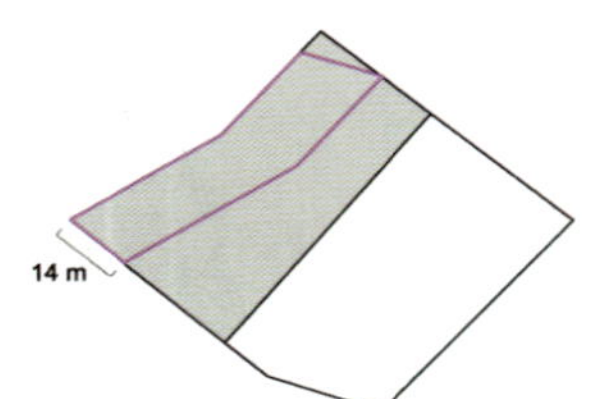


房屋里有个花园，花园里有一棵树，树上有一座房屋
THERE IS A GARDEN IN THE HOUSE, THERE IS A TREE IN THE GARDEN, THERE IS A HOSUE ON THE TREE

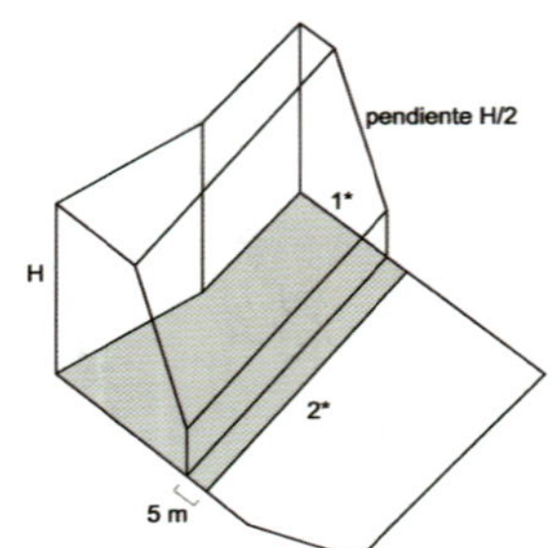


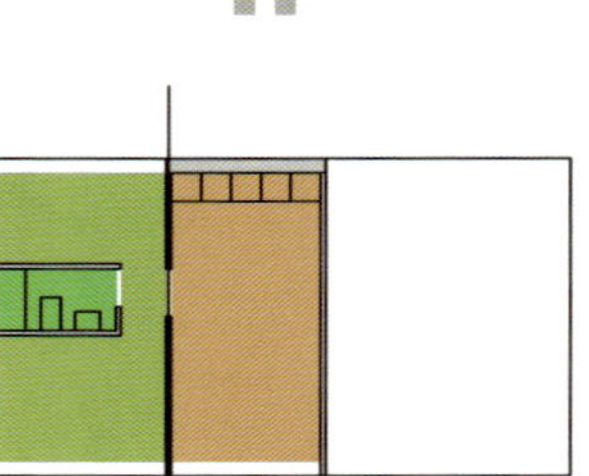

屋内核心
DOMESTIC CORE

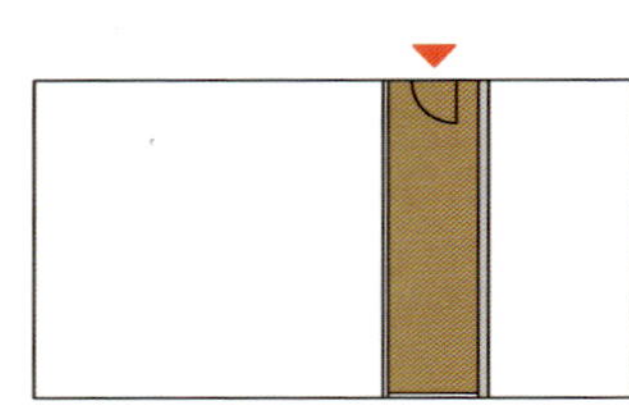
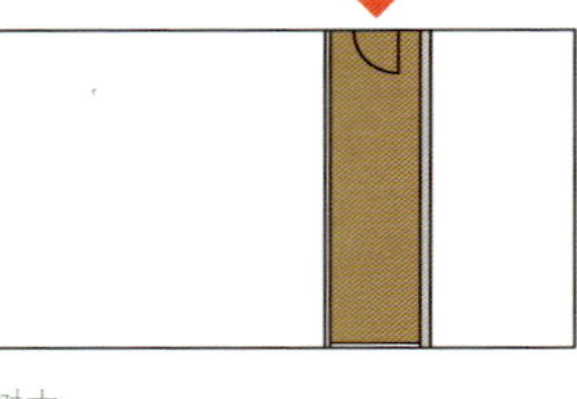

树木
THE TREE

树屋
HOUSE ON THE TREE

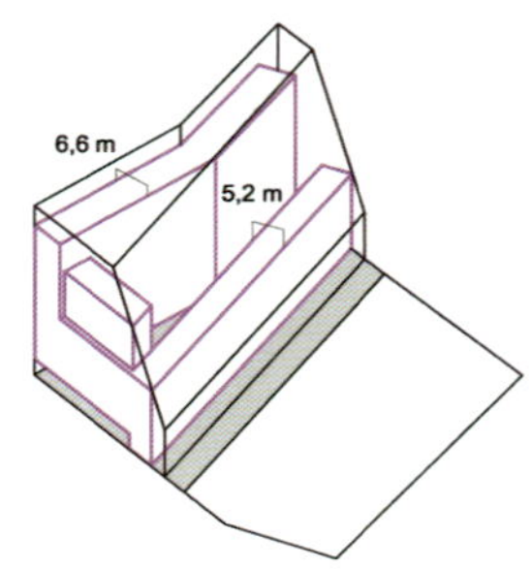


布局策略 LAYOUT STRATEGIES

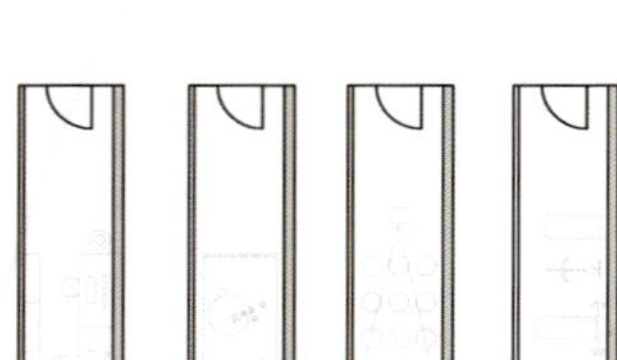
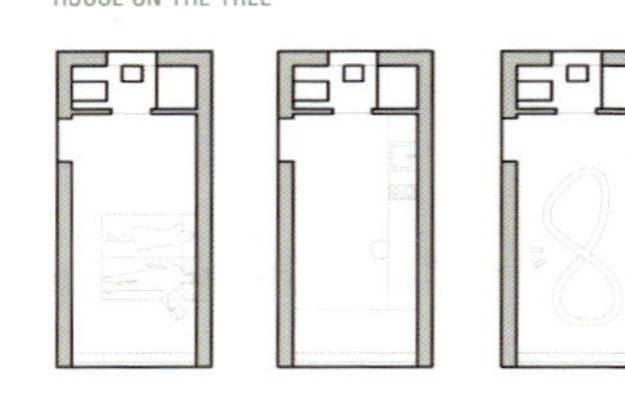
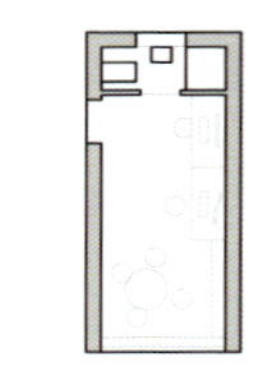
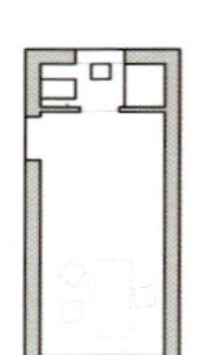
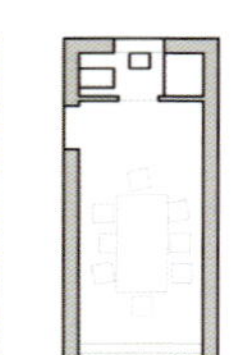

# 一百户住户

我们将修建两幢非常狭长的公寓楼：一幢6.6米宽，30米高；另一幢5.2米宽，15.5米高。如何设计能容纳100户住户，并能长时间满足其不同生活方式及不同需求的公寓？通过引入跨空间区域，打造出出入口、露台，其中，种植在其间的一棵树将这片独立空间与内部空间（树屋）分隔开。

# 100 DIFFERENT FAMILIES

We propose two very narrow bars, one of 6.6m wide and 30m high and another of 5.2m wide and 15.5m high. How can we design a few small apartments capable of accommodating over time the different characters and lifestyles of 100 different families? By introducing a cross space which serves as access and terrace - the tree, separating an independent space from the domestic core - the house on the tree.

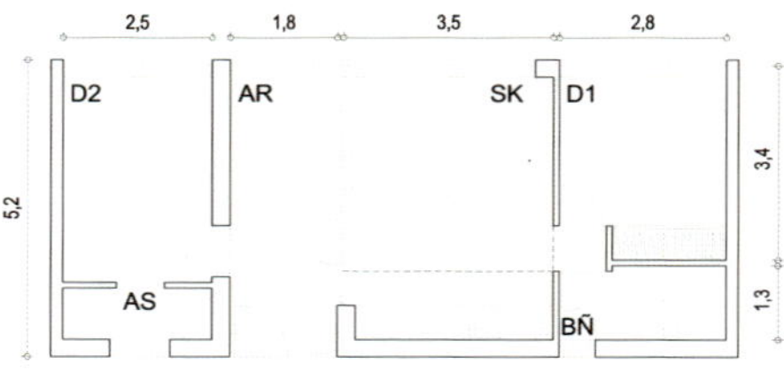


2卧室房屋单元 · 树屋 HOUSING UNIT 2 BEDROOMS · HOUSE ON THE TREE

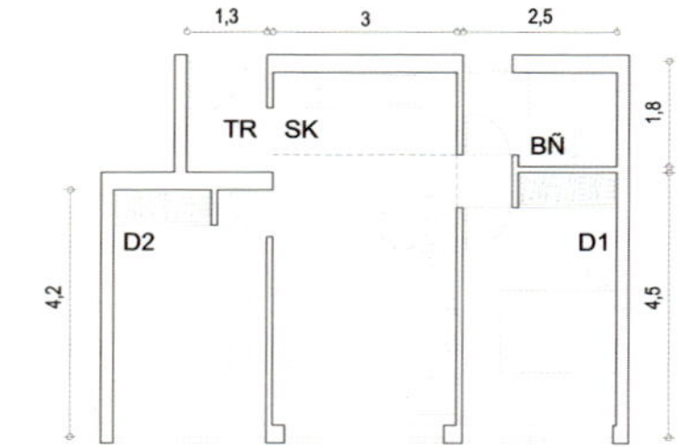


2卧室房屋单元 HOUSING UNIT 2 BEDROOMS

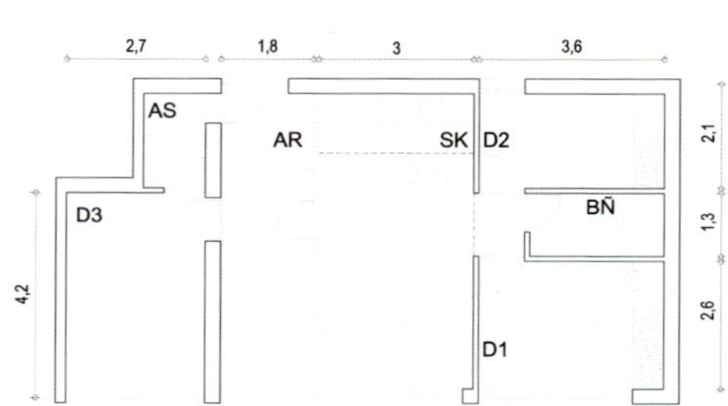


3卧室房屋单元 · 树屋 HOUSING UNIT 3 BEDROOMS · HOUSE ON THE TREE

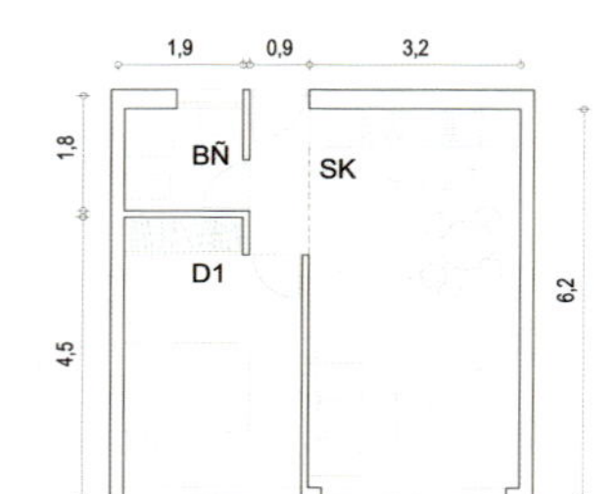


1卧室房屋单元 HOUSING UNIT 1 BEDROOM

地块位置 SITE PLAN

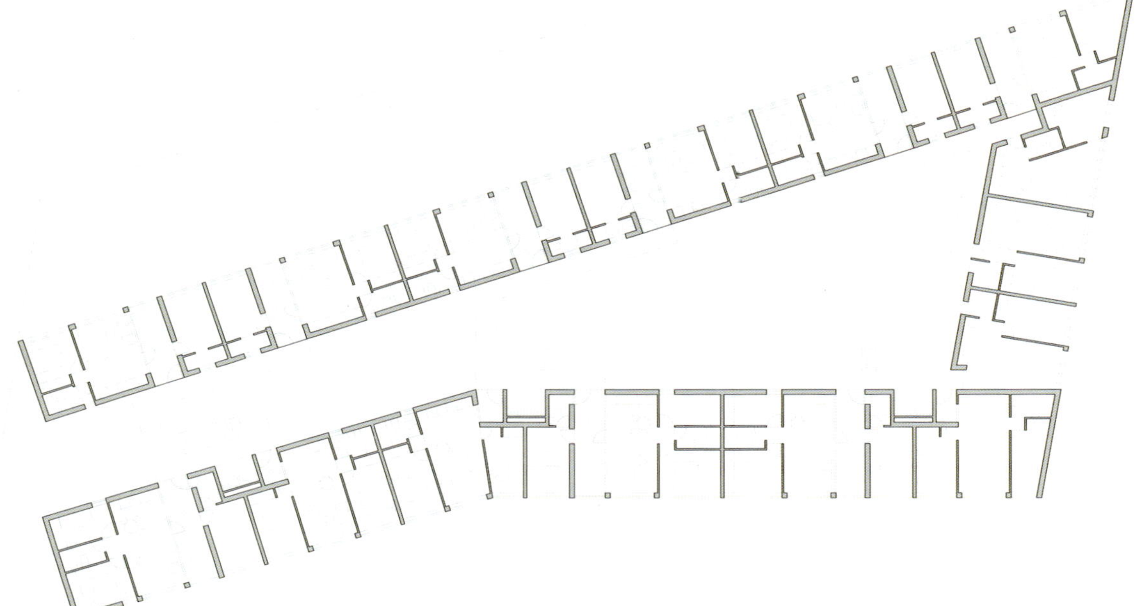

标准层平面图 二、三层 TYPICAL FLOOR PLANS 1,2

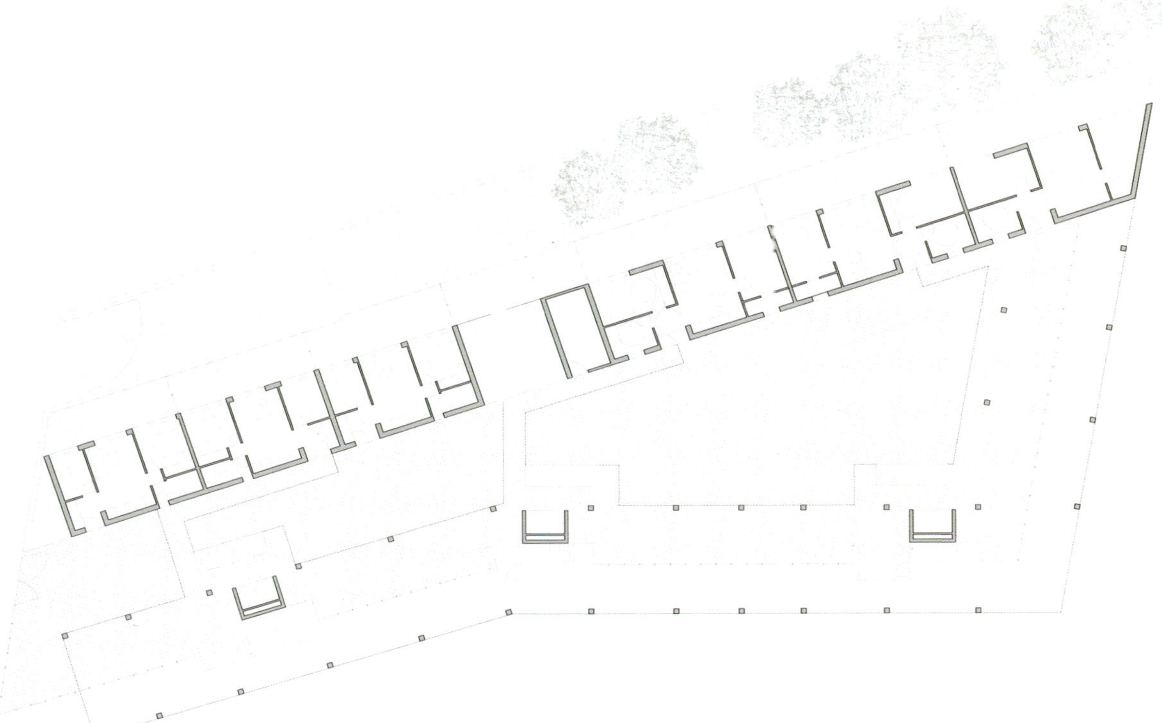

底层平面图 GROUND FLOOR PLAN LEVEL 0

# Luis Arredondo (建筑师)

第二提名奖 second mention

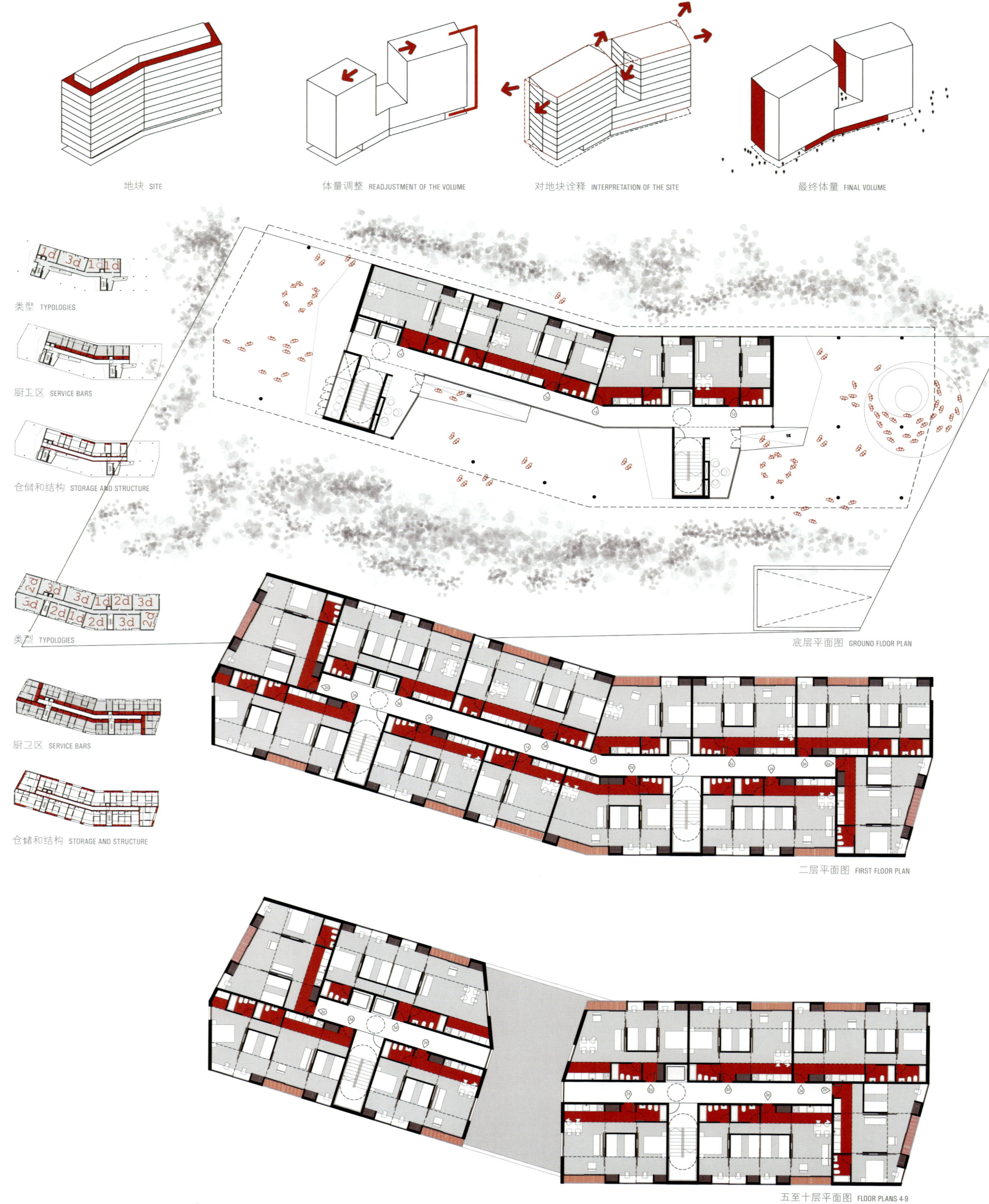

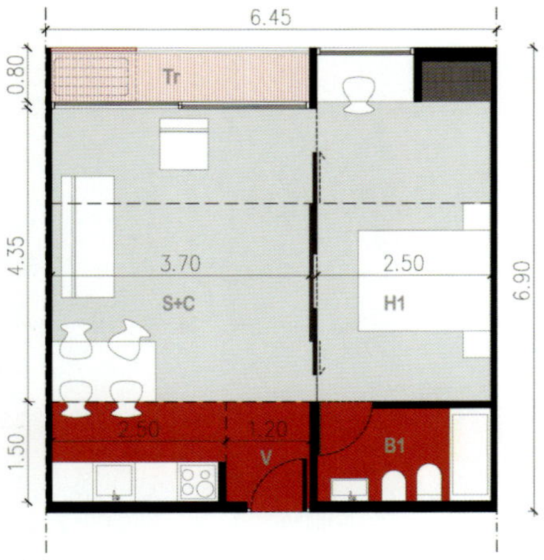

1卧室 1 BEDROOM

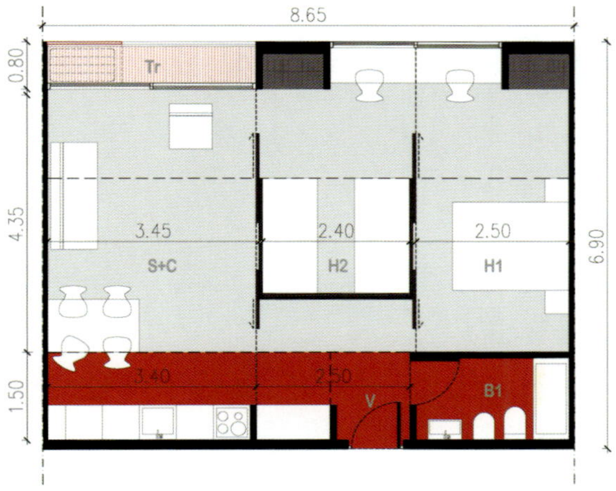

2卧室 2 BEDROOMS

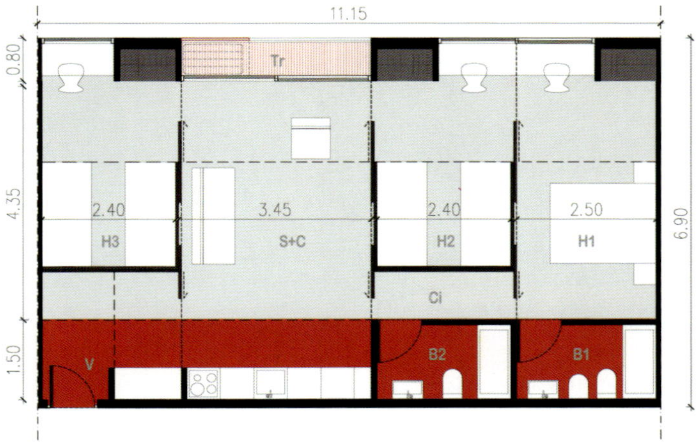

3卧室 3 BEDROOMS

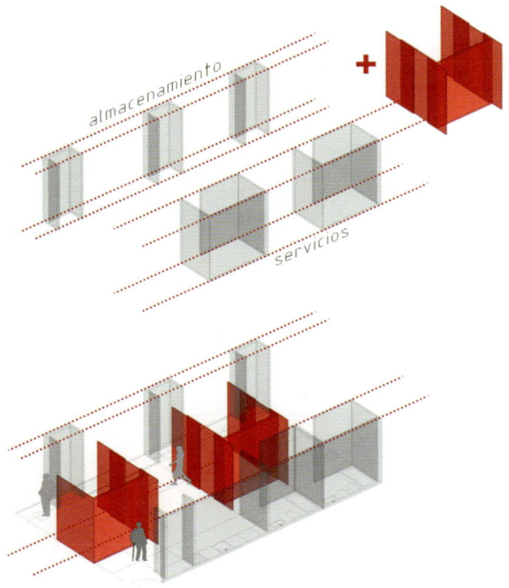

# 两个相互平行的区域

# TWO PARALLEL SERVICE BARS

我们将该建筑设计为前后各一排的公寓。为了使各住宅单元通透性更好，减少屏蔽效应，同时提高各住宅单元质量的双重目的，我们设法将整个地块分隔成两个相互独立的区域。

We propose a building with two vertical cores and apartments on both sides. In order to achieve the dual purpose of allowing permeability inside the building, removing the screen effect and the improvement of the quality of the housing units, we work on the volume separating it into two separate bodies.

家具室 FURNITURE ROOM

万用型房间 THE ROOM AS A WILD CARD

一个纵向的起居室 A LONGITUDINAL LIVING ROOM

0123HOP

Fernando Araujo Fuster · Ana Dolado Cosín (建筑师)

合作 (c) Nuria Heras Donoso

第四提名奖 fourth mention

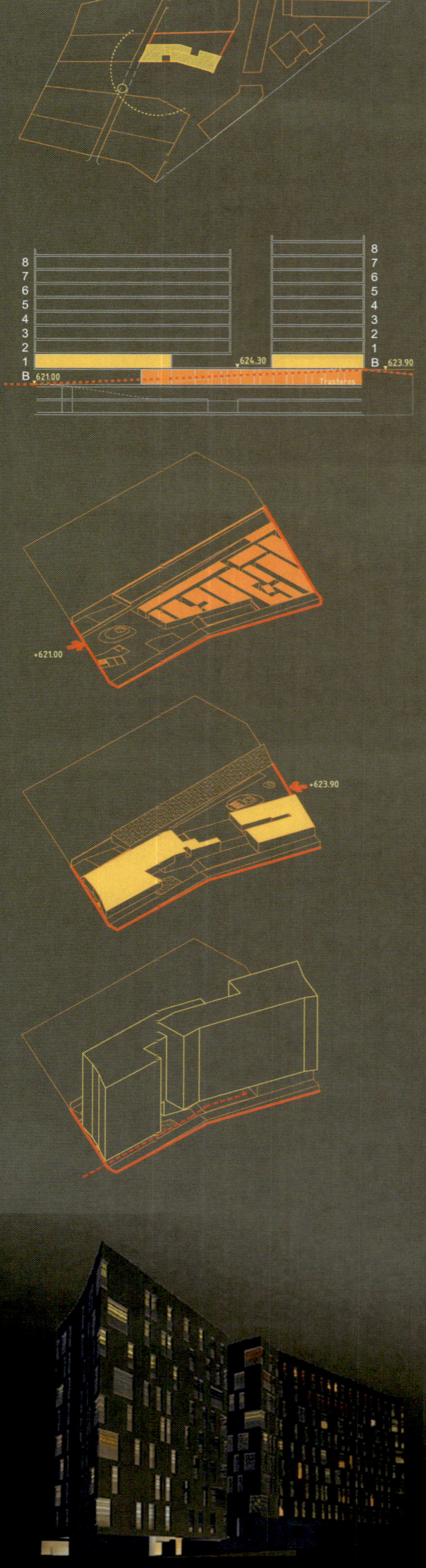

## 两种结构

该项目利用不规则的布局，根据地形将底层分成两层，各设置一个出入口，分别位于左右两侧。这种连续性的建筑表皮与地块东南面边缘连成一行。该区域的西北面是封闭式的，可巧妙地降低狭窄区域的高度。

## TWO TRAYS

The project takes advantage of the uneven situation, developing the ground floor as topography in two trays that offer opposite access in two different levels. The continuous skin-façade is aligned with the southeast edge of the plot.The volume is closed to the north-west, strategically reducing its height where the plot is narrow.

标准层平面图 TYPICAL FLOOR PLAN

1卧室房屋单元 HOUSING UNIT 1 BEDROOM
2卧室房屋单元 HOUSING UNIT 2 BEDROOMS
3卧室房屋单元 HOUSING UNIT 3 BEDROOMS
底层平面图 · 备选入口 GROUND FLOOR PLAN · ALTERNATIVE ACCESS
底层平面图 · 主入口 GFOUND FLOOR PLAN · MAIN ACCESS
地下层平面图 UNDERGROUND FLOOR PLANS

MIRADOR

# SOMOS arquitectos (建筑师事务所)

Luis Burriel · Pablo Fernandez Lewicki · Jose Antonio Tallon (建筑师)

第五提名奖 fifth mention

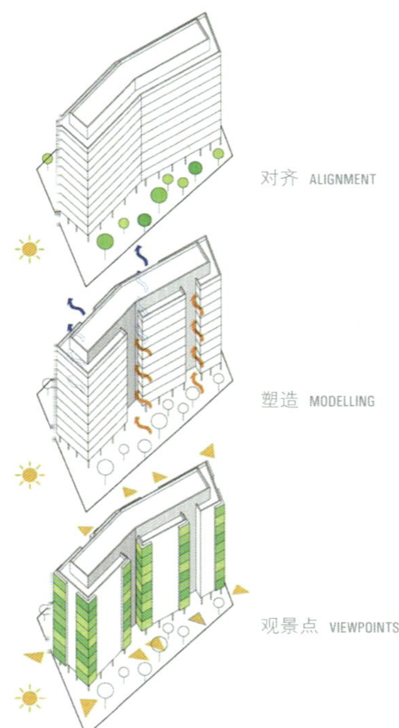


## 直接照明

我们将封闭式的建筑进行分隔，这样能让这个环形区域以及各个住宅单元采光好，景观佳。我们扩大公开区域的利用率，同时保持其紧凑性，这样可将整块的建筑进行分隔，以打破建筑表面的单一性。

## DIRECT LIGHTING

We look for the fragmentation of the hermetic block, so that we can introduce the light and views into circulation areas, as well as inside the housing units. We dilate the occupation of the houses that open to views while they keep compacted. This breaks down the monolithic block breaking the plane of the façade.

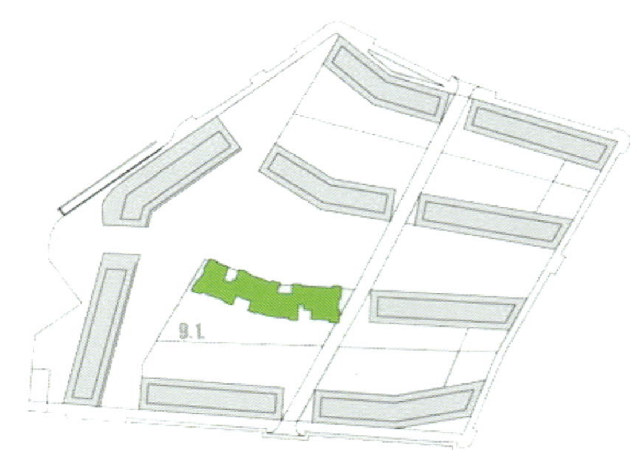


地块位置 SITE PLAN

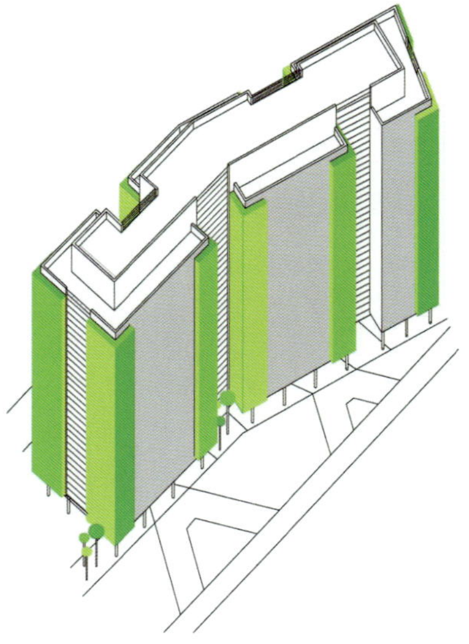

房间：晚间卧室区域，砂浆保温外墙
ROOMS: NIGHT AREA; MORTAR FAÇADE WITH THERMAL INSULATION

观景点：起居室；绿草及金属栅栏
VIEWPOINTS: LIVING ROOMS; GLASS AND METALLIC LATTICE

走道：阳极氧化铝板
CORRIDORS; ANODIZED ALUMINIUM SHEETS

横向剖面图 CROSS SECTION

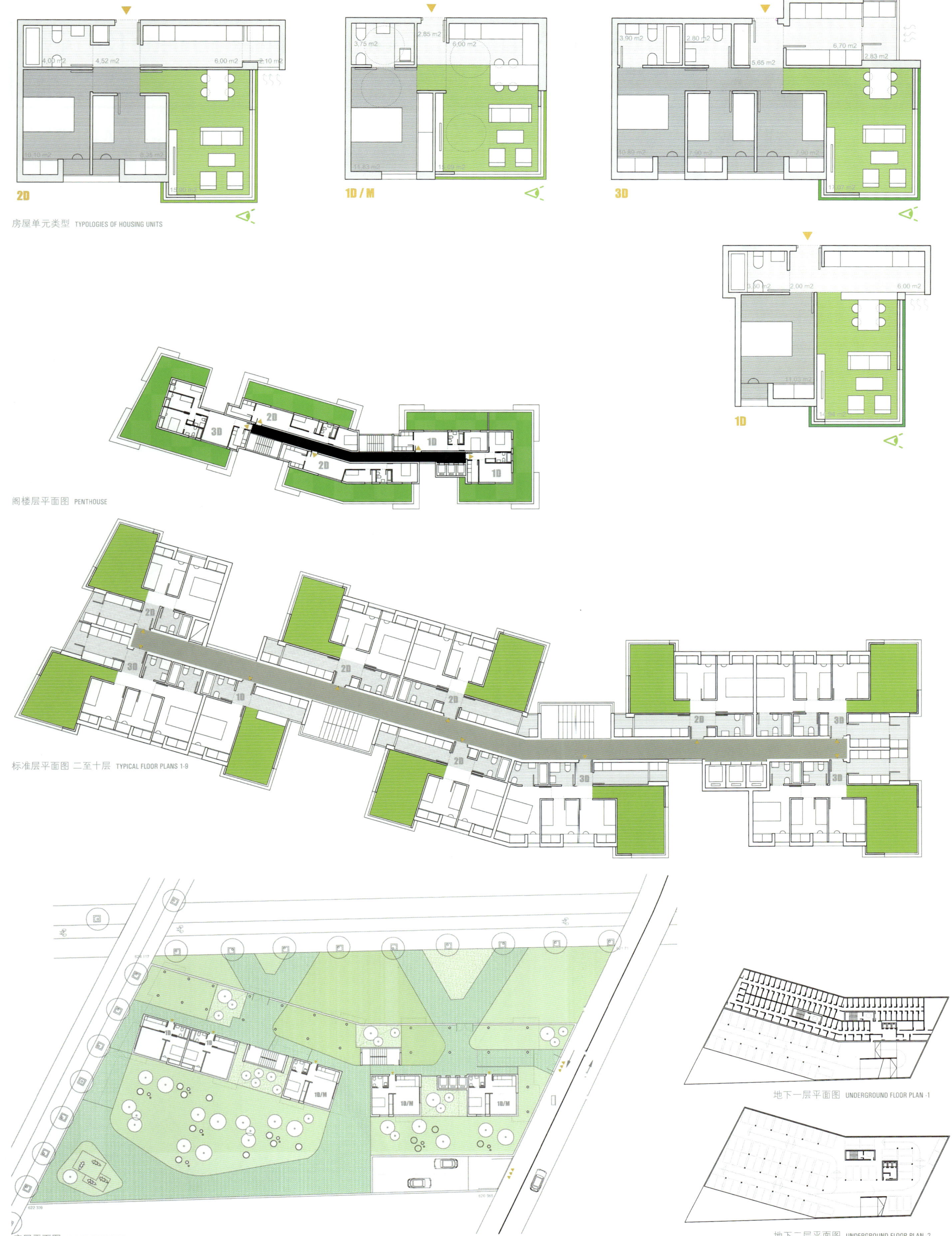

房屋单元类型 TYPOLOGIES OF HOUSING UNITS

阁楼层平面图 PENTHOUSE

标准层平面图 二至十层 TYPICAL FLOOR PLANS 1-9

底层平面图 GROUND FLOOR PLAN

地下一层平面图 UNDERGROUND FLOOR PLAN -1

地下二层平面图 UNDERGROUND FLOOR PLAN -2

MIRADAS

## Ramon Gónzalez · Fernando Pancorbo · Mathias Schuette (建筑师)

中标 winner

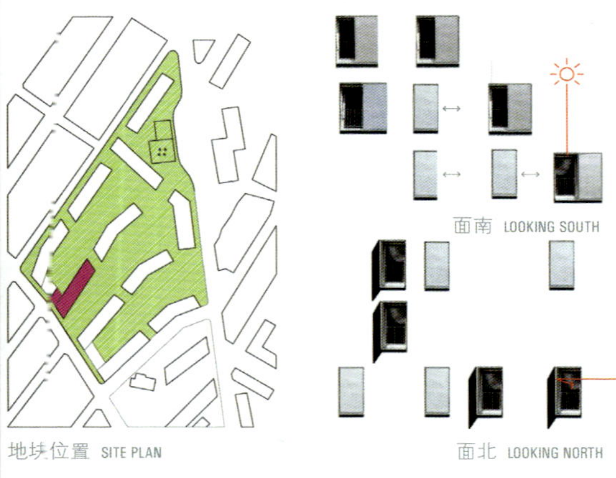


地块位置 SITE PLAN

面北 LOOKING NORTH

# 相向的建筑表面

可使用基石解决地块的纵向斜坡问题，并可打造地下储物空间，使得出入更加方便。两个建筑表面朝向相反，均呈模糊的明暗浮雕效果，与窗户的鲜亮颜色形成对比，使陶制外观更富有色彩与活力。

# OPPOSITE FAÇADES

The longitudinal slope of the plot is resolved with a plinth that houses the underground storage areas and makes the access easier. The façades respond differently to opposite directions. Both show a subtle chiaroscuro relief accentuated by the brightness of the front shutters which turn the ceramic façade into a vibrant and lively area.

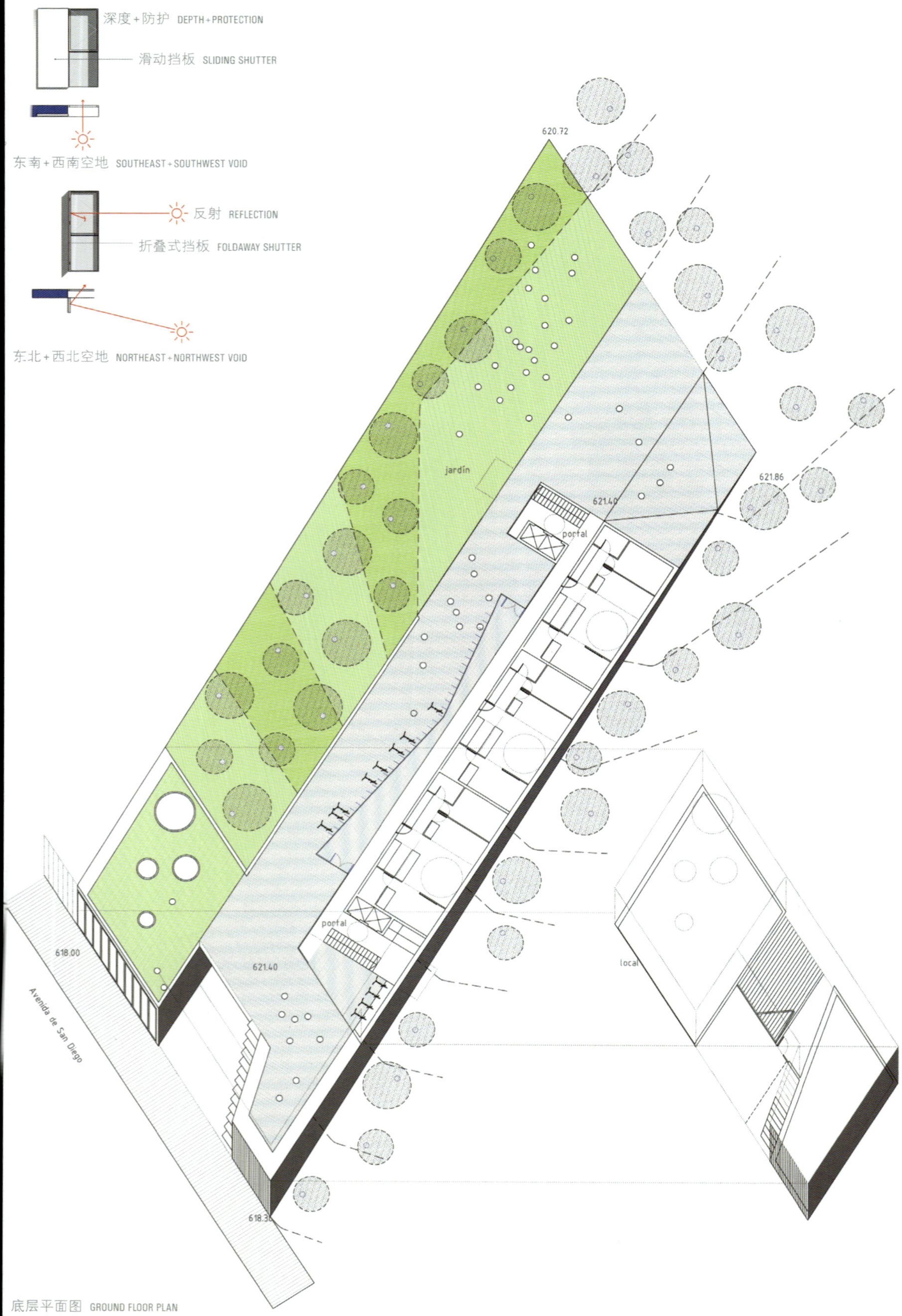

深度+防护 DEPTH+PROTECTION

滑动挡板 SLIDING SHUTTER

东南+西南空地 SOUTHEAST+SOUTHWEST VOID

反射 REFLECTION

折叠式挡板 FOLDAWAY SHUTTER

东北+西北空地 NORTHEAST+NORTHWEST VOID

620.72

jardín

621.40

portal

621.86

618.00

621.40

portal

local

Avenida de San Diego

618.3

底层平面图 GROUND FLOOR PLAN

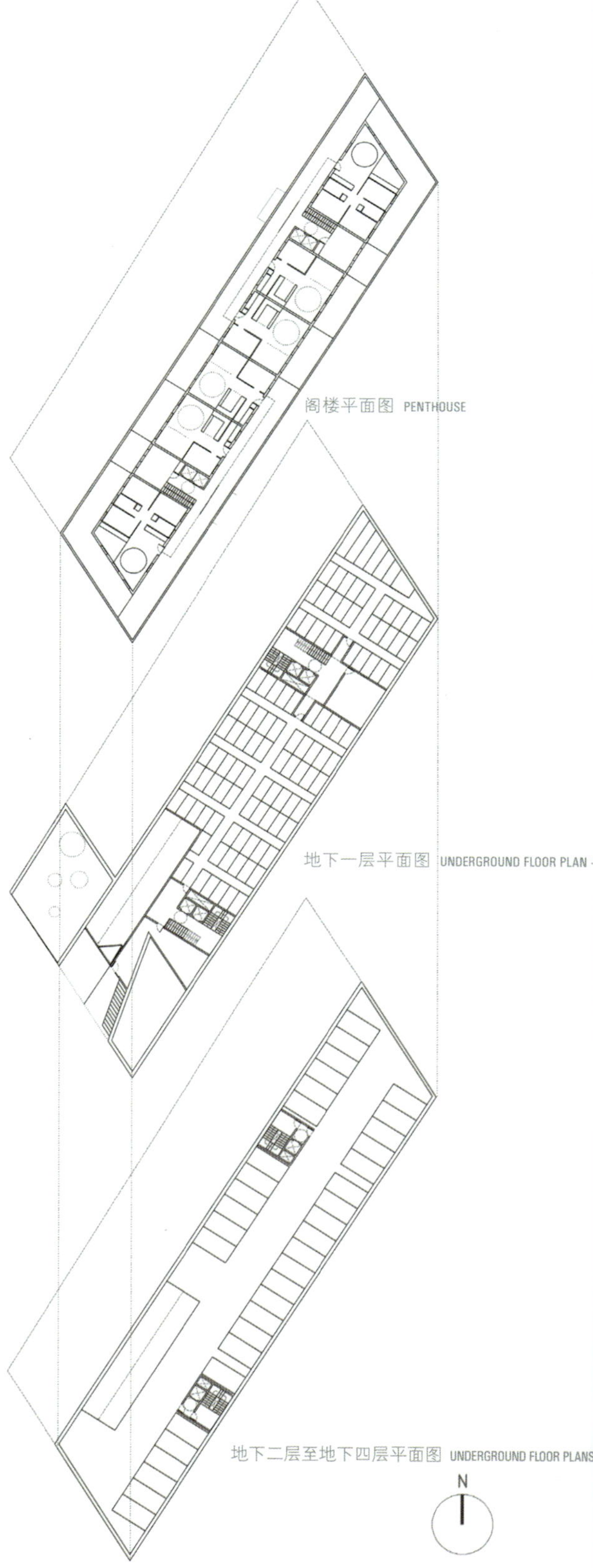


阁楼平面图 PENTHOUSE

地下一层平面图 UNDERGROUND FLOOR PLAN -1

地下二层至地下四层平面图 UNDERGROUND FLOOR PLANS

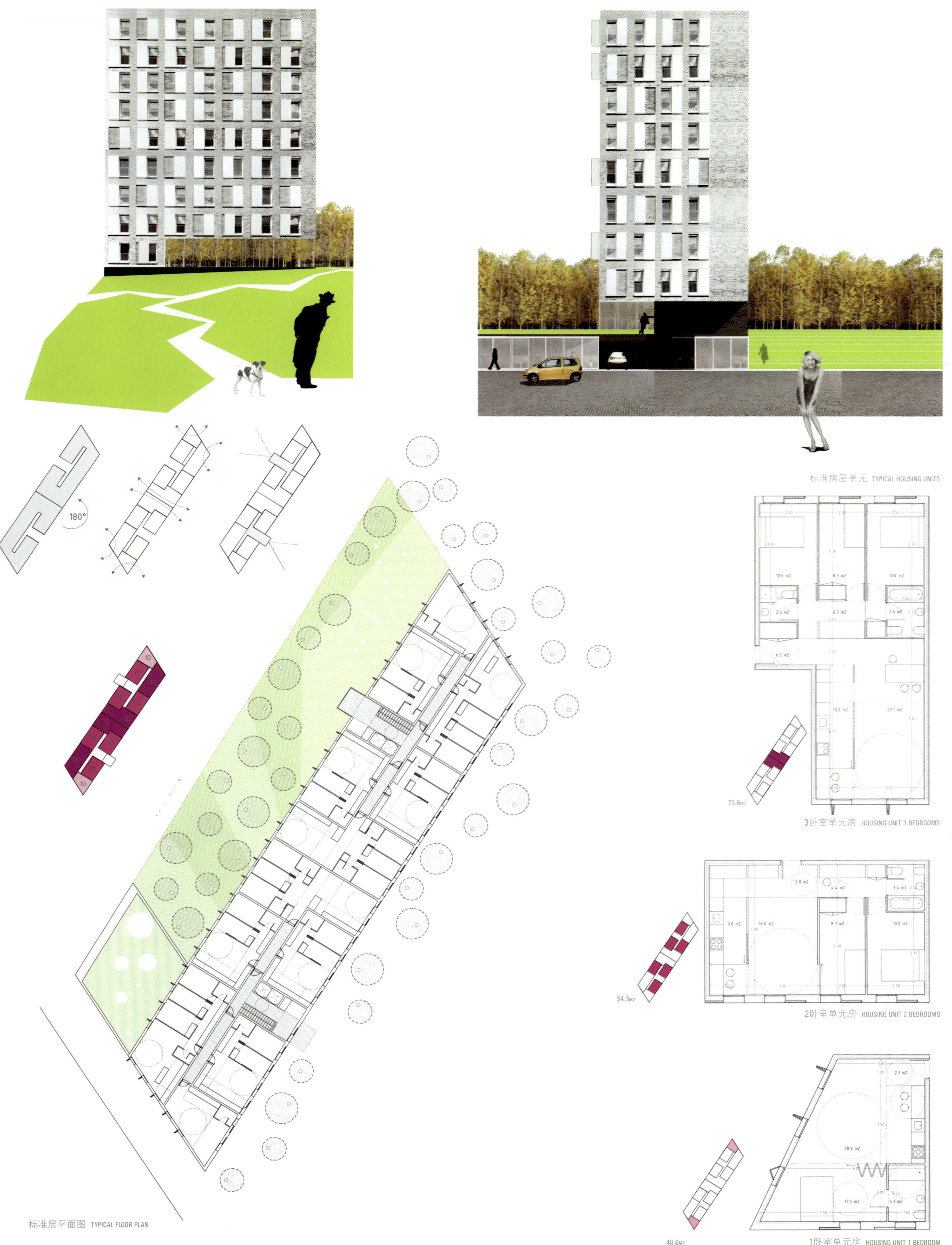
180°
标准房屋单元 TYPICAL HOUSING UNITS
79.6M2
3卧室单元房 HOUSING UNIT 3 BEDROOMS
54.3M2
2卧室单元房 HOUSING UNIT 2 BEDROOMS
40.6M2
1卧室单元房 HOUSING UNIT 1 BEDROOM
标准层平面图 TYPICAL FLOOR PLAN

# Santiago Cifuentes Barrio (建筑师)

第一提名奖 first mention

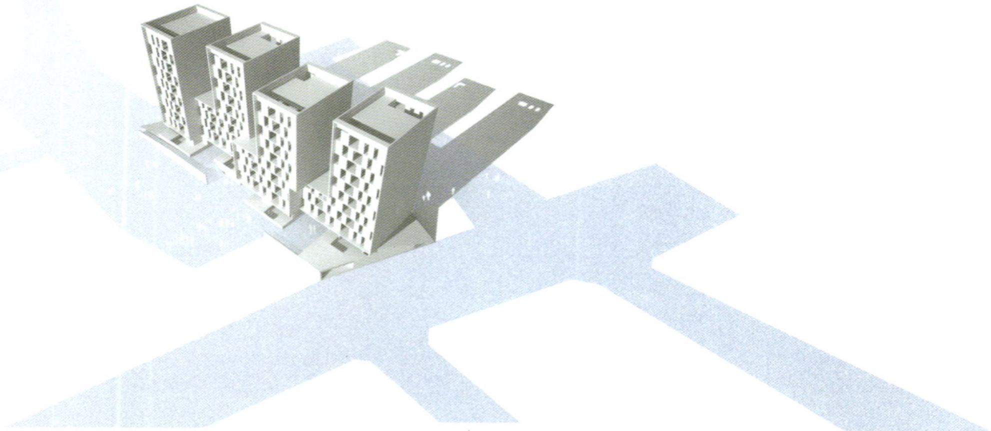

vistas desde la terraza mirador

TERRAZA

CALLE PUERTO DE LA BONAIGUA

平面图 FLOOR PLAN LEVEL +19.95

平面图 FLOOR PLAN LEVEL +5.70

底层平面图 GROUND FLOOR PLAN

ACCESO APARCAMIENTO

AVENIDA DE SAN DIEGO

VIAL DE COEXISTENCIA

+25.65

+22.80

+19.95 / +17.10 / +14.25

+11.40

+8.55

+5.70 / +2.85

+0.00

-2.85

-5.55

-8.25

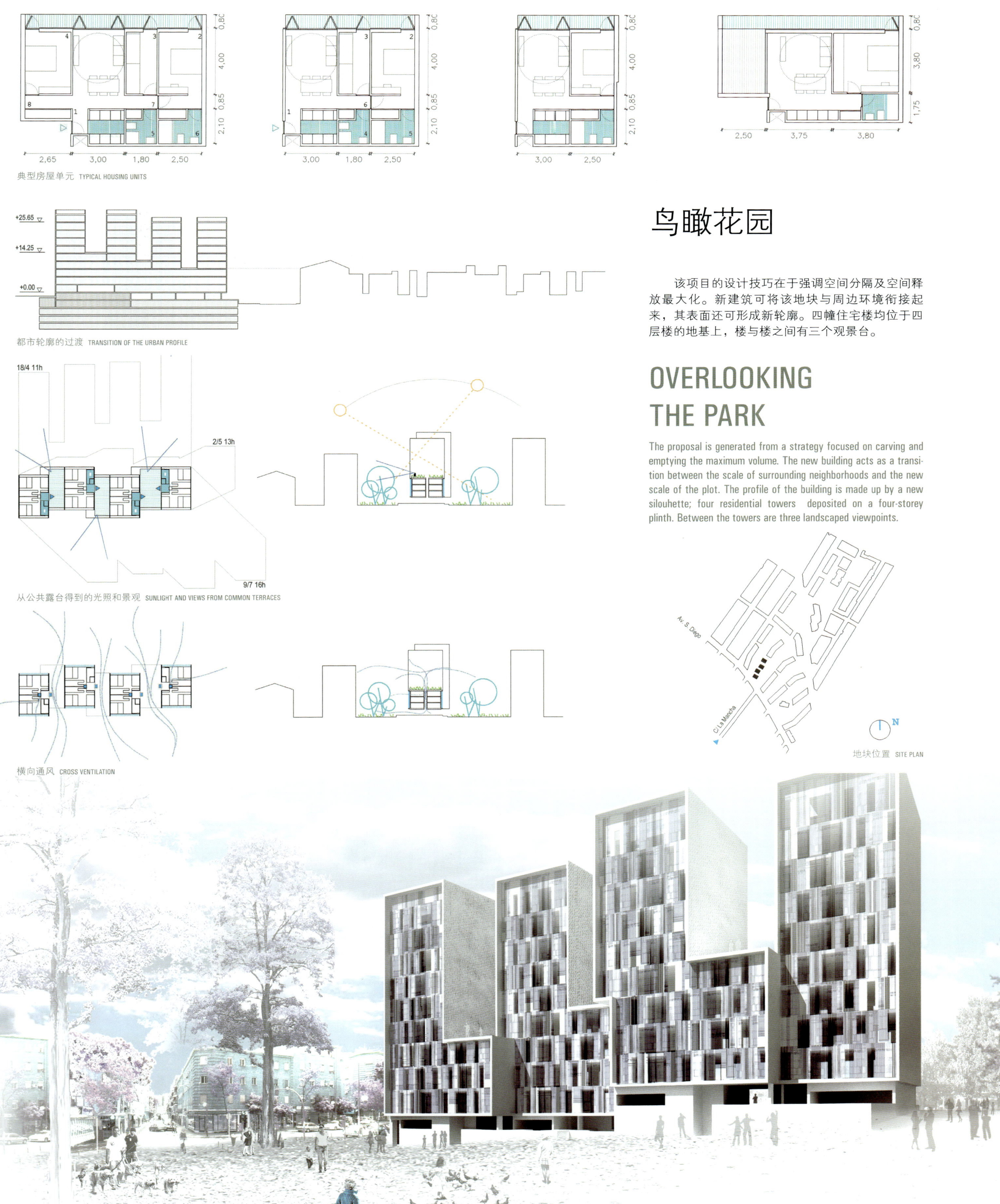


典型房屋单元 TYPICAL HOUSING UNITS

都市轮廓的过渡 TRANSITION OF THE URBAN PROFILE

从公共露台得到的光照和景观 SUNLIGHT AND VIEWS FROM COMMON TERRACES

横向通风 CROSS VENTILATION

地块位置 SITE PLAN

# 鸟瞰花园

该项目的设计技巧在于强调空间分隔及空间释放最大化。新建筑可将该地块与周边环境衔接起来，其表面还可形成新轮廓。四幢住宅楼均位于四层楼的地基上，楼与楼之间有三个观景台。

# OVERLOOKING THE PARK

The proposal is generated from a strategy focused on carving and emptying the maximum volume. The new building acts as a transition between the scale of surrounding neighborhoods and the new scale of the plot. The profile of the building is made up by a new silouhette; four residential towers deposited on a four-storey plinth. Between the towers are three landscaped viewpoints.

1401011

# Olalquiaga Arquitectos Rafael Olalquiaga Soriano · Pablo Olalquiaga Bescós · Alfonso Olalquiaga Bescós (建筑师)

合作 (c) Javier Morales Luchena

第二提名奖 second mention

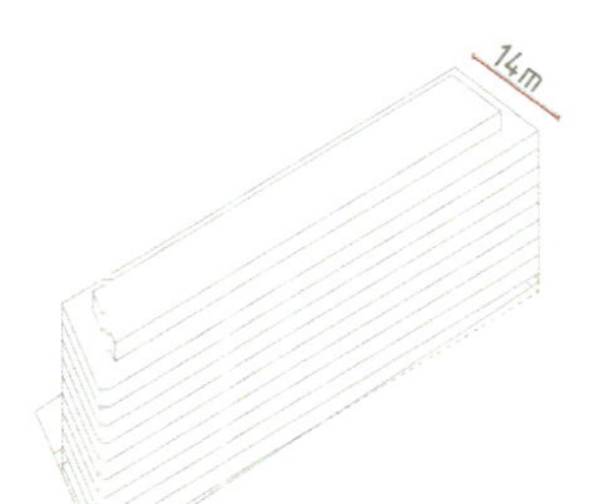

高度：十层+阁楼 HEIGHT: 9 STOREYS+PENTHOUSE

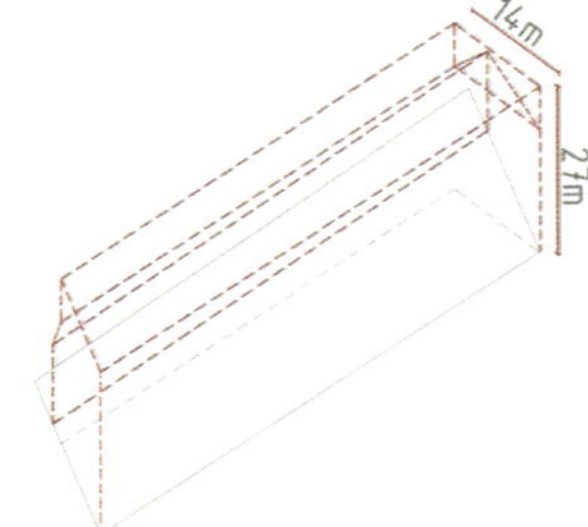


最大体量 MAXIMUM VOLUME

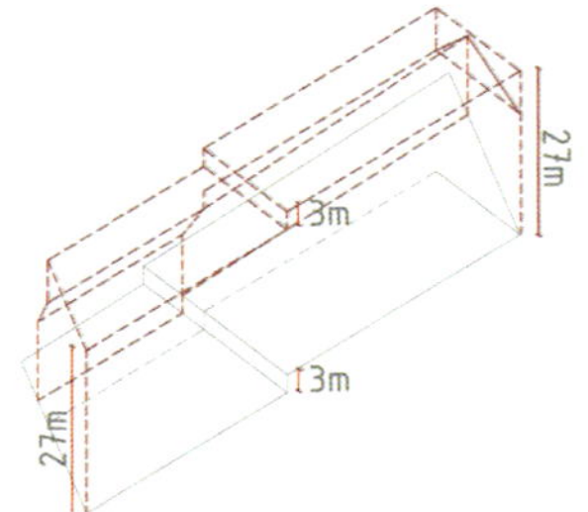


三米沉降 3 METRE DROP

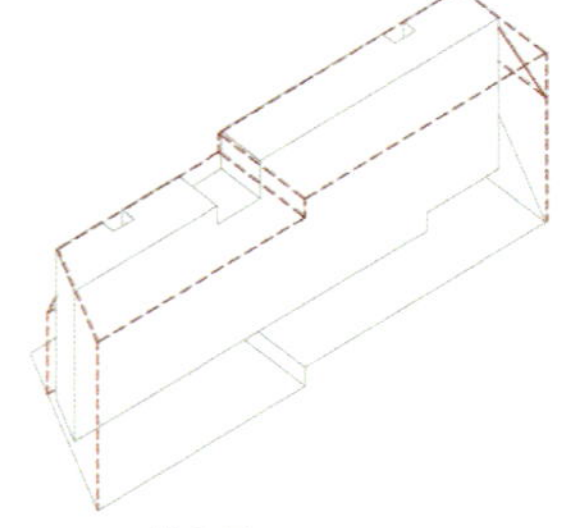

A楼十层 BUILDING A: 9 STOREYS

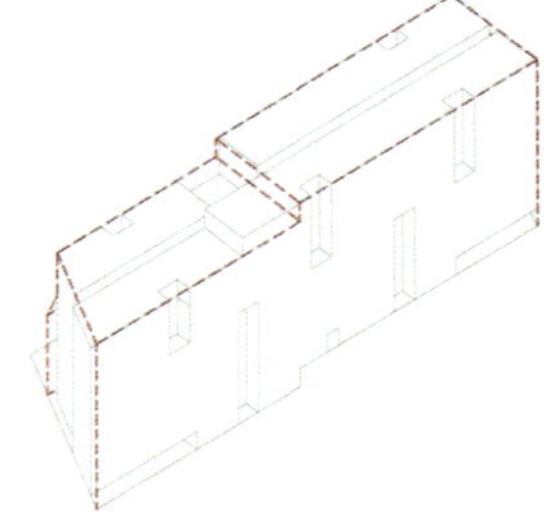

B楼十层 BUILDING B: 9 STOREYS

## 两幢建筑

我们的设计意图：

–在该地块上打造最大的开放空间和绿色空间；

–充分利用街道两侧的落差。

通道走廊的两端均设出入口，以改善自然采光，扩展外部视野。

## TWO BUILDINGS

We propose:

- To obtain the maximum open and green space in the plot.
- To take advantage of the level drop in opposite streets.

An access corridor spatially opens at the ends to improve natural lighting and exterior views.

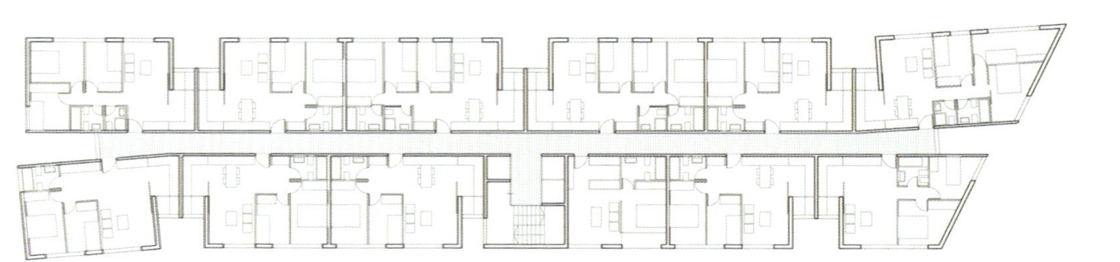

标准层平面图 TYPICAL FLOOR PLAN

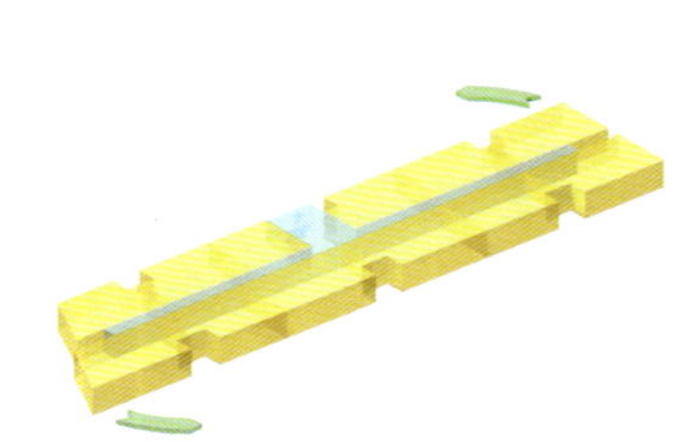

标准解决方案 TYPICAL SOLUTION

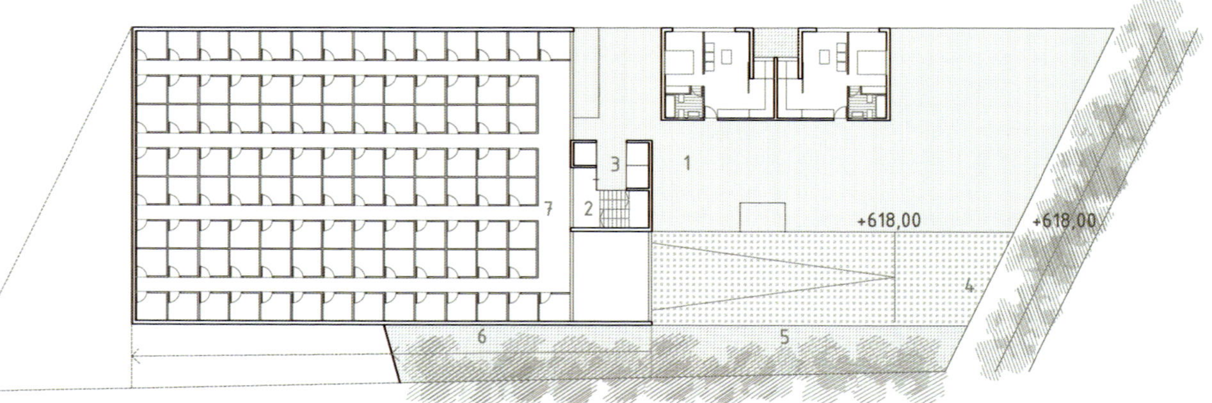


门廊底层平面图·临街层 PORCH GROUND FLOOR PLAN · STREET LEVEL

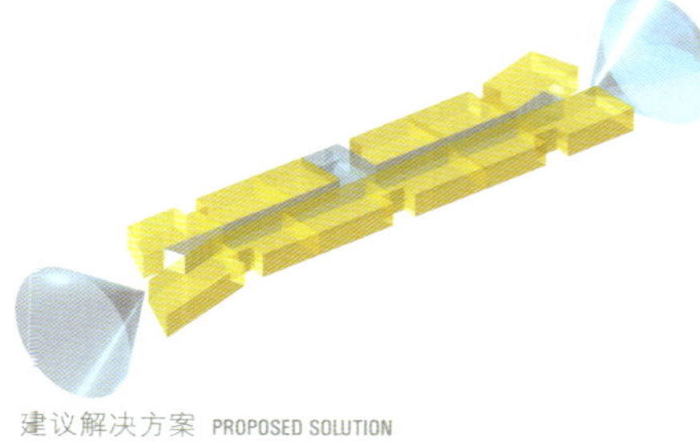

建议解决方案 PROPOSED SOLUTION

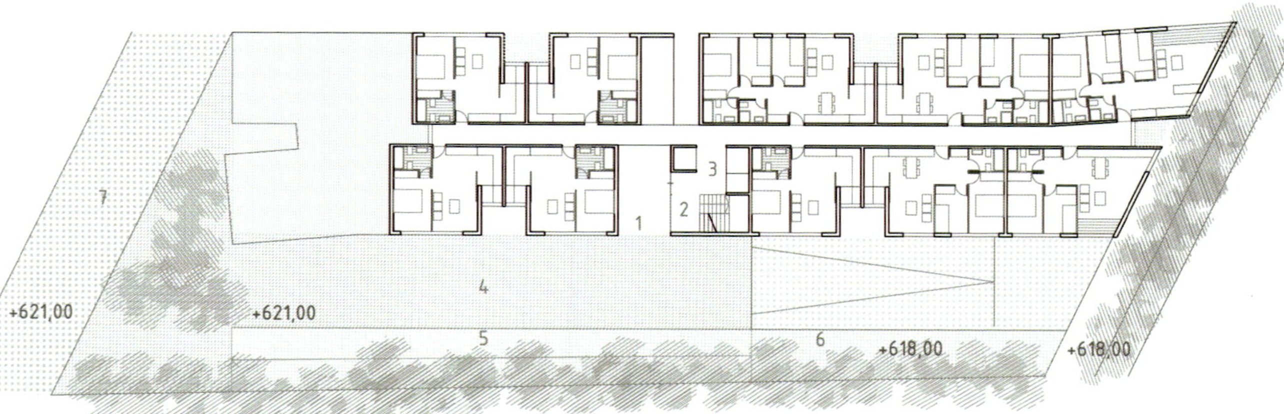


底层平面图·行人通道层 GROUND FLOOR PLAN · PEDESTRIAN PATHWAY LEVEL

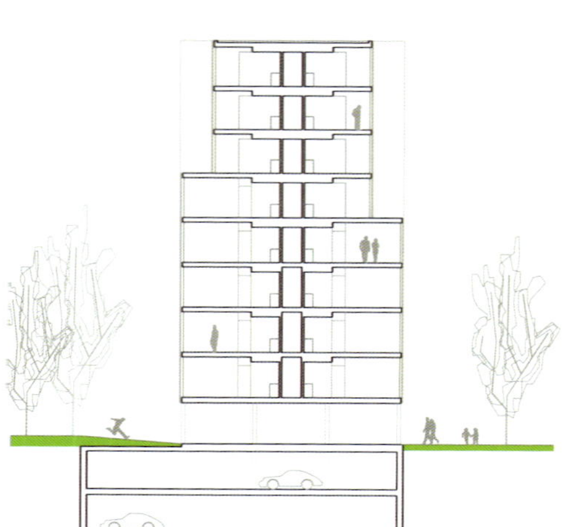

横向剖面图 CROSS SECTION

西南立面图 SOUTHWEST ELEVATION

B楼
BUILDING B
同时通向两楼的入口区
ACCESS AREAS TO BOTH BUILDINGS
A楼
BUILDING A
3卧室 3 BEDROOMS 65.20M2
1卧室 1 BEDROOM 40.20M2
2卧室 2 BEDROOMS 52.20M2 • 外墙细部 FAÇADE DETAIL

PATIOSV

# Estudio Cano Lasso (建筑师事务所)

第三提名奖 third mention

地块位置 SITE PLAN

## 精简

各套公寓朝向不同的公共区域，所以采光和通风俱佳。该部分公共区域位于内外两排公寓之间，相对私密，如同格栅一般可遮风挡雨，为公寓遮阴，是一块受保护的私密活动空间。

## SLENDERNESS

Each apartment opens to various open spaces, so all its dependencies can light and ventilate. They are spaces of transition between interior and exterior. Areas of privacy, which act as the same time as climatic filters. Their behavior is similar to a lattice. It produces shade inside the apartment and allows its use, as a protected and private outdoor space.

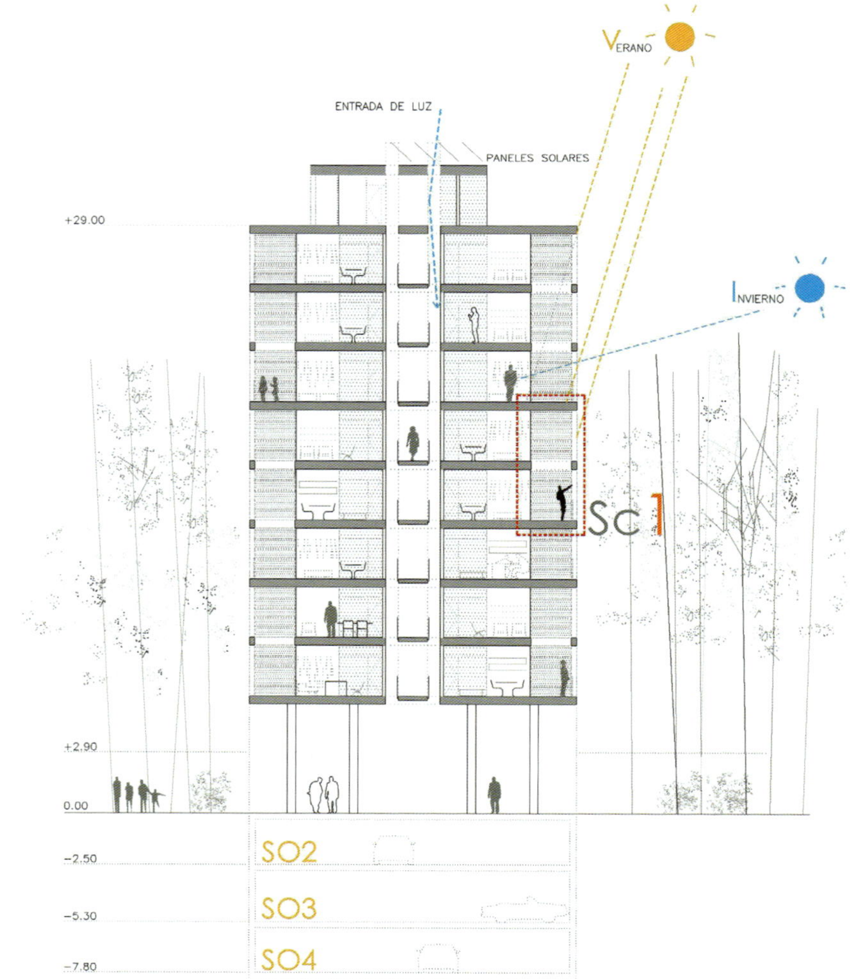


东西剖面图 · 能效 EAST-WEST SECTION · ENERGY EFFICIENCY

1卧室 1 BEDROOM 40.33M2

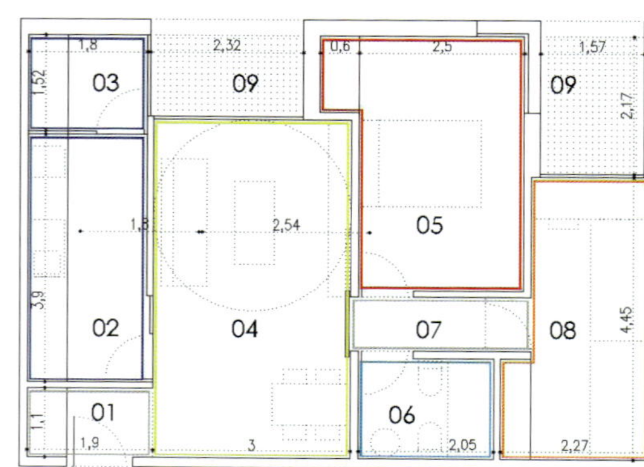

2卧室 2 BEDROOMS 55.79M2

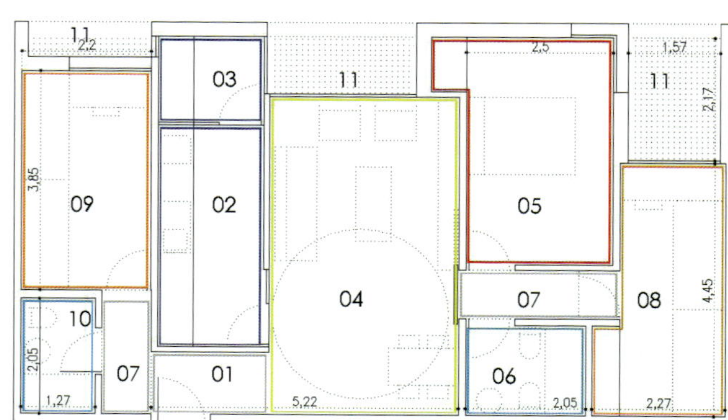

3卧室 3 BEDROOMS 70.59M2

东北立面图 NORTHEAST ELEVATION

正立面图 MAIN ELEVATION

临街立面图 STREET ELEVATION

功能布局图 PROGRAM SQUEME

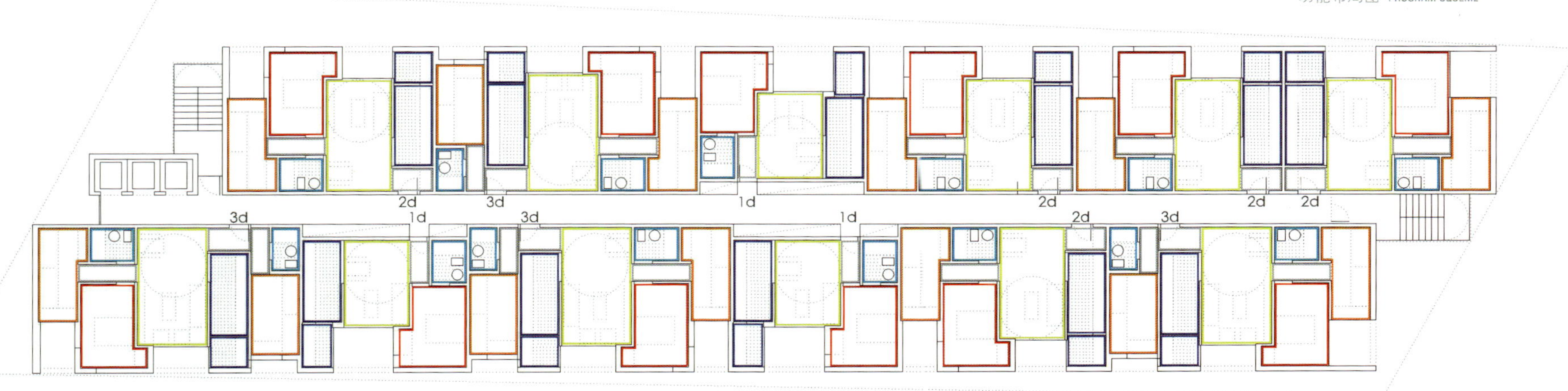


二至九层标准层平面图 TYPICAL FLOOR PLANS 1-8

功能布局图 PROGRAM SQUEME

底层平面图 GROUND FLOOR PLAN

GEA (建筑师事务所)
Enrique Martínez · Ignacio Marqués · Israel Belloso · Felix Aramburu
合作 (c) Lara Muñoz Carlos Torres Marta Macho Carmen Hernández Vanesa Gutierrez
第四提名奖 fourth mention

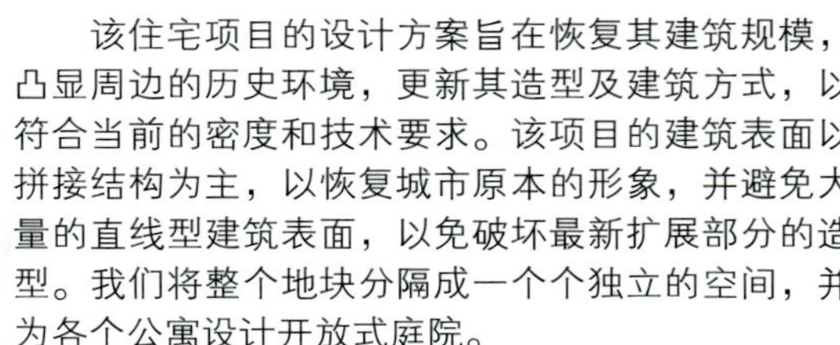

# 间隙空间

该住宅项目的设计方案旨在恢复其建筑规模，凸显周边的历史环境，更新其造型及建筑方式，以符合当前的密度和技术要求。该项目的建筑表面以拼接结构为主，以恢复城市原本的形象，并避免大量的直线型建筑表面，以免破坏最新扩展部分的造型。我们将整个地块分隔成一个个独立的空间，并为各个公寓设计开放式庭院。

# INTERSTITIAL SPACE

The proposed residential project aims to recover the scale of the volumes that define the historical surroundings, updating the construction and shape to the current density and technology. The project is defined by the fragmentation of the façade, which seeks to recover a domestic urban image, avoiding the massive linear façades that have distorted the new extensions. We seek to break up this great item on separate elements, to create the open courtyards of each apartment.

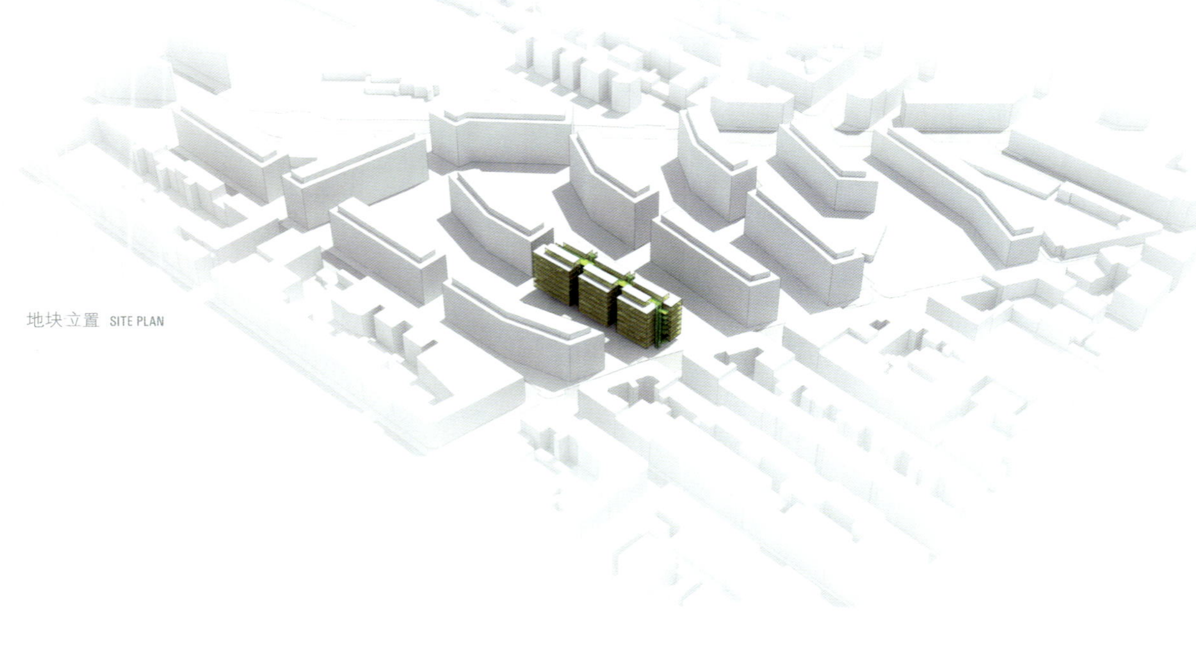

地块立置 SITE PLAN

标准层平面图 TYPICAL FLOOR PLAN

正立面图 MAIN ELEVATION

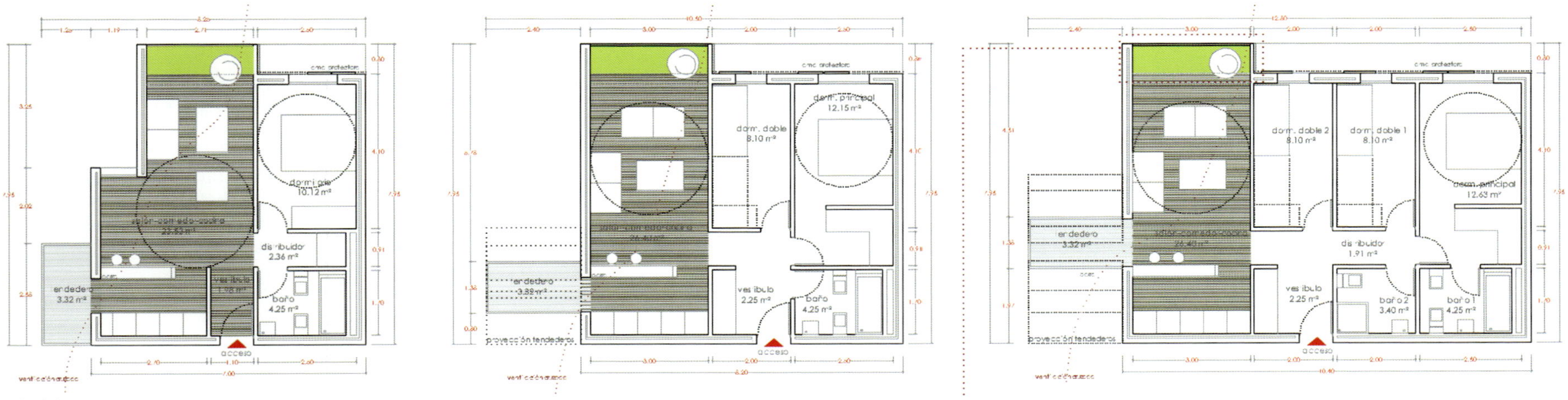

标准房屋单元 TYPICAL HOUSING UNITS

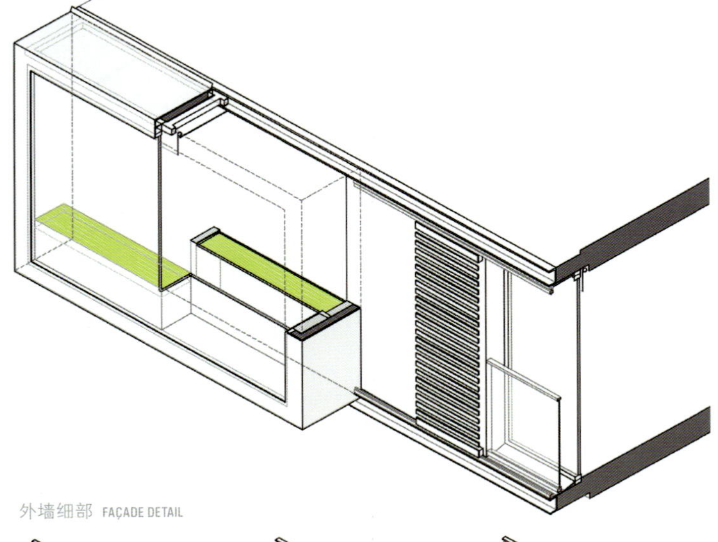

外墙细部 FAÇADE DETAIL

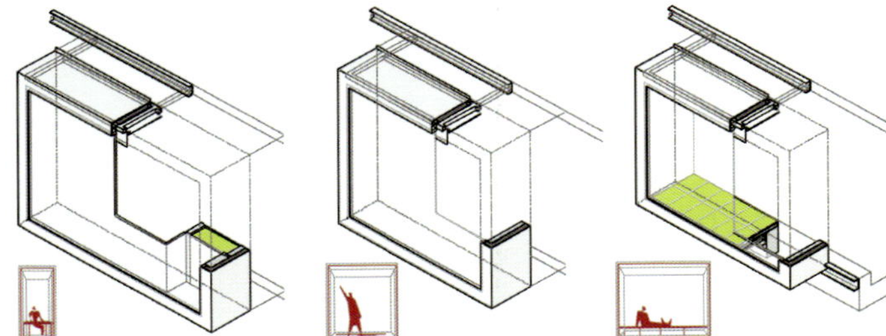

构框内的变化 VARIATION IN THE FRAME

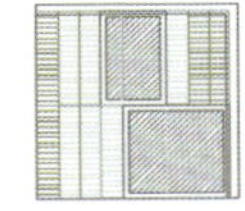

前外墙类型1
FRONT FAÇADE TYPE 1

侧外墙类型2
SIDE FAÇADE TYPE 2

后外墙类型3
BACK FAÇADE TYPE 3

# Estudio UNTERCIO (建筑师事务所)

Marina del Mármol · Daniel Bergman · Mauro Bravo · Miguel Herraiz (建筑师)

中标 winner

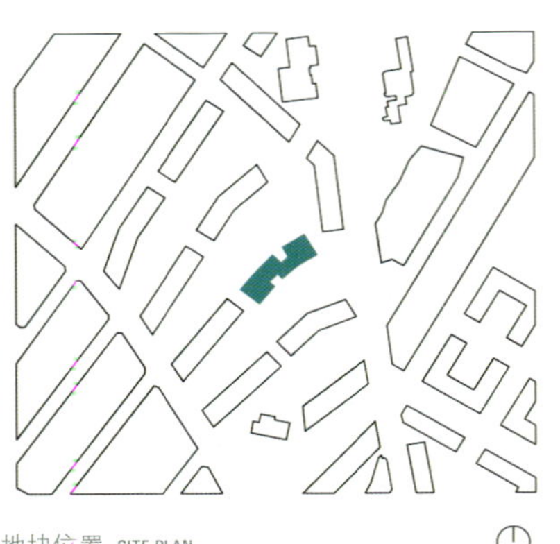

地块位置 SITE PLAN

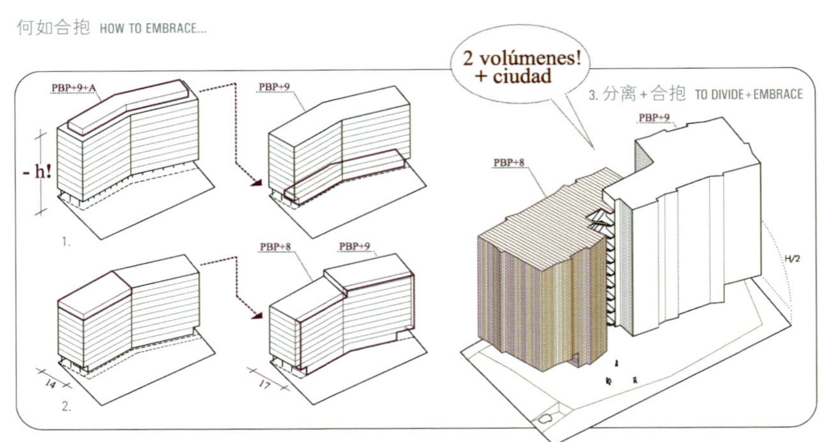


## 贴近城市

为了增进与城市的联系，我们决定以减小建筑规模来提高其美观度。因此，我们将其设计成高度限定且外观多样化的建筑。通过采用不同的方式打造这两个开放式的庭院，使其外观呈现多样化。最终，两个庭院相互连接，建筑规模减小。

## RELATIVE TO THE CITY

To enhance the relationship with the city, we decided to reduce the scale of the building looking for a pleasant appearance. Thus, we designed a building with a limited height and fragmented. We produce a sensation of diversity designing two open courtyards with opposite finishes in the façades of the two volumes. The result is a reduction in the scale of the building inside the city appearing as two volumes which embrace.

交错楼梯 INTERTWINED STAIRS

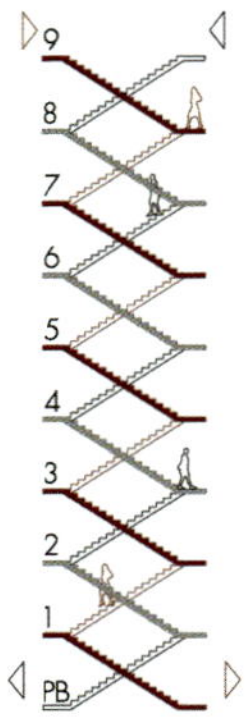

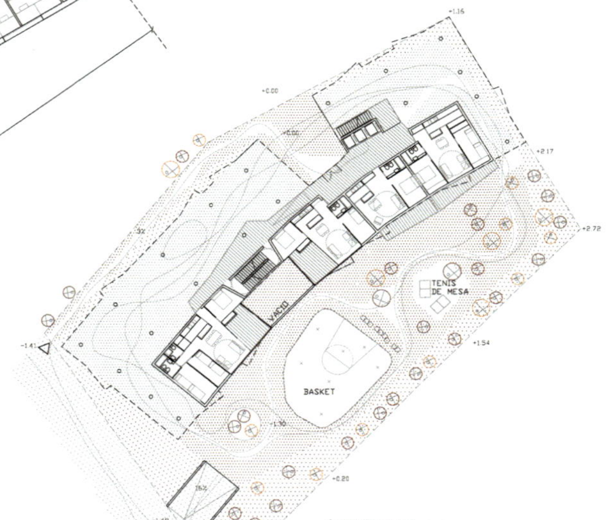

底层平面图 GROUND FLOOR PLAN

二至九层平面图 FLOOR PLANS 1-8

地下一、二层平面图 UNDERGROUND FLOOR PLANS -1,-2

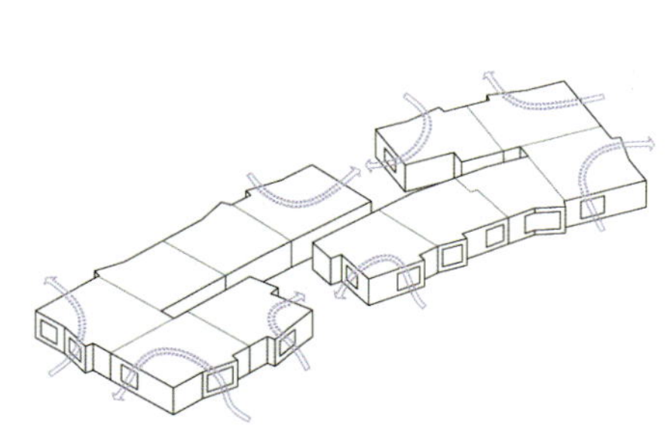

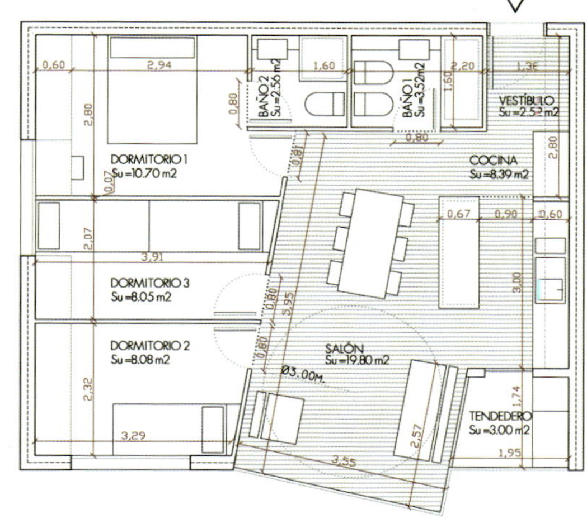

类型1 3卧室 TYPE 1. 3 BEDROOMS. 65.12M2

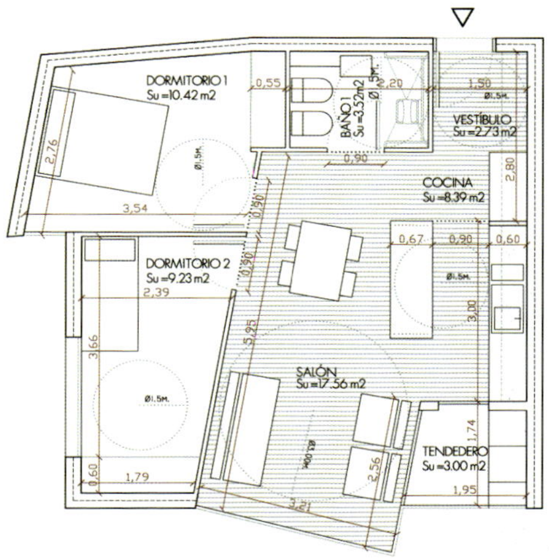

类型2 2卧室 TYPE 2. 2 BEDROOMS. 53.35M2

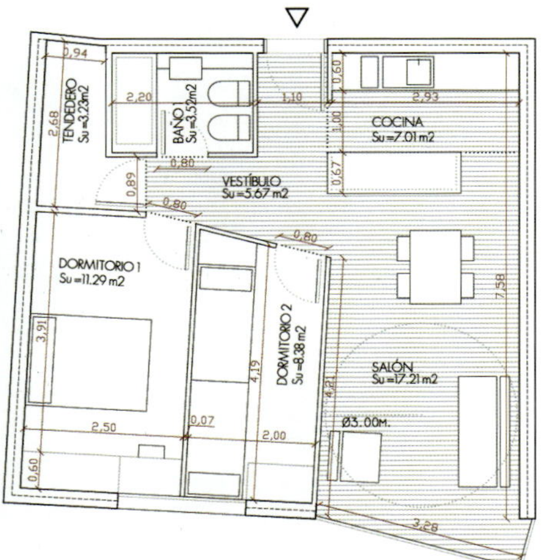

类型3 2卧室 TYPE 3. 2 BEDROOMS. 54.69M2

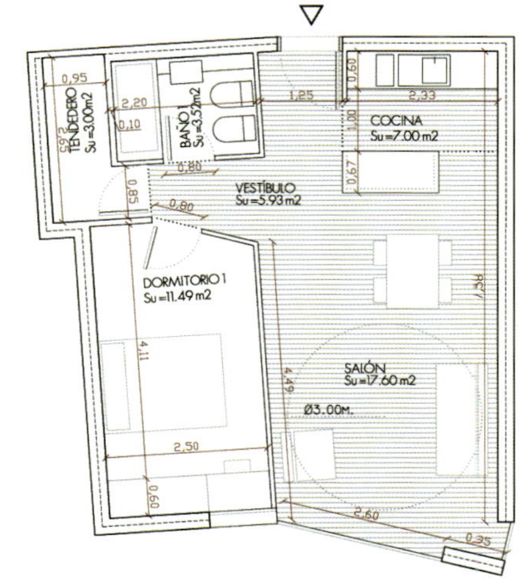

类型4 1卧室 TYPE 4. 1 BEDROOM. 47.04M2

# Ana Dolado Cosín · Fernado Araujo Fuster (建筑师)

合作 (c) Nuria Heras Donoso
入围短名单 shortlisted

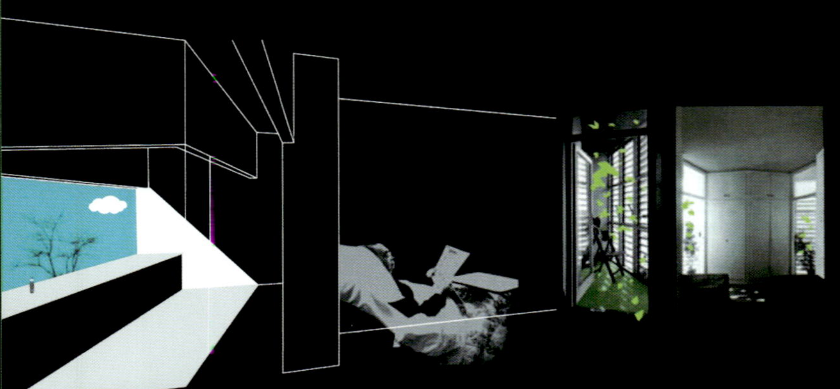

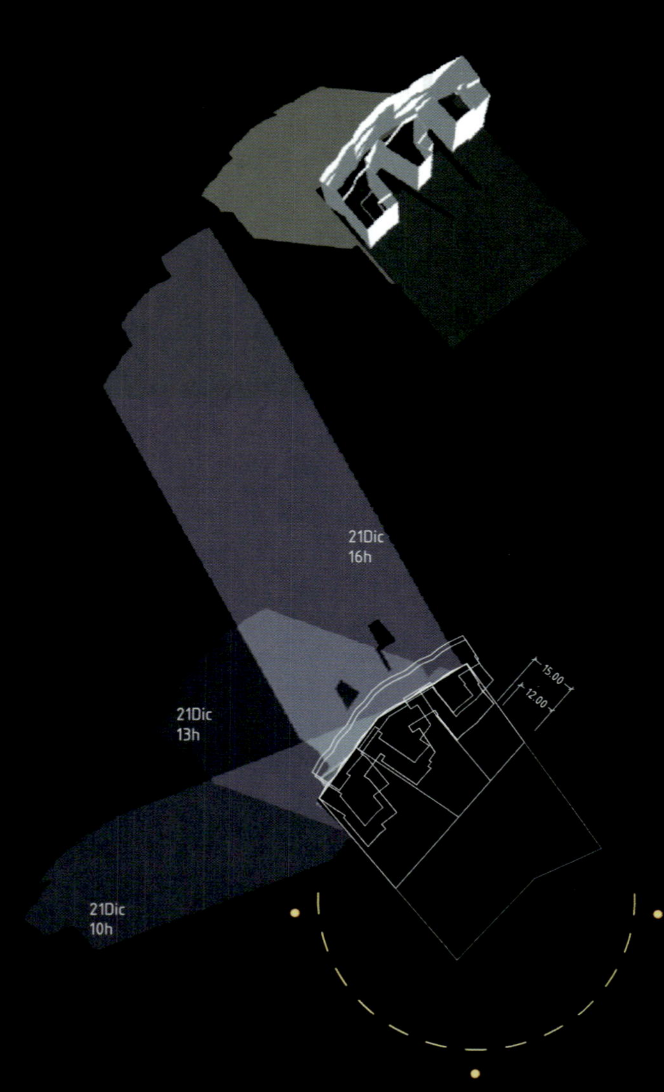

底层平面图 GROUND FLOOR PLAN

## 梳子结构

建筑表面朝向西北，迂回折叠，以便更好地通风和采光。建筑底层将分隔成三个区域，由斜坡连接，打造成景观层。景观层之上供残疾人士使用的核心通道和公寓，形成一个空间，呈扇形延伸至内部的自由空间。

## AS A COMB

The façade-skin adjusts to the northwest edge of the plot, and folds over itself looking for ventilation and lighting. The ground floor is developed as a landscape with three trays that accompany the slope. Above them, 3 communication cores and the apartments for the handicapped people are carved to set the void which fans out into the interior free space.

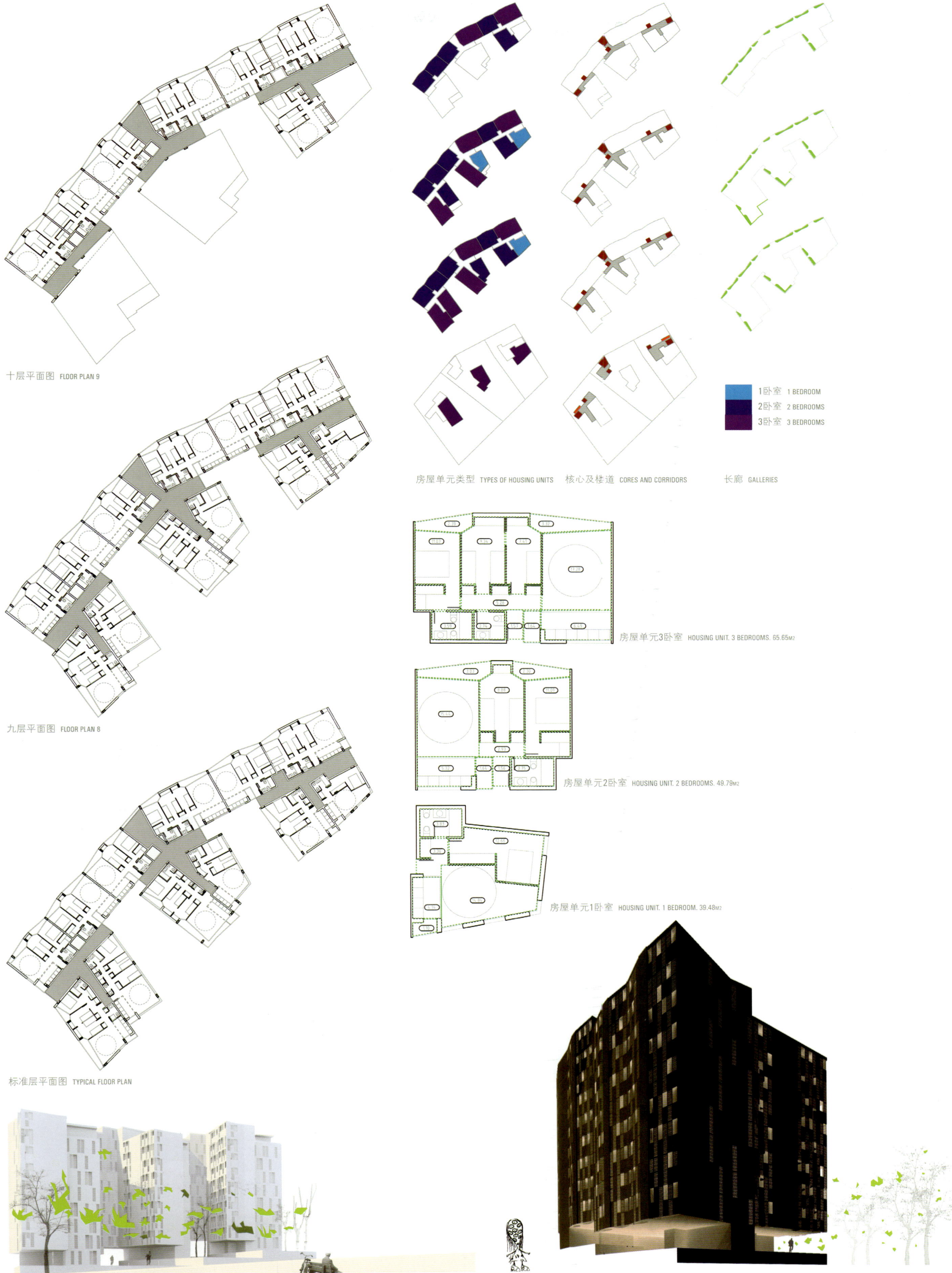

十层平面图 FLOOR PLAN 9
九层平面图 FLOOR PLAN 8
标准层平面图 TYPICAL FLOOR PLAN
房屋单元类型 TYPES OF HOUSING UNITS
核心及楼道 CORES AND CORRIDORS
长廊 GALLERIES
1卧室 1 BEDROOM
2卧室 2 BEDROOMS
3卧室 3 BEDROOMS
房屋单元3卧室 HOUSING UNIT. 3 BEDROOMS. 65.65M2
房屋单元2卧室 HOUSING UNIT. 2 BEDROOMS. 49.79M2
房屋单元1卧室 HOUSING UNIT. 1 BEDROOM. 39.48M2

AIRE LIBRE

# Juan Cortés Pedrosa (建筑师)

入围短名单 shortlisted

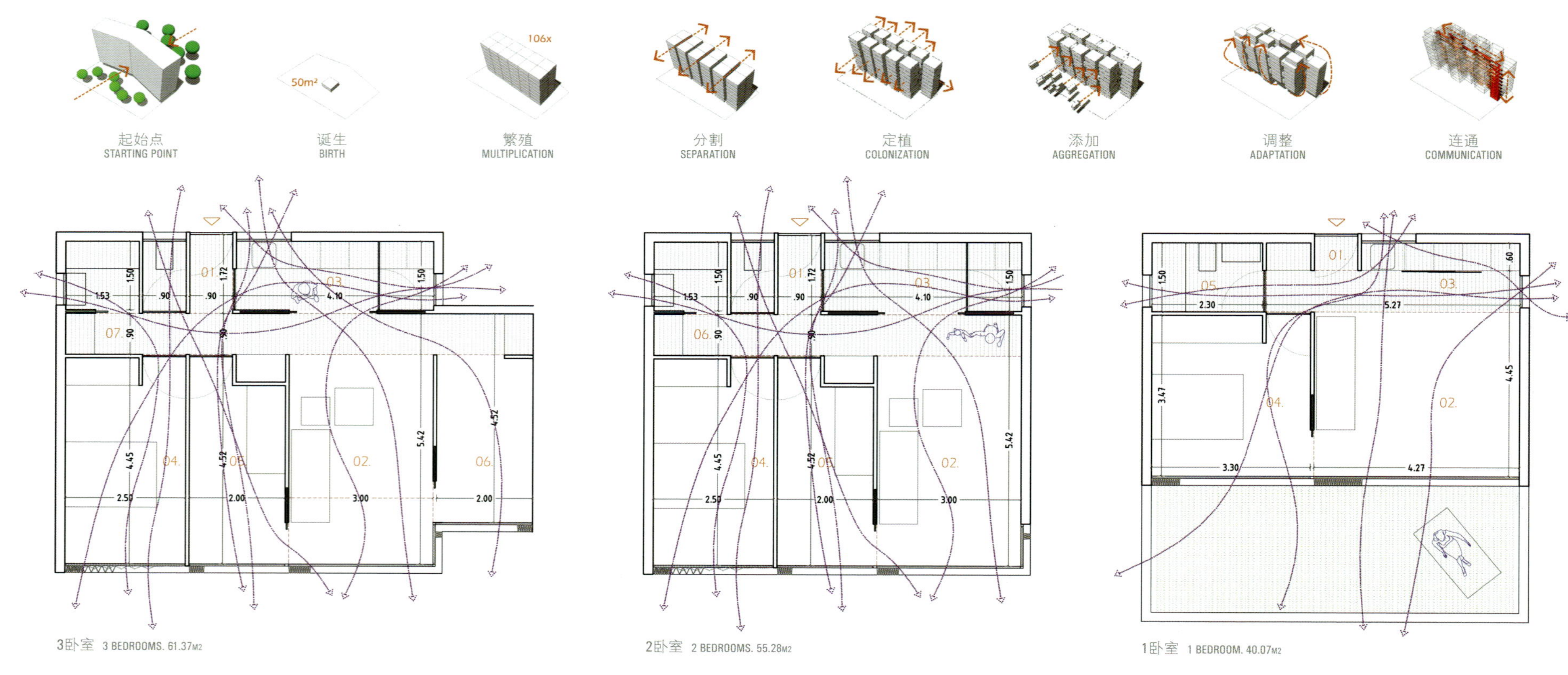


3卧室 3 BEDROOMS. 61.37M2

2卧室 2 BEDROOMS. 55.28M2

1卧室 1 BEDROOM. 40.07M2

## 建筑内的“大自然”

目前很有必要构建一个灵活性更强、渗透性更高、公共空间更佳的新城市体系。为此，我们努力寻求一个更好的方式，将居住区周边的公共空间连接起来：一个大型的公共空间可作为社交中心，并根据未来的需要，通过小型通道与外部相连，以扩大住宅单元。其间的空隙有助于入口处的通风和采光。

## NATURE INSIDE THE BUILDING

It is necessary to propose a new urban system more permeable, flexible and with better public spaces. Therefore, we seek for a new way of grouping individual cells surrounding residential communal spaces: a large open central space which acts as communications space, connected to the outside by means of small holes which also promote the extension of housing units depending on future needs. These gaps serve to introduce air and light in the entrance areas.

efecto chimenenea (verano)
cierre corredero traslúcido
nivel_10
nivel_09
ventilación cruzada (viviendas)
cubierta vegetal (aislamiento)
nivel_08
cierre corredero traslúcido
nivel_07
ganancias térmicas (invierno)
arbustos de hoja caduca
protección solar (lamas verticales)
cierre plegable traslúcido
nivel_06
nivel_05
espacios en sombra (verano)
captación de brisas (verano)
nivel_04
captación de brisas (verano)
evapotranspiración de vegetación
nivel_03
protección solar (lamas verticales)
nivel_02
captación de brisas (verano)
ventilación cruzada (viviendas)
nivel_01
evapotranspiración de vegetación
entrada de aire caliente (verano)
nivel_00
enfriamiento por vaporización

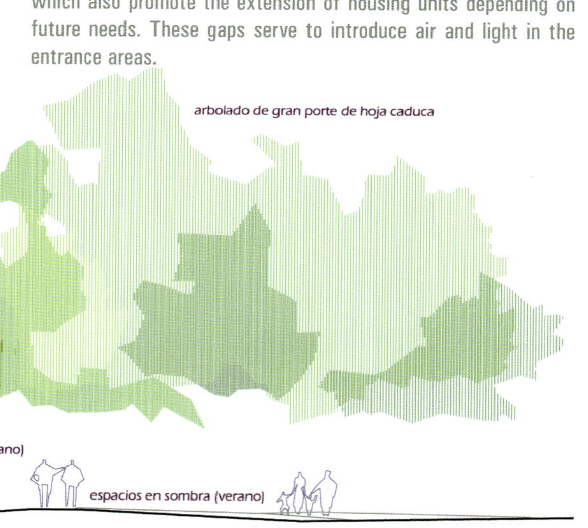


横向剖面图 CROSS SECTION

七层平面图 FLOOR PLAN 6

底层平面图 GROUND FLOOR PLAN

东南立面图 SOUTHEAST ELEVATION

# eslava y tejada arquitectos (建筑师事务所)

Clara Eslava + MiguelTejada (建筑师)

入围短名单 shortlisted

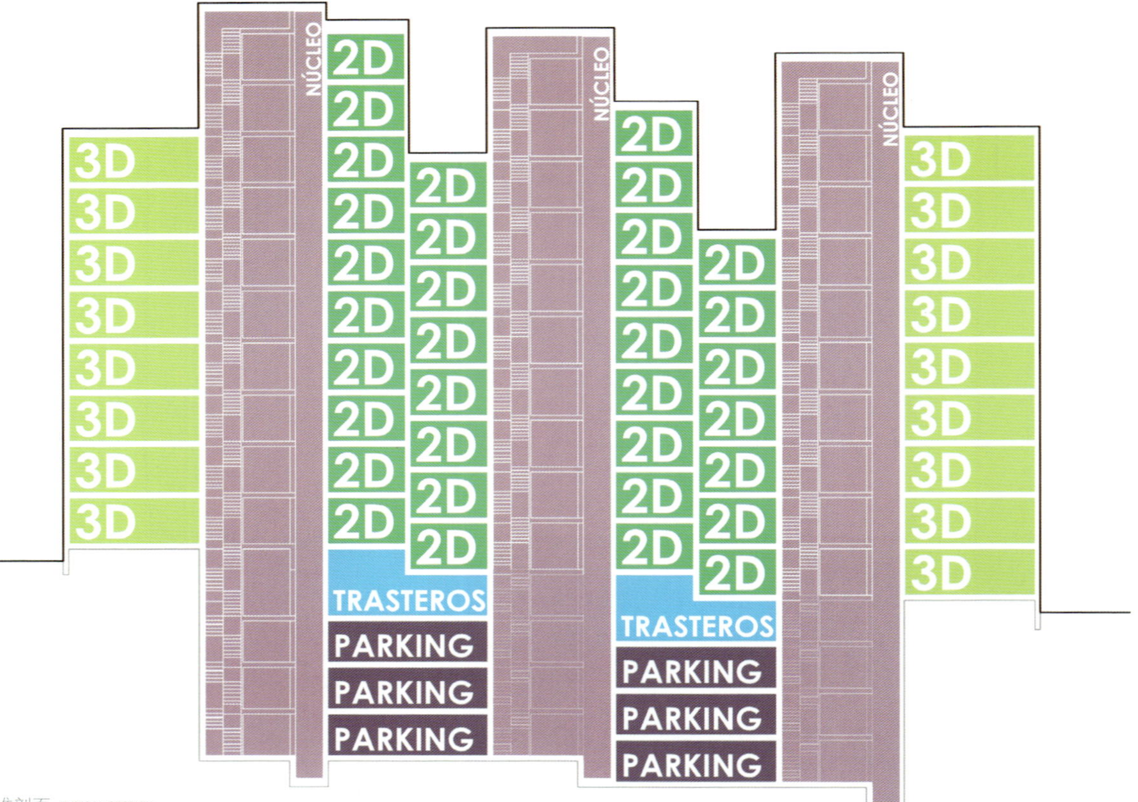


标准剖面 TYPICAL SECTION

## 分离的空间

住宅单元为大楼的组成部分，各个相连，形成分散的楼群，是未来“生态社区”绿地的“大型通道”。该楼群似巨大的折叠式屏幕，并不区分主、次建筑表面，同时，楼群的层叠结构与地块间的间隙相融合。

## BROKEN VOLUME

The housing units are components of a tower that blends with the next, forming a fragmented block, a “pathway of giants” for the green spaces for future eco-district. The block is a large folding screen without differentiating between the main and back front, while it staggers, adapting to the existing gap between the ends of the plot.

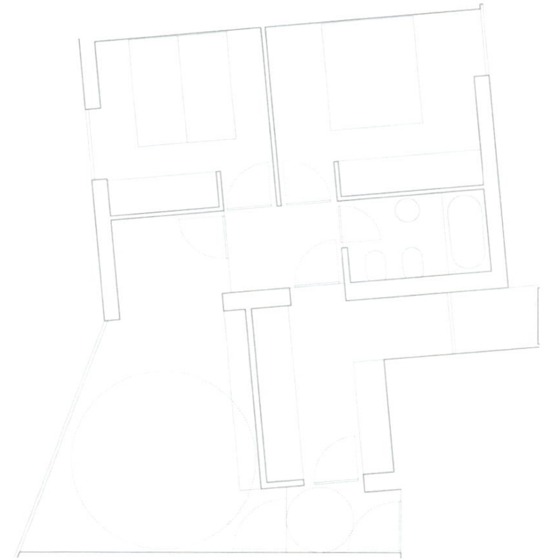

2卧室 2 BEDROOMS. 53.22$M^2$

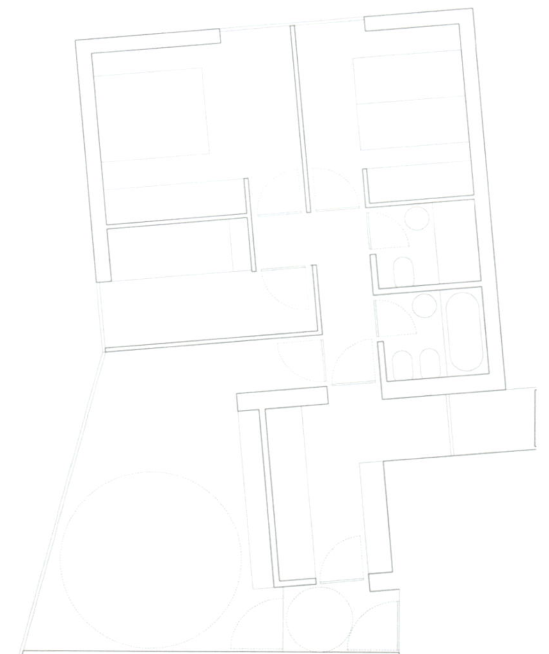

3卧室 3 BEDROOMS. 65.00$M^2$

十层平面图 FLOOR PLAN 9

1卧室 1 BEDROOM. 40.02M2

八层平面图 FLOOR PLAN 7

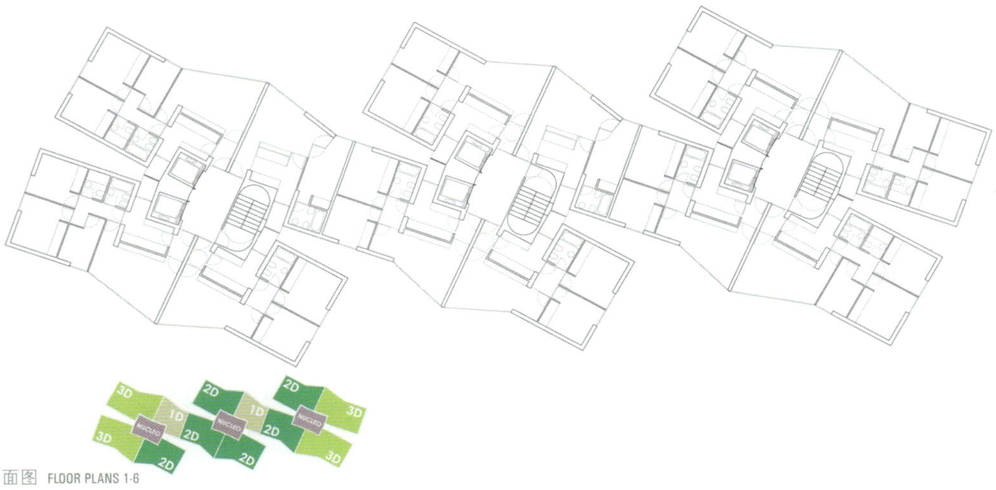

二至七层平面图 FLOOR PLANS 1-6

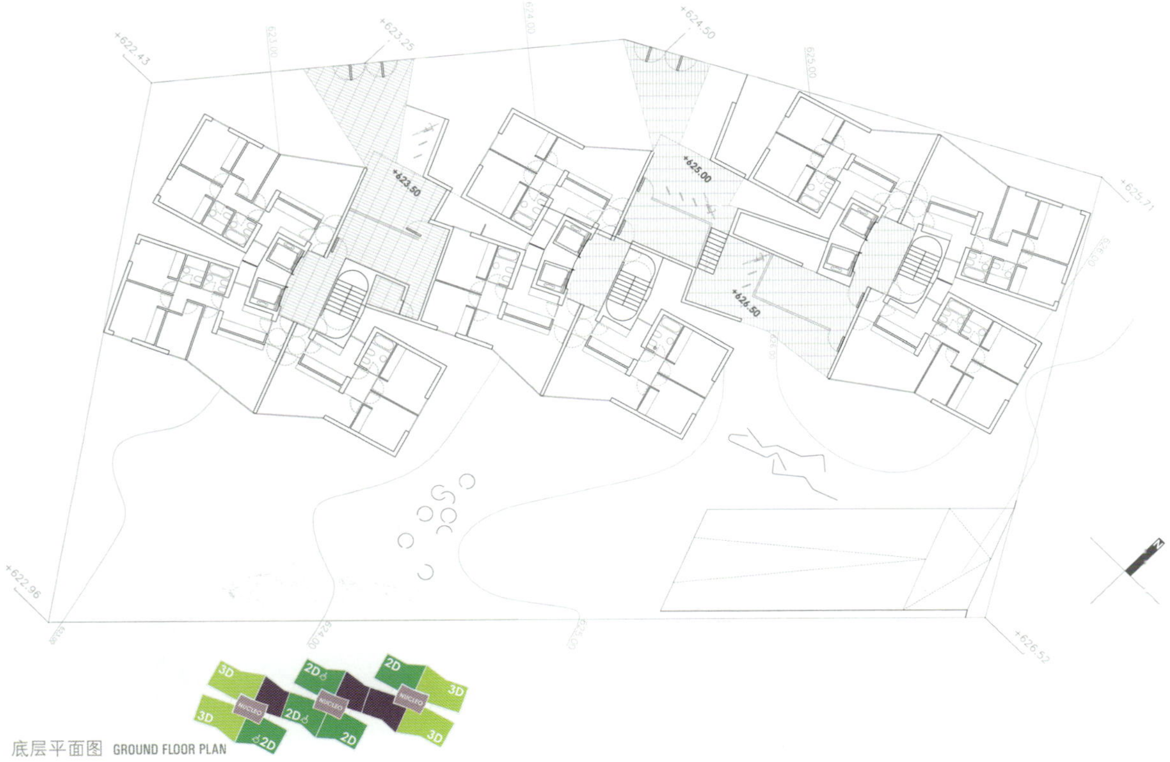

底层平面图 GROUND FLOOR PLAN

Alejandro Gómez · Begoña López · Juliane Haider (建筑师)

中标 winner

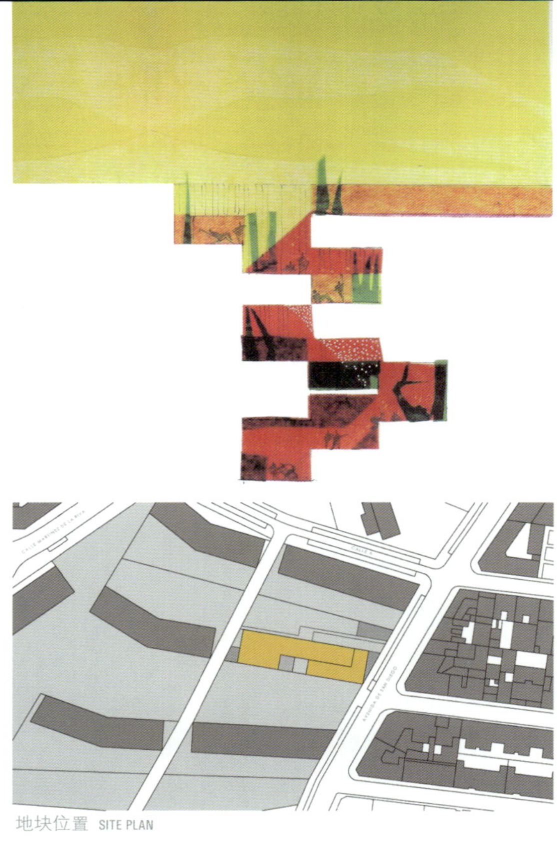

地块位置 SITE PLAN

# 伸出阳台之外

我们设计的公寓楼具有统一的垂直核心通道，且各层均带有空中花园。这些区域的设置可促进公寓内部邻里间的沟通，并增进社区与外部绿化空间之间的和谐。通过相连接的双层设计，各空中花园得以两两相连，成为一体。

# LEAN OUT OF THE BALCONIES

We propose a block of flats with a single vertical core of communications and a hanging garden on each floor. These spaces promote an internal communication between neighbours and the rapprochement between neighbours and green exterior spaces. The gardens will be interconnected through linked double heights.

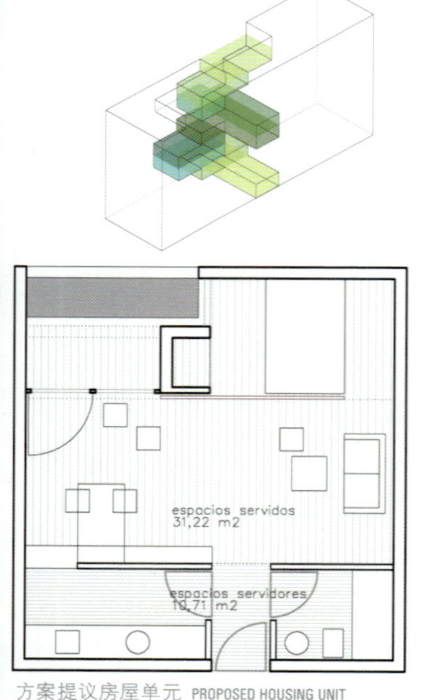


方案提议房屋单元 PROPOSED HOUSING UNIT

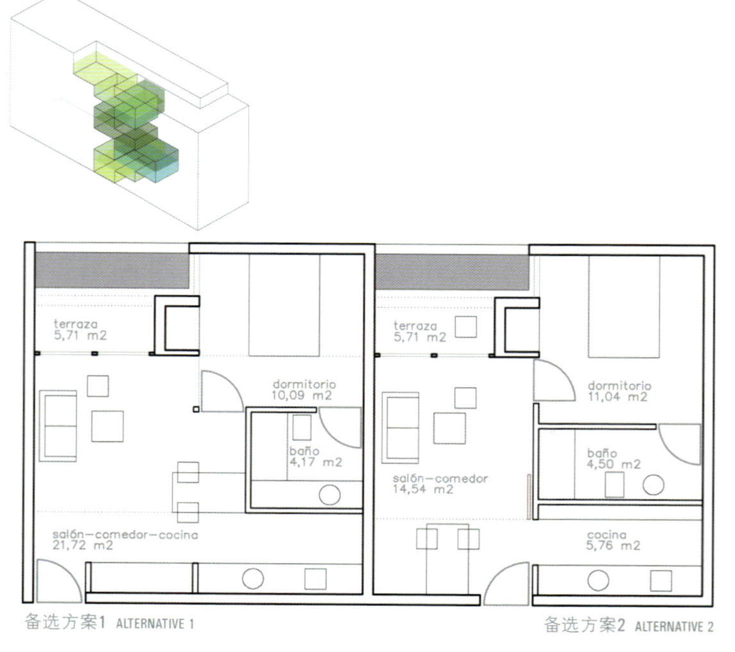


备选方案1 ALTERNATIVE 1

备选方案2 ALTERNATIVE 2

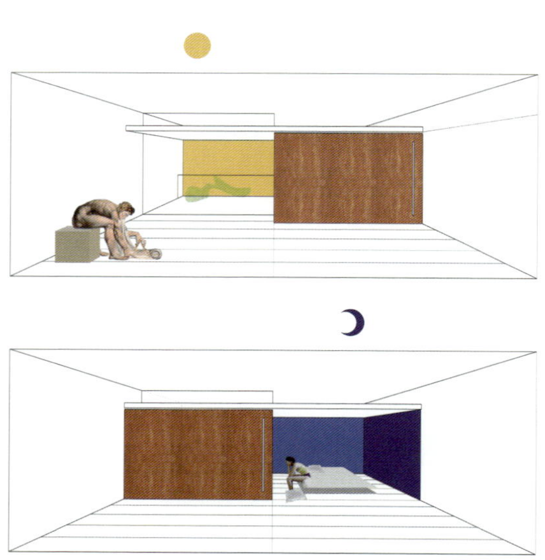

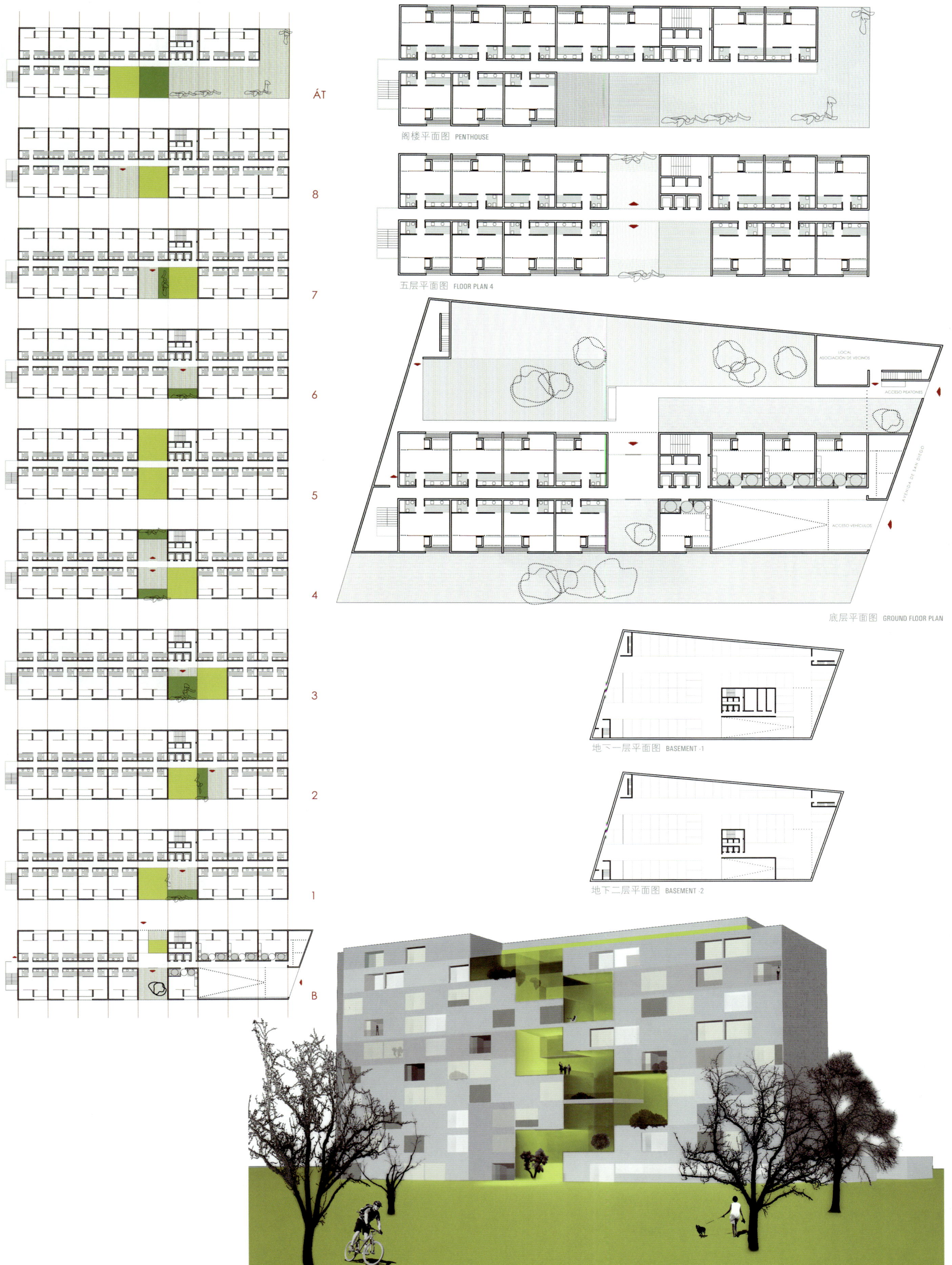
ÁT
8
7
6
5
4
3
2
1
B
阁楼平面图 PENTHOUSE
五层平面图 FLOOR PLAN 4
底层平面图 GROUND FLOOR PLAN
地下一层平面图 BASEMENT -1
地下二层平面图 BASEMENT -2

ABCDEF7

# Eugenio Muñoz Pérez (建筑师)

合作 (c) Daniela Careaga Varela · Alberto Bustamante Domínguez

第一提名奖 first mention

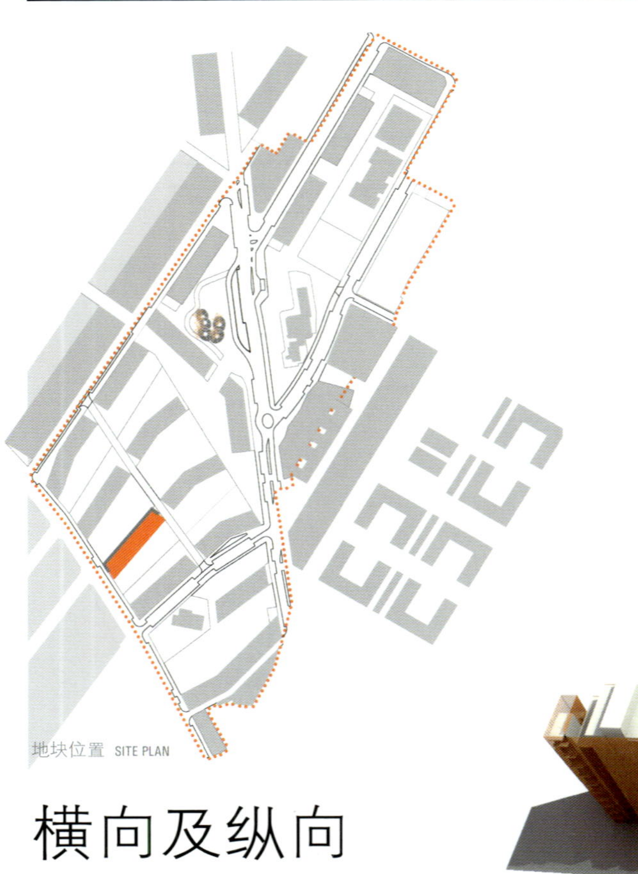

地块位置 SITE PLAN

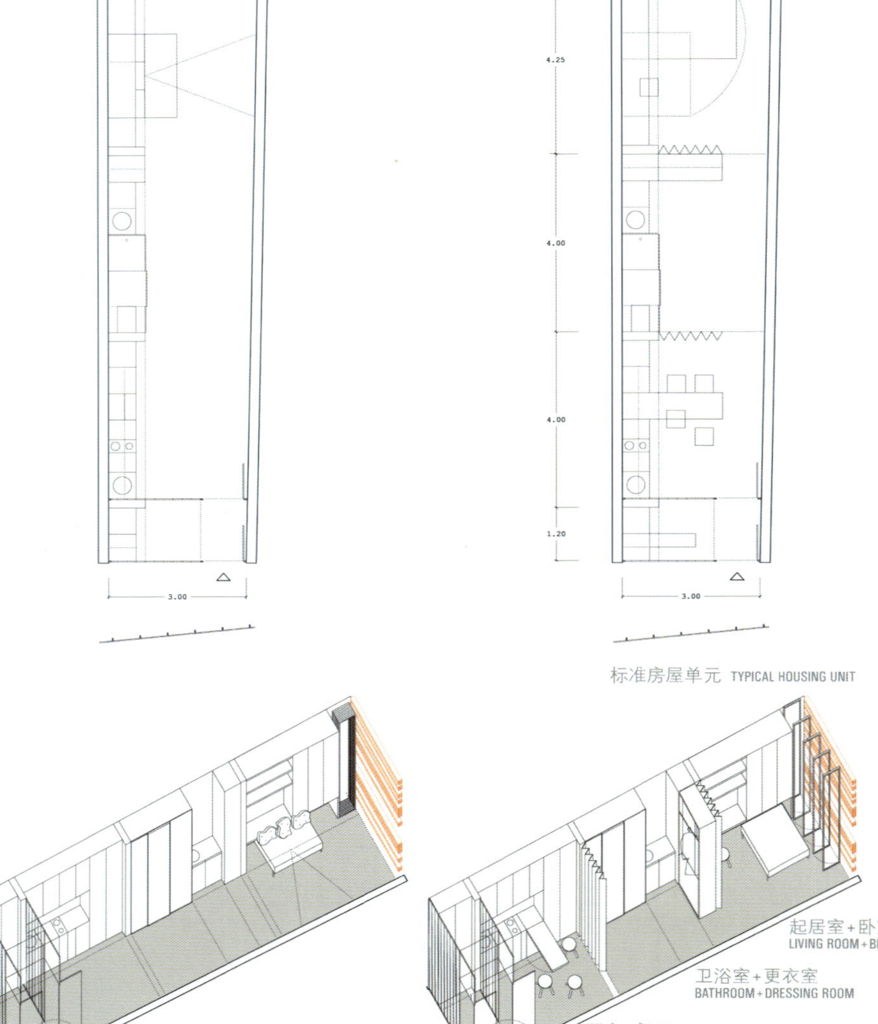

标准房屋单元 TYPICAL HOUSING UNIT

起居室+卧室 LIVING ROOM+BEDROOM

卫浴室+更衣室 BATHROOM+DRESSING ROOM

厨房+餐厅 KITCHEN+DINNING ROOM

## 横向及纵向延伸

整个方案通过重复建造同一样式的住宅单元找到了一个简单的解决办法。将纵向住宅单元打造成双面朝向，使其对流通风，并可经走廊进出，由此形成自然、简单、合理的建筑结构。根据该地块的地形特点，我们还将各个单元设计成了回转结构，使其与圣地亚哥大道相协调。如果将这些单元楼设计得样式各异，那么视觉上它们将会纵向重重叠加，使整个建筑群显得极不紧凑。

## HORIZONTAL AND VERTICAL MOVEMENT

The whole proposal is solved by a simple solution, generated from the repetition of a single type of housing unit. A rational, simple and natural structure made up by longitudinal housing units with a double orientation and cross ventilation to be accessed via a corridor. A slight twist on all units allows the building to fit inside the plot aligning with Avenue San Diego. Once these groups of houses achieve independence, they slide vertically on one another, and the building loses compactness.

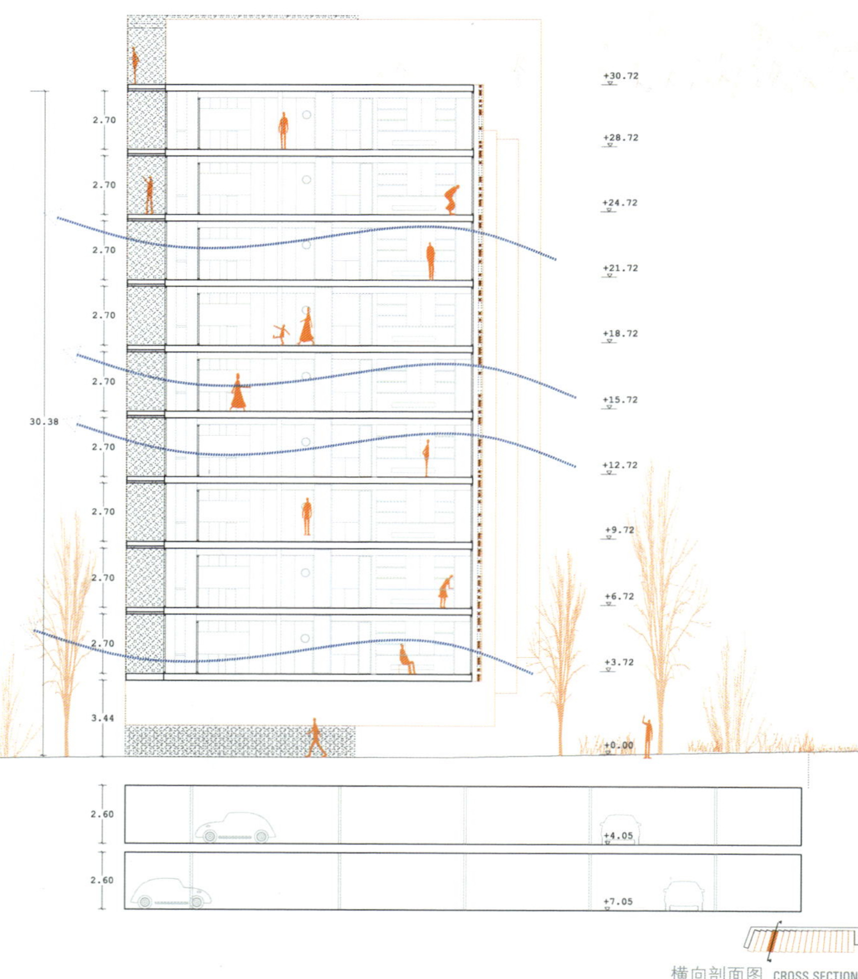

横向剖面图 CROSS SECTION

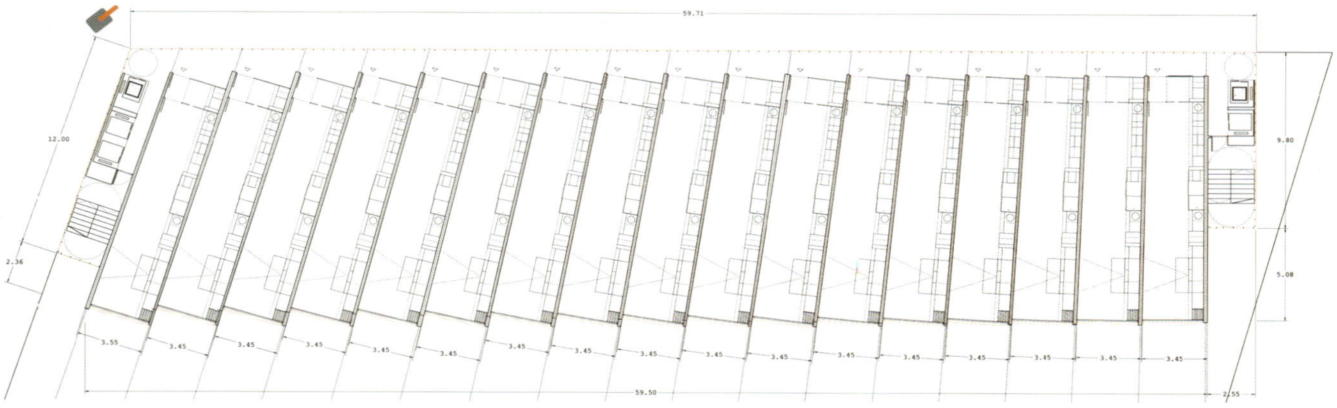

ACCESSO

LOCAL ASOCIACIÓN
DE VECINOS
216,97 m2

ACCESSO

ACCESO
GARAJE

618.71

622.46

619.17

622.98

标准层平面图 TYPICAL FLOOR PLAN

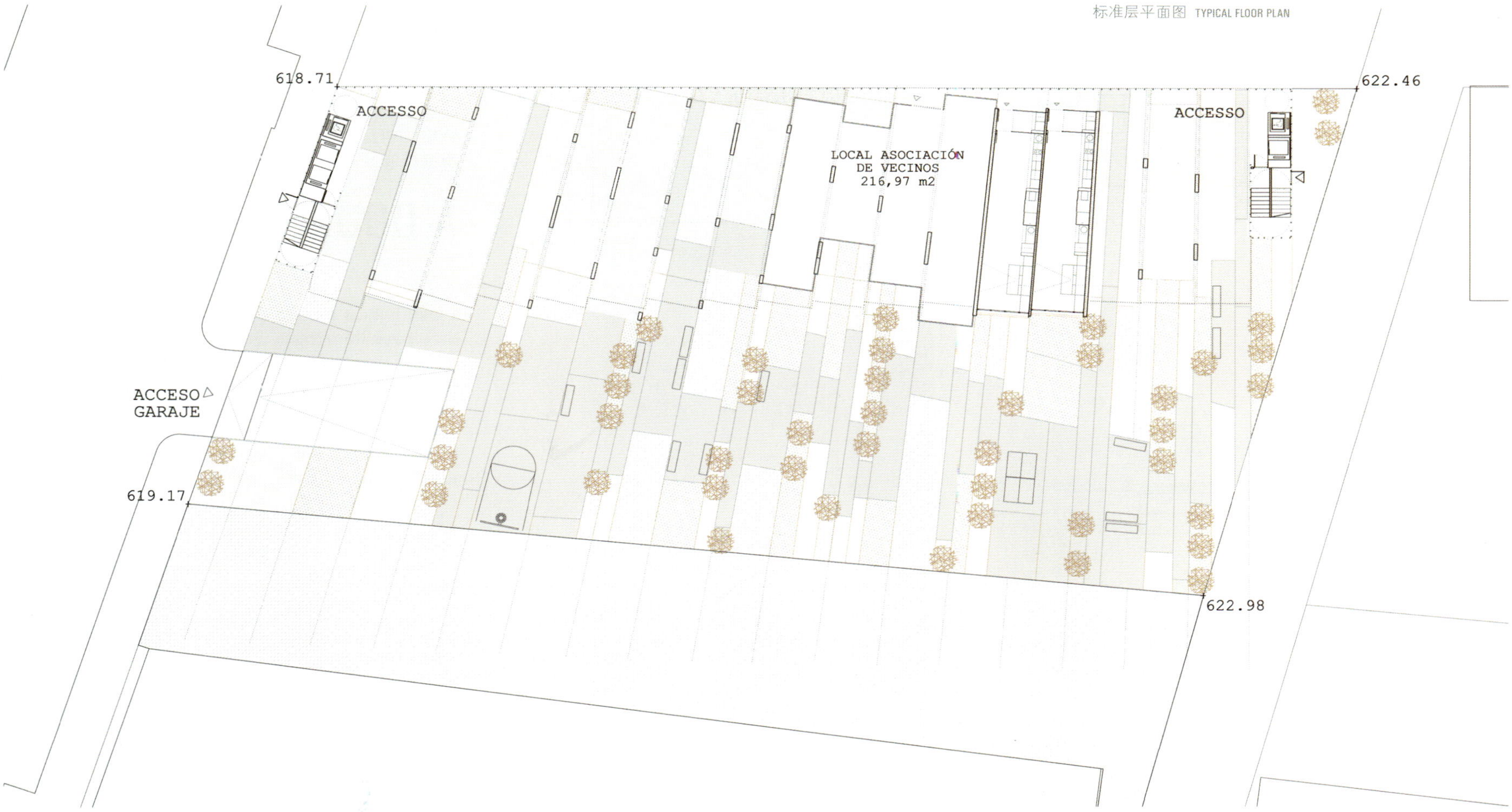

底层平面图 GROUND FLOOR PLAN

OOOTEAM

# Fernando Maniá Buono · Santiago Cifuentes Barrio Martina Sibona Guevara (建筑师)

第二提名奖 second mention

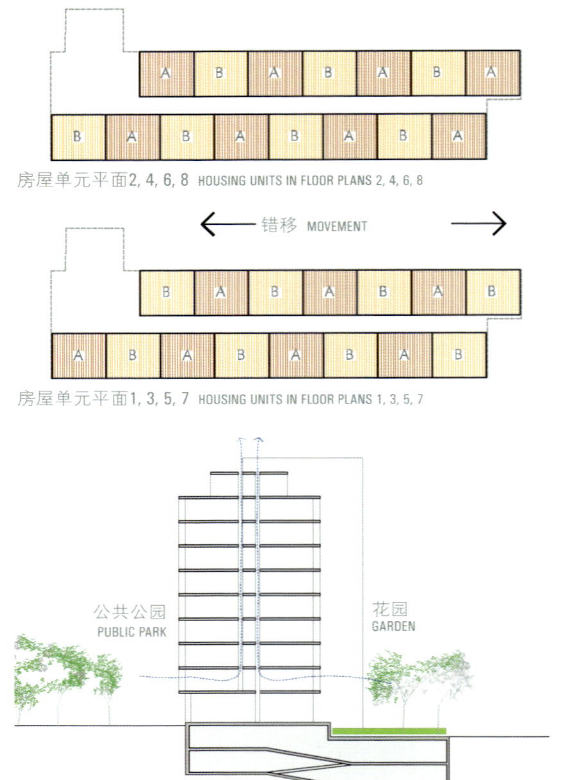


房屋单元平面2, 4, 6, 8 HOUSING UNITS IN FLOOR PLANS 2, 4, 6, 8

房屋单元平面1, 3, 5, 7 HOUSING UNITS IN FLOOR PLANS 1, 3, 5, 7

横向剖面图 CROSS SECTION

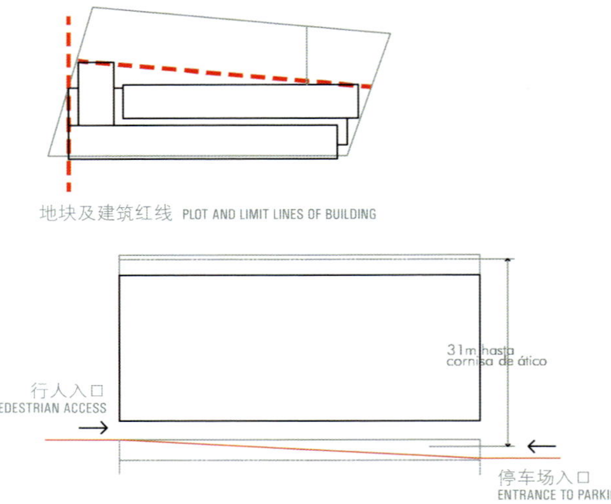


地块及建筑红线 PLOT AND LIMIT LINES OF BUILDING

## 双层建筑表面

公园边上有两幢相连的大楼互成直角，将其连接处打造为该区域的入口。该架构使得建筑群和公园完美融合，连为一体。考虑到该建筑的结构规划及其有限的面积，我们欲将建筑外观作最大化设计。我们设计的住宅单元灵活性很强。80厘米宽的悬壁可打造为观景台或花房；冬天，封闭起来还可用来保暖。

## A DOUBLE SKIN

A linear block at the edge of the park and another perpendicular block shape the entrance, linking two intersecting pieces. The structural grid adapts, perfectly modulated, to housing and parking. Given the organization of the building, and considering housing is of limited size, we propose maximum development possible for the façades. We have designed highly flexible housing units. The 80cm-cantilever is a panoramic terrace, a greenhouse, and in winter, closed, it will serve as an energy sensor.

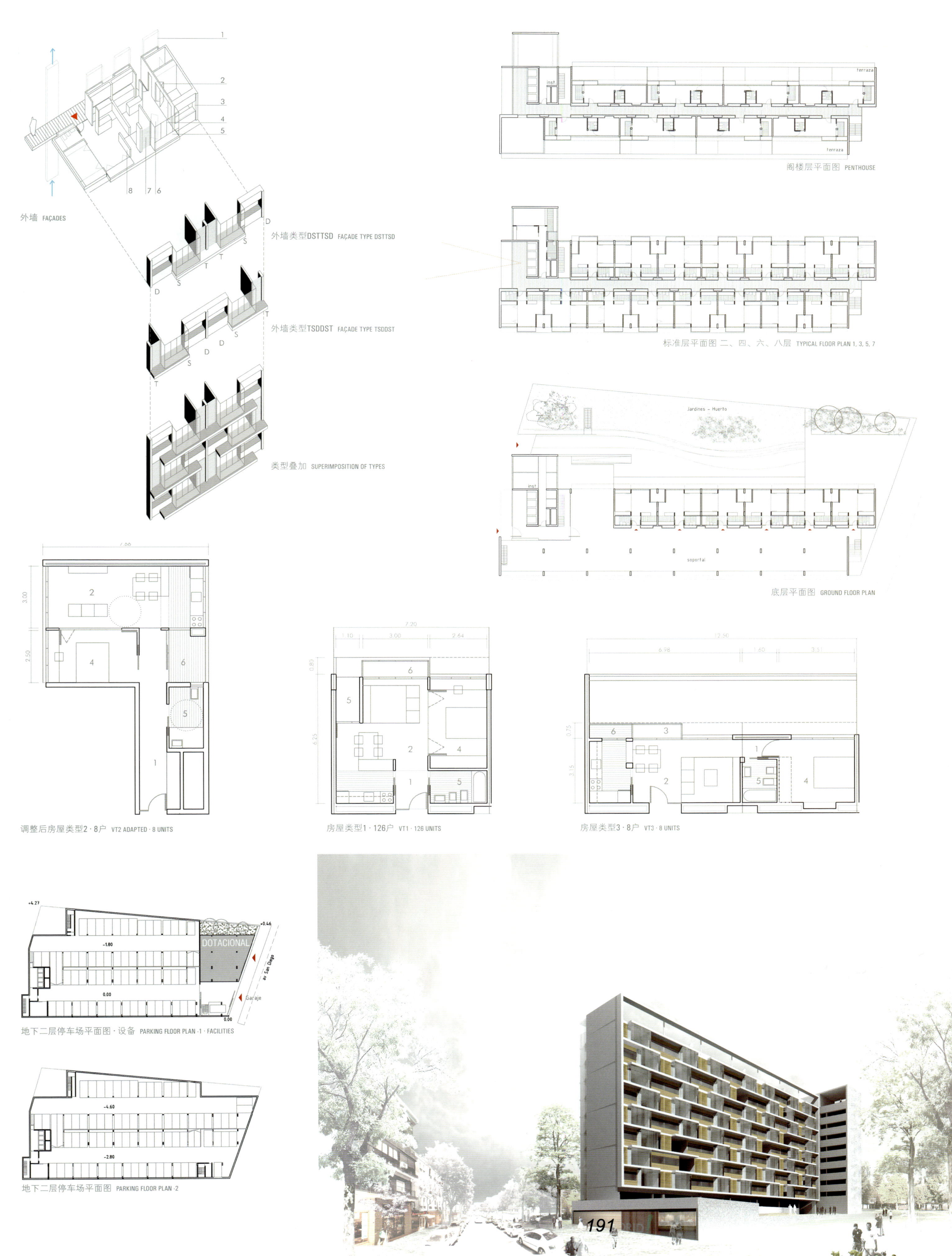


外墙 FAÇADES

外墙类型DSTTSD FAÇADE TYPE DSTTSD

外墙类型TSDDST FAÇADE TYPE TSDDST

类型叠加 SUPERIMPOSITION OF TYPES

阁楼层平面图 PENTHOUSE

标准层平面图 二、四、六、八层 TYPICAL FLOOR PLAN 1, 3, 5, 7

底层平面图 GROUND FLOOR PLAN

调整后房屋类型2・8户 VT2 ADAPTED・8 UNITS

房屋类型1・126户 VT1・126 UNITS

房屋类型3・8户 VT3・8 UNITS

地下二层停车场平面图・设备 PARKING FLOOR PLAN -1・FACILITIES

地下二层停车场平面图 PARKING FLOOR PLAN -2

ATALAYA

# AC Arquitectos (Carlos Miguel Iglesias + Ángel Nodar) (建筑师)

合作 (c) Marta Rama Poza · Sofía Corsini Fuhrmann

第四提名奖 fourth mention

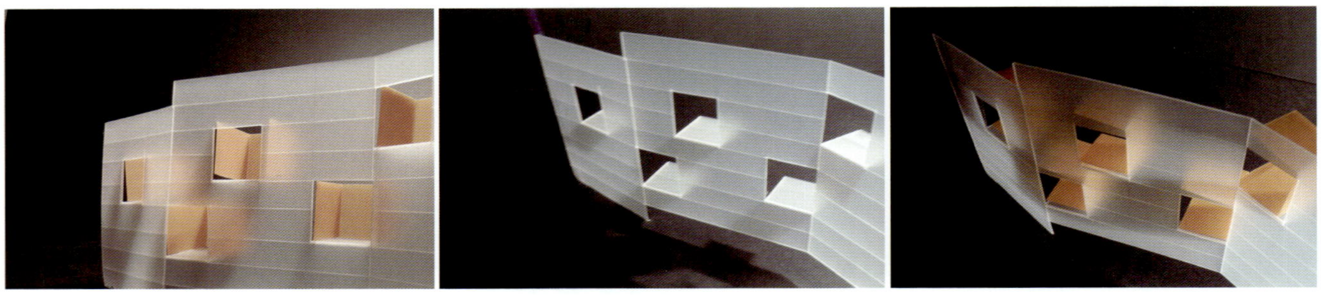

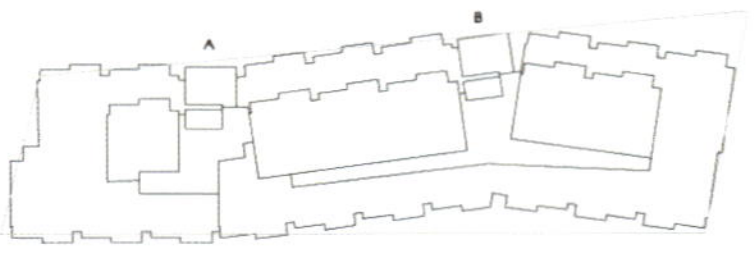

屋顶层平面 ROOF PLAN

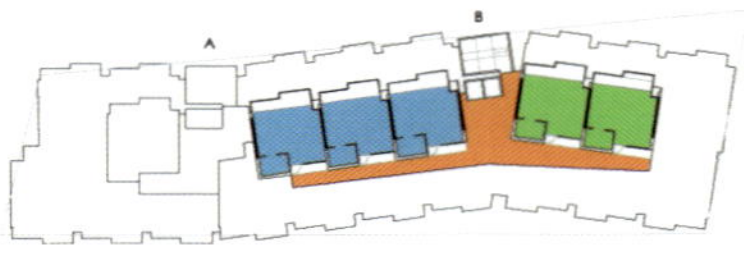

阁楼层平面 · 核心B PENTHOUSE · CORE B

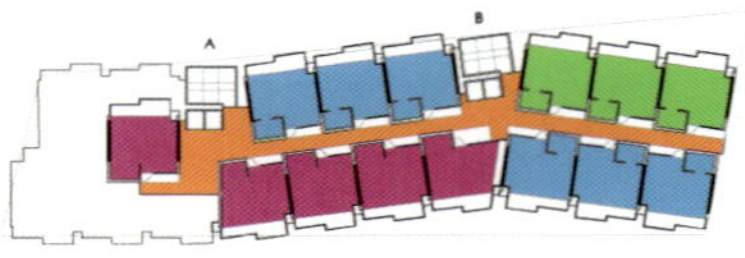

阁楼层平面 · 核心A + 九层平面 · 核心B PENTHOUSE · CORE A + FLOOR PLAN 8 · CORE B

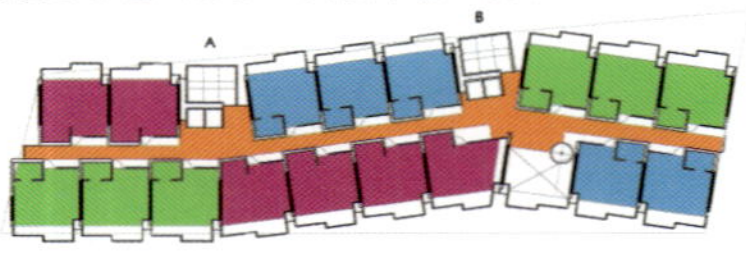

九层平面 · 核心A + 八层平面 · 核心B FLOOR PLAN 8 · CORE A + FLOOR PLAN 7 · CORE B

八层平面 · 核心A + 七层平面 · 核心B FLOOR PLAN 7 · CORE A + FLOOR PLAN 6 · CORE B

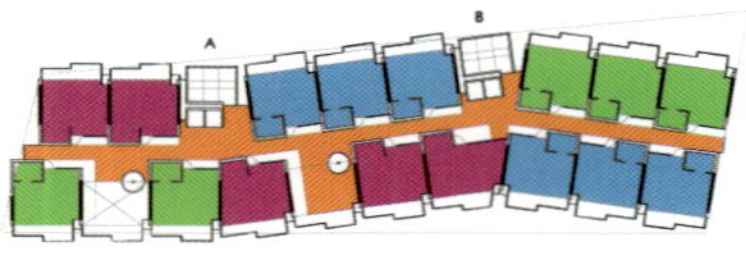

七层平面 · 核心A + 六层平面 · 核心B FLOOR PLAN 6 · CORE A + FLOOR PLAN 5 · CORE B

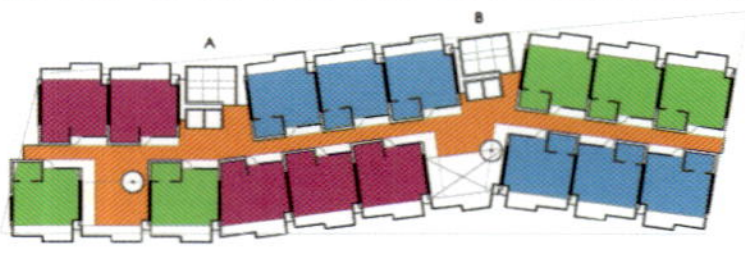

六层平面 · 核心A + 五层平面 · 核心B FLOOR PLAN 5 · CORE A + FLOOR PLAN 4 · CORE B

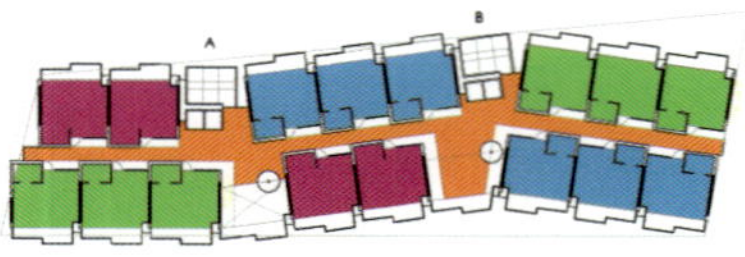

五层平面 · 核心A + 四层平面 · 核心B FLOOR PLAN 4 · CORE A + FLOOR PLAN 3 · CORE B

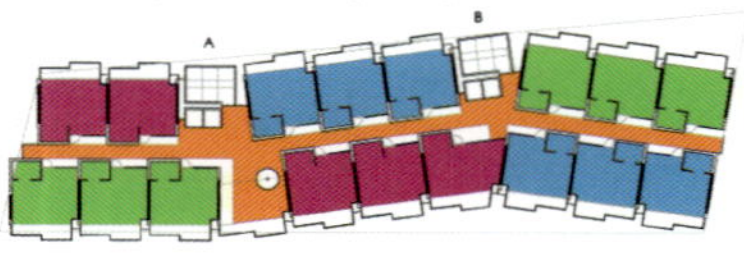

四层平面 · 核心A + 三层平面 · 核心B FLOOR PLAN 3 · CORE A + FLOOR PLAN 2 · CORE B

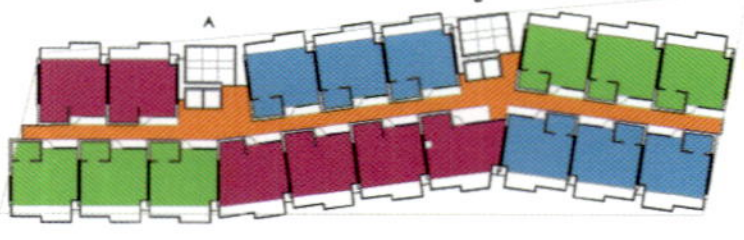

三层平面 · 核心A + 二层平面 · 核心B FLOOR PLAN 2 · CORE A + FLOOR PLAN 1 · CORE B

二层平面 · 核心A + 底层平面 · 核心B FLOOR PLAN 1 · CORE A + FLOOR PLAN 0 · CORE B

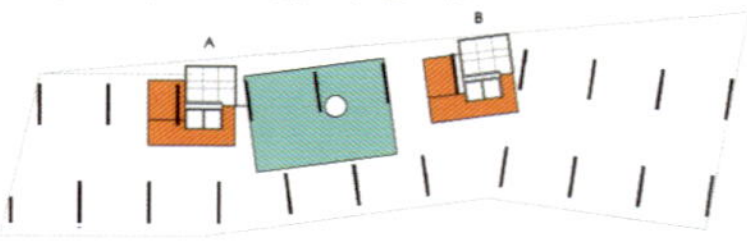

底层平面 · 核心A FLOOR PLAN 0 · CORE A

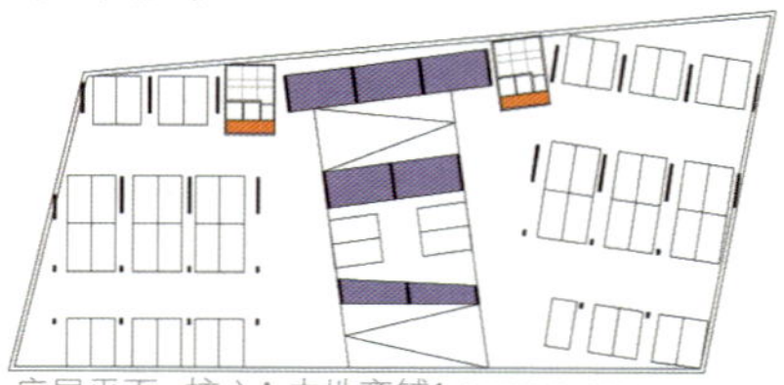

底层平面 · 核心A 本地商铺1 PLANTA 0 · CORE A. LOCAL COMERCIAL 1

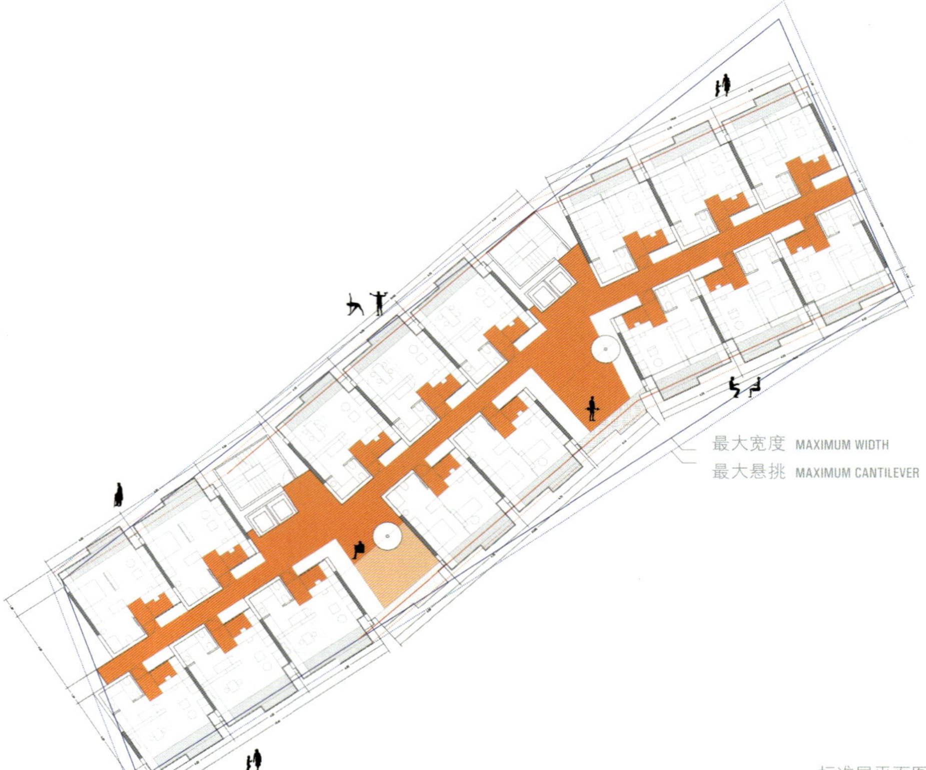

最大宽度 MAXIMUM WIDTH

最大悬挑 MAXIMUM CANTILEVER

标准层平面图 TYPICAL FLOOR PLAN

AVENIDA DE SAN DIEGO

N

底层平面图 GROUND FLOOR PLAN

南立面图 SOUTH ELEVATION

横向剖面图 CROSS SECTION

# 悬浮的空间

一条中央走廊将两个流通中心连通，这样的布局使得人们可以自由出入紧贴房屋外围的两条狭长地带。这种走廊—街道式的设计可扩展人们的视阈，同时又给大家提供了一个公共空间。五个大型双层空间使得人们站在这里犹如置身于瞭望塔之上，既可俯瞰周边，亦可瞭望天际。

# SUSPENDED VOIDS

A system based on a central corridor linked with two communication cores provides access to two peripheral strips which shape the housing. This corridor-street organizes the layout promoting views and perceptions which turns the space into a communal room. Five large-scale double height holes drill the volume as watchtowers to watch the site and scan the skyline.

类型1 家庭工作单元 TYPE 1. HOME-WORK-STUDIO UNIT

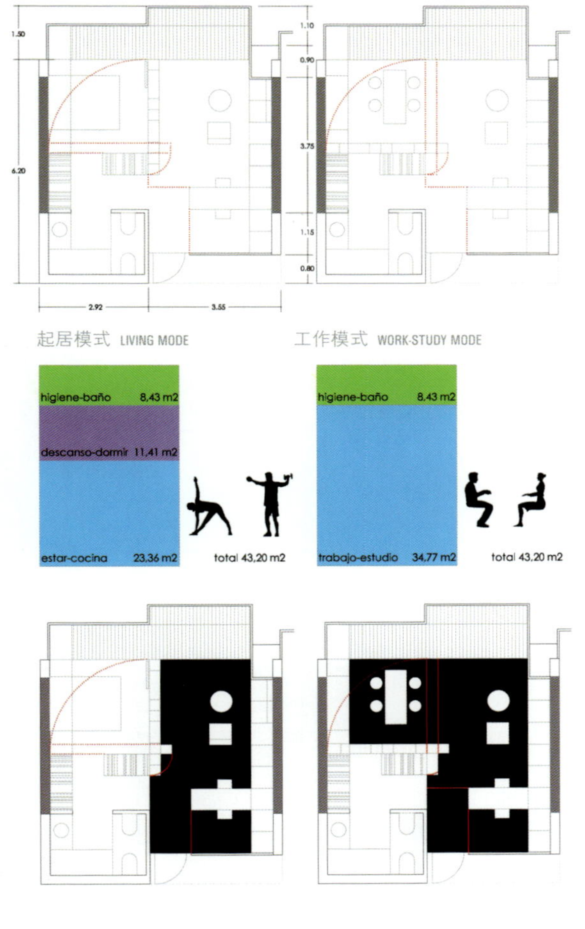


类型2 家庭照料单元 TYPE 2. CARE HOME

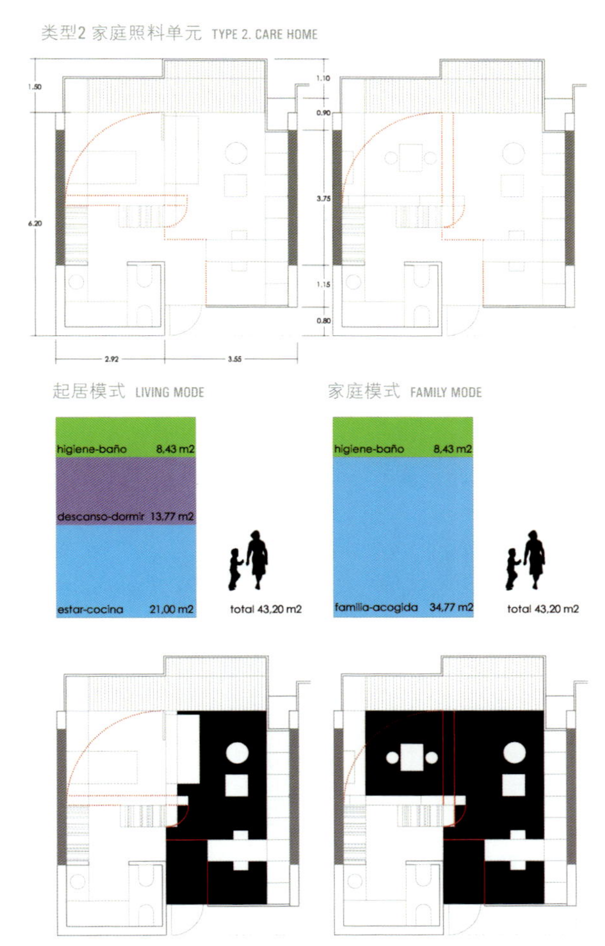


类型3 老年家庭单元 TYPE 3. ELDERLY HOME UNIT

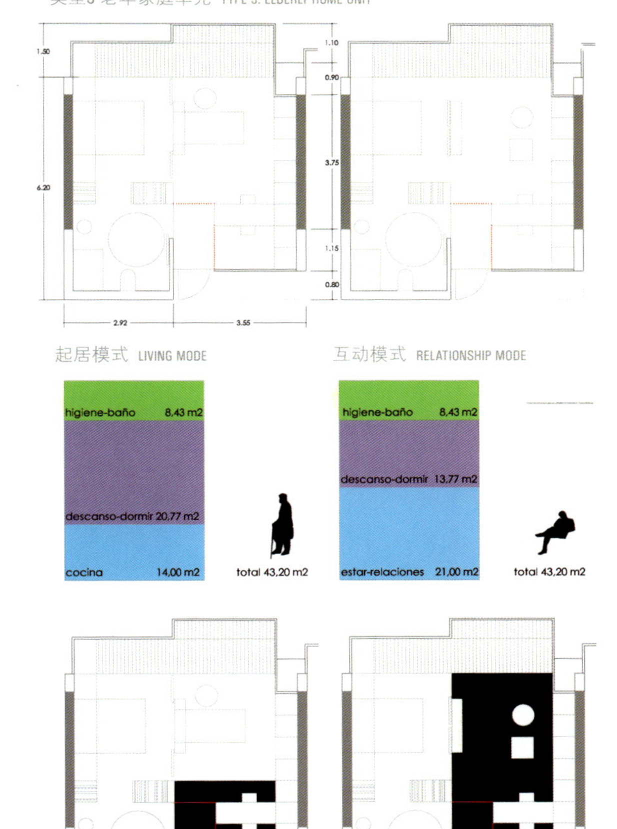

TAMICES

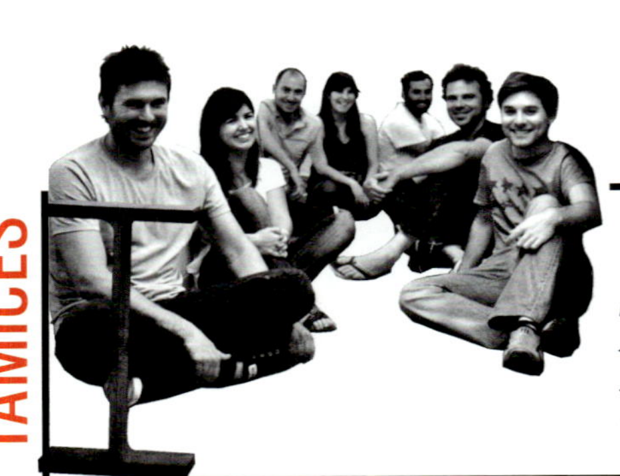

# Arturo Franco Díaz (建筑师)

合作 (c) Diego Castellanos · Yolanda Ferrero · Isabel Gil · Jesús Gallo

第五提名奖 fifth mention

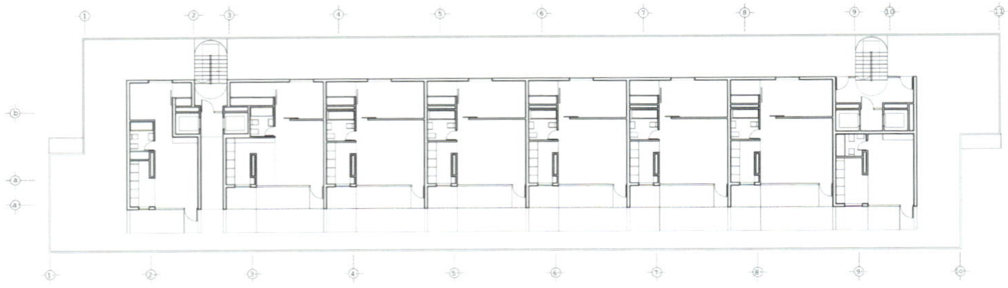

阁楼平面图 PENTHOUSE

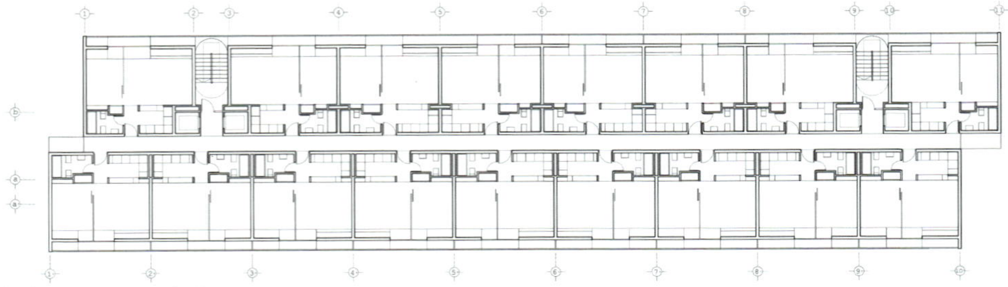

标准层平面图 三至九层 TYPICAL FLOOR PLAN 2-8

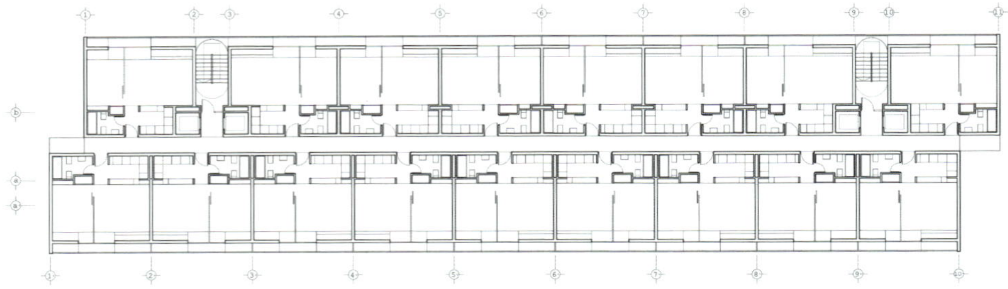

二层平面图 FLOOR PLAN 1

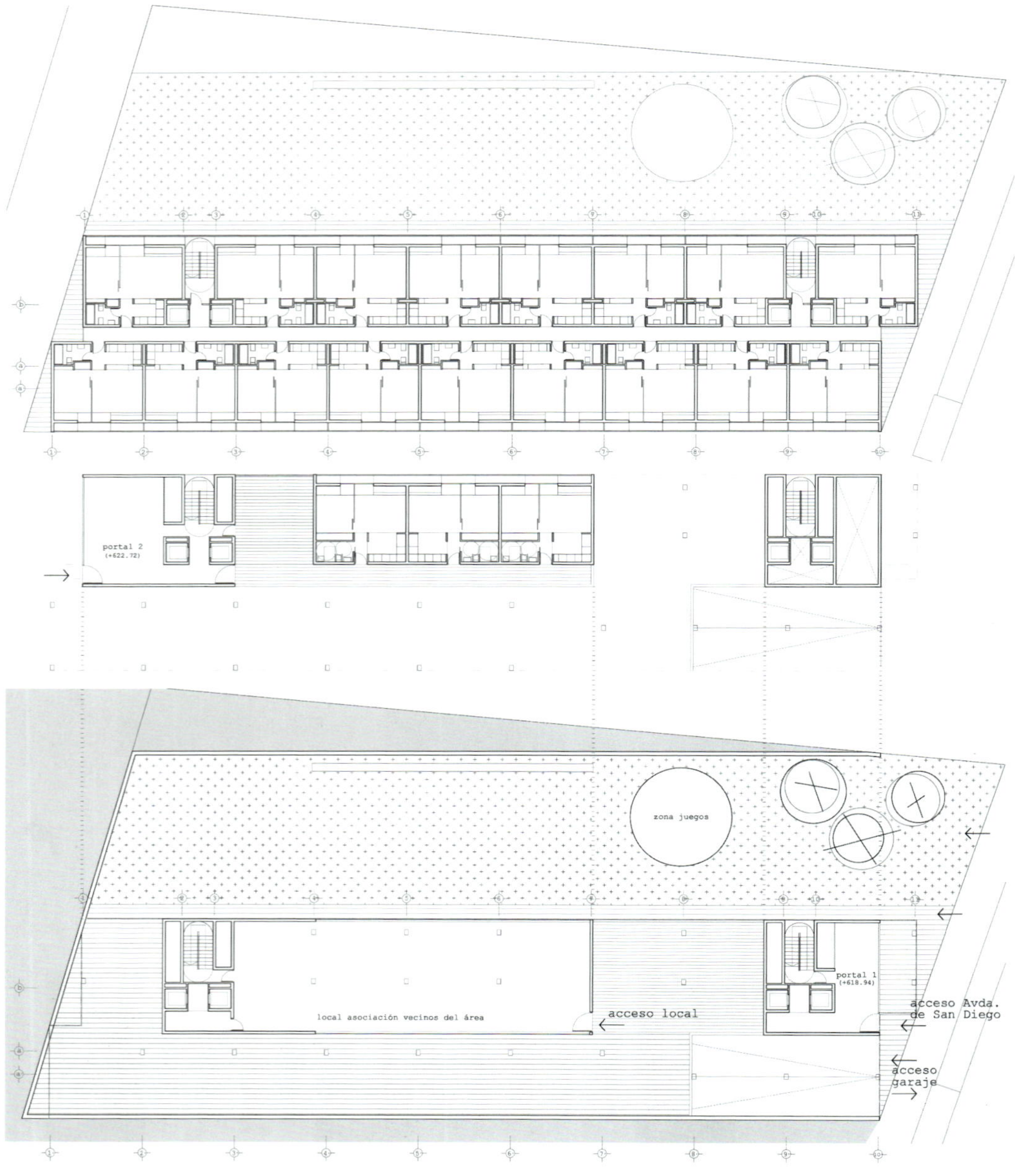

portal 2
zona juegos
local asociación vecinos del área
acceso local
portal 1
acceso Avda. de San Diego
acceso garaje


底层平面图 GROUND FLOOR PLAN

Área estancial calle de prioridad peatonal
venida de San Diego


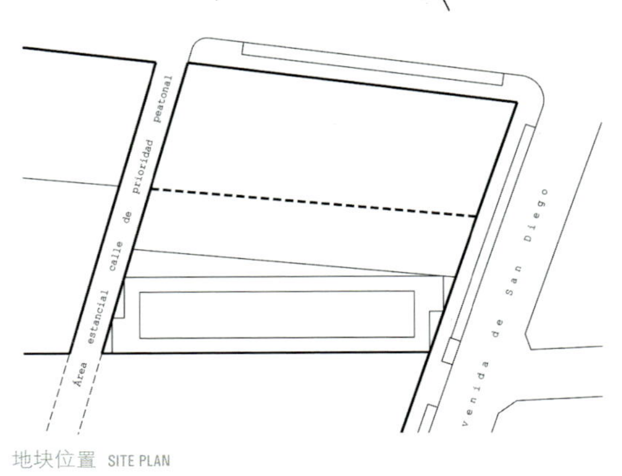

地块位置 SITE PLAN

## 传统体系

该项目主要打造简洁及严谨之风格。金属架构用以加强厂墙的牢固度，在这种情况下，该建筑从外围砖墙开始修建。建筑南、北面不用砖墙，但横向处仍保留金属架构，作为垂直圆截面钢条结构的导向。我们利用两个结构的连接，控制采光以及由里及外的视野。

## CONVENTIONAL SYSTEM

The project is a commitment to clarity and rigour. The metal system used to reinforce factory walls, in this case a brick enclosure serves as a starting point for the project. In the south and north façades, the brick disappears but the reinforcing system remains horizontally acting as a guide for another system of vertical steel rods of circular section. The link of both systems helps us to control the entrance of light and the interior-exterior views.

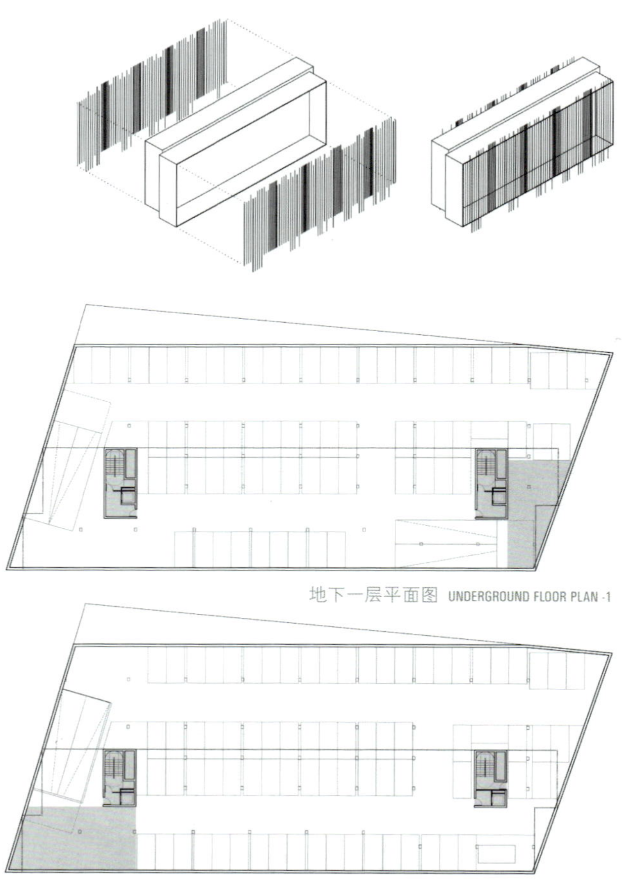

地下一层平面图 UNDERGROUND FLOOR PLAN -1

地下二层平面图 UNDERGROUND FLOOR PLAN -2

南立面细部 SOUTH FAÇADE DETAIL

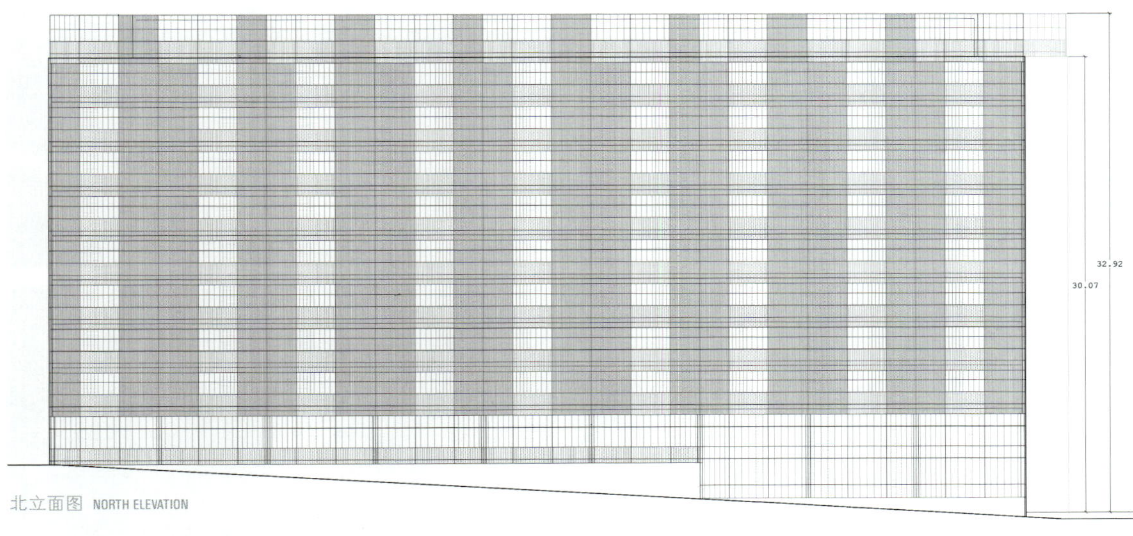

北立面图 NORTH ELEVATION

纵向剖面图 LONGITUDINAL SECTION

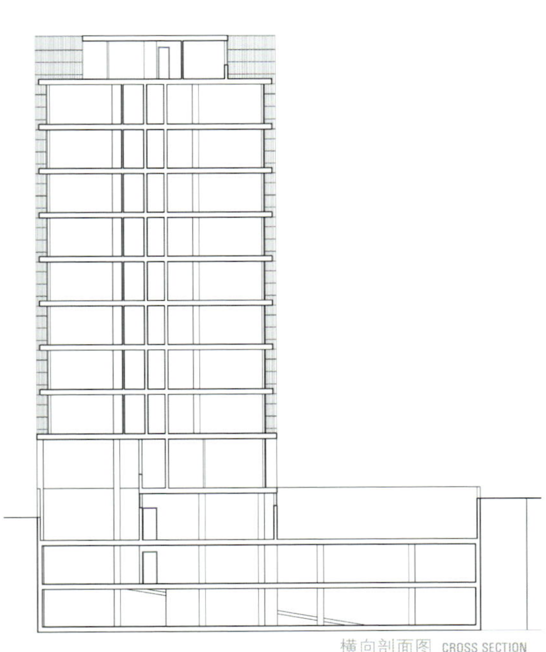
横向剖面图 CROSS SECTION

ENLAZA2

## Francisco Parrón Ortiz · Domingo Melendo Arancón (建筑师)

合作 (c) Miguel Angel Suárez · Fernando López Barrientos

第六提名奖 sixth mention

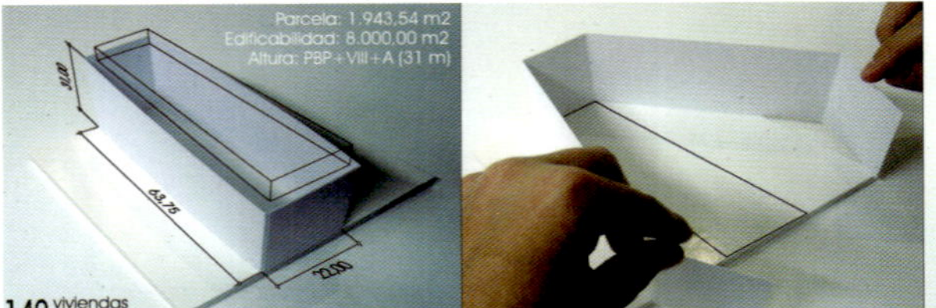

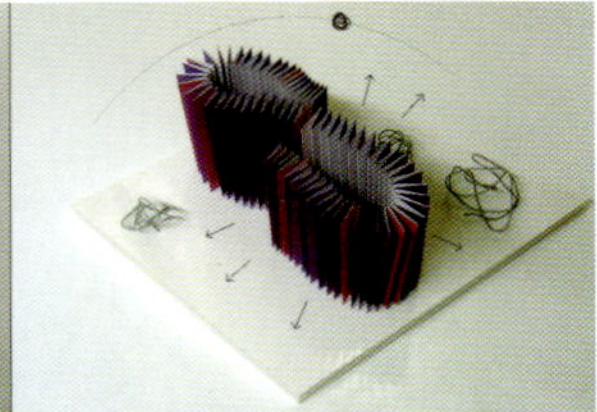

# 边缘凸起

# EXTRUSION OF THE PERIMETER

我们知道，建筑是为活动提供空间的框架。我们巧妙处理空间，并调整活动，以达到和谐共生的目的。我们的设计点着眼于生物气候因素，用其作为连接居住者、城市和气候间的纽带。

We understood the building as a frame which houses the activity (housing). We manipulate the shape and adjust the activity to obtain a symbiotic harmony between them. We design bioclimatic viewpoints, which act as a link between residents, the city and the weather.

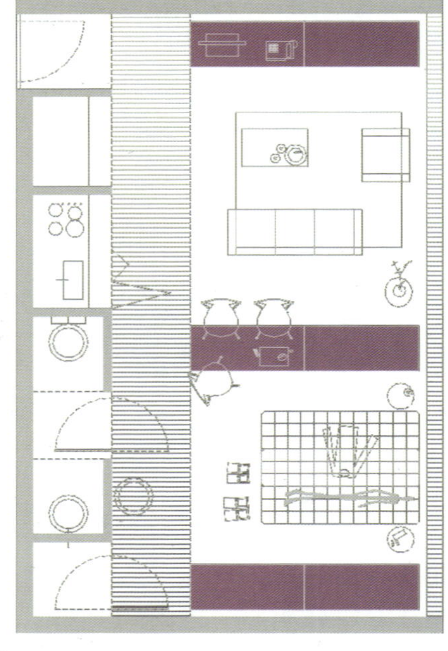

房屋单元 HOUSING UNIT

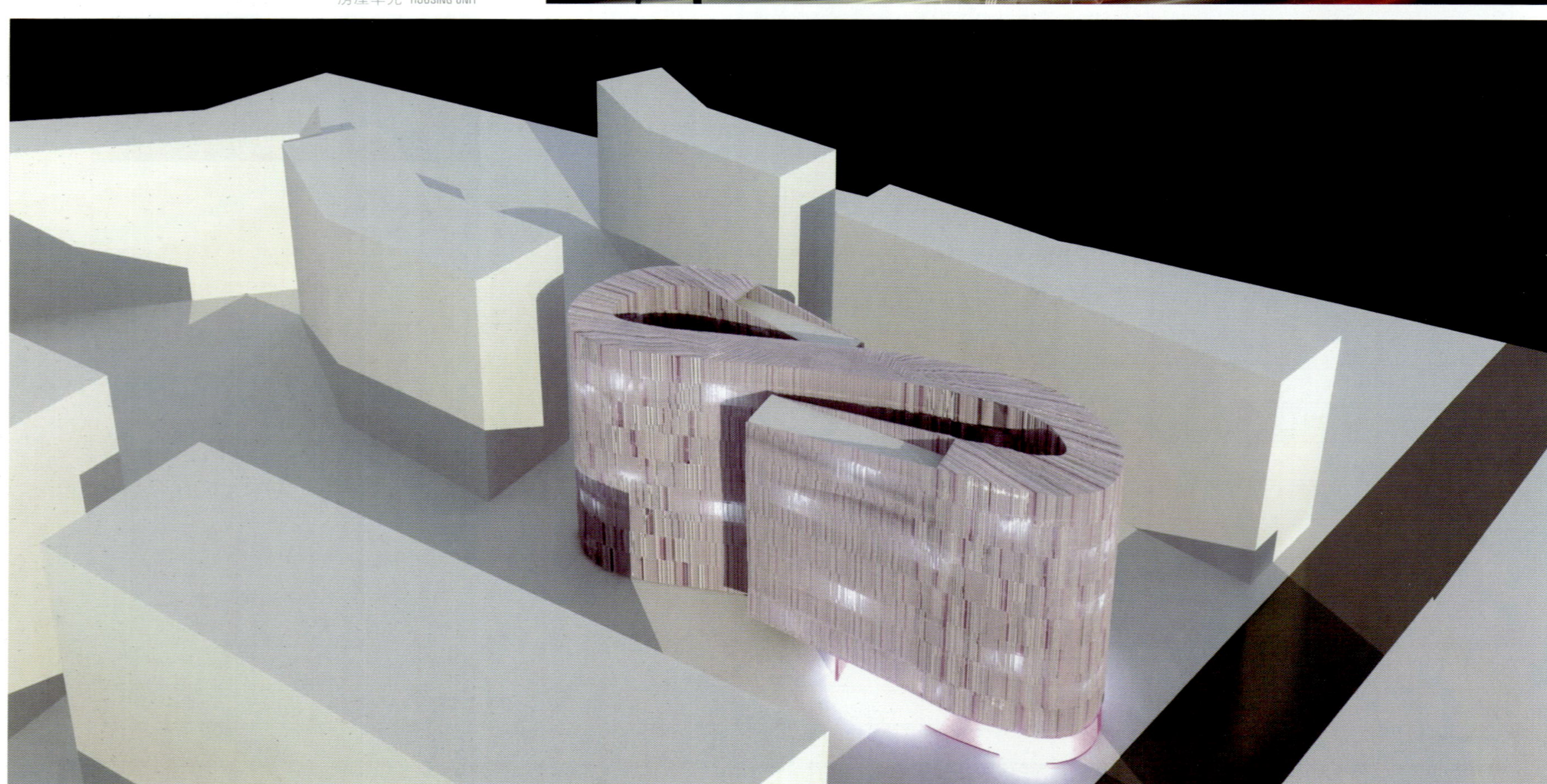

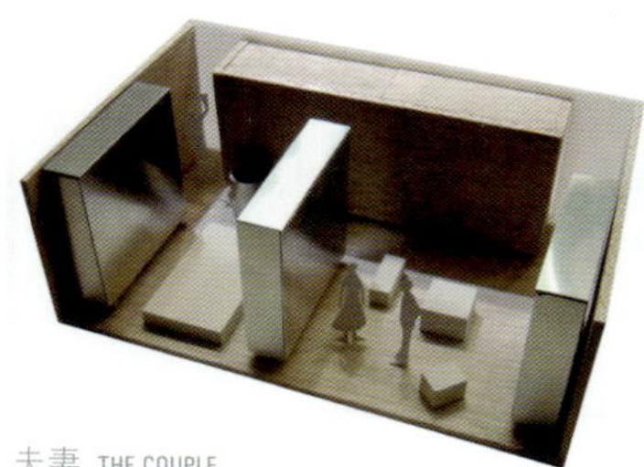

夫妻 THE COUPLE

LOFT THE LOFT

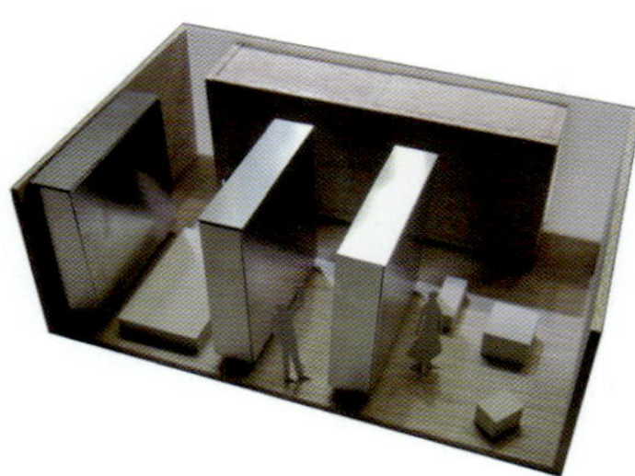

他单身 HE IS SINGLE

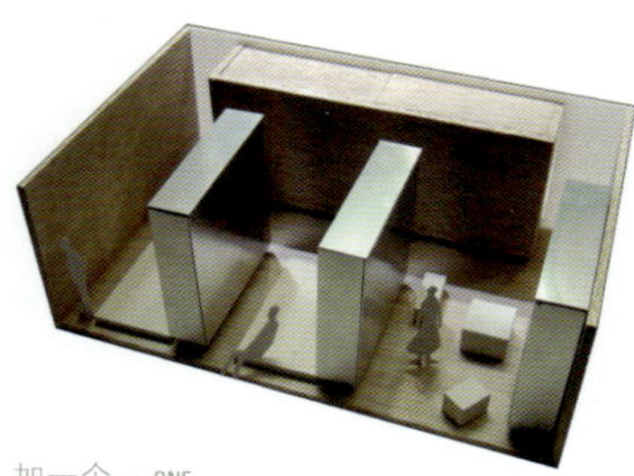

加一个 + ONE

派对 THE PARTY

社交 THE COMUNNITY

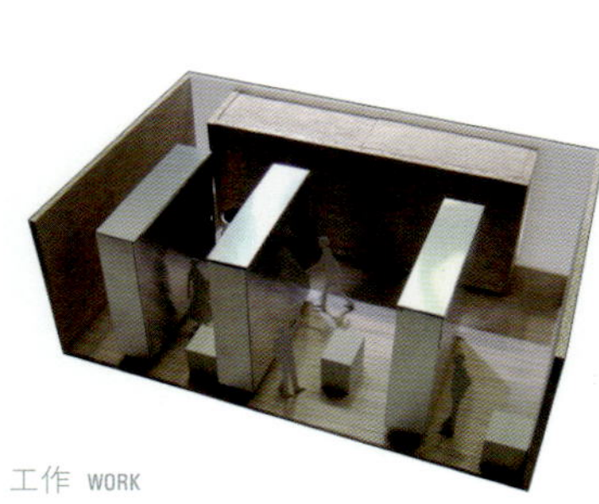

工作 WORK

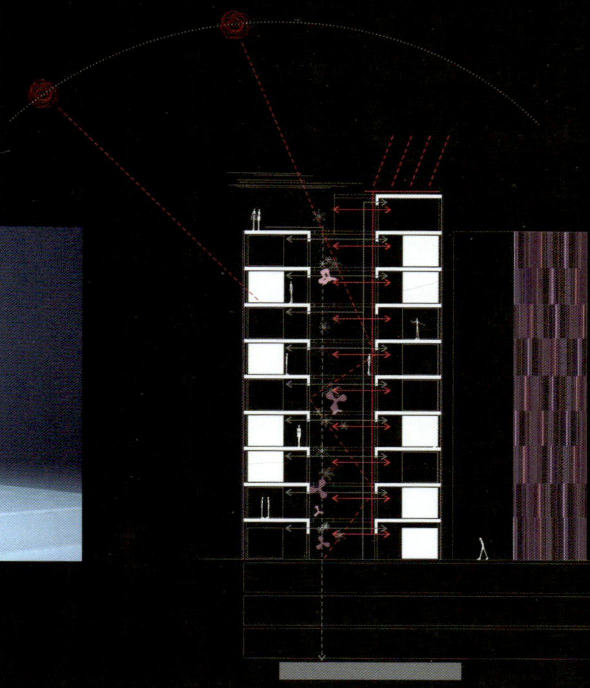

生物气候剖面 BI

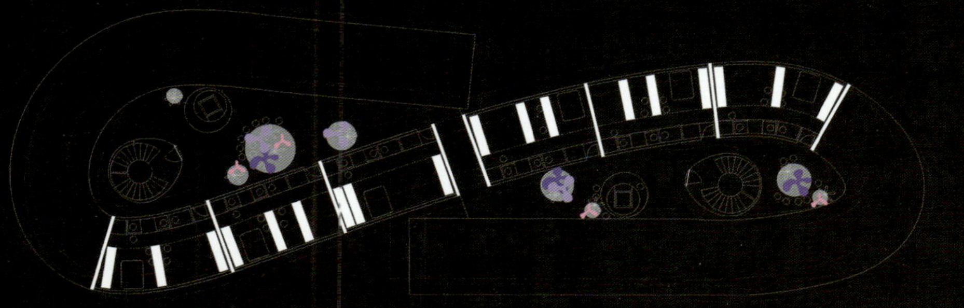

阁楼层平面

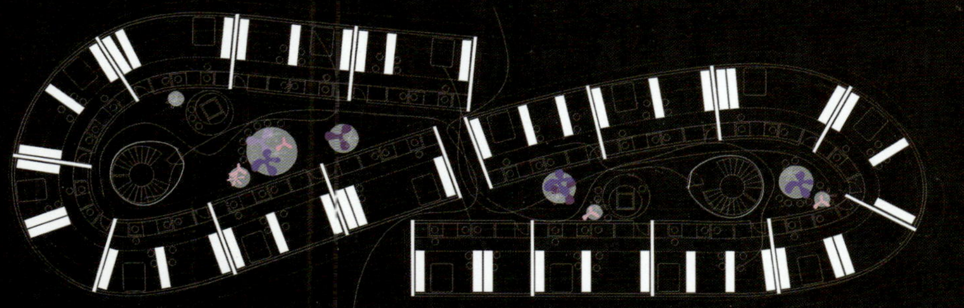

标准层平面图 T

底层平面图 G

## Estudio Cano Lasso (建筑师事务所)

Gonzalo Cano Pintos · Alfonso Cano Pintos · Diego Cano-Lasso Pintos · Beatriz Pozo Fernández · Manuel Ordóñez Abarca · Lucía Pérez Gamarra · Javier Esteban Lecumberri · Antonio Mas-Guindal · Adela Rueda Márquez de la Plata Elena Estella Pérez · Mayca Sánchez Carvajal (建筑师)

中标 winner

标准层平面图 二至十层 TYPICAL FLOOR PLANS 1-9

通风管 SOLAR CHIMNEYS

地块位置 SITE PLAN

# 圆顶

过度注重建筑“外表”是当前的一个趋势，由此涌现了大量这样的建筑设计：繁杂的外观在不同程度上掩盖了内部的真实构造。在该设计方案中，我们重新定义目标，力求建筑外表能反映建筑内部和住宅单元，并去除外部的假象。

# CIRCULAR CROWN

The obsessive preoccupation with the "skins", feature of the current scene, is resulting in a catalog of architecture, in which the dubious enclosures, camouflage and hide, with varying degrees of success, their inner reality. We claim in this proposal for the clarity of the object, the image as an expression of inner reality and housing units without masks and without artifice.

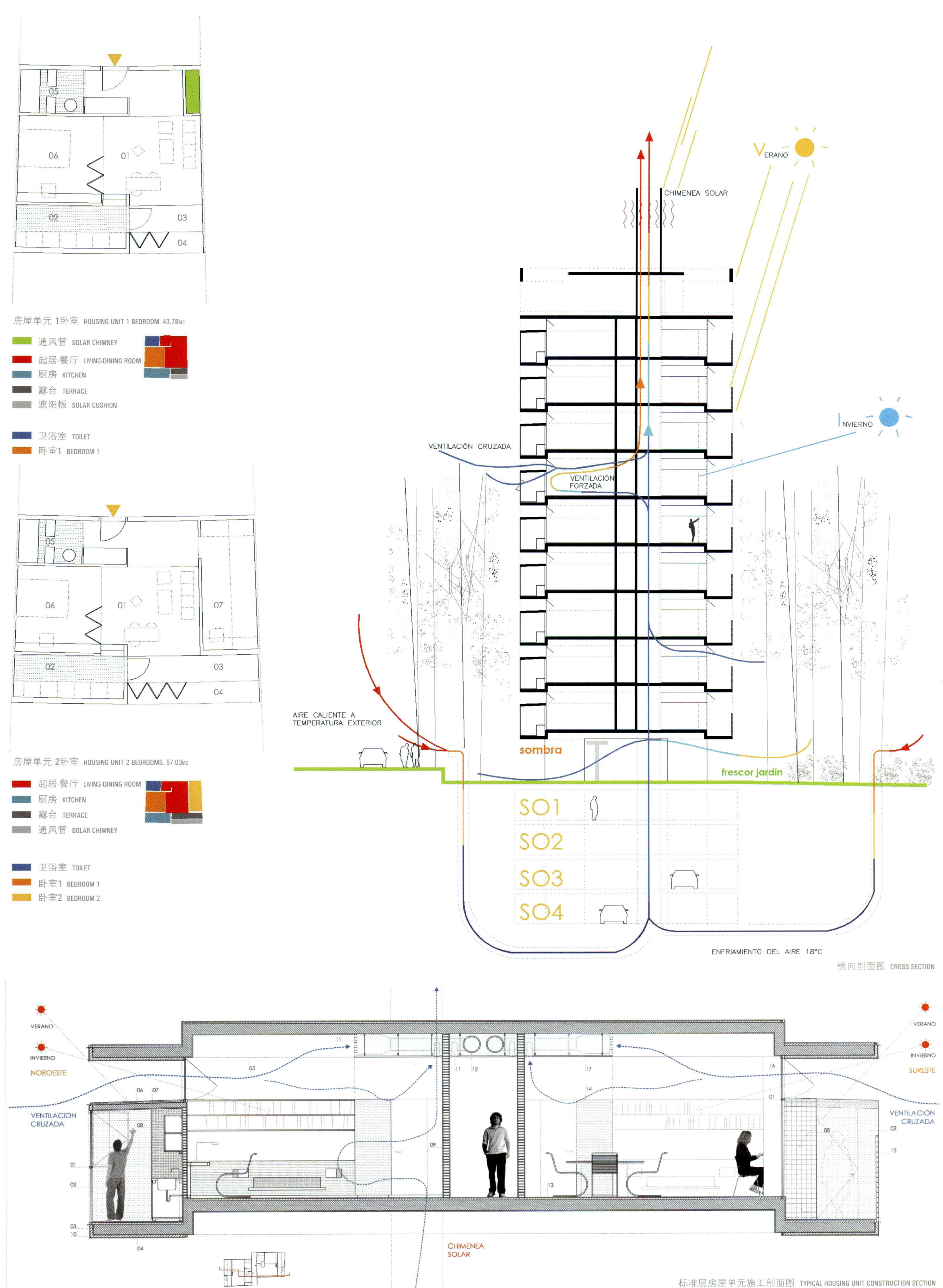
房屋单元 1卧室 HOUSING UNIT 1 BEDROOM. 43.78M2
通风管 SOLAR CHIMNEY
起居·餐厅 LIVING-DINING ROOM
厨房 KITCHEN
露台 TERRACE
遮阳板 SOLAR CUSHION
卫浴室 TOILET
卧室1 BEDROOM 1
房屋单元 2卧室 HOUSING UNIT 2 BEDROOMS. 57.03M2
起居·餐厅 LIVING-DINING ROOM
厨房 KITCHEN
露台 TERRACE
通风管 SOLAR CHIMNEY
卫浴室 TOILET
卧室1 BEDROOM 1
卧室2 BEDROOM 2
VERANO
CHIMENEA SOLAR
INVIERNO
VENTILACIÓN CRUZADA
VENTILACIÓN FORZADA
AIRE CALIENTE A TEMPERATURA EXTERIOR
sombra
frescor jardín
SO1
SO2
SO3
SO4
ENFRIAMIENTO DEL AIRE 18°C
横向剖面图 CROSS SECTION
VERANO
INVIERNO
NOROESTE
VENTILACION CRUZADA
VERANO
INVIERNO
SURESTE
VENTILACION CRUZADA
CHIMENEA SOLAR
标准层房屋单元施工剖面图 TYPICAL HOUSING UNIT CONSTRUCTION SECTION

JULIETA

# Olalquiaga Arquitectos (建筑师事务所)

Rafael Olalquiaga Soriano · Pablo Olalquiaga Bescós · Alfonso Olalquiaga Bescós (建筑师)

合作 (c) Javier Morales Luchena

第一提名奖 first mention

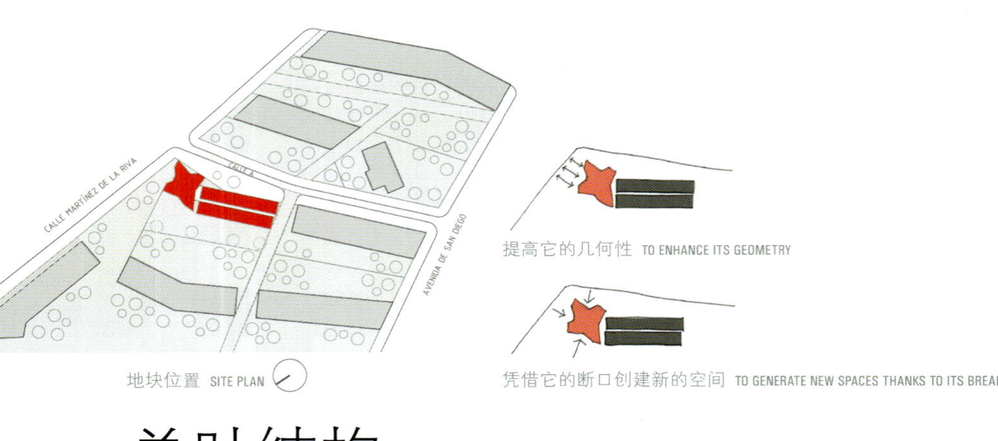


地块位置 SITE PLAN

提高它的几何性 TO ENHANCE ITS GEOMETRY

凭借它的断口创建新的空间 TO GENERATE NEW SPACES THANKS TO ITS BREAKS

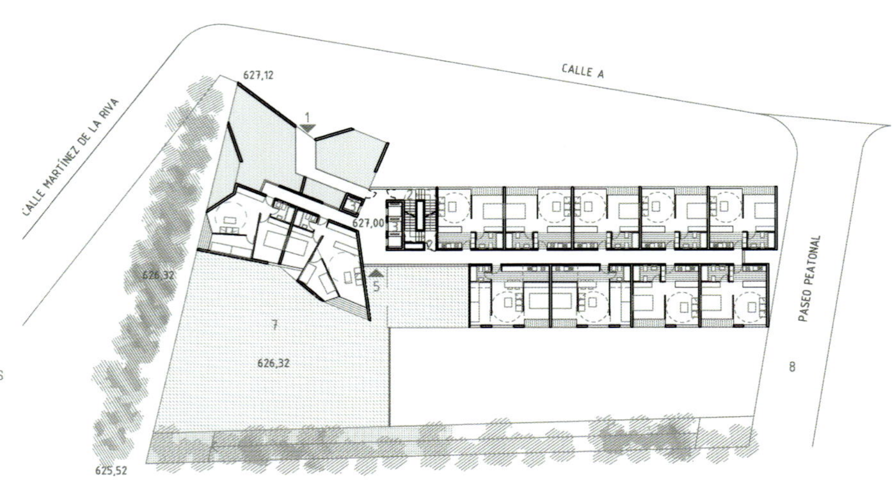


门廊底层平面图 · 大街层 PORCH GROUND FLOOR PLAN · STREET LEVEL

## 单叶结构

我们的设计主旨在于使得该建筑与周边的公园风格一体化，于是将该建筑划分为三部分，其中两部分与街道A平行，并利用曲折结构，在底层释放新的空间。

## THE SINGULAR LEAF

We propose a better integration of the building and the gardens surrounding the building. We propose to divide the building into 3 volumes, setting 2 of them parallel to street A. We propose to create new spaces with its twists on the ground floor.

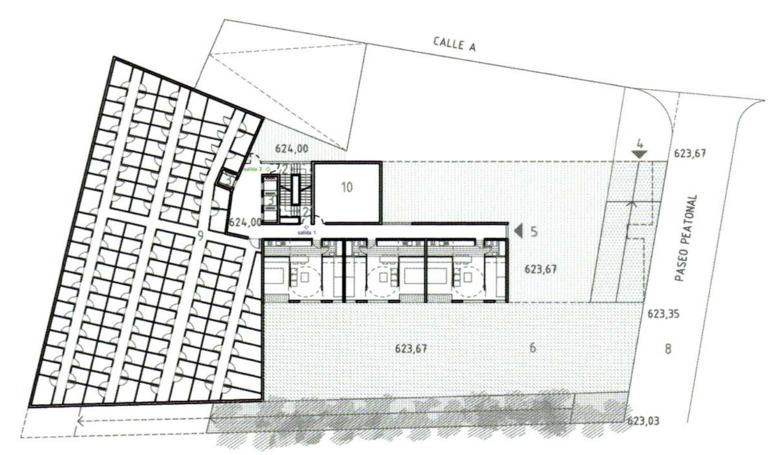


底层平面图 · 步行街层 GROUND FLOOR PLAN · PEDESTRIAN PATHWAY LEVEL

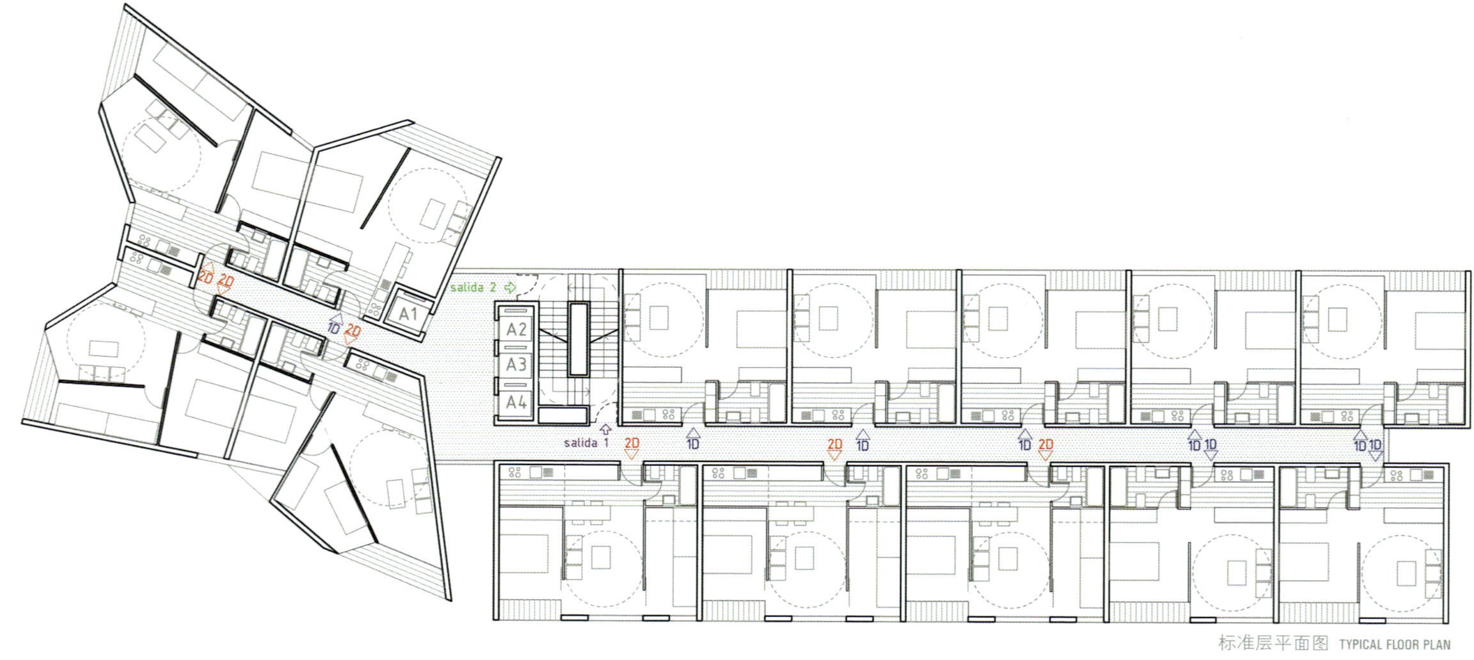


标准层平面图 TYPICAL FLOOR PLAN

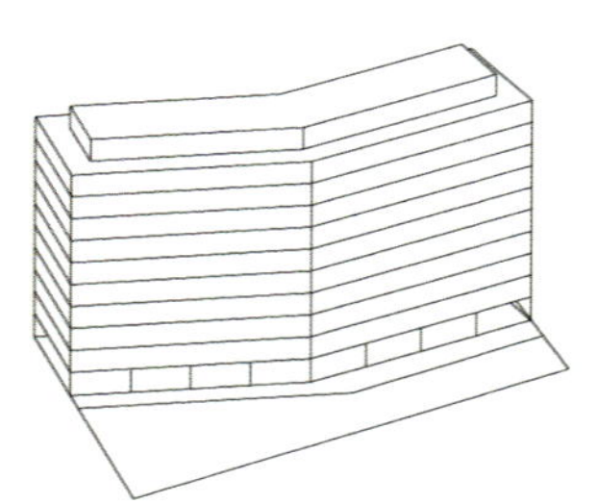

高度：十层加阁楼 HEIGHT: 9 STOREYS + PENTHOUSE

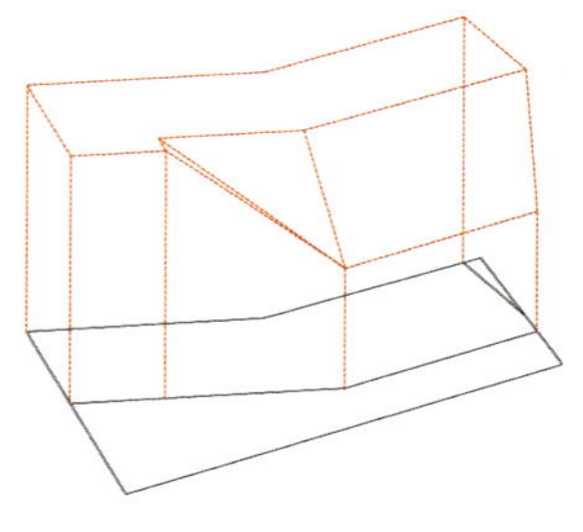

最大体量 MAXIMUM VOLUME

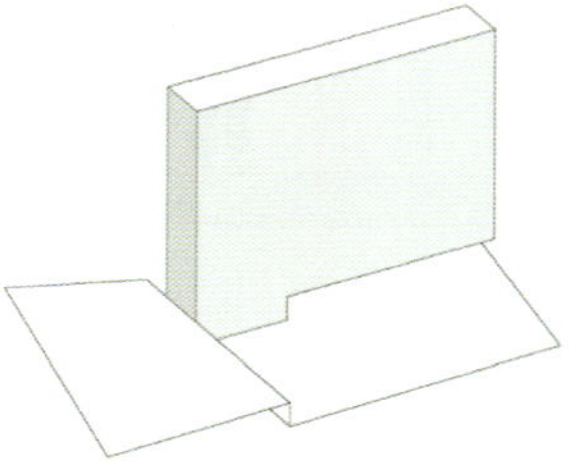

体量A：十层 VOLUME A: 9 STOREYS

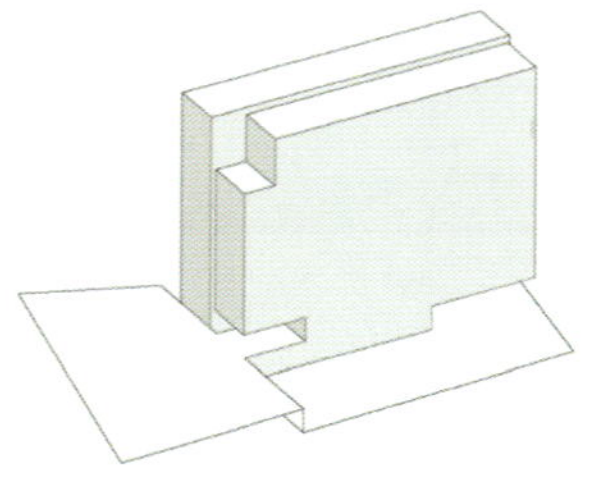

体量B：十层 VOLUME B: 9 STOREYS

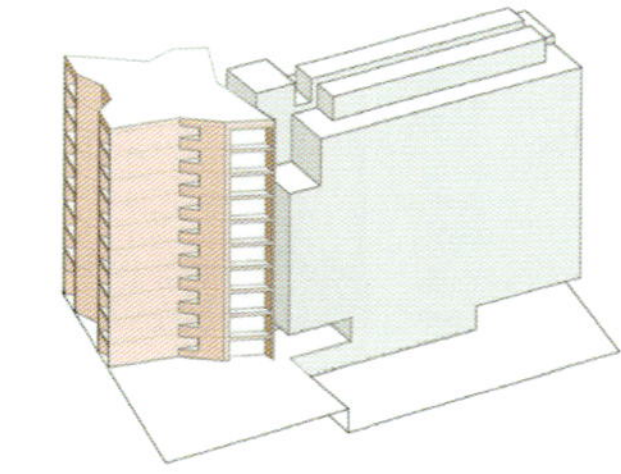

体量C：叶子 十一层 VOLUME C: THE LEAF. 10 STOREYS

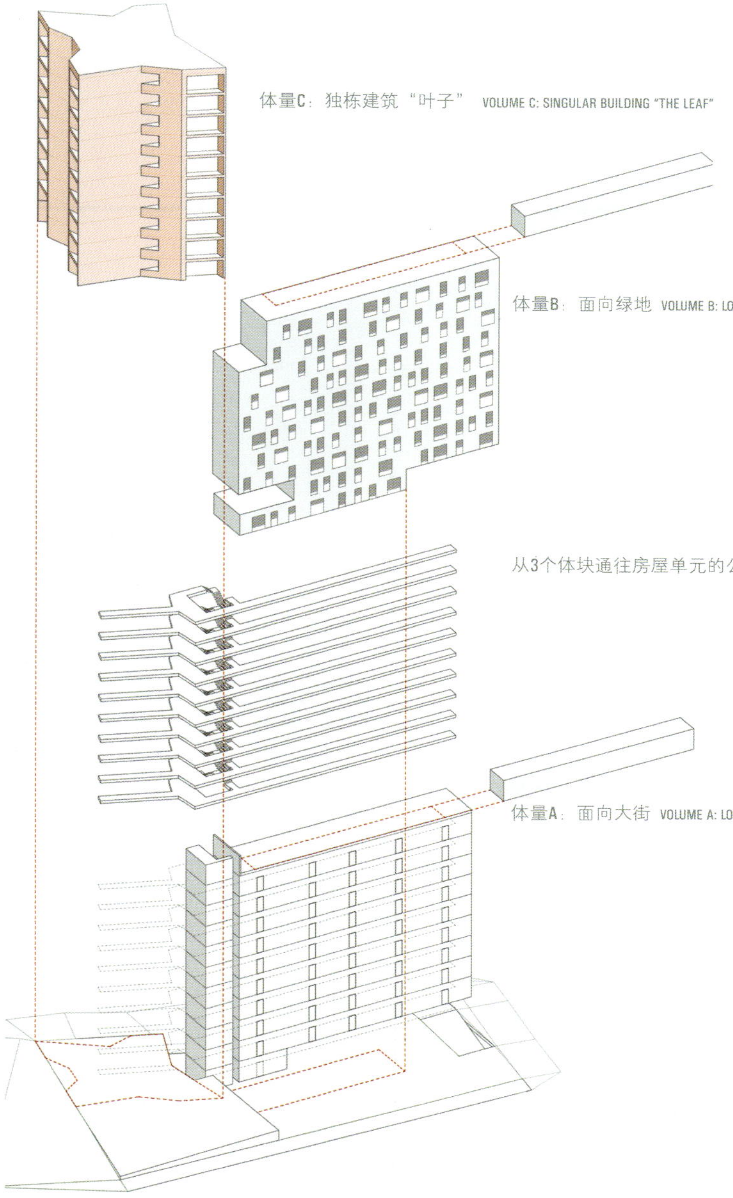

房屋单元 2卧室 HOUSING UNIT 2 BEDROOMS

体量C：独栋建筑“叶子” VOLUME C: SINGULAR BUILDING "THE LEAF"

体量B：面向绿地 VOLUME B: LOOKING TO THE GREEN AREA

从3个体块通往房屋单元的公共入口 COMMUNAL ACCESS TO HOUSING UNITS FROM THE 3 VOLUMES

体量A：面向大街 VOLUME A: LOOKING TO THE STREET

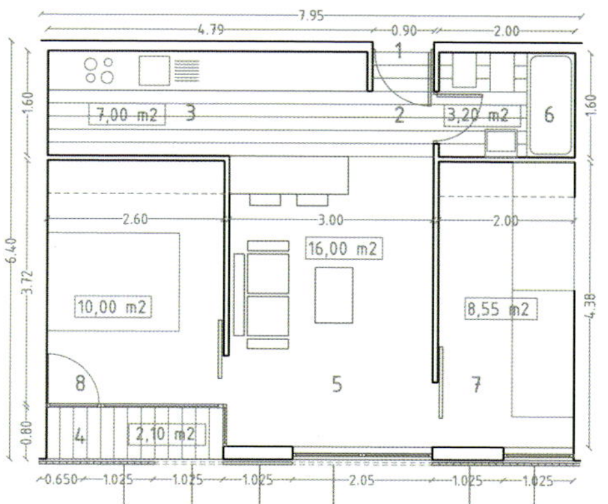

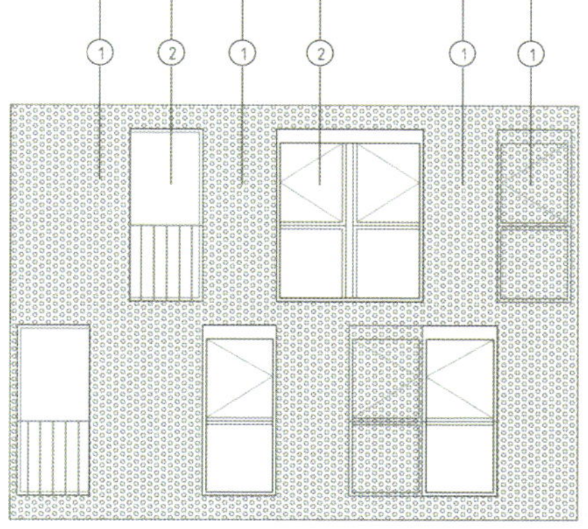

立面细部 ELEVATION DETAIL

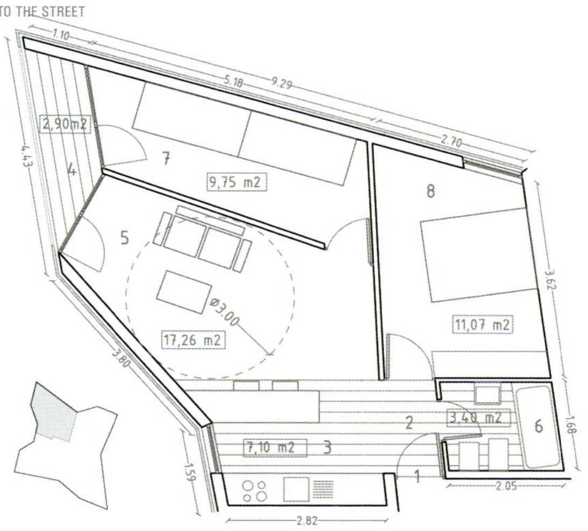

“叶子”楼房屋单元 2卧室 HOUSING UNIT 2 BEDROOMS FROM "THE LEAF"

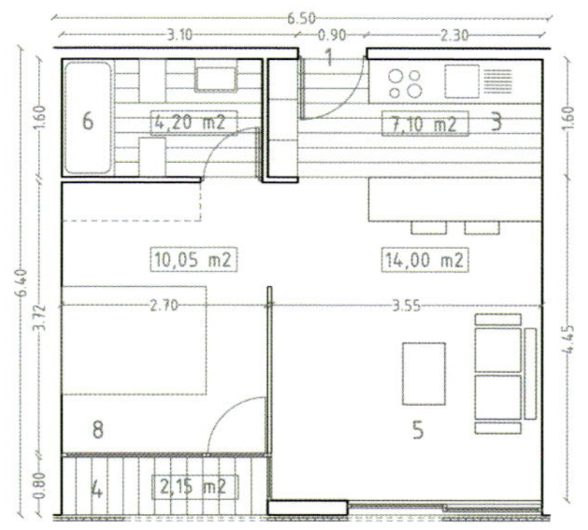

房屋单元 1卧室 HOUSING UNIT 1 BEDROOM

# ESPINOSA MORENO ARQUITECTOS (建筑师事务所)

Ana Espinosa G-Valdecasas · Álvaro Moreno Hernández · María Espinosa G-Valdecasas (建筑师)

第二提名奖 second mention

门廊底层平面图 · 大街层 PORCH GROUND FLOOR PLAN · STREET LEVEL

地块位置 SITE PLAN

## 半透明建筑表面

住房户型主要分成两类：一类是朝内深入式，厨房与卫生间位于进门处的两侧；另外一类则是向外延伸式，进门即是客厅和卧室。我们在住宅单元的布局上努力实现其灵活性最大化，并根据地形设计建筑架构，为该建筑设计两个行人入口。

## TRANSLUCENT FAÇADE

The building is divided into two bands: the first, looks inside, leaving the kitchen and bathroom on opposite sides to the entrance, the second, in contact with the exterior, is designed to the living room and bedrooms. We have looked for the maximum flexibility in the distribution of housing units. The building section is adapted to the topography of the site breaking up the ground floor. We create two pedestrian entrances to the building.

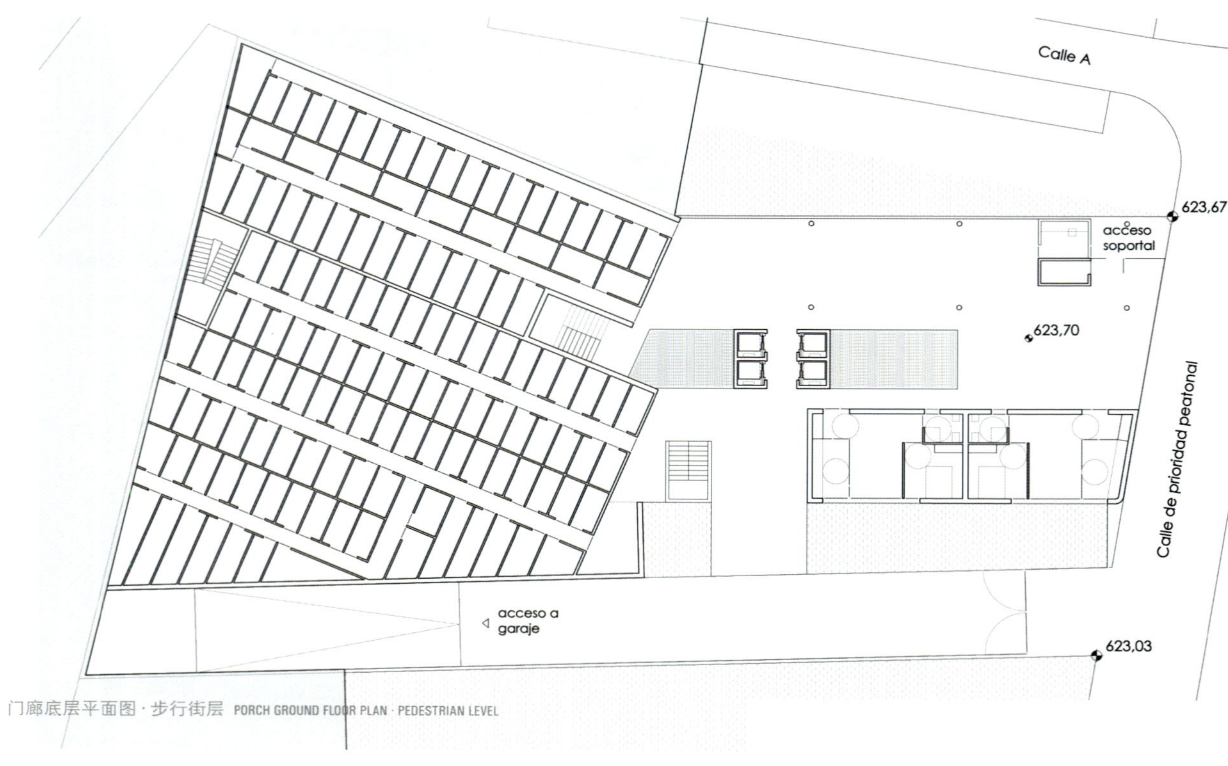


门廊底层平面图 · 步行街层 PORCH GROUND FLOOR PLAN · PEDESTRIAN LEVEL

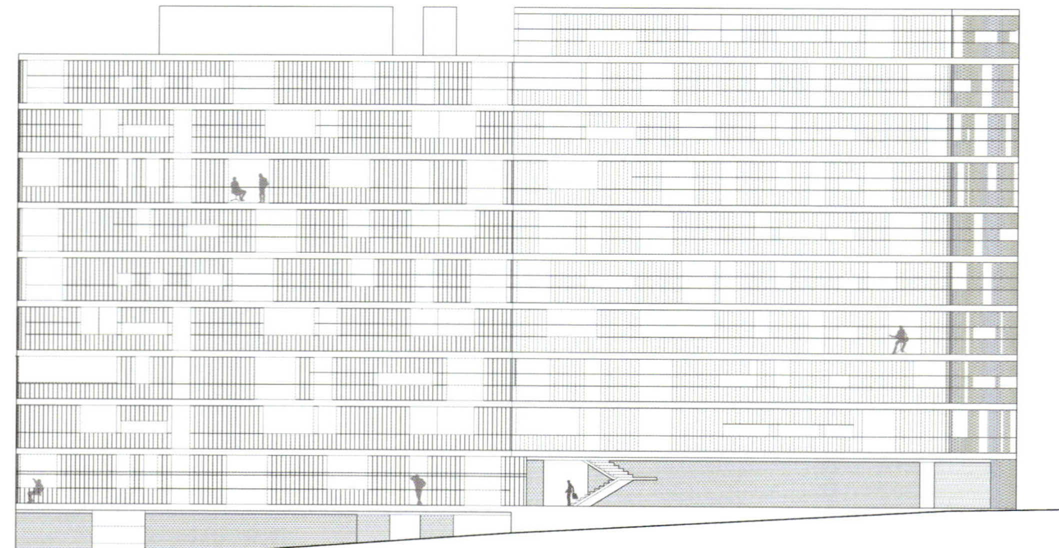

面向A大街立面图 ELEVATION TO STREET A

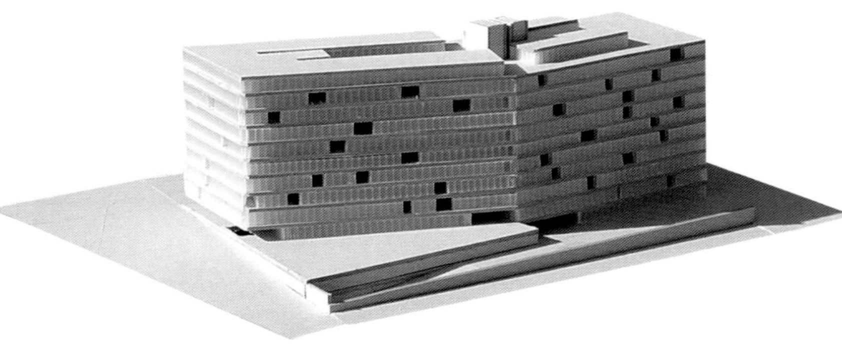

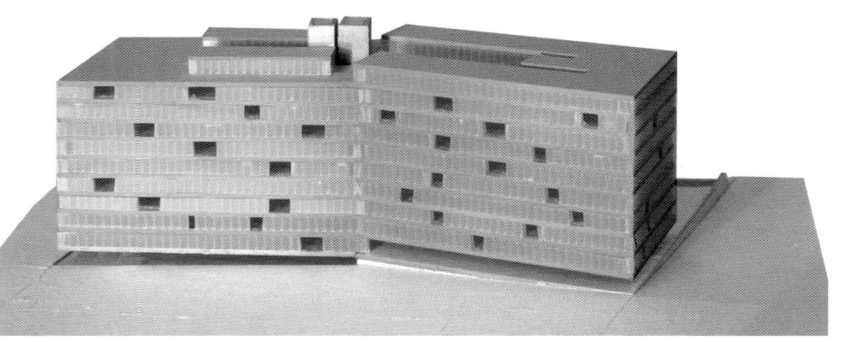

阁楼层平面图 PENTHOUSE

标准层平面图 TYPICAL FLOOR PLAN

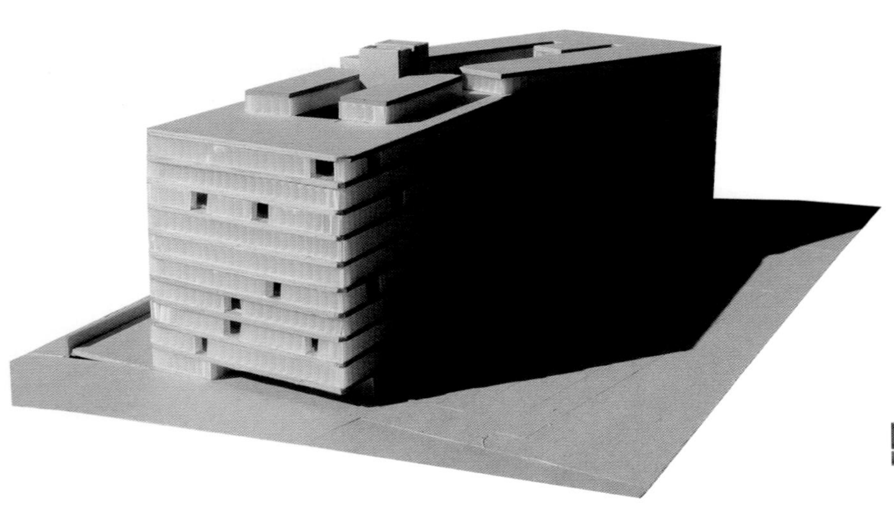

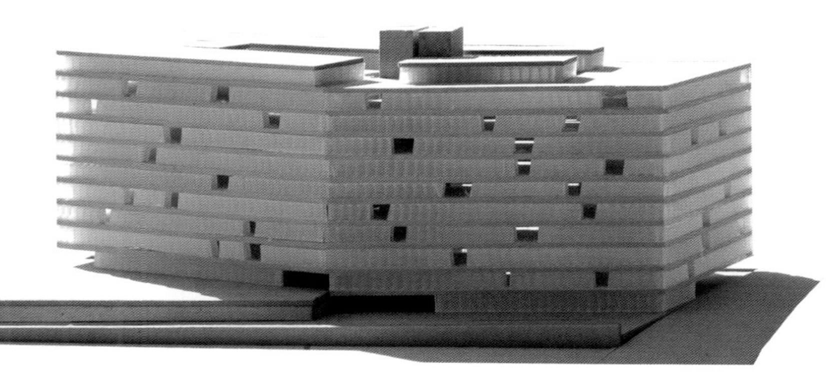

SHAKEIT

**Espegel-Fisac Arquitectos** (Carmen Espegel · Concha Fisac) (建筑师)
合作 (c) Laia Lafuente · Marcela Aragüez · Miriam Llamazares · José Manuel Calvo
第三提名奖 third mention

# 纵向的住宅单元

# LONGITUDINAL HOUSING UNITS

一个个住宅单元的聚集、叠加，构成了这个居住单元。该区域作为居住单元，其结构相当简单。我们只是将不同户型（一居室和两居室）的住宅单元融合在一起。然而，这种独特的结构极具美感，非同寻常。该设计将各住宅单元重新布局，各得其位，同时变换楼层平面图结构，使其保持合理的建筑架构。

The volume is produced from the accumulation of living cells. Its generation is simple; we just mix the housing types of one and two bedrooms and stack them. However, the set is attractively unusual. The project slides housing units, as if each one tried to match in its place, and in turn, each floor plan moves, while they maintain a construction and structural logic.

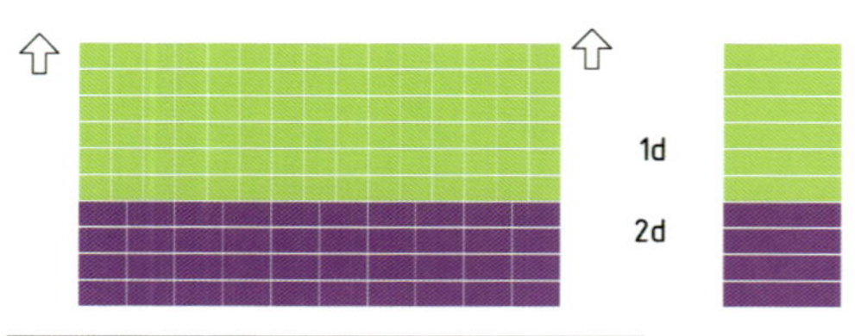


① 十层加阁楼 9 STOREYS + PENTHOUSE

② 十一层加阁楼 10 STOREYS + PENTHOUSE

③ 抖动打散 SHAKE IT shake it!!!

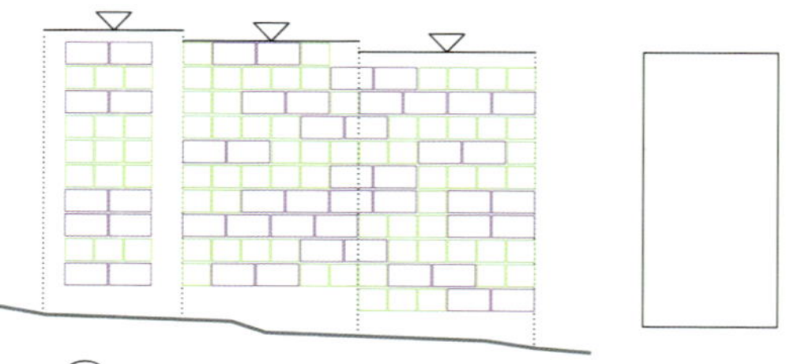

④ 最大限高 MAXIMUM PERMISSIBLE HEIGHT

⑤ 调整 ADAPTATION

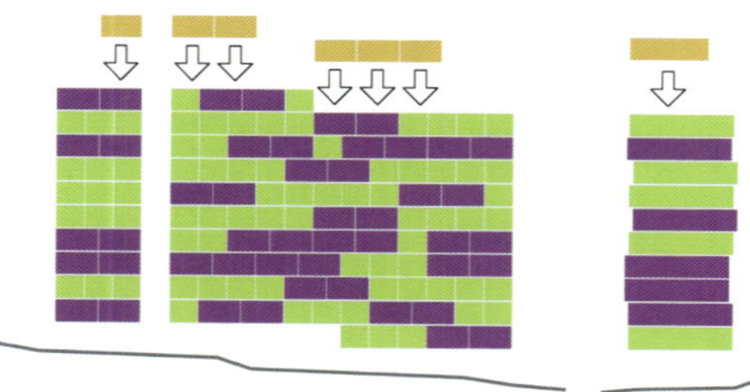

⑥ 阁楼 PENTHOUSE

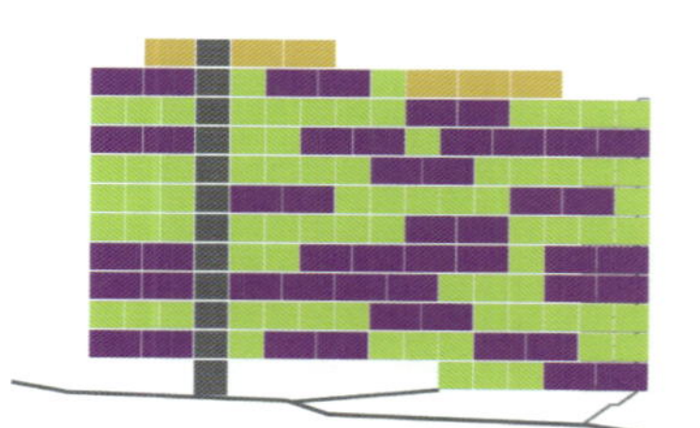

⑦ 流通核心 COMMUNICATION CORES

立面图 ELEVATION

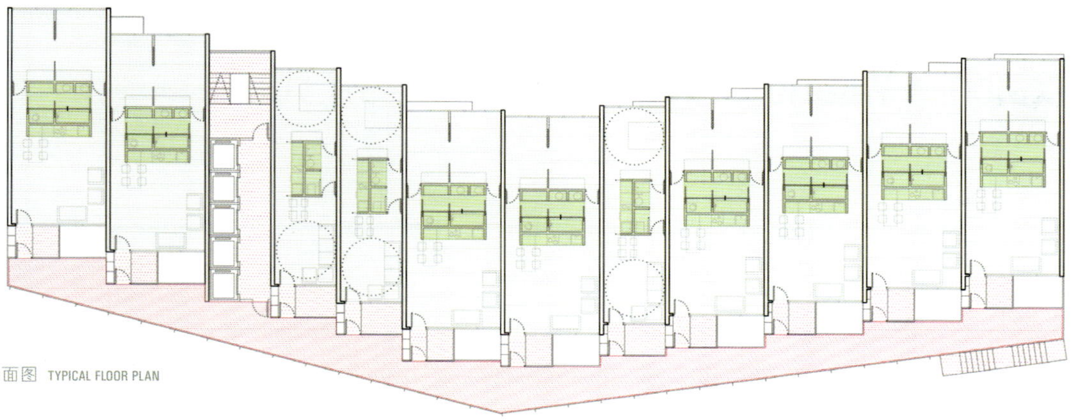

标准层平面图 TYPICAL FLOOR PLAN

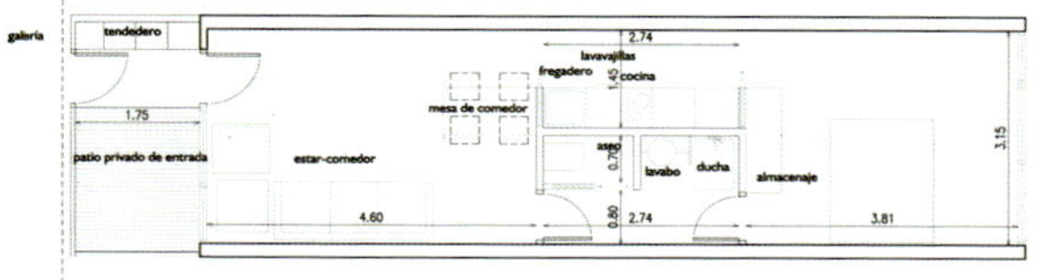

房屋单元 1卧室 HOUSING UNIT. 1 BEDROOM

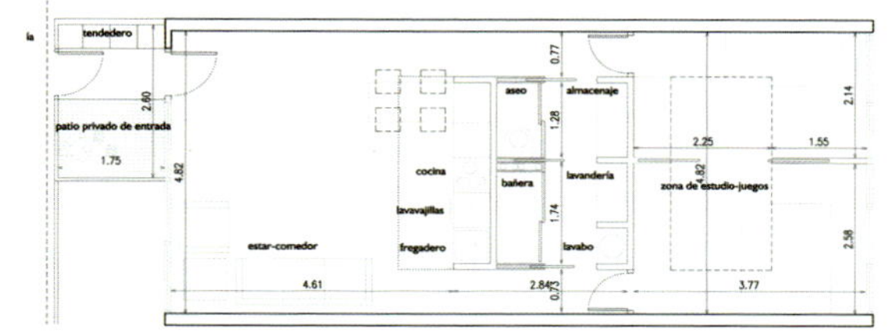

房屋单元 2卧室 HOUSING UNIT. 2 BEDROOMS

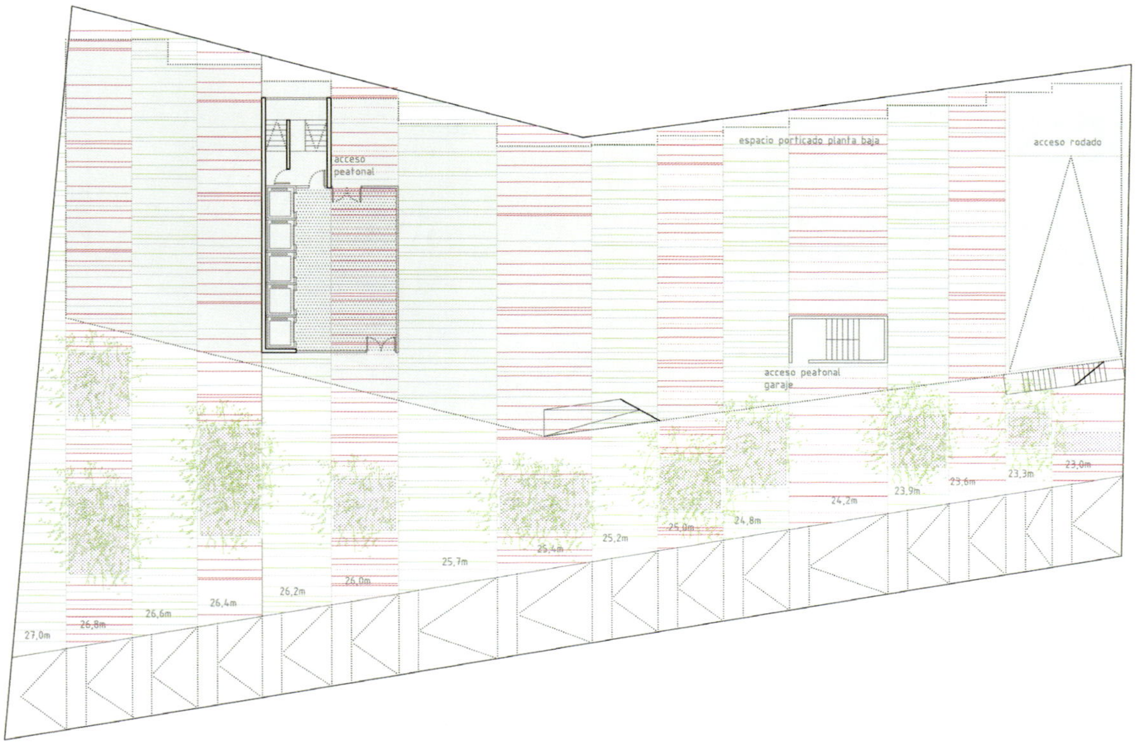

底层平面图 GROUND FLOOR PLAN

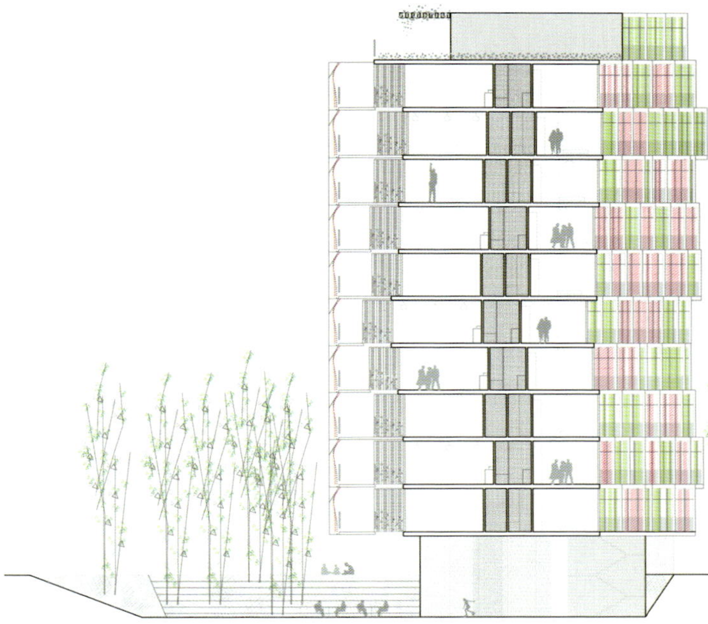

横向剖面图 CROSS SECTION

# Clara Matilde Moneo Feduchi · Valerio Canals Revilla (建筑师)

合作 (c) Lucía Martínez Martínez

第四提名奖 fourth mention

## 交错的边缘

为了使朝北的房间能够采光，使朝南和朝西的房间避免阳光直射，我们设计的建筑表面由内外两层空间构成。建筑表面为一实体空间，并非一平面。这种交错的空间延伸亦可在表面上显现出来。

## JAGGED EDGES

To try to capture the sun in apartments that look north and protect the apartments that look south and west, we propose a façade formed by incoming and outcoming volumes. The façade is planned as a volume and not as a plane with voids. This movement of jagged volumes also appears on the cover.

地块位置 SITE PLAN

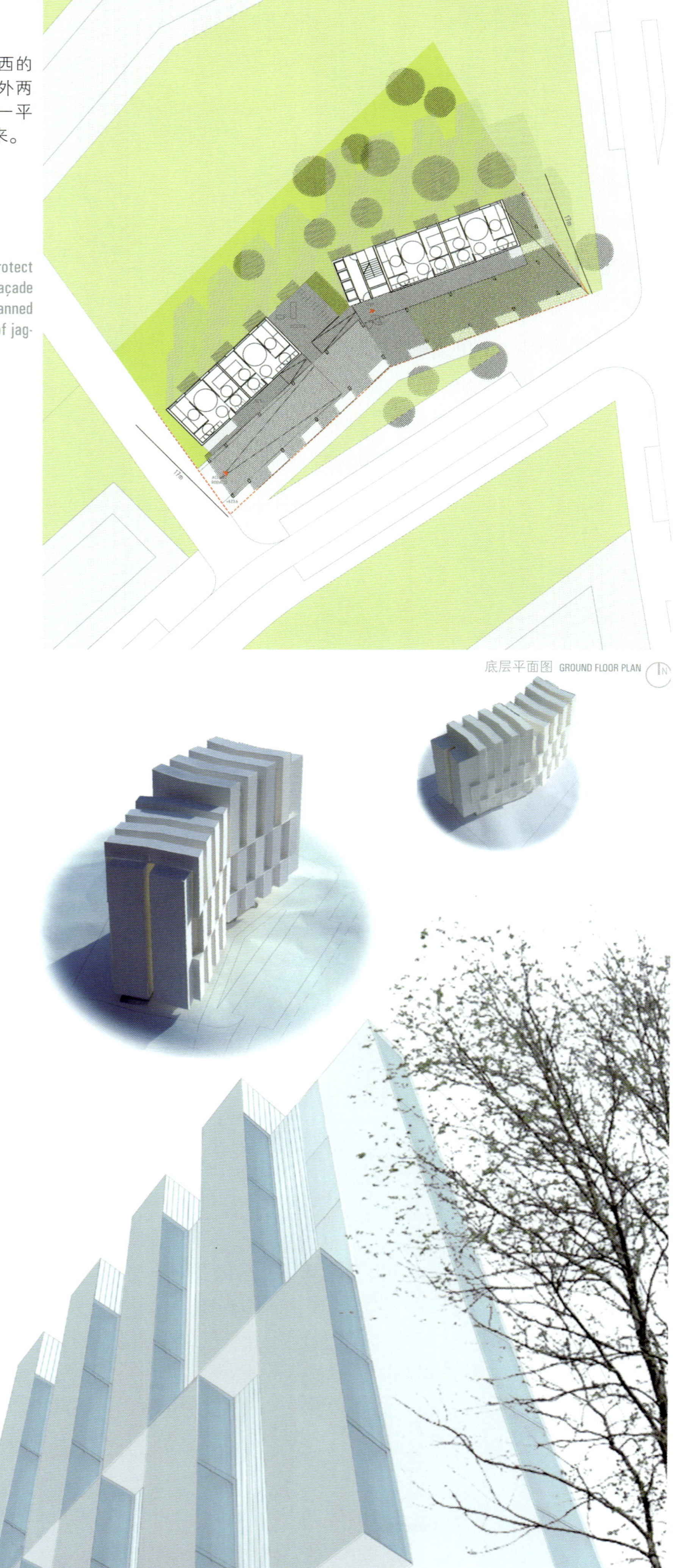

底层平面图 GROUND FLOOR PLAN

十层平面图+夹层 FLOOR PLAN 9+MEZZANINE

南外墙：U形玻璃 SOUTH VIEW: U-GLASS

地下一层平面图 UNDERGROUND FLOOR PLAN -1

地下二、三、四层平面图 UNDERGROUND FLOOR PLAN -2,3,4

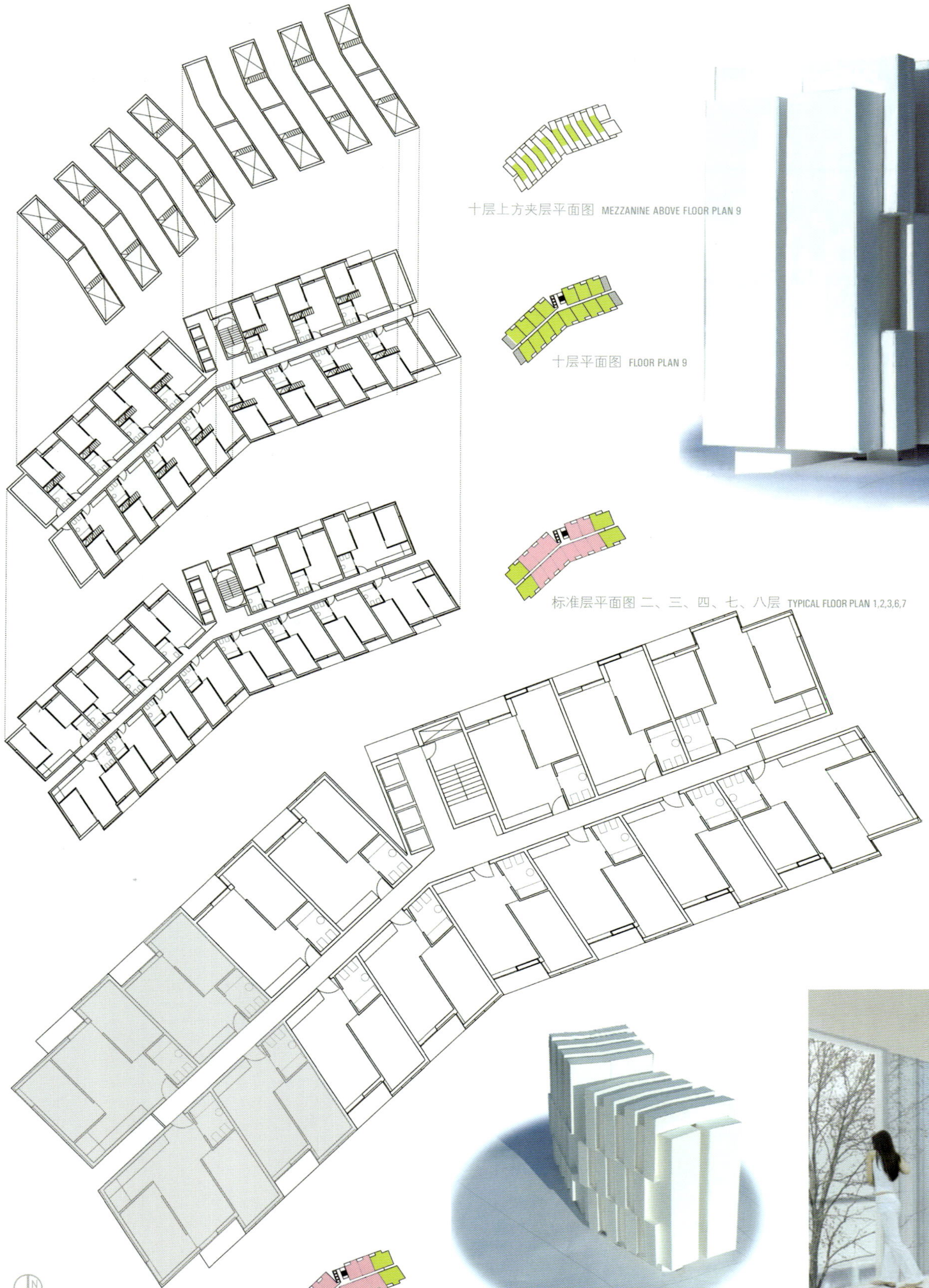

十层上方夹层平面图 MEZZANINE ABOVE FLOOR PLAN 9

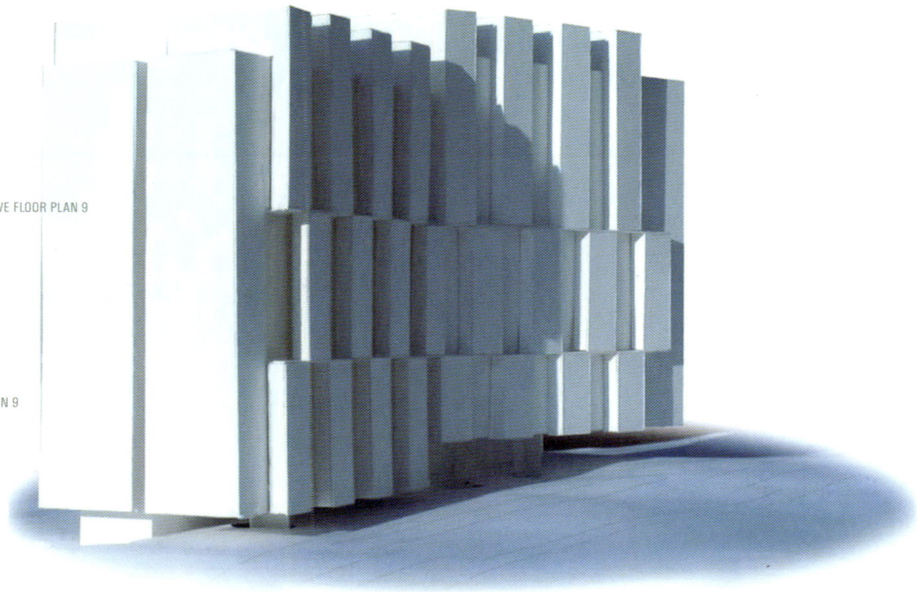

十层平面图 FLOOR PLAN 9

标准层平面图 二、三、四、七、八层 TYPICAL FLOOR PLAN 1,2,3,6,7

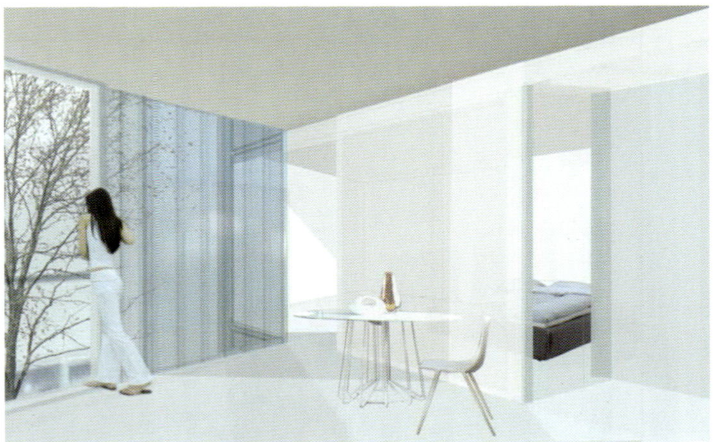

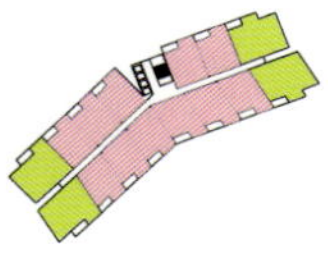

标准层平面图 五、六层 TYPICAL FLOOR PLAN 4,5

西北外墙样板 TYPOLOGIES NORTHWEST FAÇADE

东南外墙样板 TYPOLOGIES SOUTHEAST FAÇADE

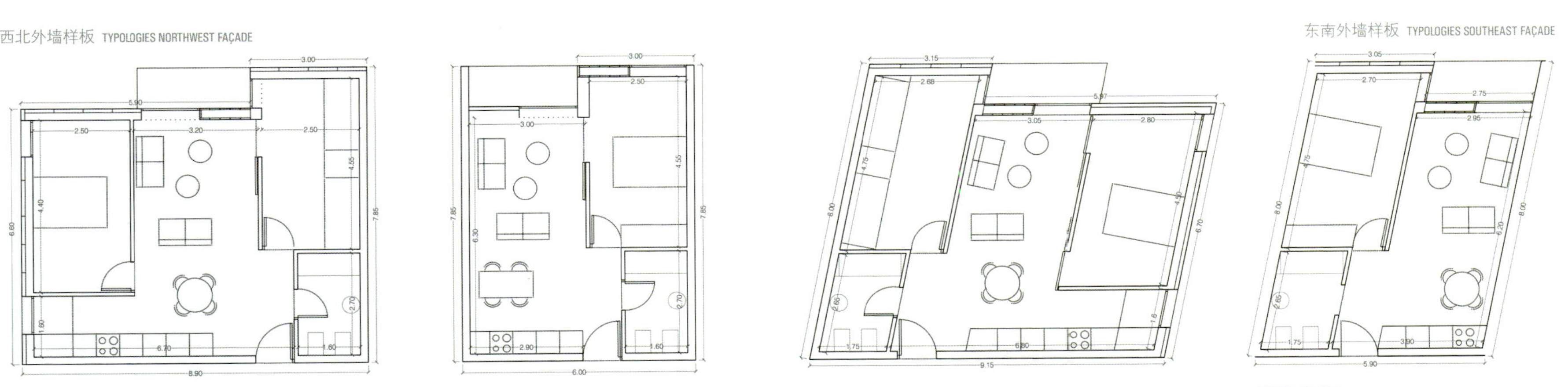

2卧室 类型A 2 BEDROOMS. TYPE A. 54.00M2

1卧室 类型B 1 BEDROOM. TYPE B. 37.50M2

2卧室 类型C 2 BEDROOMS. TYPE C. 55.00M2

2卧室 类型D 2 BEDROOMS. TYPE D. 38.00M2

TOVERDE

# Syra Abella · Jorge Muñoz · Marina Cisneros · Joaquín Mosquera (建筑师)

第五提名奖 fifth mention

底层平面图 GROUND FLOOR PLAN

## 空地

我们设计一幢规模不大的楼，面向步行区；另一幢垂直方向的楼，面向街道。我们以对角方式（沿西南方向）对建筑进行布局，建造一个走廊，将其打造成纵向延伸式花园。这样一方面给人们提供了一个交流活动的空间，另一方面则改善了这幢进深17米的楼房的采光及空气流通条件，从而阳光也能透进公共花园了。

## HOLLOW SPACE

We propose a volume with a smaller scale towards the pedestrian area and a vertical scale towards the street. We drill the block diagonally (south-west) creating a gallery housing communications (vertical garden). It promotes that light and air enter into the volume (17m deep) and sunlight into the communal garden.

阁楼层平面图 PENTHOUSE
十层平面图 FLOOR PLAN 9
九层平面图 FLOOR PLAN 8
八层平面图 FLOOR PLAN 7
七层平面图 FLOOR PLAN 6
五、六层平面图 FLOOR PLAN 4-5
四层平面图 FLOOR PLAN 3
二、三层平面图 FLOOR PLAN 1,2
planta 6
AVENIDA
Continuidad espacial
jardín - terrazas
659.8
656.8
648.3
636.9
631.2
627.0
625.5
617.0
横向平面图 CROSS SECTION

# CLarq.arquitectura (建筑师事务所)

Luis Alió Alonso · Javier García Valencia · Álvaro Marín Fernández · Sebastián Zapata Syro (建筑师)

中标 winner

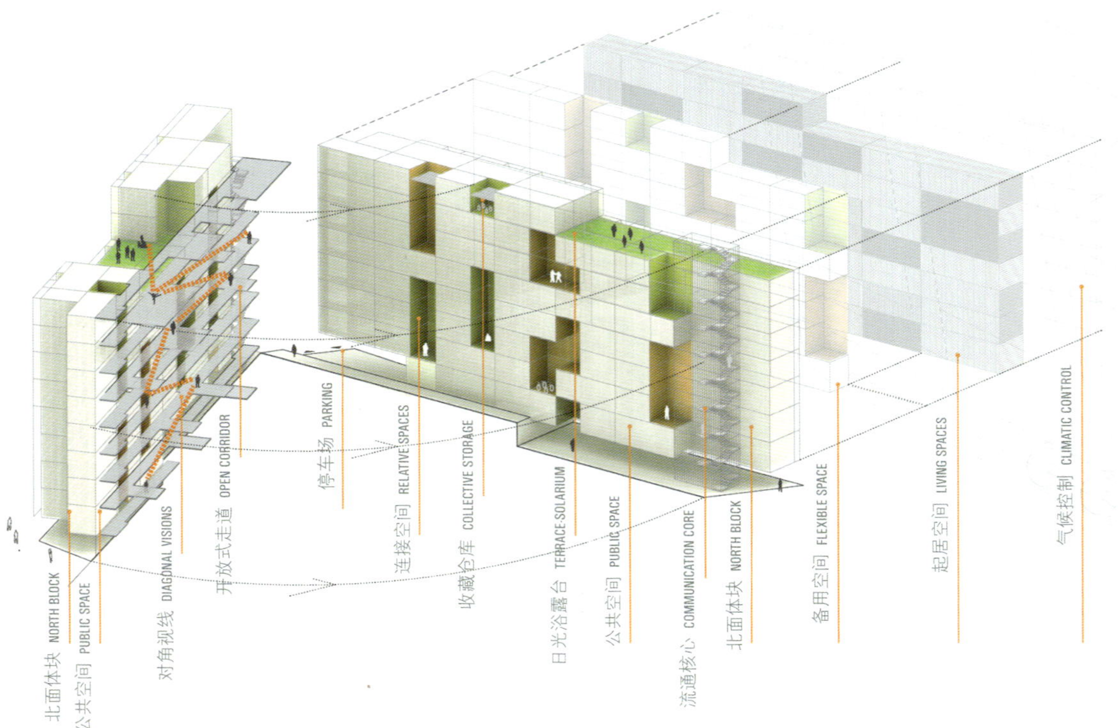

## 两幢大楼

我们的最初想法是设计两幢独立却不对称的大楼，两幢大楼的纵向侧面相接，底层顺应地形特点而设计，试图将中央的环形走廊打造为具备多重空间特性的内/外空间。"内墙"因为分散的庭院而间断，高高低低，造型各异，呈不规则状。这些空间无固定用途，可用作自行车棚，也可用作临时会议场所。

## TWO BLOCKS

Starting from the idea of two separate anti-symmetrical blocks, linked by their longitudinal side, we move their floor plans to adapt to the plot, and seek to understand the central circulation corridor as an interior/exterior space with complex spatial qualities. The "inner walls" are carved with discontinuous courtyards, irregularly distributed both in height and in plan. These spaces without a determined program can be used both to store bicycles, and for impromptu meetings.

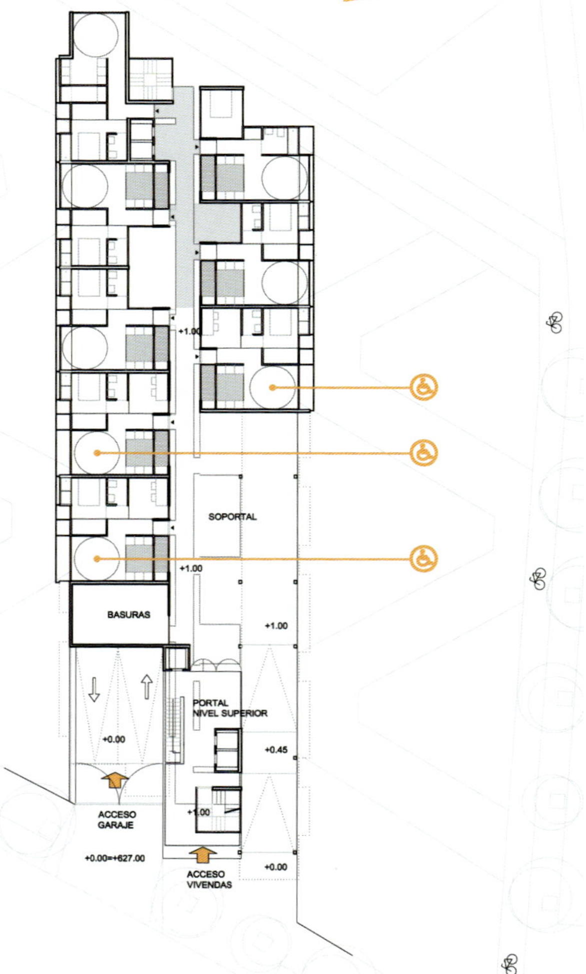


平面图 九层(B)·屋顶层(A) FLOOR PLAN 8(B) · ROOF PLAN (A)

平面图 七层(A)·八层(B) FLOOR PLAN 6(A) · 7(B)

标准层平面图 TYPICAL FLOOR PLAN

底层平面图(A)·底层平面图(B) GROUND FLOOR PLAN (A) · FLOOR PLAN 1 (B)

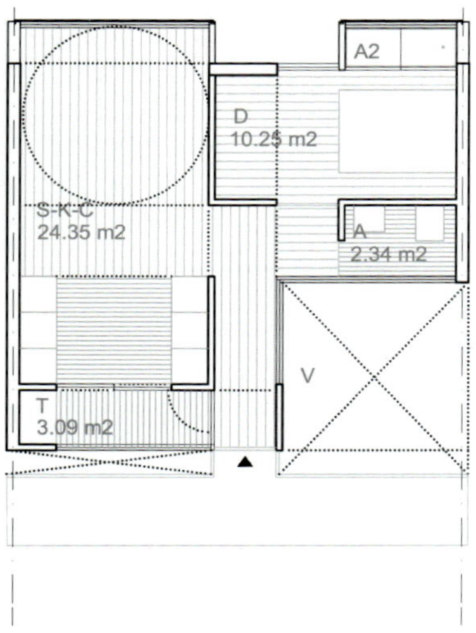

房屋单元类型1 HOUSING UNIT TYPE 1. 40.03M2

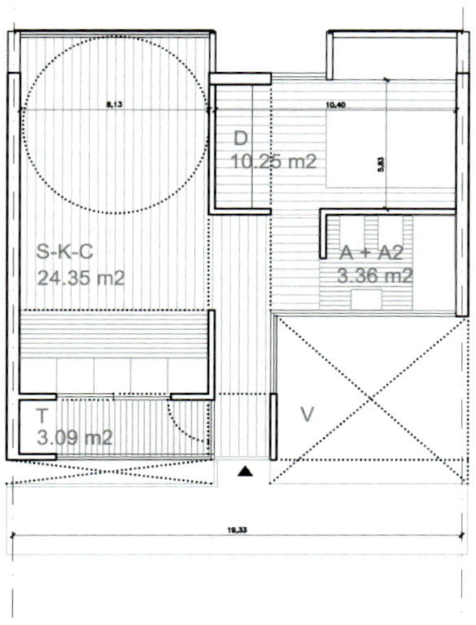

房屋单元类型1的变化 VARIATION OF HOUSING UNIT TYPE 1

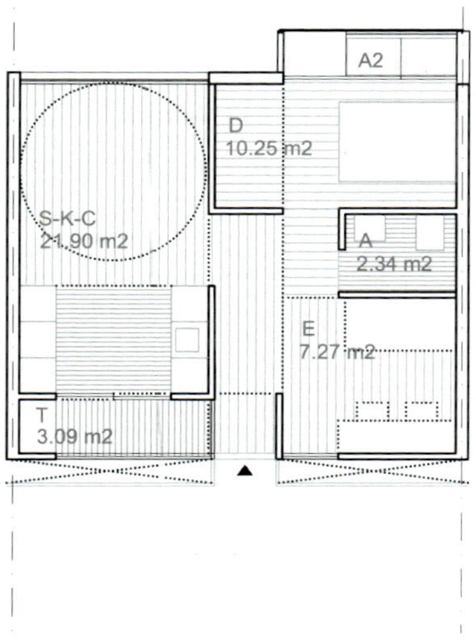

房屋单元类型2 HOUSING UNIT TYPE 2. 44.85M2

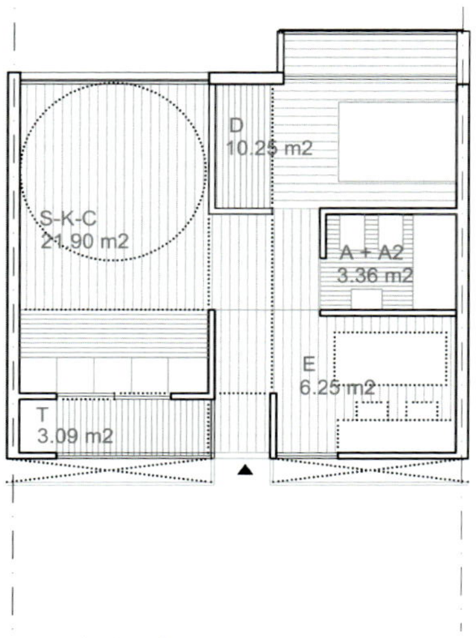

房屋单元类型2的变化 VARIATION OF HOUSING UNIT TYPE 2

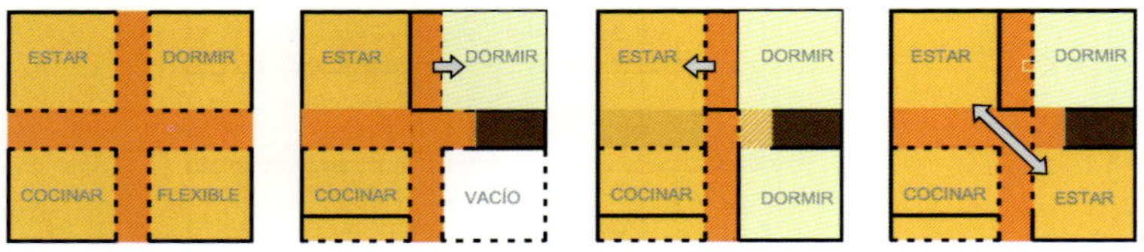

ENFRIAMIENTO EVAPORATIVO

0.00 m

横向剖面图 CROSS SECTION

43PICOS

# estudio.entresitio (建筑师事务所)

César Jiménez de Tejada · María Hurtado de Mendoza · José María Hurtado de Mendoza · Alvaro Ruiz (建筑师)

第一提名奖 first mention

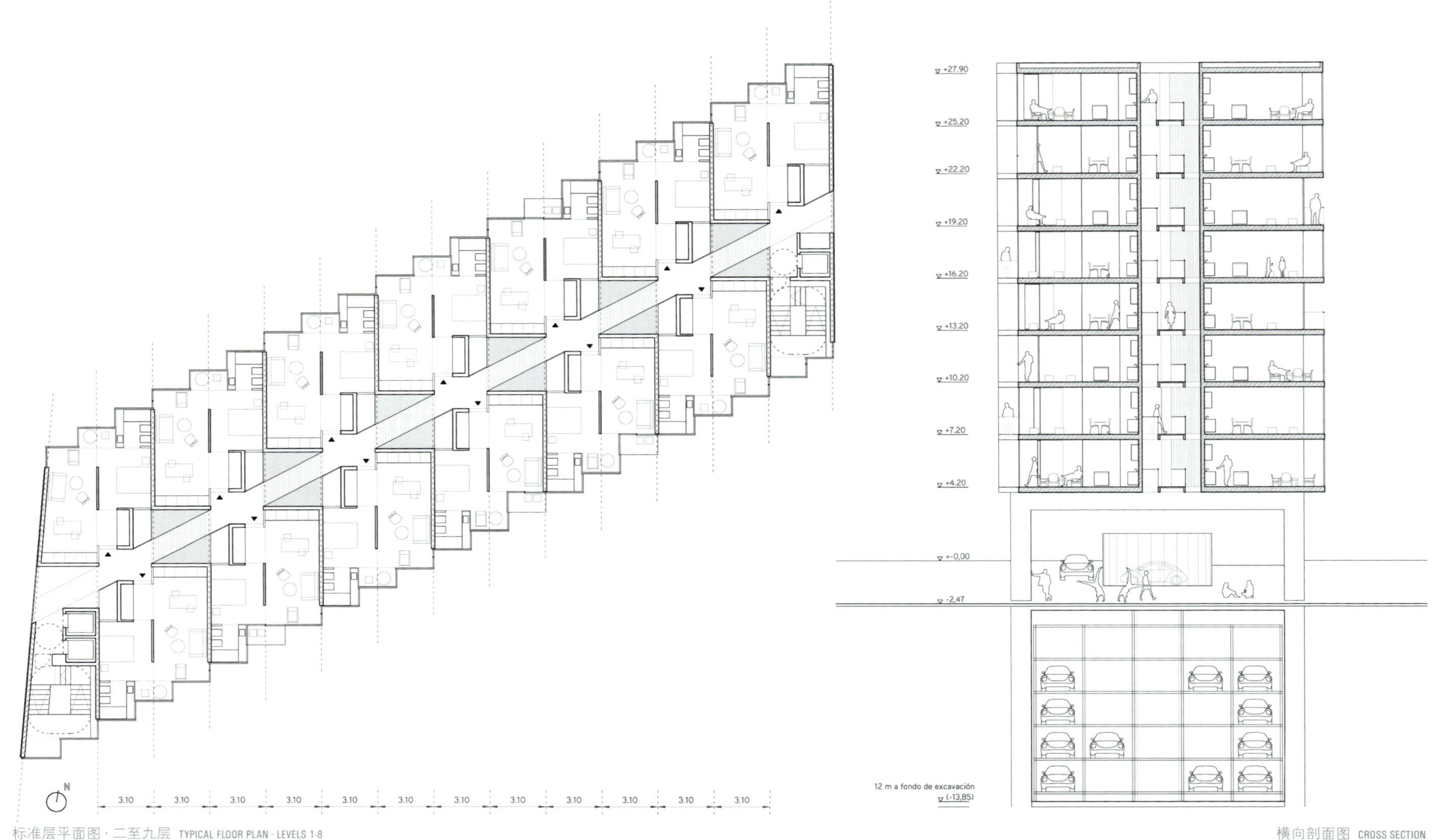


标准层平面图 · 二至九层 TYPICAL FLOOR PLAN · LEVELS 1-8

横向剖面图 CROSS SECTION

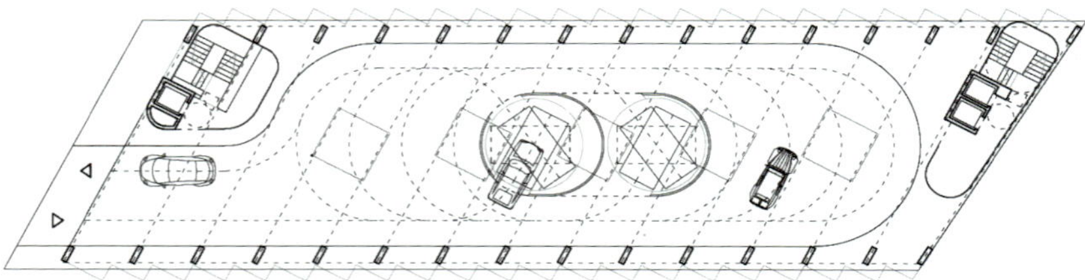

底层平面图 · 入口 GROUND FLOOR PLAN · ENTRANCE

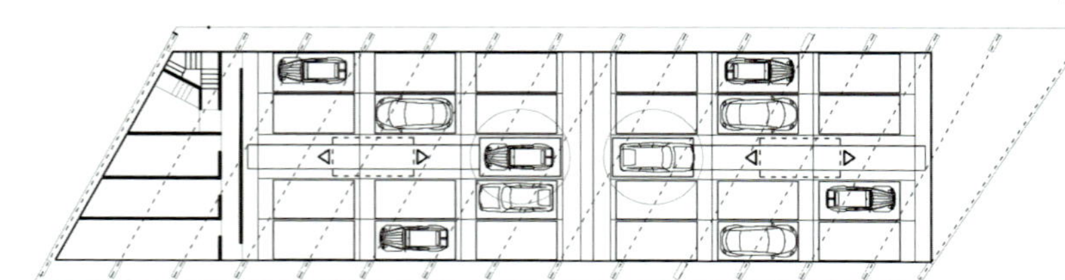

自动化停车场 AUTOMATED PARKING

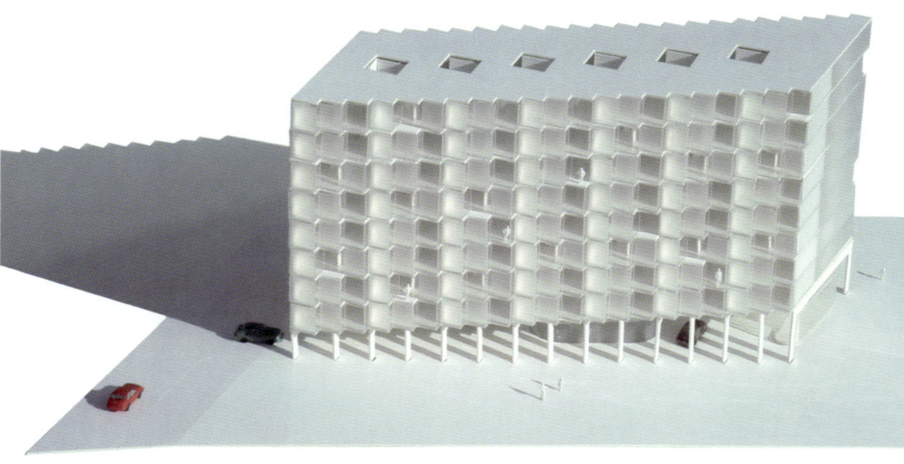

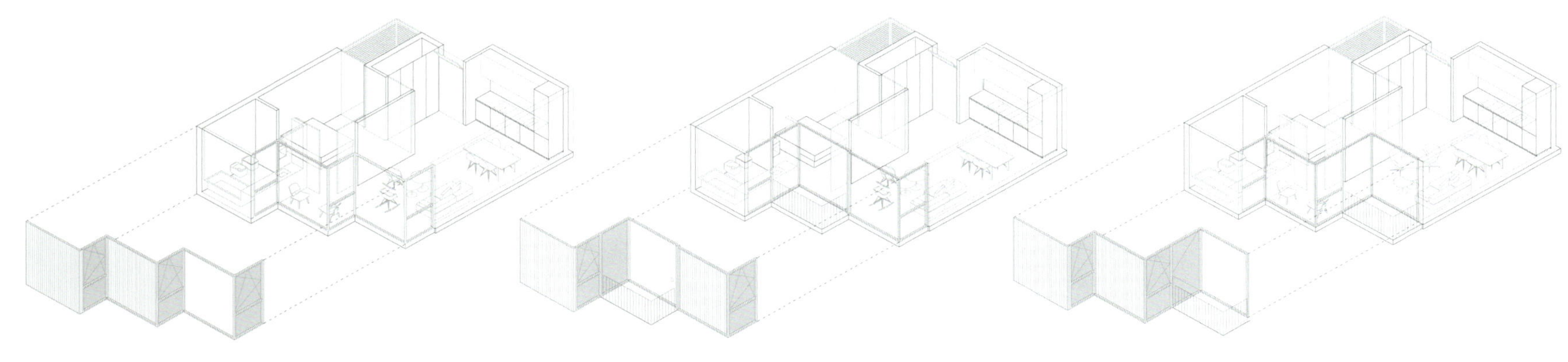

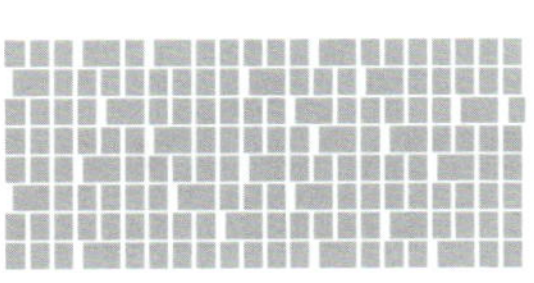
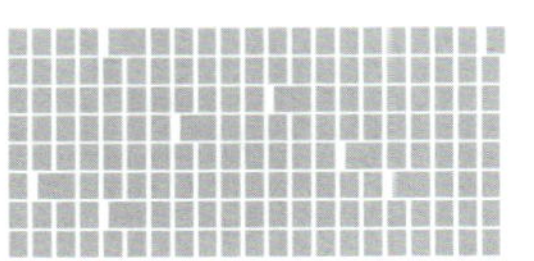

一天中不同时段中南北立面示意图 NORTH AND SOUTH FAÇADE SCHEMES AT DIFFERENT HOURS OF DAY

# 调整与重复

我们提议将入口处设计成采光、通风俱佳的纵向空间，由此打造与众不同的空间感，并实现每个住宅单元的双重通风功能。该解决方案的关键点在于将建筑外表面定义为室内与室外温度的缓冲区。

# MODULATION AND REPETITION

We propose the access area with lighted and ventilated vertical gaps as a result offer an interesting spatial experience and a double ventilation of housing units. The solution is focused on the definition of the façade as a cushion space between interior and exterior.

FACEBOX

# DJarquitectura (建筑师事务所)

Diego Jimenez López · Juana Sanchez Gómez (建筑师)

合作 (c) Jorge Salguero Ropero

第二提名奖 second mention

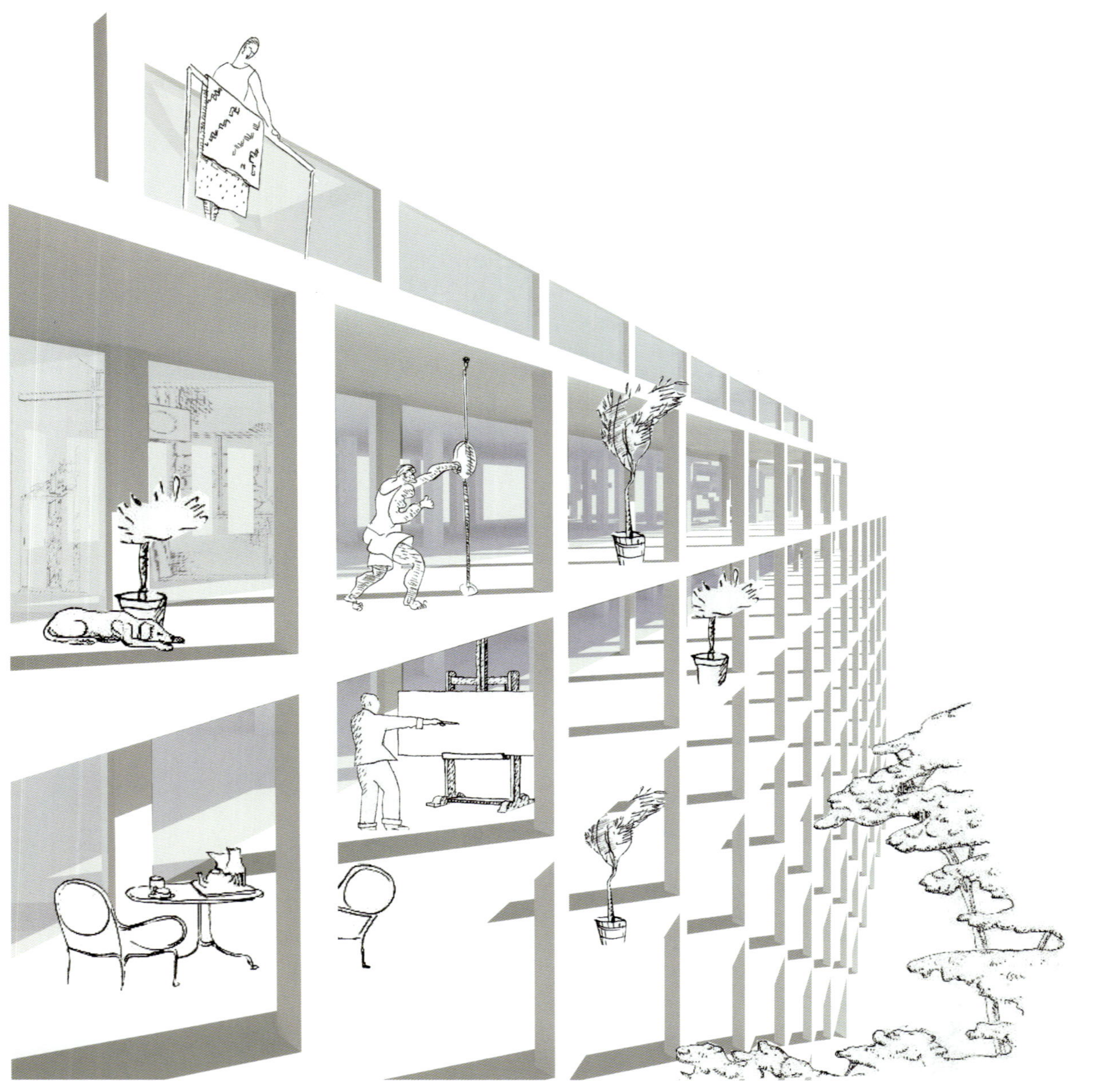

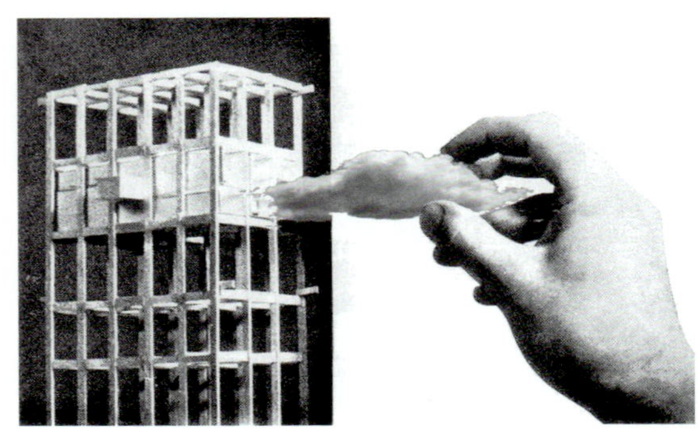

## 灵活的模块结构

我们的方案基于实现装配、户型、住户等方面的灵活性最大化而提出，可作为正式的预制方案，以加快和促进施工过程。施工中所搭的架子可还原，随时可拆走用于另一建筑工地，不造成浪费。

## FLEXIBLE MODULAR ORGANIZATION

We propose a strategy based on a system which allows for maximum flexibility (assembly, type, user...), designed as a formal prefabricated proposal, to expedite and facilitate the construction process. A reversible frame could even be removed at any time, migrating to another plot, without producing waste.

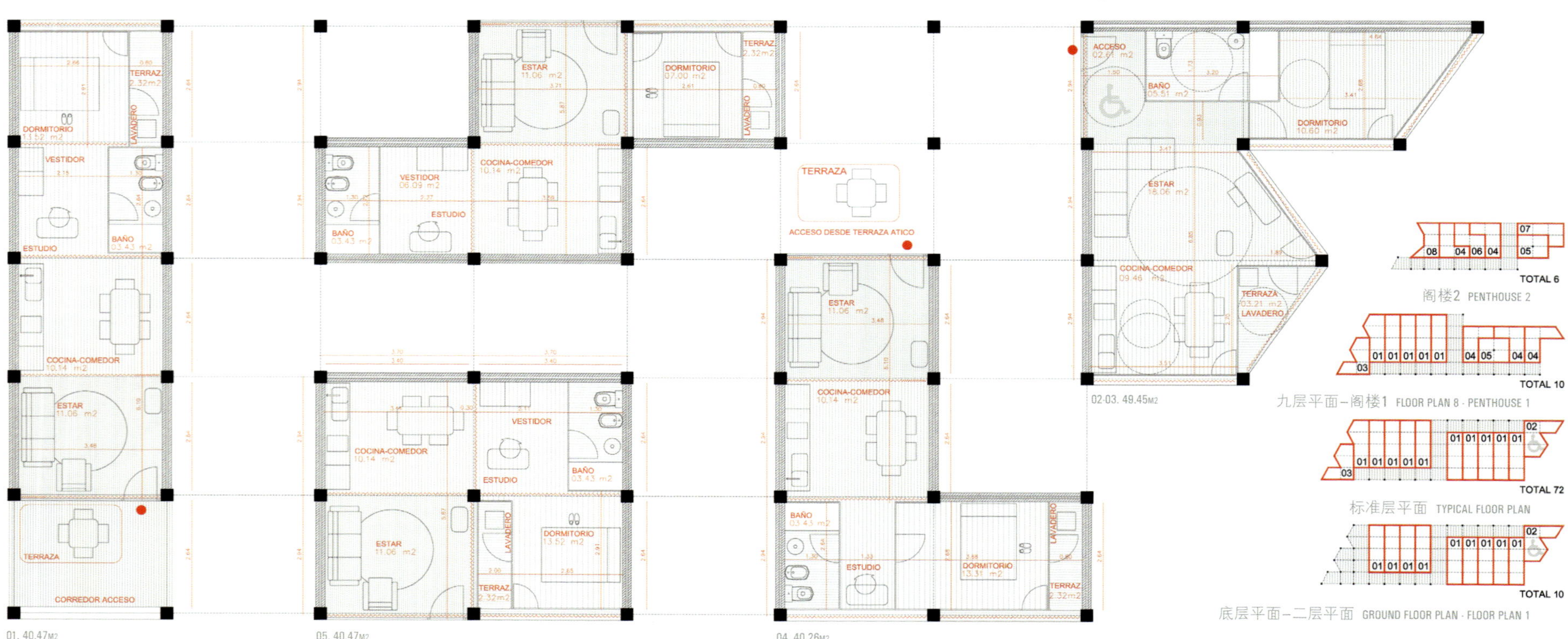


房屋单元类型 HOUSING UNITS TYPOLOGIES

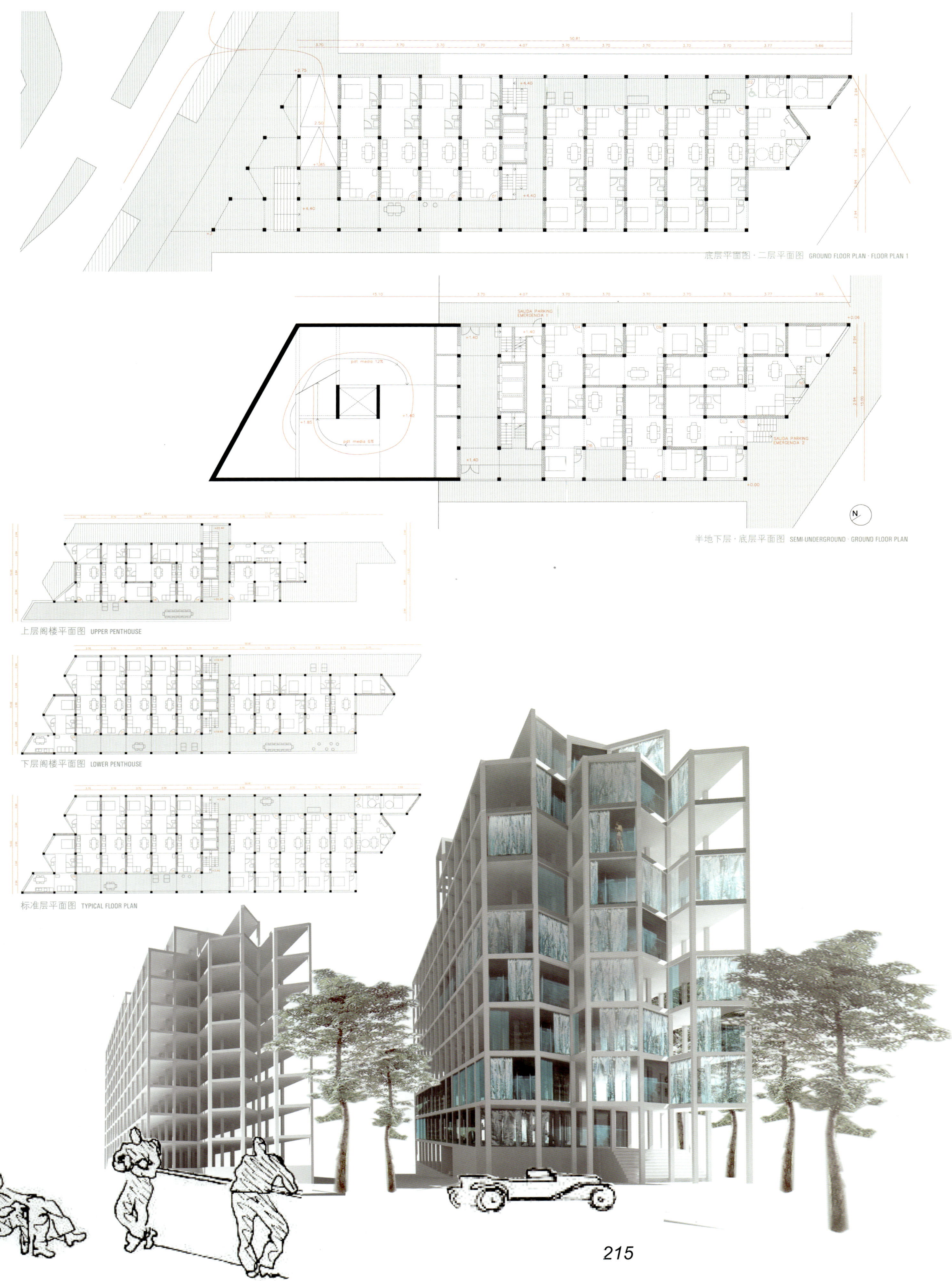

底层平面图 · 二层平面图 GROUND FLOOR PLAN · FLOOR PLAN 1

半地下层 · 底层平面图 SEMI-UNDERGROUND · GROUND FLOOR PLAN

上层阁楼平面图 UPPER PENTHOUSE

下层阁楼平面图 LOWER PENTHOUSE

标准层平面图 TYPICAL FLOOR PLAN

ZZZZZIP

# Espegel-Fisac Arquitectos (建筑师事务所)

Carmen Espegel · Concha Fisac (建筑师)
合作 (c) Laia Lafuente · Marcela Aragüez · Miriam Llamazares

第三提名奖 third mention

## 韵律

每个公寓都配备一个阳台，便于阳光照进屋里。厨房与客厅相连，空间已经最小化了，各种家具几乎挤占了半个厨房，而卫生间空间大，方便使用。阳台则布局用心，利于对流通风。

## RYTHM

Each apartment has a terrace that advances towards the façade to catch the sun. The kitchen, connected to the living area, is minimized and is half-buried by pieces of furniture, while the bathroom is provided with a spatiality that encourages its use. A central gallery promotes cross ventilation thanks to the intentional layout of the terraces.

max.30m

原始状态 STARTING POINT

max.34m

方案提议 PROPOSAL

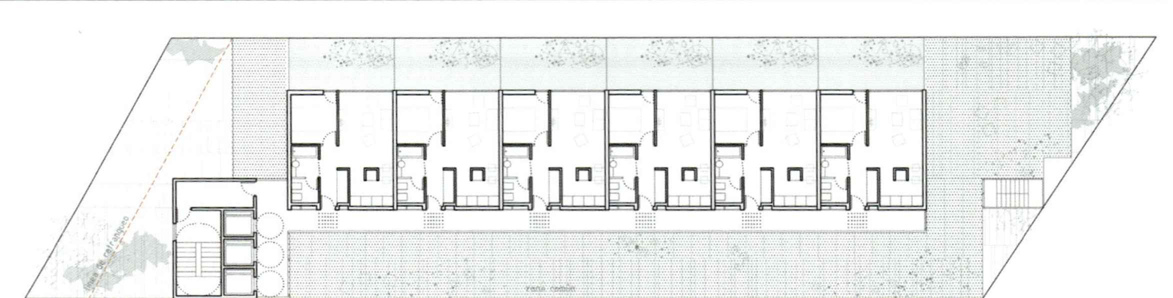

阁楼层平面图 PENTHOUSE

标准层平面图 TYPICAL FLOOR PLAN

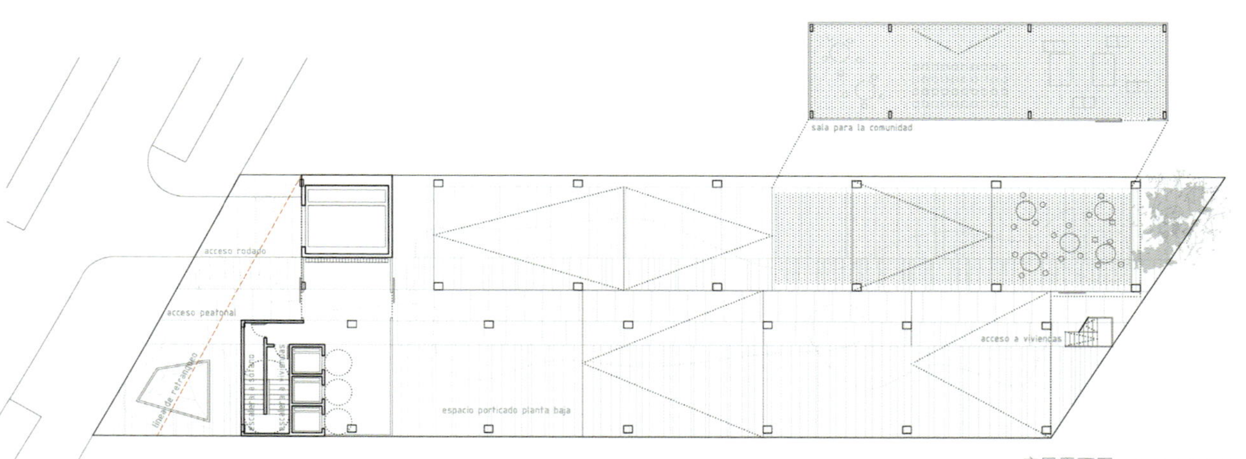


底层平面图 GROUND FLOOR PLAN

南立面 SOUTH ELEVATION

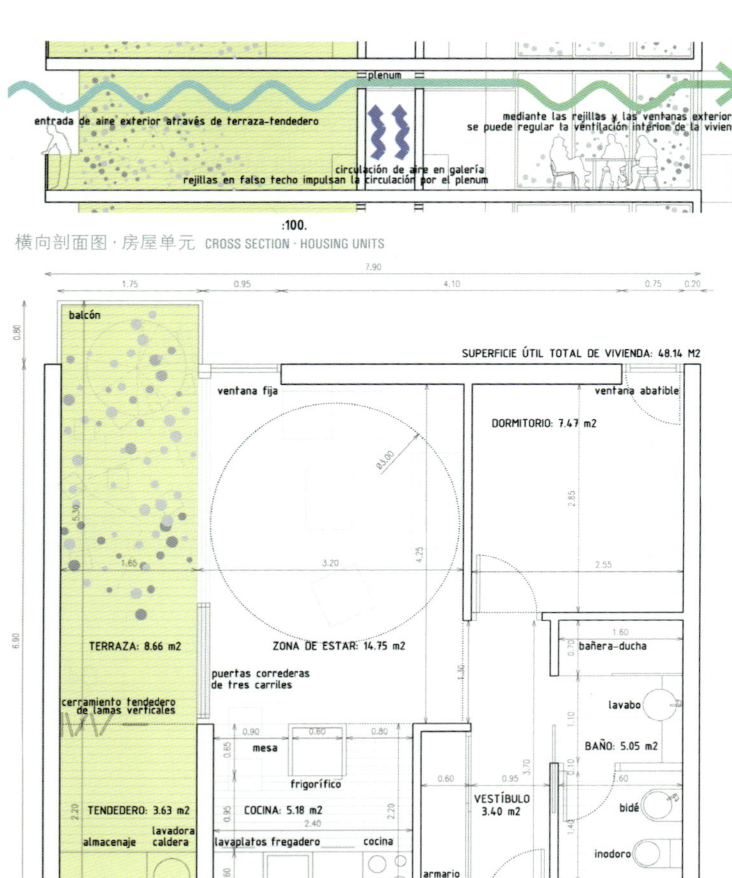


横向剖面图 · 房屋单元 CROSS SECTION · HOUSING UNITS

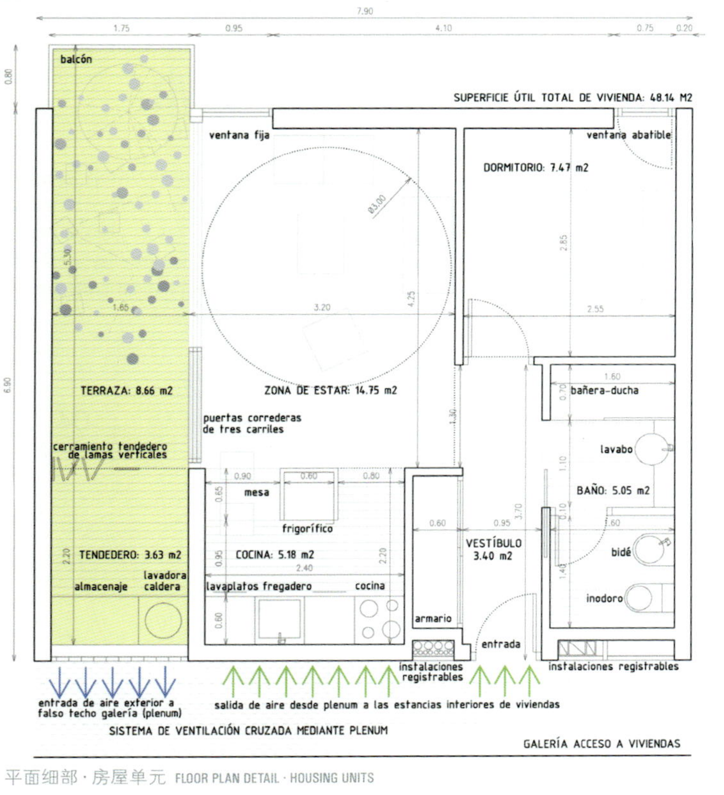


平面细部 · 房屋单元 FLOOR PLAN DETAIL · HOUSING UNITS

房屋单元内部 · 冬季 INTERIOR OF HOUSING UNIT · WINTER

房屋单元内部 · 夏季 INTERIOR OF HOUSING UNIT · SUMMER

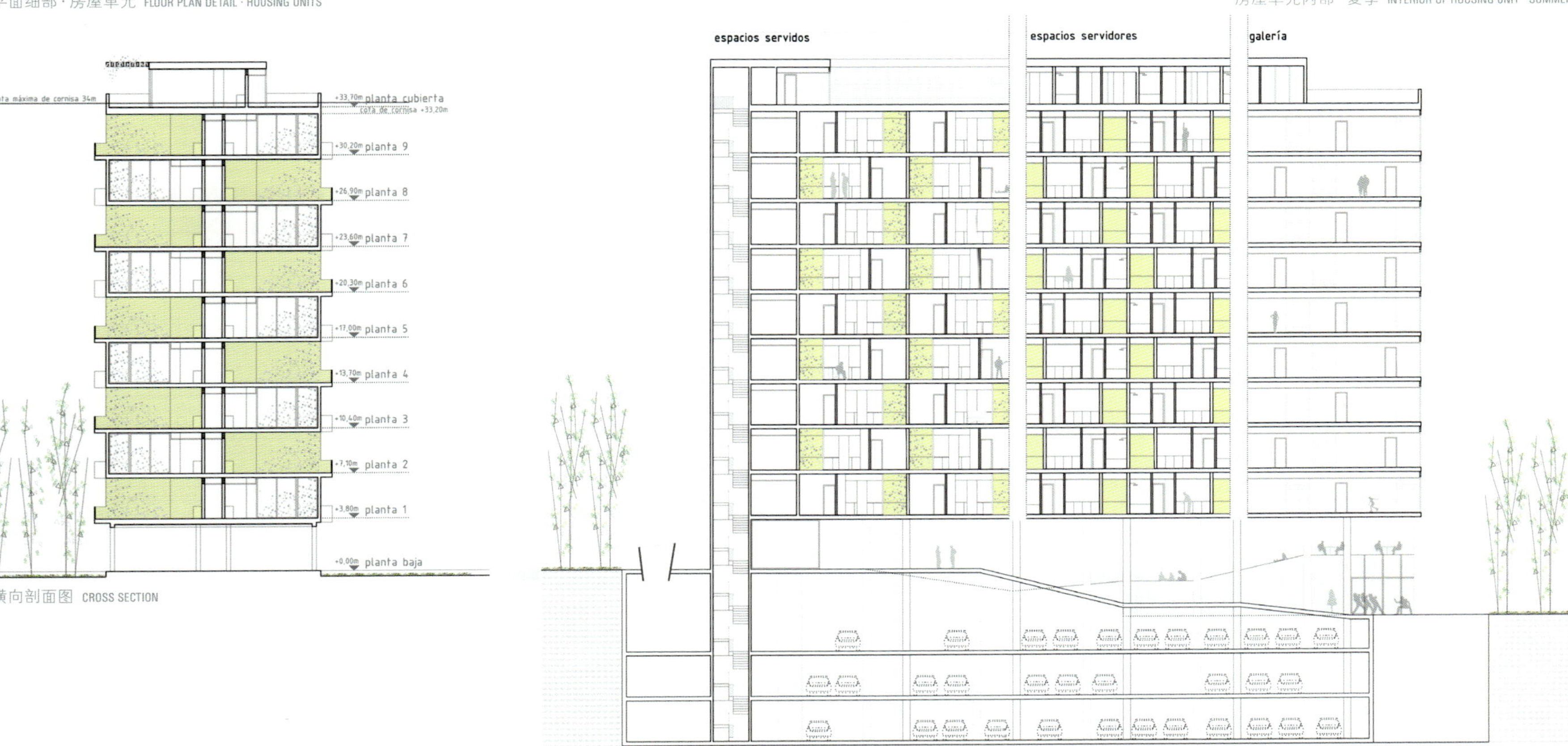


横向剖面图 CROSS SECTION

纵向剖面图 LONGITUDINAL SECTION

OOOAIRE

# Estudio Cano Lasso (建筑师)

Gonzalo Cano Pintos · Alfonso Cano Pintos · Diego Cano-Lasso Pintos · Beatriz Pozo Fernández · Manuel Ordóñez Abarca · Lucía Pérez Gamarra · Javier Esteban Lecumberri · Antonio Mas-Guindal · Adela Rueda Márquez de la Plata · Elena Estella Pérez· Mayca Sánchez Carvajal (建筑师)

第五提名奖 fifth mention

地块位置 SITE PLAN

## 大型建筑体量

公寓生动活泼、简洁敞亮、色调明快，连通公寓和外界的走廊可作为室内外气温的缓冲区。走廊嵌于建筑外表面与内屋之间，外有玻璃窗保护，避免夏日正午阳光的直射。这些公寓虽然朝向单一，但仍有自然风对流，这主要得益于太阳能烟囱。

## A GREAT VOLUME

Lively, simple and bright apartments open to the outside through a gallery, which serves as a cushion of climatic control. The gallery is tucked into the plane of the façade so that its glasses keep it protected from the summer sun at noon. Despite they are apartments with a single orientation; they have natural cross-ventilation. This is achieved through solar chimneys.

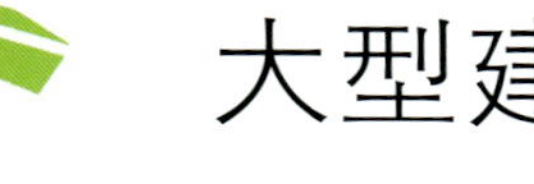

功能布局图 PROGRAM SCHEME

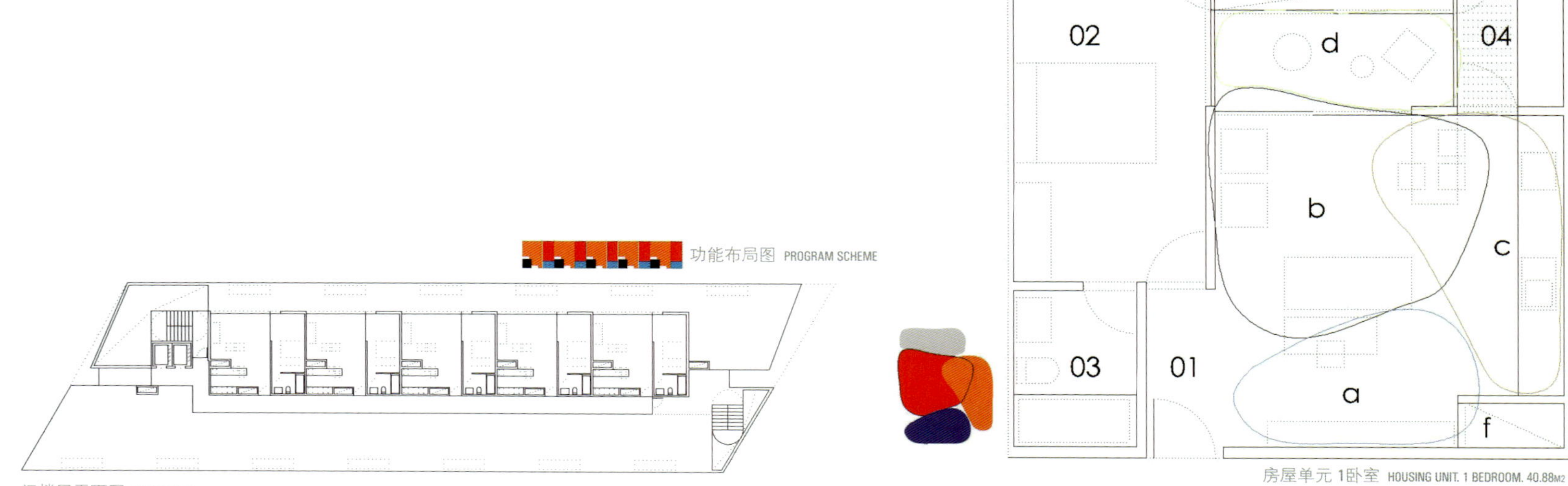


阁楼层平面图 PENTHOUSE

房屋单元 1卧室 HOUSING UNIT. 1 BEDROOM. 40.88m2

功能布局图 PROGRAM SCHEME

标准层平面图 · 二至九层 TYPICAL FLOOR PLANS · LEVELS 1-8

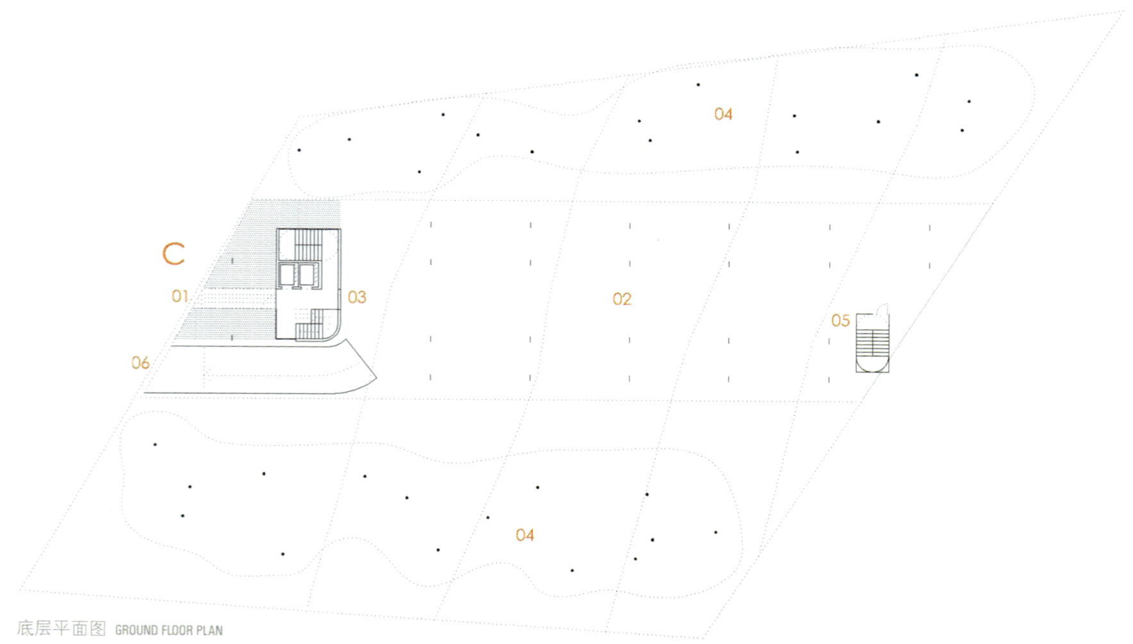


底层平面图 GROUND FLOOR PLAN

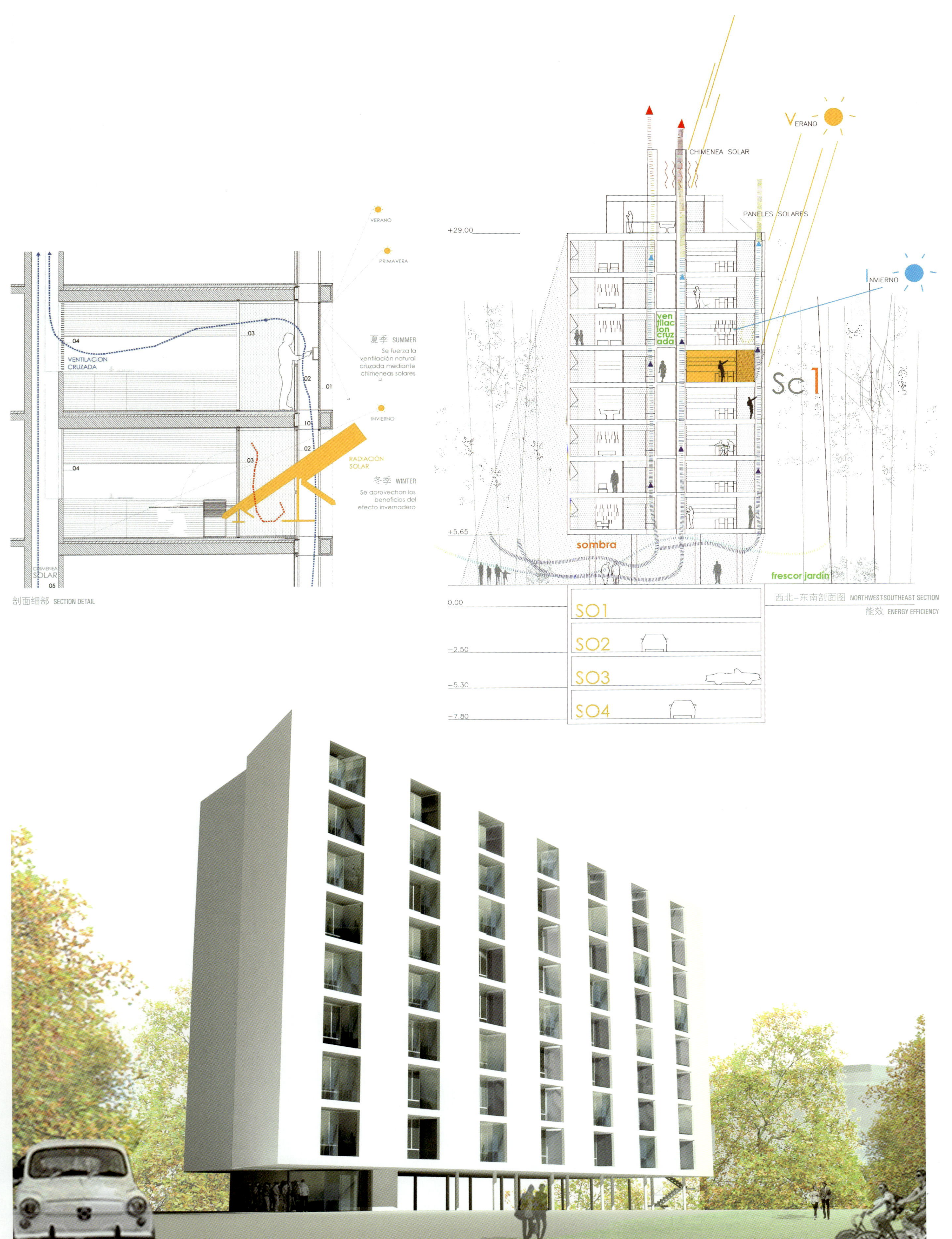

剖面细部 SECTION DETAIL

西北-东南剖面图 NORTHWEST-SOUTHEAST SECTION

能效 ENERGY EFFICIENCY

# Estudio UNTERCIO (建筑师事务所)

Marina del Mármol · Daniel Bergman · Mauro Bravo · Miguel Herraiz (建筑师)

中标 winner

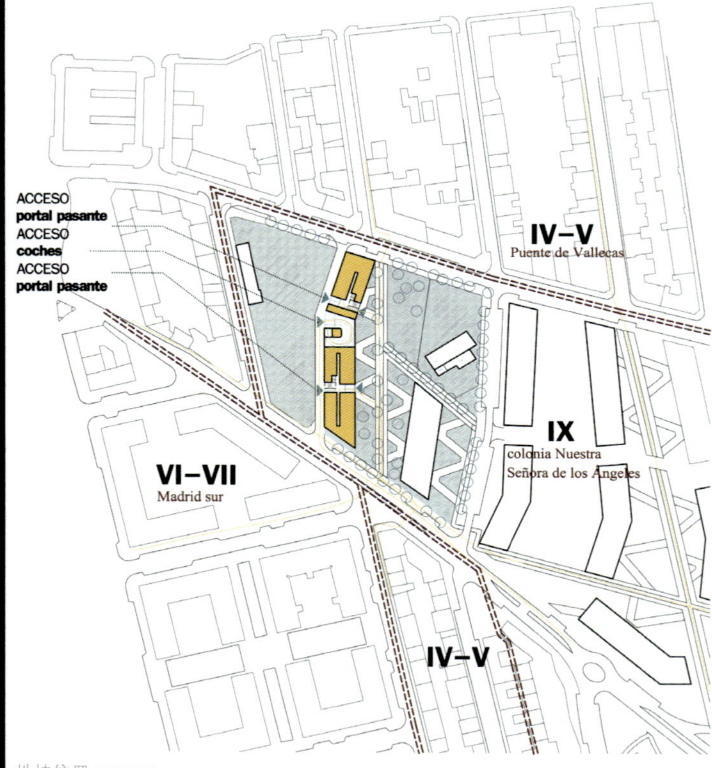


地块位置 SITE PLAN

## 阶梯结构

该设计方案将屋顶和屋顶的小棚屋分开设计，以调整建筑的高度，适应现有城市的布局。城市规划可通过两种不同的高度得以实现：一面临街——温暖、紧凑，并根据透明和非透明的外墙板进行调整，这是“建筑师的表面”；另一面是“社区的表面”——建筑物采用后缩结构，形成空隙，使阶梯结构更具活力。

## A SUCCESSION OF TERRACES

The proposal adjusts its height to the existing city by stepping the roof and the design of the penthouse as a differentiated body. The urban adaptation is expressed through two different elevations. One of its façades scales the street, it is warm and close, modulated on the basis of transparent and opaque cladding panels, "the façade of the architects", and another, "the façade of the neighbors", set back, with a freer arrangement of holes, where life emerges in the terraces.

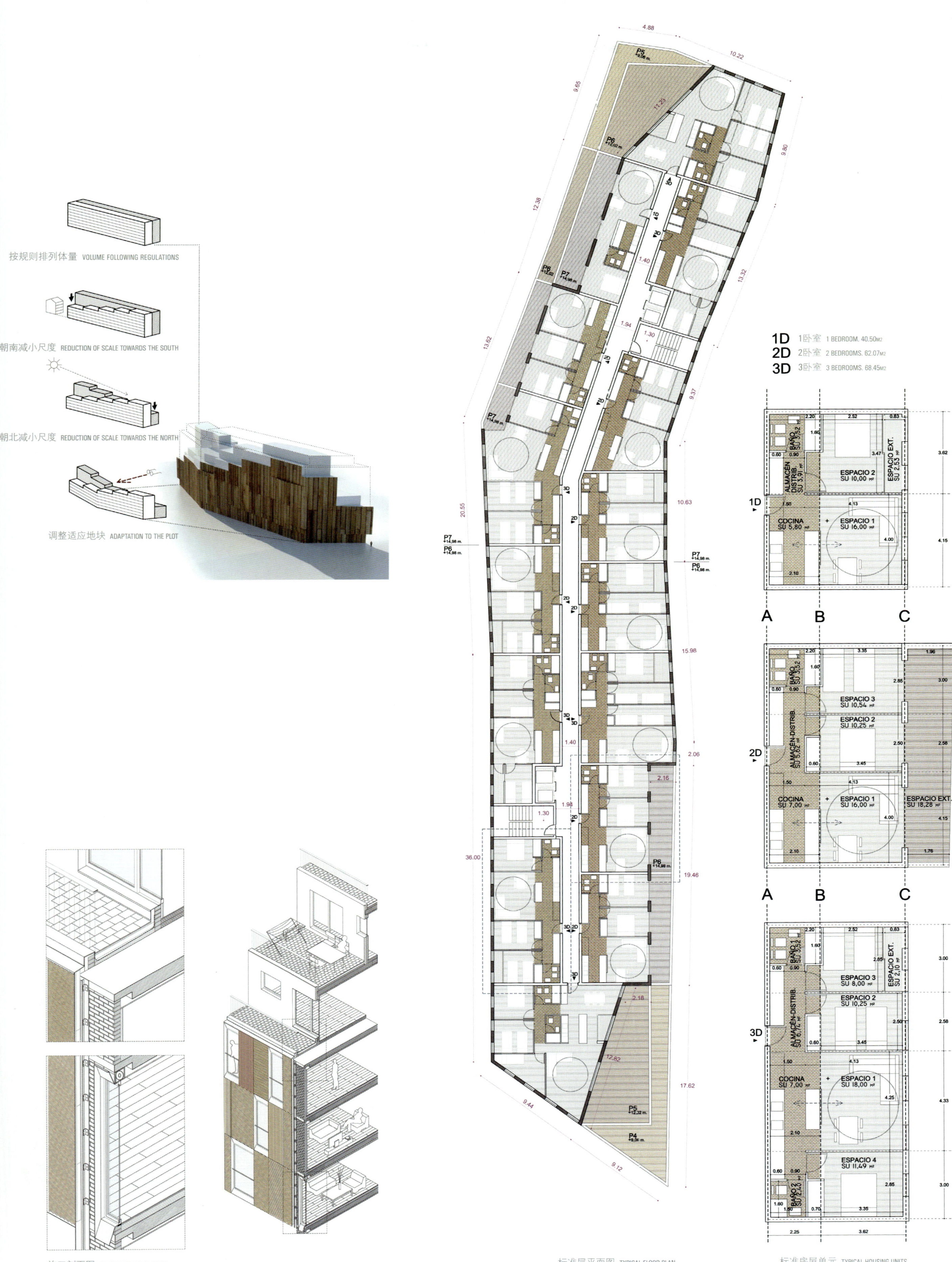


施工剖面图 CONSTRUCTION SECTION

标准层平面图 TYPICAL FLOOR PLAN

标准房屋单元 TYPICAL HOUSING UNITS

ENGAWAA

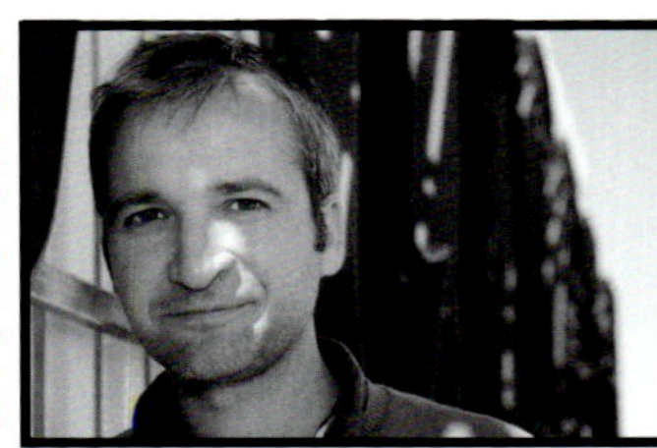

# José María Sánchez García (建筑师)

合作 (c) Enrique García-Margallo · Rafael Fernández · Laura Rojo · Marta Cabezón · Mafalda Ambrósio · Mariló Sánchez

第一提名奖 first mention

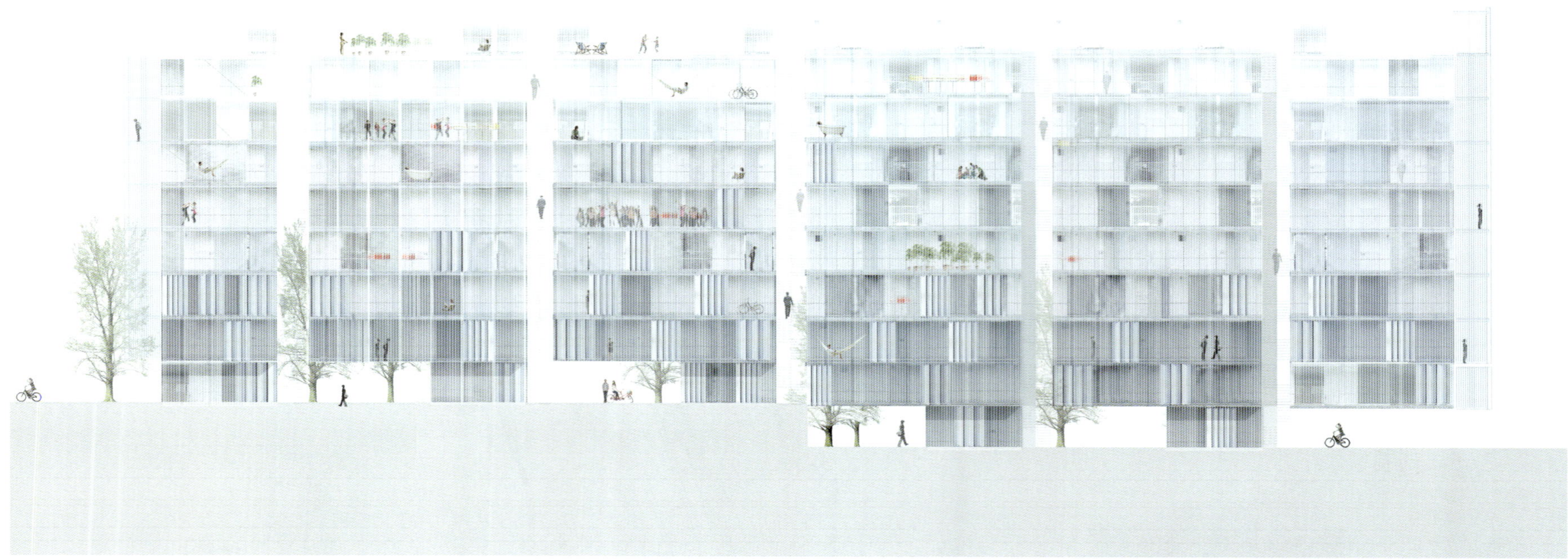

立面图 ELEVATION

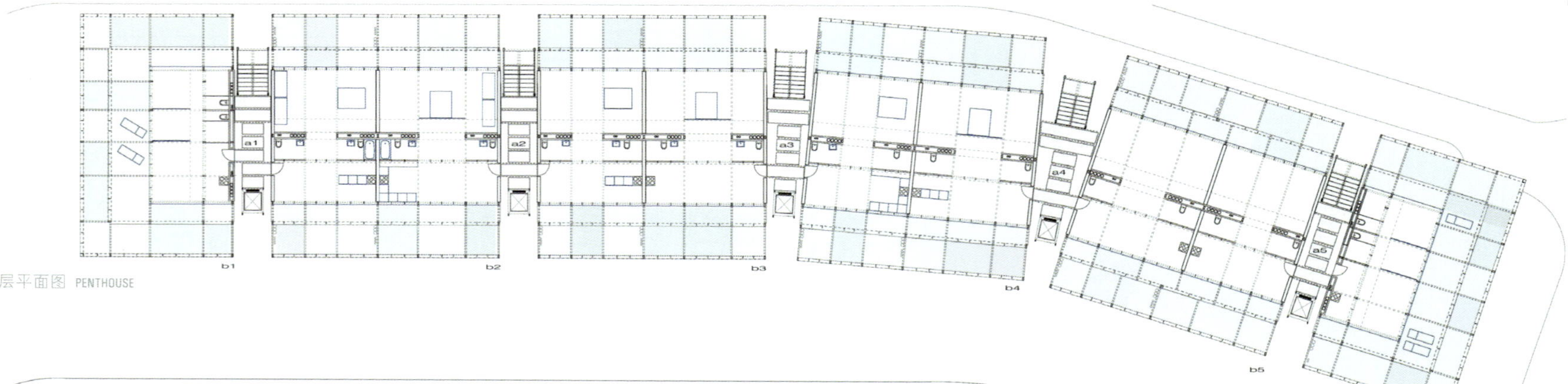

阁楼层平面图 PENTHOUSE

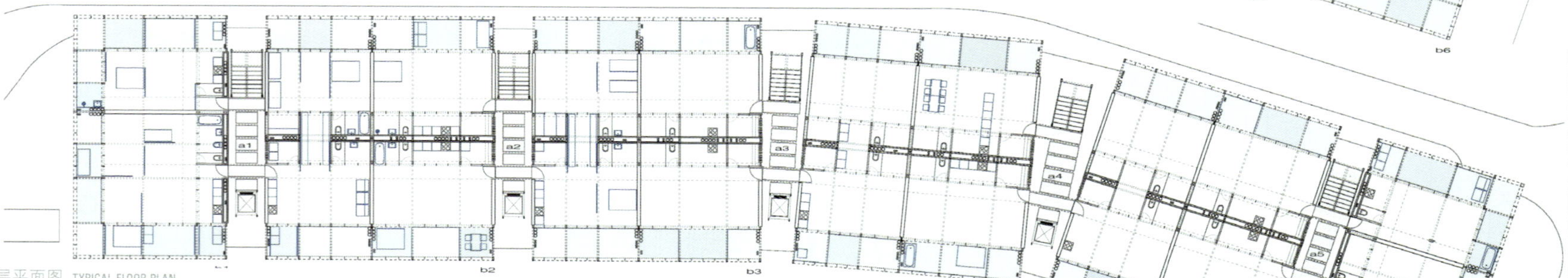

标准层平面图 TYPICAL FLOOR PLAN

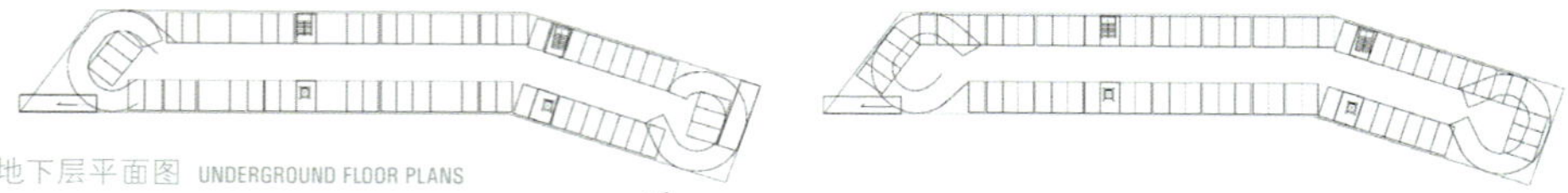

地下层平面图 UNDERGROUND FLOOR PLANS

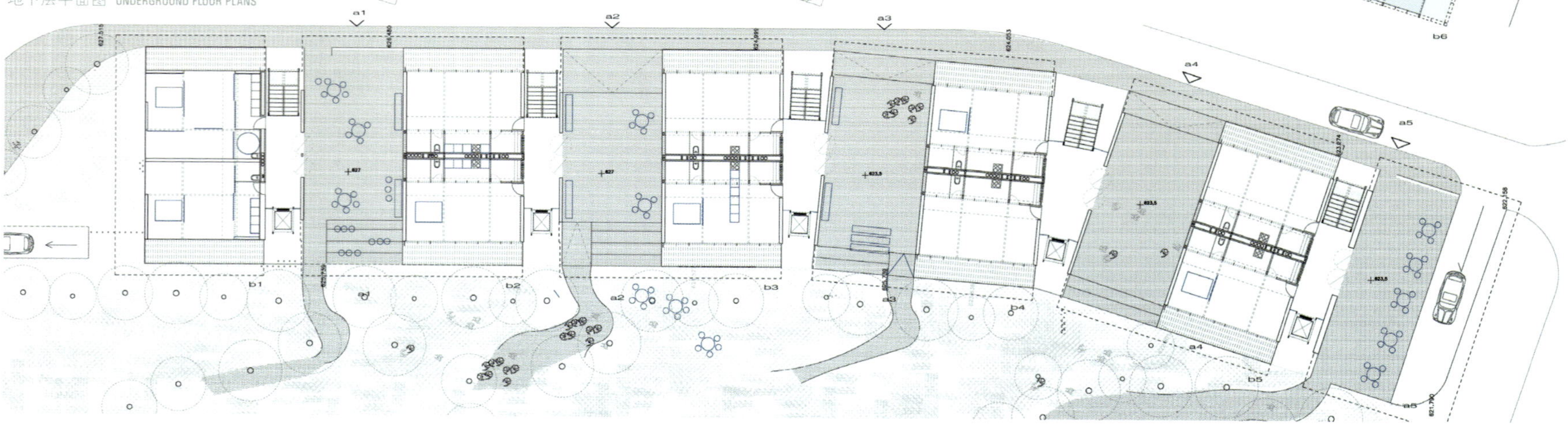

底层平面图 GROUND FLOOR PLAN

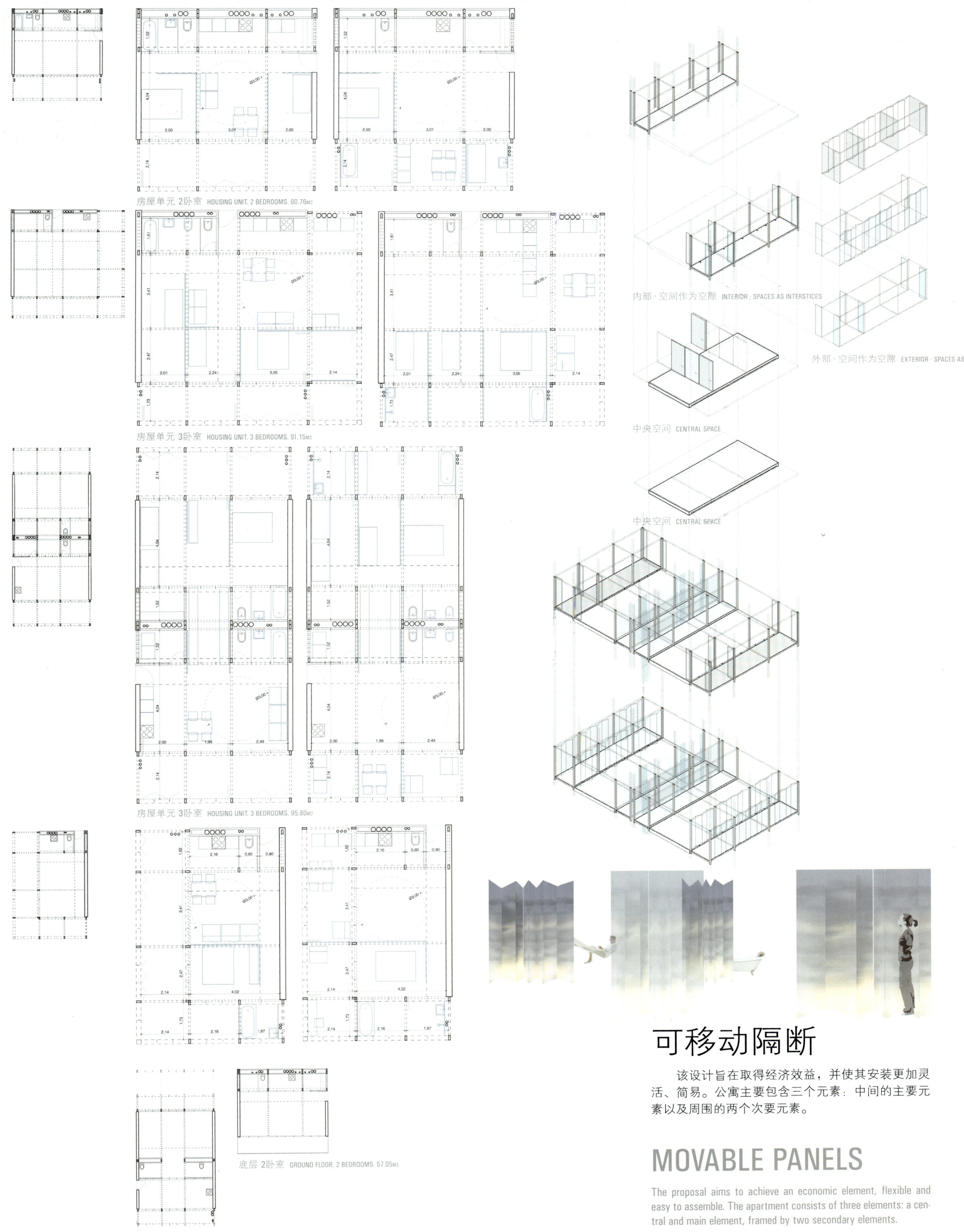


# 可移动隔断

该设计旨在取得经济效益，并使其安装更加灵活、简易。公寓主要包含三个元素：中间的主要元素以及周围的两个次要元素。

# MOVABLE PANELS

The proposal aims to achieve an economic element, flexible and easy to assemble. The apartment consists of three elements: a central and main element, framed by two secondary elements.

ELTELON

# Olalquiaga Arquitectos (建筑师事务所)

Rafael Olalquiaga Soriano · Pablo Olalquiaga Bescós · Alfonso Olalquiaga Bescós (建筑师)
合作 (c) Javier Morales Luchena

并列第二提名奖 second mention ex-aequo

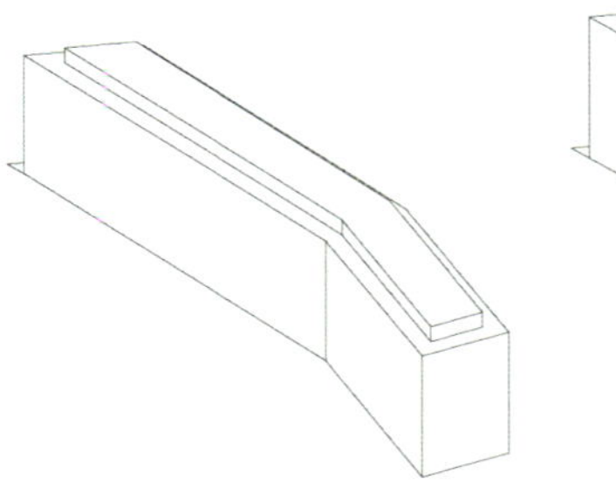
高度：九层加阁楼 HEIGHT: 8 LEVELS+PENTHOUSE

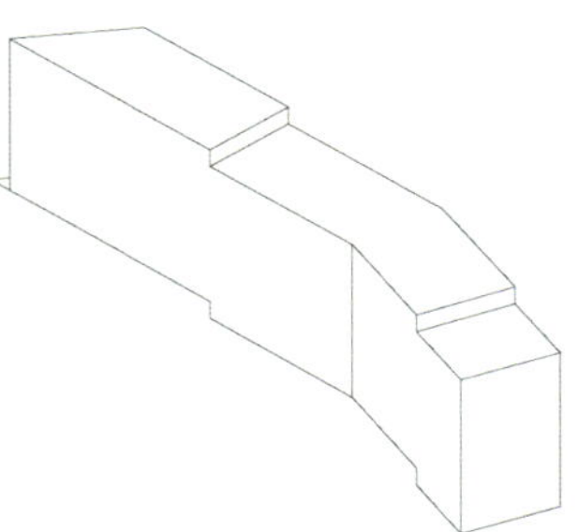
错移 STAGGER

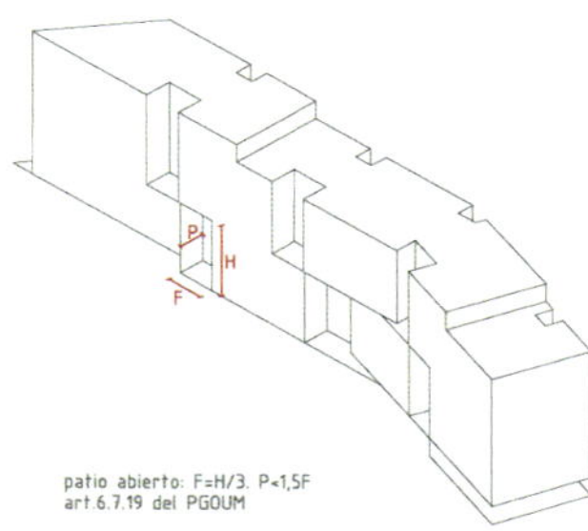

掏空建筑 TO HOLLOW THE BUILDING

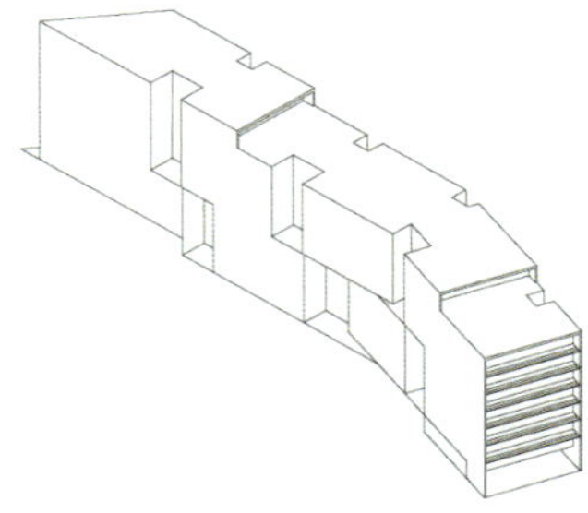
挖空后墙面 TO EMPTY THE BACK FAÇADES

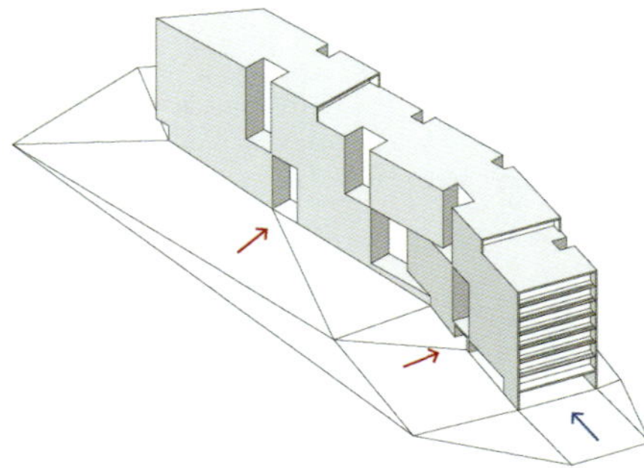
最后的体量 FINAL VOLUME
→ 行人入口 PEDESTRIAN ACCESS
→ 汽车入口 CAR ACCESS

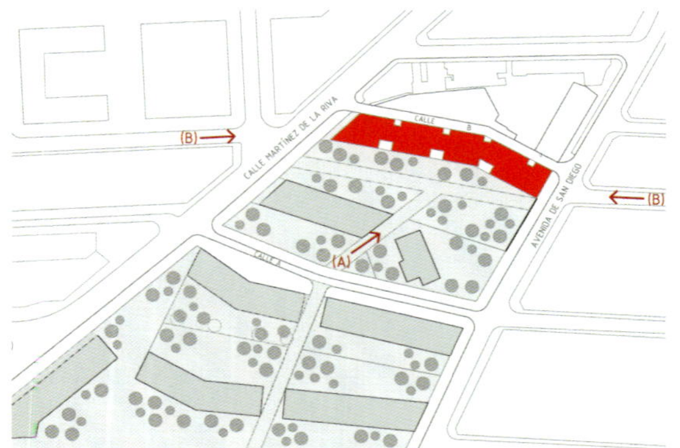
地块位置 SITE PLAN

## 断裂的空间

场地的长度和形状，以及端部6米长的下跌式结构，是帮助我们对建筑进行布局的因素。这幢建筑最后将两条大街相融合，层次分明，将整个空间分隔成一系列两两相连的空间区域。

## A BROKEN VOLUME

The length and shape of the site, together with its 6m drop at its ends, are factors that help us to configure the latest building that embraces the two main streets as a final stage. The building staggers and breaks into a series of intertwined volumes.

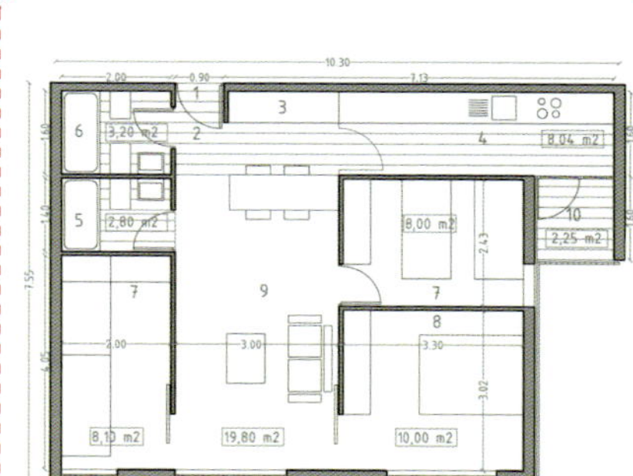
3卧室 3 BEDROOMS. 65.20M2

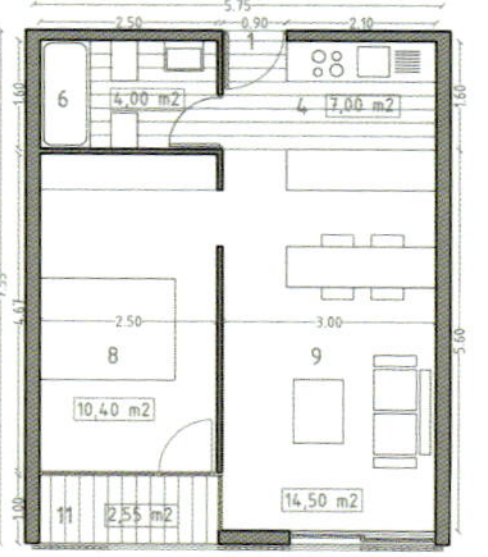
1卧室 1 BEDROOM. 41.25M2

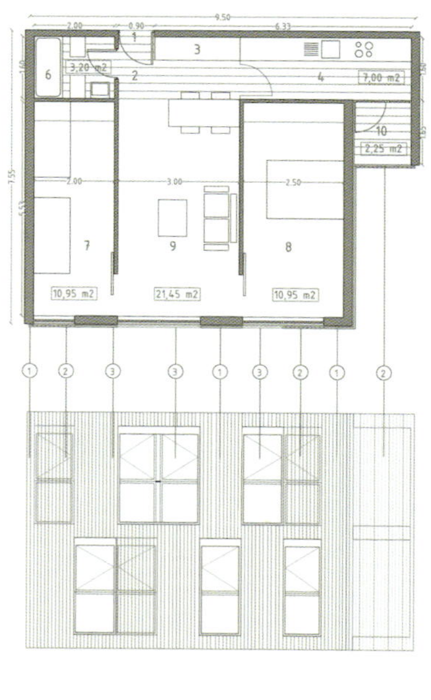
2卧室 2 BEDROOMS. 52.20M2

立面细部 ELEVATION DETAIL

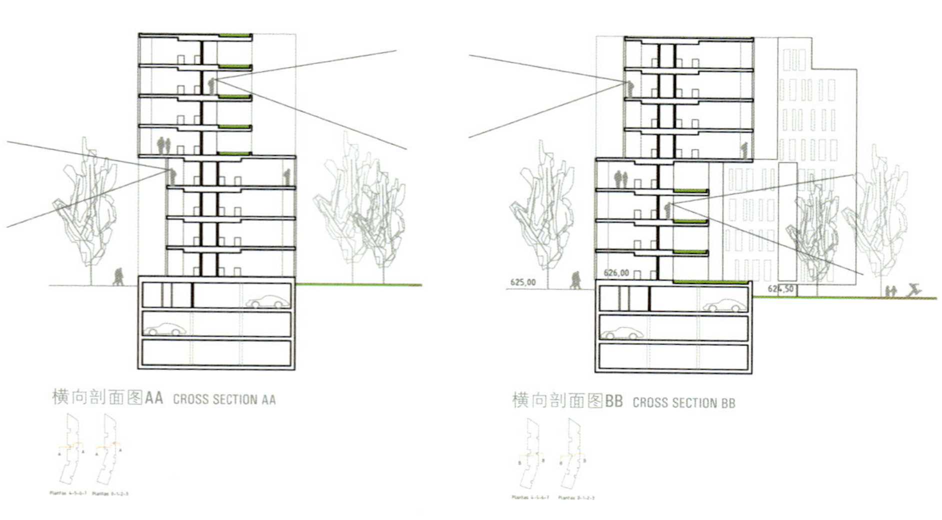
横向剖面图AA CROSS SECTION AA

横向剖面图BB CROSS SECTION BB

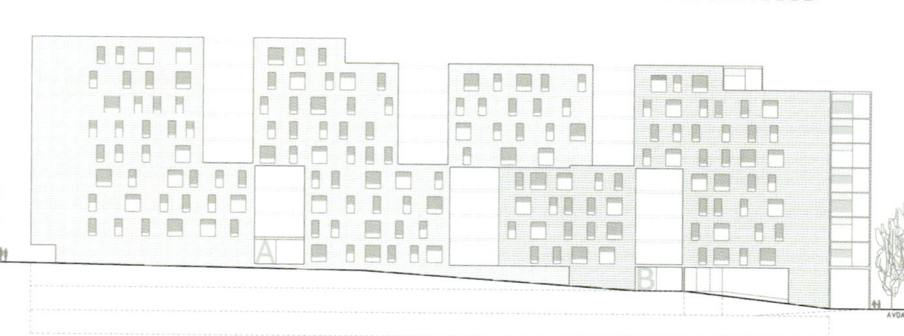
临街立面图 STREET ELEVATION

临公园立面图 PARK ELEVATION

底层平面图 GROUND FLOOR PLAN +626

二层平面图 FIRST FLOOR PLAN +629.5

五层平面图 FLOOR PLAN 4

8UM3RAN

# Estudio Cano Lasso (建筑师)

Gonzalo Cano Pintos · Alfonso Cano Pintos · Diego Cano-Lasso Pintos · Beatriz Pozo Fernández · Manuel Ordóñez Abarca · Lucía Pérez Gamarra · Javier Esteban Lecumberri · Antonio Mas-Guindal · Adela Rueda Márquez de la Plata · Elena Estella Pérez· Mayca Sánchez Carvajal (建筑师)

并列第二提名奖 second mention ex-aequo

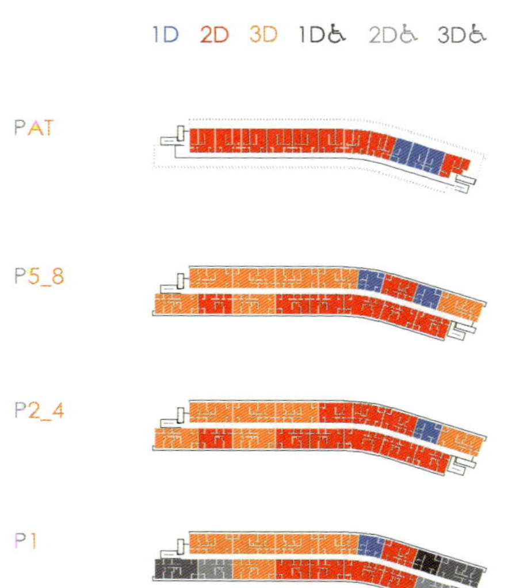


## 弧线形

借鉴“现代建筑运动”的最佳作法，并以准确性和紧凑性为指导原则，我们设计了一幢八层楼的建筑。通过清晰、平稳的调整实现准确性。紧凑性则通过基本建筑单元的系统性分类而实现。住宅单元的三个户型（一居室、两居室和三居室）可借助移动隔断作调整，以适应白天和夜晚的不同用途。

## IT CURVES SLIGHTLY

Following the example of the best models of the Modern Movement, we design an 8-storey block, with the guidelines of accuracy and compactness. The accuracy is achieved through a clear and rhythmic modulation. The compactness is achieved through a systematic typology using the growth of the basic housing unit. The 3 types of housing units, 1, 2 and 3 bedrooms, are adaptable for its use during the day and the night by movable panels.

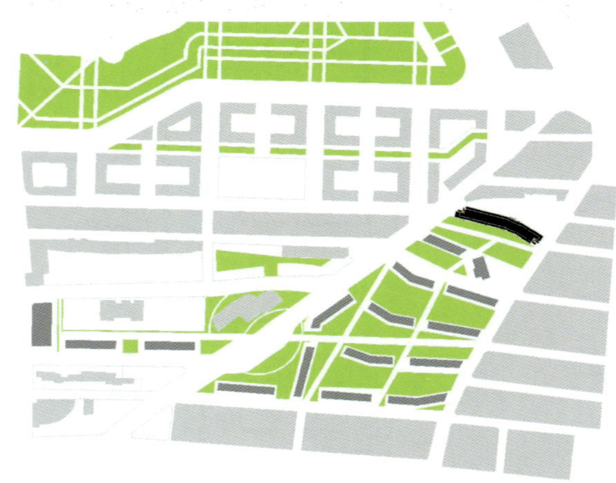

地块位置 SITE PLAN

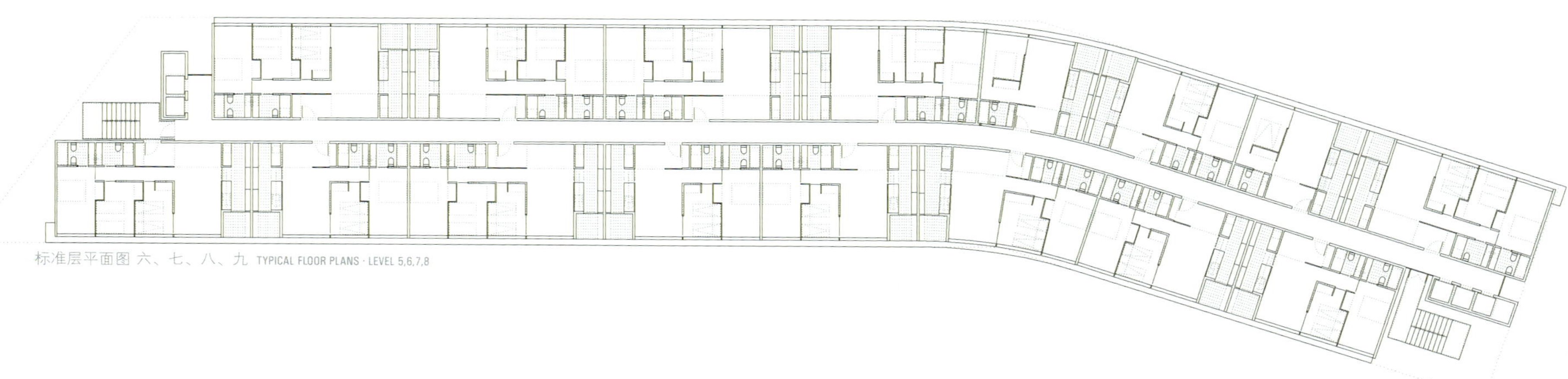

标准层平面图 六、七、八、九 TYPICAL FLOOR PLANS - LEVEL 5,6,7,8

06
01
02
05
04
03
房屋单元 1卧室 HOUSING UNIT, 1 BEDROOM, 42.18M2
09
07
01
06
02
08
05
04
03
房屋单元 2卧室 HOUSING UNIT, 2 BEDROOMS, 57.04M2
09
07
01
06
02
10
08
05
04
03
房屋单元 3卧室 HOUSING UNIT, 3 BEDROOMS, 71.77M2
白天 DAY
夜晚 NIGHT
白天 DAY
夜晚 NIGHT
白天 DAY
夜晚 NIGHT
SO1
SO2
SO3
SO4
正立面图 MAIN ELEVATION

SOMBRAS

# LUMO ARQUITECTOS (建筑师事务所)

Javier García · Alfredo Rodríguez (建筑师)

并列第二提名奖 second mention ex-aequo

西北立面图 NORTHWEST ELEVATION

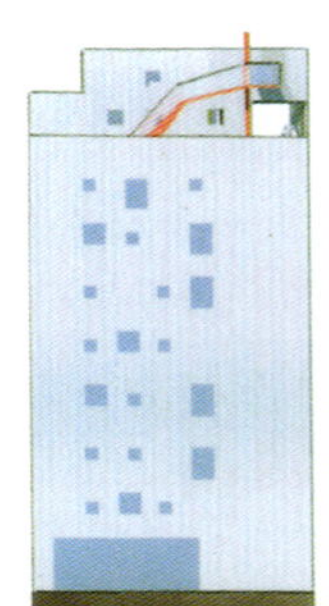

西南立面图 SOUTHWEST ELEVATION

东南立面图 SOUTHEAST ELEVATION

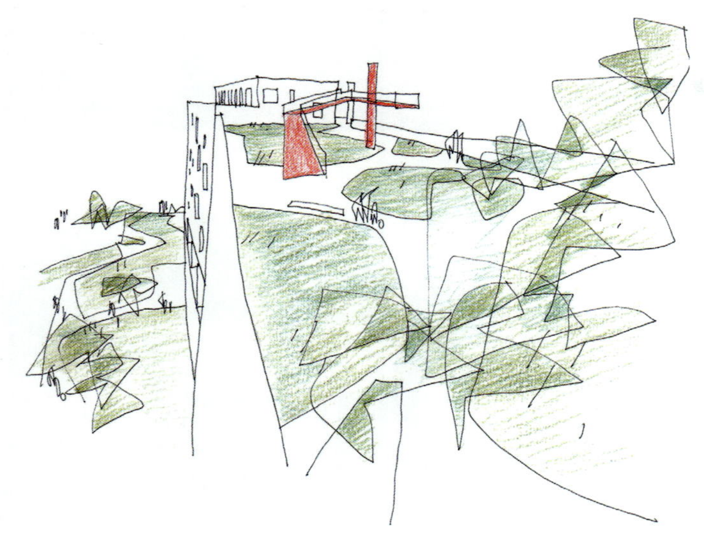

# 微缩庭院

两种不同的建筑外观：尾向朝北的传统形式，根据朝向在建筑外表面使用大小不一的窗户；双面朝南的建筑外观——传统型和过滤结构型，中间为通向房间的通道。通道带有露台和过滤型结构，形成一条内部“街道”，通风良好，植物等又可遮挡南面的阳光，生动活泼，色彩丰富，赋予城市一种丰富多彩、变幻莫测的形象。

# MICRO-COURTYARDS

Two distinct façades: Conventional north ends and façade, with windows of varying size appropriate to their orientation. Double south façade, one is conventional and the other is a filter, separated by the access corridor to the apartments. The access corridor to the apartments, with their patios and their filters, generates an internal street to meet, ventilated, with vegetation, protected from the south sun, full of fun, colorful, promoting a varied and changing image to the city.

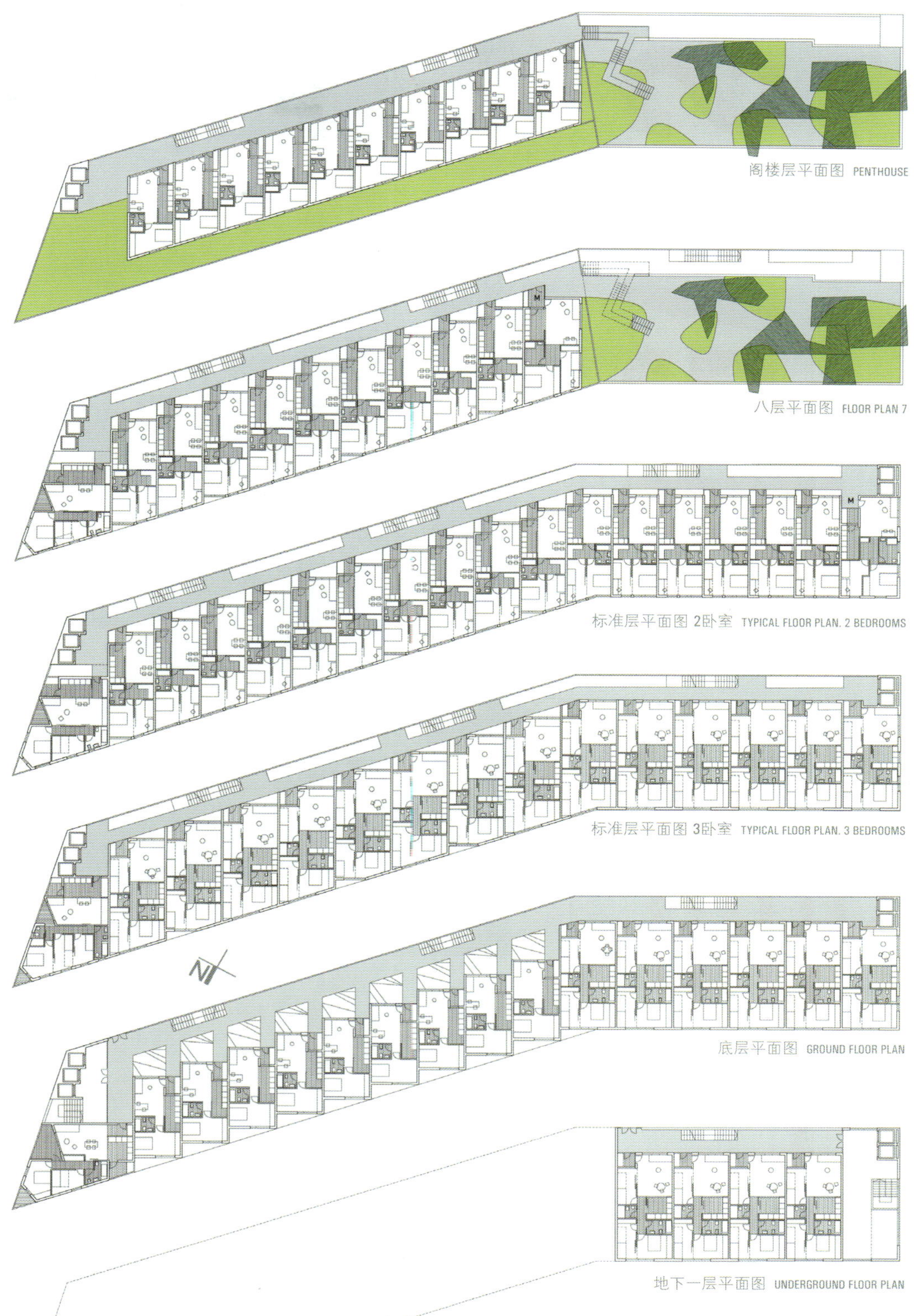


标准房屋单元 TYPICAL HOUSING UNITS

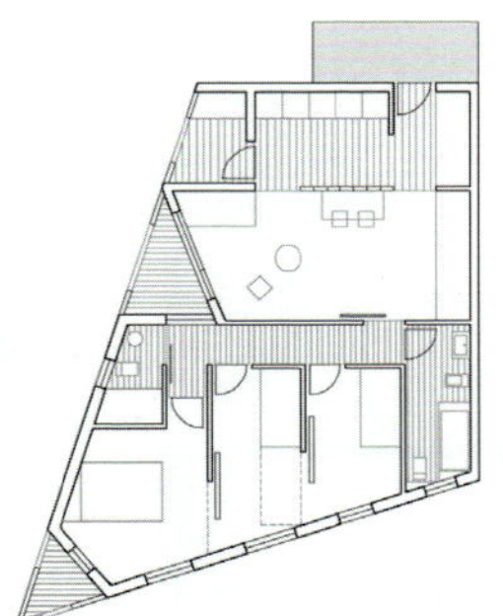

3卧室 类型1 3 BEDROOMS T1. 70.00M2

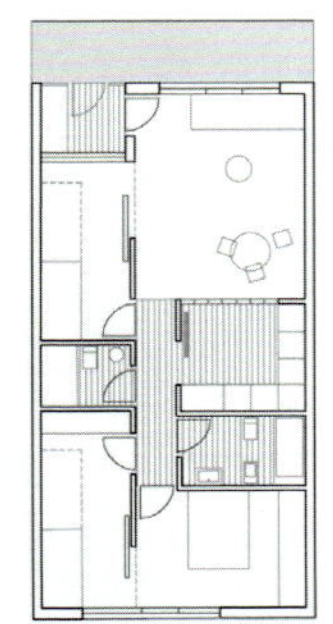

3卧室 类型2 3 BEDROOMS T2. 66.37M2

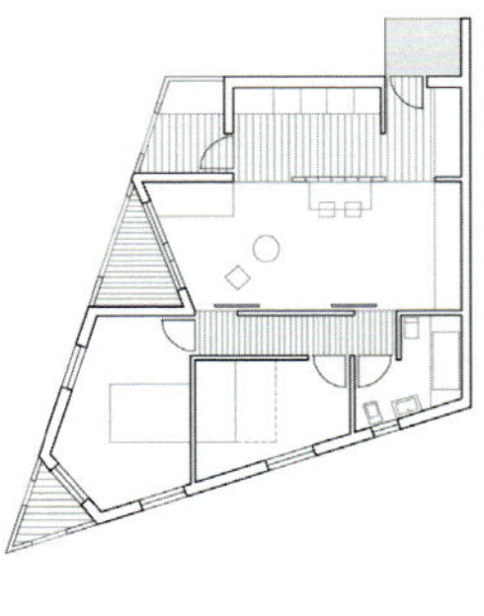

2卧室 类型1 2 BEDROOMS T1. 60.00M2

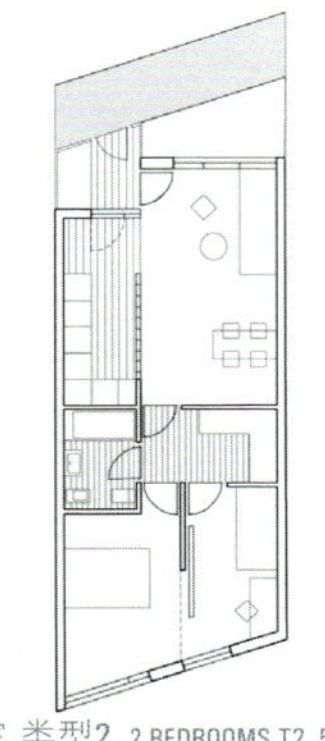

2卧室 类型2 2 BEDROOMS T2. 53.01M2

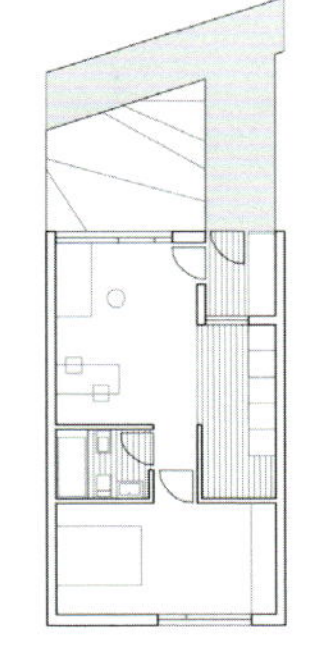

1卧室 类型1 1 BEDROOM T1. 40.38M2

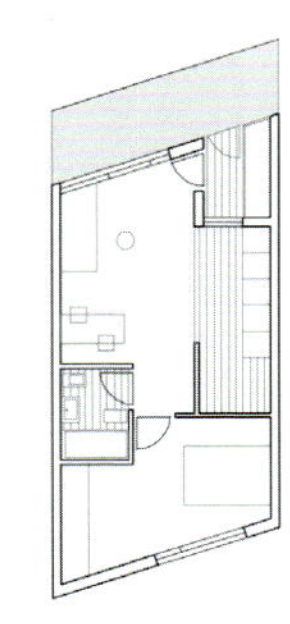

1卧室 类型2 1 BEDROOM T2. 40.47M2

## VHOM音乐学校·维也纳
## Vienna House of Music VHOM · Austria

并列一等奖 FIRST PRIZE EX-AEQUO
Sónia Santos
Diogo Monteiro
Guilherme Rodrigues
Universidade de Évora (大学), 葡萄牙 Portugal

在第三届ArchMedium学生竞赛中，维也纳音乐之家建筑设计竞赛的参赛者遇到了一个难题：该建筑选址位于奥地利维也纳一个欧洲历史中心的公园内。此次竞赛的评审团决定颁发两个一等奖，而这两位一等奖获得者在设计时必须采取完全不同的标准和方法。

During the third **ArchMedium contest of ideas for students**, the participants of the Vienna House of Music competition confronted a difficult job: a plot within a park in a European historical centre, in Vienna, Austria. The Jury decided to award 2 first prizes, who had radically different criteria and ways to respond to the same exercise.

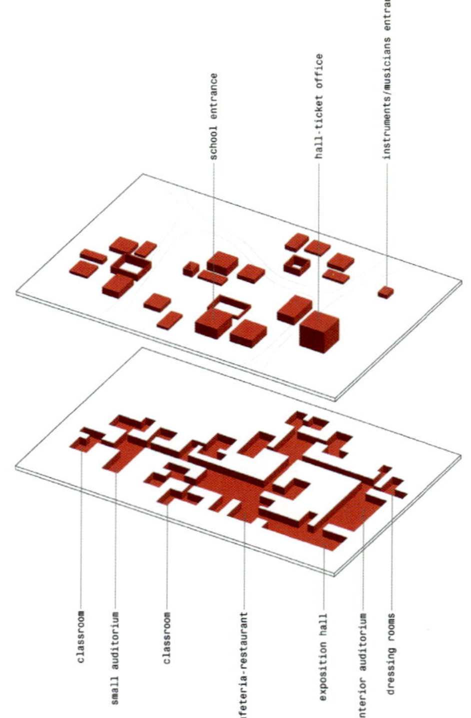

剖面图 SECTION

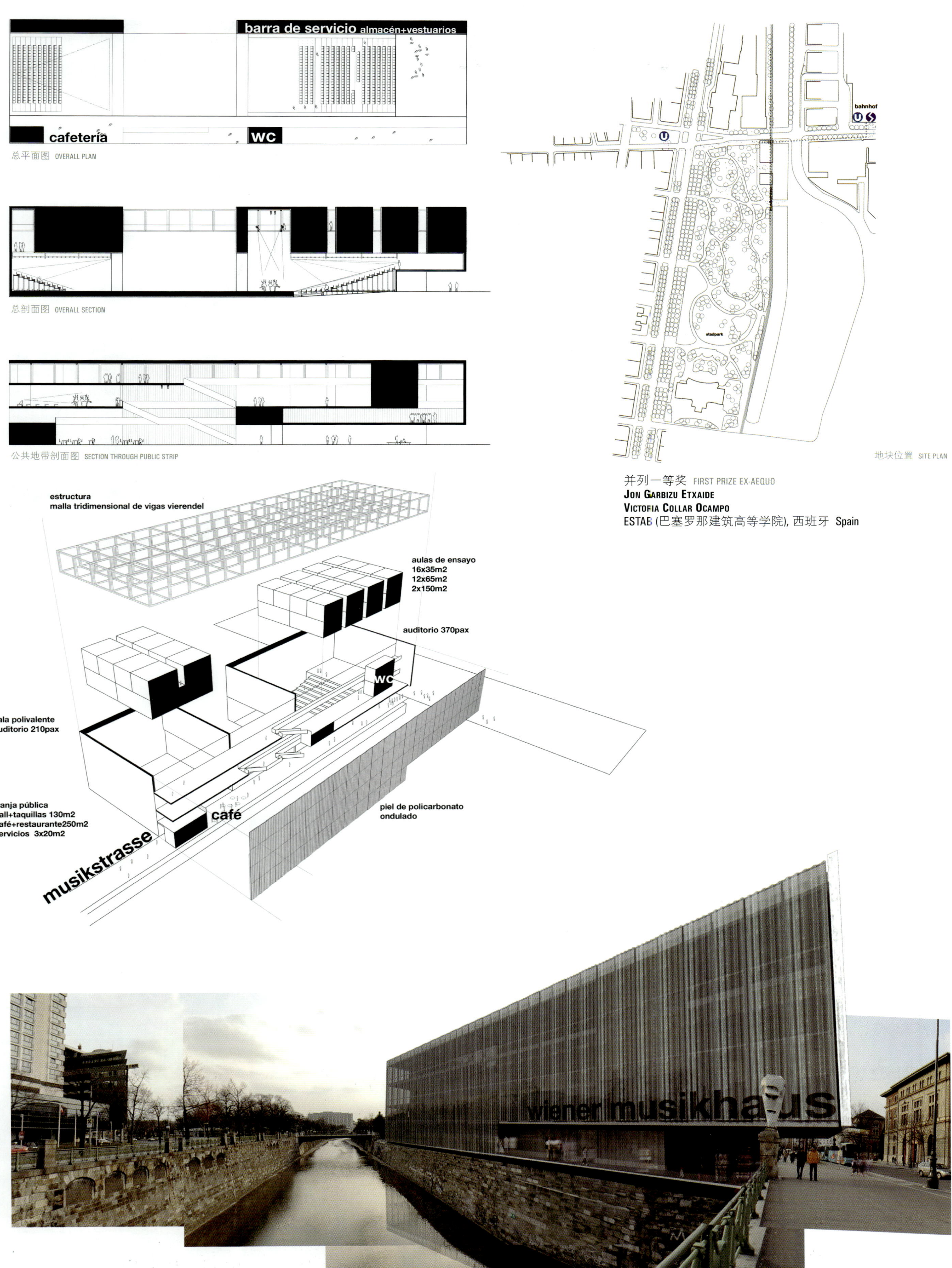


总平面图 OVERALL PLAN

总剖面图 OVERALL SECTION

公共地带剖面图 SECTION THROUGH PUBLIC STRIP

地块位置 SITE PLAN

并列一等奖 FIRST PRIZE EX-AEQUO

**Jon Garbizu Etxaide**
**Victoria Collar Ocampo**
ESTAB (巴塞罗那建筑高等学院), 西班牙 Spain

## Alzheimer疗养院 · 贝纳温特，扎莫拉
## Day Centre for Alzheimer's patients · Spain

有了一家金融机构的拨款资助，位于贝纳文特（萨莫拉）的老年痴呆患者新中心将有望建成。该项目于一年多前由Ruben Garcia Rubio和Juanes Enrique Martin承接。

The construction of the new center for Alzheimer's patients in Benavente (Zamora) will be possible thanks to a grant from a financial institution. The project was awarded to **Ruben Garcia Rubio and Juanes Enrique Martin**, more than one year ago.

概念 CONCEPT

上层平面图 UPPER FLOOR PLAN

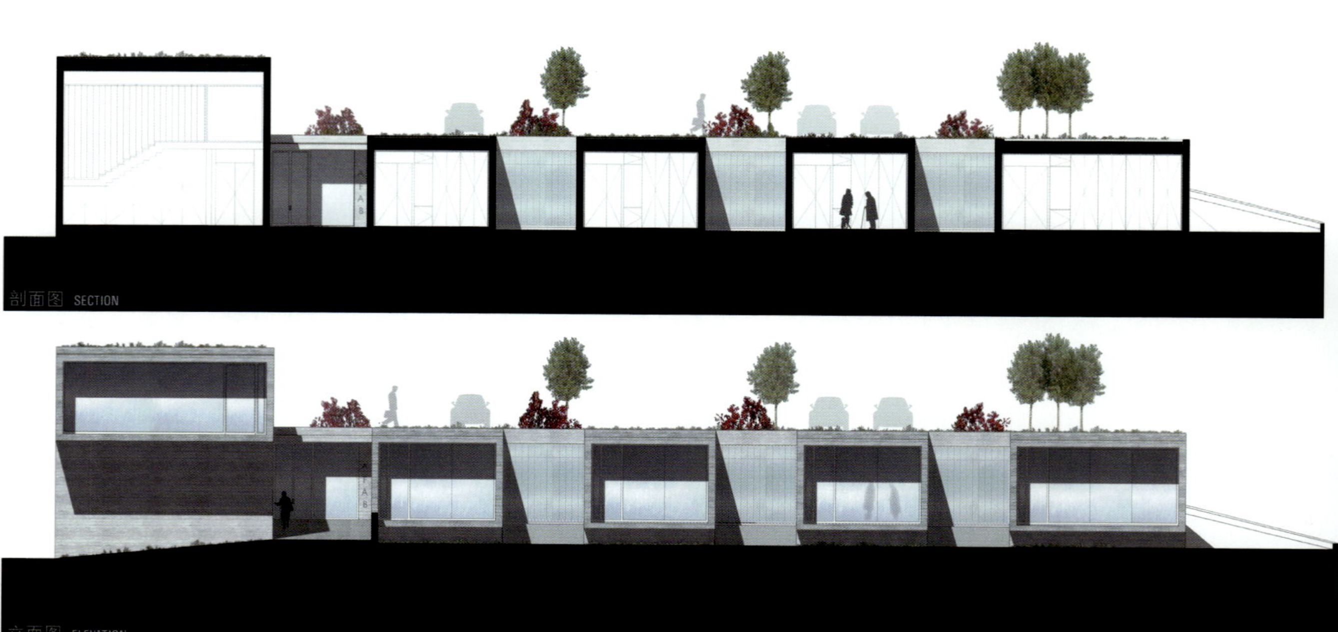

剖面图 SECTION

立面图 ELEVATION

下层平面图 LOWER FLOOR PLAN

# 新水族馆·巴塔米
# New Aquarium · Georgia

亨宁·拉森建筑事务所设计了位于格鲁吉亚共和国海港城市巴统的新水族馆，并因其优秀的设计理念而荣获一等奖。除了亨宁·拉森建筑事务所之外，德国的建筑公司Drei Architekten和两家美国的建筑公司——PJA建筑事务所和Pryor & Morrow建筑事务所也参与了此次大赛的角逐。占地2000平方米的水族馆将在以往的港水族馆原址上进行修建。

**Henning Larsen Architects** has won first prize for its significant design concept for a new aquarium in the seaport of Batumi, the Republic of Georgia. In addition to Henning Larsen Architects, the German architecture company Drei Architekten and the two American companies PJA Architects and Pryor & Morrow Architects were in the running to win the competition. The 2,000 m2 aquarium will replace the previous aquarium of the port.

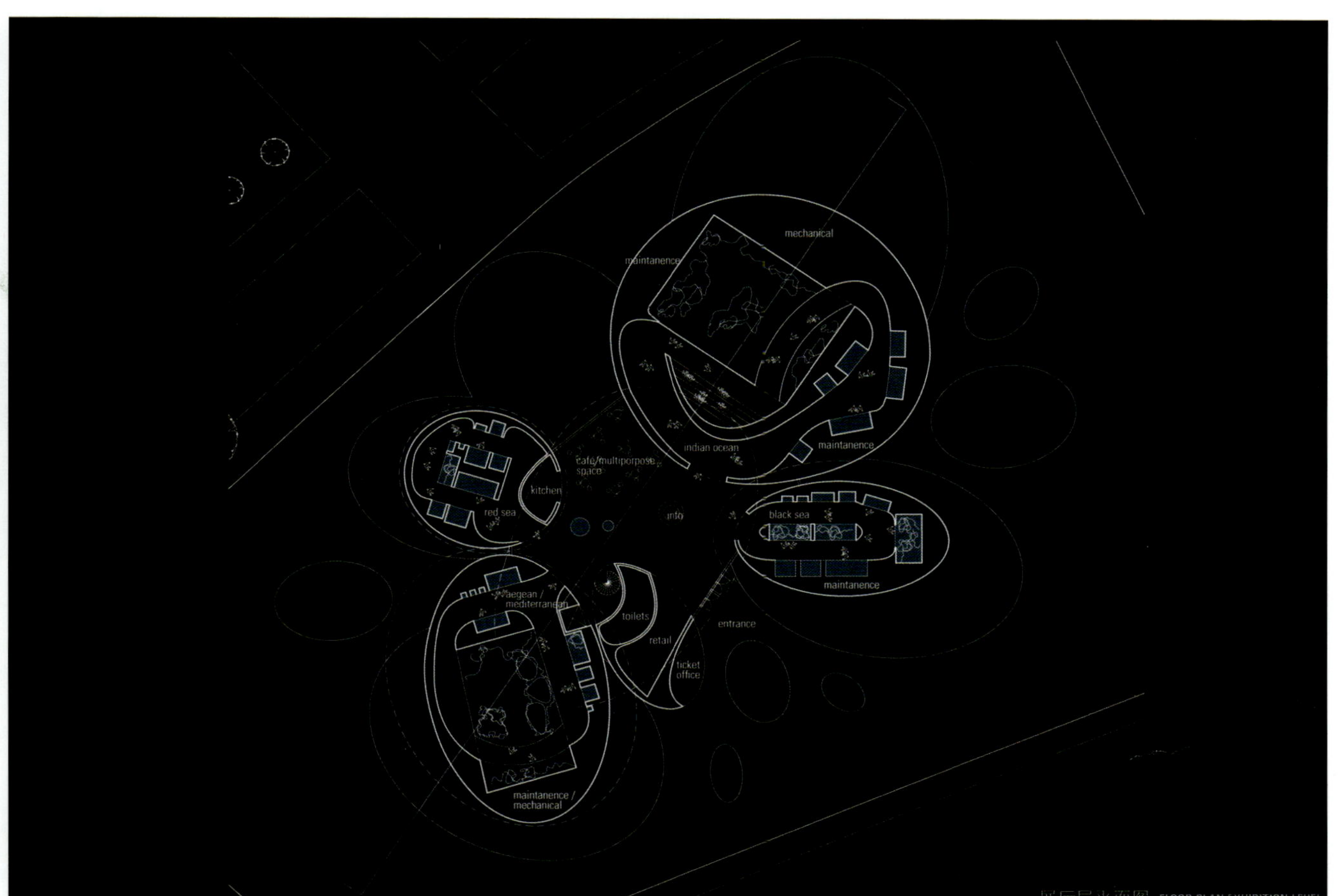


展厅层平面图 FLOOR PLAN EXHIBITION LEVEL

南立面图 SOUTH ELEVATION

东立面图 EAST ELEVATION

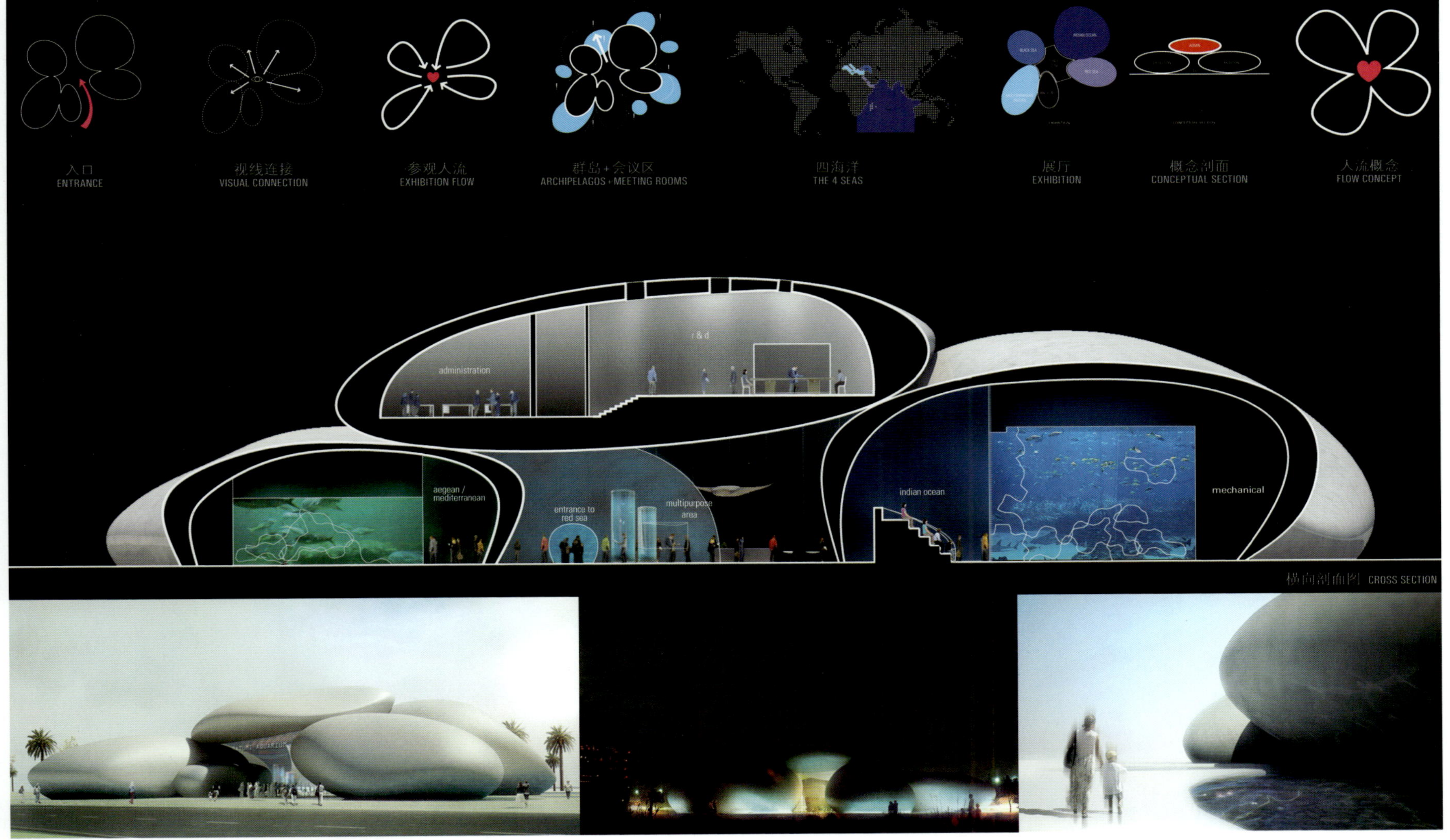


横向剖面图 CROSS SECTION

# 住宅综合体 · 贾不里亚
## Housing Complex · Kuwait

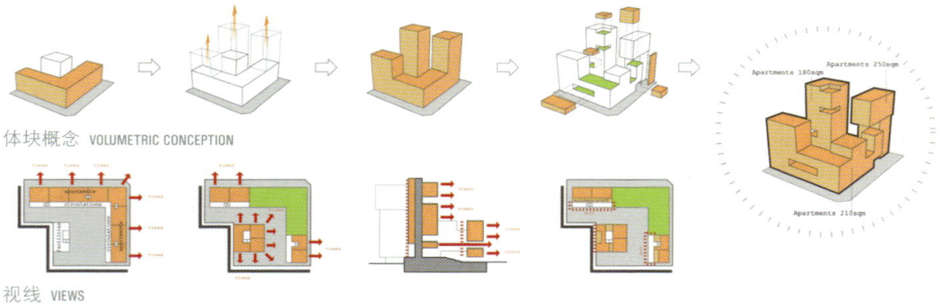

体块概念 VOLUMETRIC CONCEPTION

视线 VIEWS

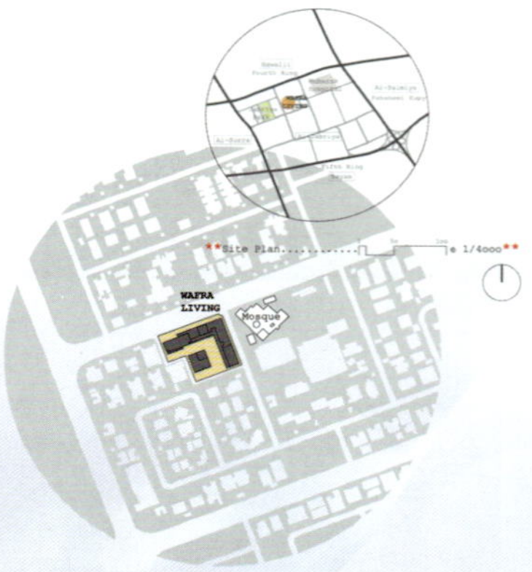

地块位置 SITE PLAN

沃夫拉生活区位于贾卜里派，周围是住宅楼和商业设施，包括一幢不临街的高层建筑以及一幢沿街的L形建筑，成为该地的一个新型地标。

Located in Jabriya, surrounded by residential buildings and commercial facilities, the proposed design for the Wafra Living complex, which consists of a high rise building set back from the street and an L-shaped building defining the street edge, is conceived as a new type of landmark

一等奖 FIRST PRIZE

**AGi ARCHITECTS (建筑师事务所)**
Nasser Bader Abulhasan · Joaquin Perez-Goicoechea
team: Salvador Cejudo · Gwenola Kergall · Daniel Munoz · Stefania Rendinelli · Carmen Sagredo Bruno Gomes · Jose Del Campo · Lucia Sanchez Sharifa Alshalfan · Hanan Alkouh · Robert Varghese Babu Abraham · Moyra Montoya · Nicolas Martin
(建筑师)

平面图 +6 FLOOR PLAN LEVEL +6

平面图 +9 FLOOR PLAN LEVEL +9

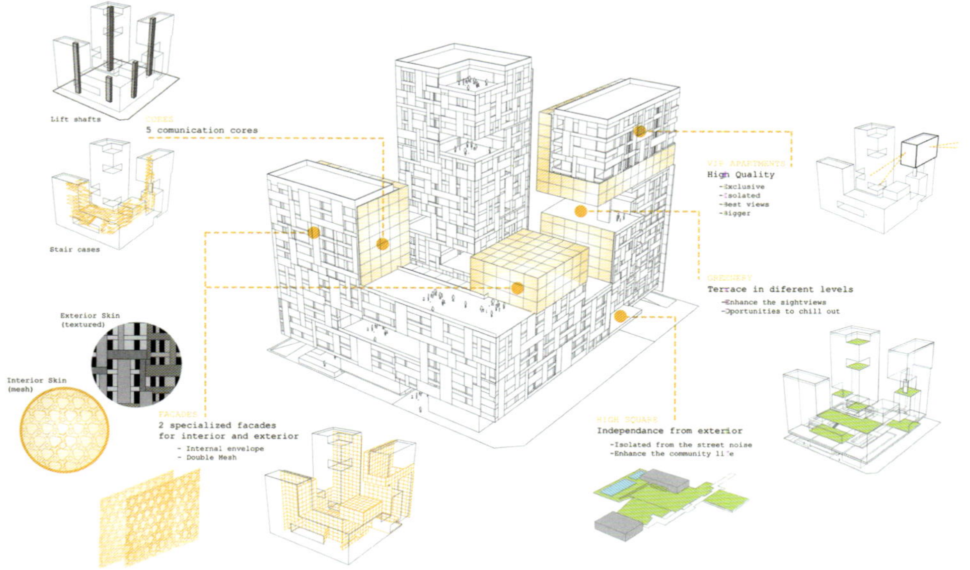

Lift shafts

CORES
5 comunication cores

Stair cases

VIP APARTMENTS
High Quality
-Exclusive
-Isolated
-Best views
-Bigger

GREENERY
Terrace in diferent levels
-Enhance the sightviews
-Oportunities to chill out

Exterior Skin (textured)

Interior Skin (mesh)

FACADES
2 specialized facades for interior and exterior
- Internal envelope
- Double Mesh

HIGH SQUARE
Independance from exterior
-Isolated from the street noise
-Enhance the community life

# 图书预购

## 4 本

*包括：*
*4本240页图书*

| 880¥ | 含税及邮资 |
|---|---|
| 800¥ | 学生优惠价<br>含税及邮资 |

我想从第几辑开始购买未来建筑系列图书

预购信息

姓名

身份证号码

单位

地址

邮编

电话

电邮

腾讯

更多信息请联系：

未来建筑中国联络办公室 ***future*** arquitecturas China
杭州市文晖路303号交通大厦11楼
邮编：310014
电话 +86 85303277 手机 +86 13706505166

**china@arqfuture.com**

**QQ 860464402**

日期 签名

***www.arqfuture.com***

*2012年有效*

# *pre-order*

## 4 books

*includes:*
*4 books of 240 pages*

| 880¥ | taxes and shipping costs included in China |
|---|---|
| 800¥ | special prize for students<br>taxes and shipping costs included in China |

I wish to pre-order ***future China*** architecture books starting from number

*pre-order information*

NAME

ID NUMBER

COMPANY

ADDRESS

ZIP CODE

PHONE

E-MAIL

QQ

For more information, please contact:

***future*** arquitecturas China
Zhejiang Communication Group Tower 11st floor
Wenhui Road 303, 310014, Hangzhou, China.
tel: +86 85303277 cell: +86 13706505166

**china@arqfuture.com**

**QQ 860464402**

DATE SIGNATURE

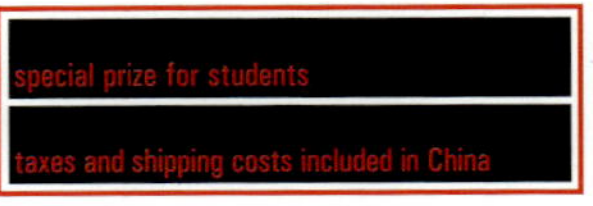

***www.arqfuture.com***

valid only for year *2012*

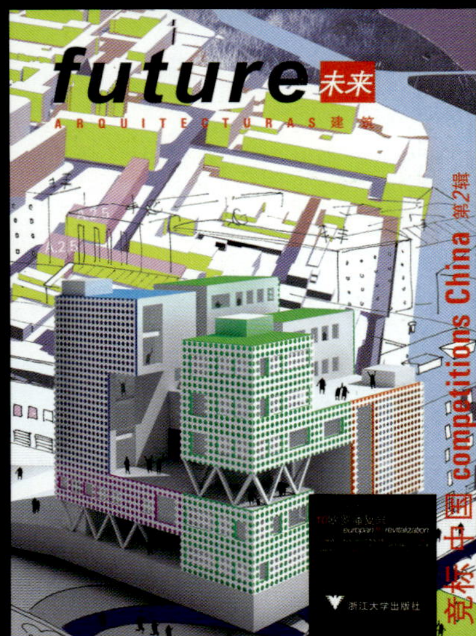

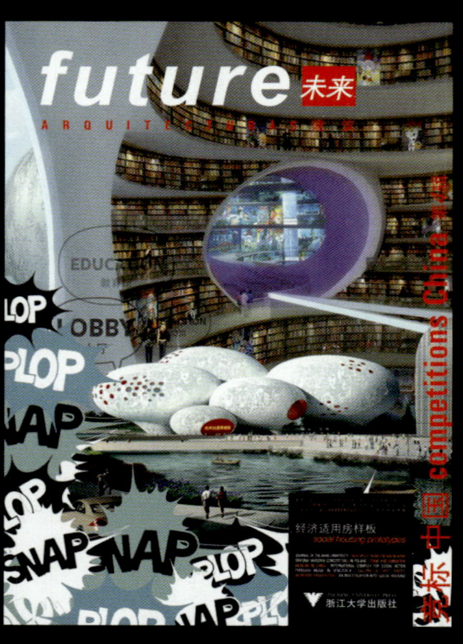